Theophil Eicher, Siegfried Hauptmann
The Chemistry of Heterocycles

[

Further Reading from Wiley-VCH

Fuhrhop, J.-H., Li, G.
Organic Synthesis, 3. Ed.
2003. 3-527-30272-7 (Hardcover)
3-527-30273-5 (Softcover)

Schmalz, H.-G., Wirth, T. (Eds.)
Organic Synthesis Highlights V
2003. 3-527-30611-0

Nicolaou, K. C., Snyder S. A.
Classics in Total Synthesis II
2003. 3-527-30685-4 (Hardcover)
3-527-30684-6 (Softcover)

Green, M. M., Wittcoff, H. A.
Organic Chemistry Principles and Industrial Practice
2003. 3-527-30289-1

Theophil Eicher, Siegfried Hauptmann
in Collaboration with Andreas Speicher

The Chemistry of Heterocycles

Structure, Reactions, Syntheses,
and Applications

Second, Completely Revised, and Enlarged Edition

Translated by Hans Suschitzky and Judith Suschitzky

WILEY-VCH

WILEY-VCH GmbH & Co. KGaA

Authors

Professor Dr. Theophil Eicher
University of the Saarland
Am Botanischen Garten 1
D-66123 Saarbrücken
Germany

Professor Dr. Siegfried Hauptmann
Naunhofer Strasse 137
D-04299 Leipzig
Germany

PD Dr. Andreas Speicher
Department of Chemistry
University of the Saarland
D-66041 Saarbrücken
Germany

Translators
Professor Dr. Hans Suschitzky and Mrs. Judith
Suschitzky
Department of Chemistry and Applied Chemistry
University of Salford
Salford M5 4WT
United Kingdom

This book was carefully produced. Nevertheless, authors and publisher do not warrant the information contained therein to be free of errors. Readers are advised to keep in mind that statements, data, illustrations, procedural details or other items may inadvertently be inaccurate.

Library of Congress Card No.: applied for

British Library Cataloging-in-Publication Data:

A catalogue record for this book is available from the British Library

Bibliographic information published by

Die Deutsche Bibliothek

Die Deutsche Bibliothek lists this publication in the Deutsche Nationalbibliografie; detailed bibliographic data is available in the

Internet at <http://dnb.ddb.de>.

© 2003 WILEY-VCH GmbH & Co. KGaA, Weinheim

Printed in the Federal Republic of Germany

Printed on acid-free paper

Printing   Strauss Offsetdruck GmbH, Mörlenbach

Bookbinding   Großbuchbinderei J. Schäffer GmbH & Co. KG, Grünstadt

ISBN   3-527-30720-6

Dedicated to Ursula and Gundela

Foreword

The heterocycles constitute the largest group of organic compounds and are becoming ever more important in all aspects of pure and applied chemistry. The monograph, *The Chemistry of Heterocycles – Structure, Reactions, Syntheses and Applications*, is a comprehensive survey of this vast field. The discussion is supported by numerous lucid diagrams and the extensive reaction schemes are supported by relevant and up-to-date references. Aromatic and nonaromatic heterocycles are treated according to increasing ring size under six defined headings. Thus, information can be easily located and compared. Natural occurance, synthetic aspects, as well as modern applications of many heterocyclic compounds in the chemical and pharmaceutical industries are also described.

This book will no doubt prove to be an invaluable reference source. It is eminently for advanced undergraduate and graduate students of chemistry, and of related subjects such as biochemistry and medicinal chemistry. It also provides an important aid to professional chemists, and teachers of chemistry will find it most useful for lecture preparation. It will surely find a place on the bookshelf of university libraries and in the laboratories of scientists concerned with any aspect of heterocyclic chemistry.

Hans Suschitzky, University of Salford

Preface

Of the more than 20 million chemical compounds currently registered, about one half contain heterocyclic systems. Heterocycles are important, not only because of their abundance, but above all because of their chemical, biological and technical significance. Heterocycles count among their number many natural products, such as vitamins, hormones, antibiotics, alkaloids, as well as pharmaceuticals, herbicides, dyes, and other products of technical importance (corrosion inhibitors, antiaging drugs, sensitizers, stabilizing agents, etc.).

The extraordinary diversity and multiplicity of heterocycles poses a dilemma: What is to be included in an introductory book on heterocyclic chemistry which does not aim to be an encyclopaedia? This difficulty had to be resolved in a somewhat arbitrary manner. We decided to treat a representative cross section of heterocyclic ring systems in a conventional arrangement. For these heterocycles, structural, physical and spectroscopic features are described, and important chemical properties, reactions and syntheses are discussed. Synthesis is consequently approached as a retrosynthetic problem for each heterocycle, and is followed by selected derivatives, natural products, pharmaceuticals and other biologically active compounds of related structure type, and is concluded by aspects of the use in synthesis and in selected synthetic transformations. The informations given are supported by references to recent primary literature, reviews and books on experimental chemistry. Finally, a section of "problems" and their solutions – selected in a broad variety and taken mainly from the current literature – intends to deepen and to extend the topics of heterocyclic chemistry presented in this book.

The book is designed for the advanced student and research worker, and also for the industrial chemist looking for a survey of well-tried fundamental concepts as well as for information on modern developments in heterocyclic chemistry. The contents of this book can also serve as a basis for the design of courses in heterocyclic chemistry. Above all, however, we intend to demonstrate that general chemical principles of structure, reactivity and synthesis can be elucidated by using examples from the chemistry of heterocycles.

Text and diagrams were produced with the Word for Windows and ChemWindow packages, respectively, in the Desktop Publishing program.

We are indebted to Prof. Dr. H. Becker, Prof. Dr. R. W. Hartmann, Prof. Dr. U. Kazmaier and Prof. Dr. L. F. Tietze for valuable advice and encouragement. Special thanks are due to Mrs. Ch. Altmeyer for her excellent assistance and cooperativeness in preparing the camera-ready version of this book. We also thank Dr. E. Westermann and the staff of the editorial office of Wiley VCH for their collaboration and understanding.

Saarbrücken and Leipzig, April 2003

Theophil Eicher   Siegfried Hauptmann

Contents

Contents

Abbreviations and Symbols

mp	melting point	de	diastereoisomeric excess
bp	boiling point	%	percentage
ca.	circa	°C	degrees centigrade
cf.	compare	Δ	thermal
cf. p	see page	$h\nu$	photochemical
MO	molecular orbital	dil	dilute
INN	international nonproprietary name	concd	concentrated
IR	infrared spectrum	ref.	reference
cm^{-1}	wave number	$\Delta H^{\neq}$	activation enthalpy (kJ mol^{-1})
UV	ultraviolet spectrum	rfl.	heated under reflux
λ	wavelength	r.t.	room temperature
ε	molar extinction coefficient	et al.	and other authors
^{1}H NMR	proton resonance spectrum	nm	nanometer (10^{-9} m)
^{13}C NMR	^{13}C resonance spectrum	pm	picometer (10^{-12} m)
δ	chemical shift ($\delta_{TMS} = 0$)		
ppm	parts per million (10^{-6})		
ee	enantiomeric excess		

Ac	acetyl
Ar	aryl
Boc	*tert*-butoxycarbonyl
Bn	benzyl
Bz	benzoyl
n-Bu	*n*-butyl
sec-Bu	*sec*-butyl
tert-Bu	*tert*-butyl
Et	ethyl
Me	methyl
Mes	mesyl (methanesulfonyl)
Ph	phenyl
i-Pr	isopropyl
n-Pr	*n*-propyl
Tos	tosyl (*p*-toluenesulfonyl)

DABCO	1,4-diazabicyclo[2.2.2]octane
DMF	dimethylformamide
DMSO	dimethyl sulfoxide
DDQ	2,3-dichloro-5,6-dicyano-1,4-benzoquinone
DBU	1,8-diazabicyclo[5.4.0]undec-7-ene
HMPT	hexamethylphosphoric triamide
LDA	lithiumdiisopropylamide
LiTMP	lithium-2,2,6,6-tetramethylpiperidide
MOM	methoxymethyl
NBS	*N*-bromosuccinimide
NCS	*N*-chlorosuccinimide
PPA	polyphosphoric acid
TBAF	tetra-*n*-butylammonium fluoride
THF	tetrahydrofuran
TMEDA	*N,N,N',N'*-tetramethylethylenediamine
TMS	trimethylsilyl
TosMIC	(*p*-toluenesulfonyl)methylisocyanide

1   The Structure of Heterocyclic Compounds

Most chemical compounds consist of molecules. The classification of such chemical compounds is based on the structure of these molecules, which is defined by the type and number of atoms as well as by the covalent bonding within them. There are two main types of structure:

— The atoms form a chain – aliphatic (acyclic) compounds
— The atoms form a ring – cyclic compounds

Cyclic compounds in which the ring is made up of atoms of one element only are called isocyclic compounds. If the ring consists of C-atoms only, then we speak of a carbocyclic compound, e.g.:

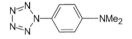

(4 - dimethylaminophenyl) pentazole
isocyclic

cyclopenta - 1,3 - diene
isocyclic und carbocyclic

Cyclic compounds with at least two different atoms in the ring (as ring atoms or members of the ring) are known as heterocyclic compounds. The ring itself is called a heterocycle. If the ring contains no C-atom, then we speak of an inorganic heterocycle, e.g.:

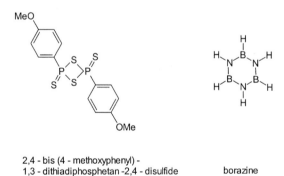

2,4 - bis (4 - methoxyphenyl) -
1,3 - dithiadiphosphetan -2,4 - disulfide
(Lawesson - Reagent)

borazine

If at least one ring atom is a C-atom, then the molecule is an organic heterocyclic compound. In this case, all the ring atoms which are not carbon are called heteroatoms, e.g.:

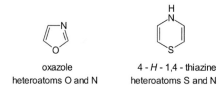

oxazole
heteroatoms O and N

4 - H - 1,4 - thiazine
heteroatoms S and N

In principle, all elements except the alkali metals can act as ring atoms.

Along with the type of ring atoms, their total number is important since this determines the ring size. The smallest possible ring is three-membered. The most important rings are the five- and six-membered heterocycles. There is no upper limit; there exist seven-, eight-, nine- and larger-membered heterocycles.

Although inorganic heterocycles have been synthesized, this book limits itself to organic compounds. In these, the N-atom is the most common heteroatom. Next in importance are O- and S-atoms. Heterocycles with Se-, Te-, P-, As-, Sb-, Bi-, Si-, Ge-, Sn-, Pb- or B-atoms are less common.

To determine the stability and reactivity of heterocyclic compounds, it is useful to compare them with their carbocyclic analogues. In principle, it is possible to derive every heterocycle from a carbocyclic compound by replacing appropriate CH_2 or CH groups by heteroatoms. If one limits oneself to monocyclic systems, one can distinguish four types of heterocycles as follows:

- *Saturated heterocycles (heterocycloalkanes), e.g.:*

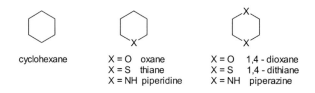

cyclohexane

X = O   oxane
X = S   thiane
X = NH piperidine

X = O   1,4 - dioxane
X = S   1,4 - dithiane
X = NH piperazine

In this category, there are no multiple bonds between the ring atoms. The compounds react largely like their aliphatic analogues, e.g. oxane (tetrahydropyran) and dioxane behave like dialkyl ethers, thiane and 1,4-dithiane like dialkyl sulfides, and piperidine and piperazine like secondary aliphatic amines.

- *Partially unsaturated systems (heterocycloalkenes), e.g.:*

cyclohexene

X = O 3,4 - dihydro - 2H - pyran
X = S
X = NH

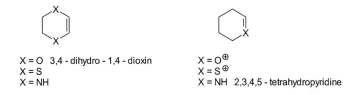

X = O   3,4 - dihydro - 1,4 - dioxin   X = O⊕
X = S   X = S⊕
X = NH   X = NH   2,3,4,5 - tetrahydropyridine

If the multiple bonds are between two C-atoms of the ring, as, for instance, in 3,4-dihydro-2*H*-pyran, the compounds react essentially like alkenes or alkynes. The heteroatom can also be involved in a double bond. In the case of X = O⁺, the compounds behave like oxenium salts, in the case of X = S⁺, like sulfenium salts, and in the case of X = N, like imines (azomethines).

- *Systems with the greatest possible number of noncumulated double bonds (heteroannulenes)*, e.g.:

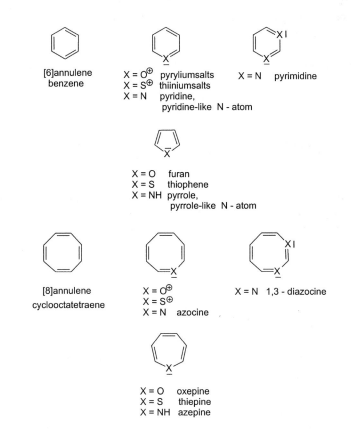

[6]annulene   X = O⊕ pyryliumsalts   X = N   pyrimidine
benzene   X = S⊕ thiiniumsalts
                   X = N   pyridine,
                            pyridine-like N - atom

X = O   furan
X = S   thiophene
X = NH   pyrrole,
         pyrrole-like N - atom

[8]annulene   X = O⊕   X = N   1,3 - diazocine
cyclooctatetraene X = S⊕
                   X = N   azocine

X = O   oxepine
X = S   thiepine
X = NH   azepine

From the annulenes, one can formally derive two types of heterocycles:

— systems of the same ring size, if CH is replaced by X
— systems of the next lower ring size, if HC=CH is replaced by X.

In both cases, the resulting heterocycles are iso-π-electronic with the corresponding annulenes, i.e. the number of π-electrons in the ring is the same. This is because in the pyrylium and thiinium salts, as well as in pyridine, pyrimidine, azocine and 1,3-diazocine, each heteroatom donates one electron pair to the conjugated system and its nonbonding electron pair does not contribute. However, with furan, thiophene, pyrrole, oxepin, thiepin and azepine, one electron pair of the heteroatom is incorporated into the conjugated system (delocalization of the electrons). Where nitrogen is the heteroatom, this difference can be expressed by the designation *pyridine-like N-atom* or *pyrrole-like N-atom.*

• *Heteroaromatic systems*

This includes heteroannulenes, which comply with the HÜCKEL rule, i.e. which possess $(4n + 2)$ π-electrons delocalized over the ring. The most important group of these compounds derives from [6]annulene (benzene). They are known as *heteroarenes*, e.g. furan, thiophene, pyrrole, pyridine, and the pyrylium and thiinium ions. As regards stability and reactivity, they can be compared to the corresponding benzenoid compounds [1].

The antiaromatic systems, i.e. systems possessing $4n$ delocalized electrons, e.g. oxepin, azepine, thiepin, azocine, and 1,3-diazocine, as well as the corresponding annulenes, are, by contrast, much less stable and very reactive.

The classification of heterocycles as heterocycloalkanes, heterocycloalkenes, heteroannulenes and heteroaromatics allows an estimation of their stability and reactivity. In some cases, this can also be applied to inorganic heterocycles. For instance, borazine (see p 1), a colourless liquid, bp 55°C, is classified as a heteroaromatic system.

[1] P. v. Rague Schleyer, H. Jiao, *Pure Appl.. Chem.* **1996**, *68*, 209;
     Chem. Rev. **2001**, *101*, 1115;
     C. W. Bird, *Tetrahedron* **1998**, *54*, 10179;
     T. M. Krygowski, M. K. Cyranski, Z. Czarnocki, G. Häfelinger,
     A. R. Katritzky, *Tetrahedron* **2000**, *56*, 1783.

2 Systematic Nomenclature of Heterocyclic Compounds

Many organic compounds, including heterocyclic compounds, have a *trivial name*. This usually originates from the compounds occurrence, its first preparation or its special properties.

Structure	Trivial name	Systematic name
	ethylene oxide	oxirane
	pyromucic acid	furan - 2 - carboxylic acid
	nicotinic acid	pyridine - 3 - carboxylic acid
	coumarin	2*H* - chromen - 2 - one

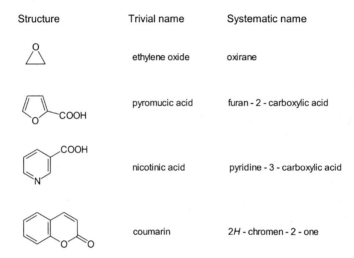

The derivation of the *systematic name* of a heterocyclic compound is based on its structure. Nomenclature rules have been drawn up by the IUPAC Commission and these should be applied when writing theses, dissertations, publications and patents. These rules are listed in section R-2 of the most recent IUPAC 'Blue Book' together with worked examples (H.R.Panico, W.H.Powell, J.-C.Richer *A Guide to IUPAC Nomenclature of Organic Compounds, Recommendations 1993*; Blackwell Scientific: Oxford, 1993; the previous IUPAC Blue Book: J.Rigandy, S.P.Klesney *Nomenclature of Organic Chemistry*; Pergamon: Oxford, 1979).

The IUPAC rules are not given in detail here, rather instructions are given for formulating systematic names with appropriate reference to the Blue Book.

Every heterocyclic compound can be referred back to a *parent ring system*. These systems have only H-atoms attached to the ring atoms. The IUPAC rules allow two nomenclatures. The *Hantzsch–Widman nomenclature* is recommended for three- to ten-membered heterocycles. For larger ring heterocycles, *replacement nomenclature* should be used.

2.1   Hantzsch–Widman Nomenclature

- *Type of heteroatom*

The type of heteroatom is indicated by a prefix according to Table 1. The sequence in this table also indicates the preferred order of prefixes (*principle of decreasing priority*).

Table 1   Prefixes to indicate heteroatoms

Element	Prefix	Element	Prefix
O	oxa	Sb	stiba
S	thia	Bi	bisma
Se	selena	Si	sila
Te	tellura	Ge	germa
N	aza	Sn	stanna
P	phospha	Pb	plumba
As	arsa	B	bora
		Hg	mercura

- *Ring size*

The ring size is indicated by a suffix according to Table 2. Some of the syllables are derived from Latin numerals, namely ir from tri, et from tetra, ep from hepta, oc from octa, on from nona, ec from deca.

Table 2   Stems to indicate the ring size of heterocycles

Ring Size	Unsaturated	Saturated
3	irene[a]	irane[b]
4	ete	etane[b]
5	ole	olane[b]
6A[c]	ine	ane
6B[c]	ine	inane
6C[c]	inine	inane
7	epine	epane
8	ocine	ocane
9	onine	onane
10	ecine	ecane

[a]   The stem irine may be used for rings containing only N.

[b]   The traditional stems 'irine', 'etidine' and 'olidine' are preferred for N-containing rings and are used for saturated heteromonocycles having three, four or five ring members, respectively.

[c]   The stem for six-membered rings depends on the least preferred heteroatom in the ring, that immediately preceding the stem. To detemine the correct stem for a structure, the set below containing this least-preferred heteroatom is selected.

6A: O, S, Se, Te, Bi, Hg;   6B: N, Si, Ge, N, Pb;   6C: B, P, As, Sb

- *Monocyclic systems*

The compound with the maximum number of noncumulative double bonds is regarded as the parent compound of the monocyclic systems of a given ring size. The naming is carried out by combining one or more prefixes from Table 1 with a suffix from Table 2. If two vowels succeed one another, the letter a is omitted from the prefix, e.g. azirine (not azairine).

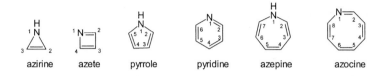

| azirine | azete | pyrrole | pyridine | azepine | azocine |

Note that trivial names are permitted for some systems, e.g. pyrrole, pyridine. Permitted trivial names can be found in the latest IUPAC Blue Book pp 166-172; if a trivial name is permitted then it should be used.

Partly or completely saturated rings are denoted by the suffixes according to Table 2. If no ending is specified the prefixes dihydro-, tetrahydro-, etc. should be used.

| 2,3-dihydropyrrole | pyrrolidine | 1,4 - dihydropyridine | piperidine (hexahydropyridine) |

- *Monocyclic systems, one heteroatom*

The numbering of such systems starts at the heteroatom.

- *Monocyclic systems, two or more identical heteroatoms*

The prefixes di-, tri-, tetra-, etc., are used for two or more heteroatoms of the same kind. When indicating the relative positions of the heteroatoms, the principle of the lowest possible numbering is used, i.e. the numbering of the system has to be carried out in such a way that the heteroatoms are given the lowest possible set of locants:

1,2,4 - triazole (not 1,3,5 -triazole)   pyrimidine (1,3 - diazine, not 1,5 - diazine)

In such a numerical sequence, the earlier numbers take precedence, e.g. 1,2,5 is lower than 1,3,4.

- *Monocyclic systems, two or more different heteroatoms*

For heteroatoms of different kinds, prefixes are used in the order in which they appear in Table 1, e.g. thiazole, not azathiole; dithiazine, not azadithiine. The heteroatom highest in Table 1 is allocated the 1-position in the ring. The remaining heteroatoms are assigned the smallest possible set of number locants:

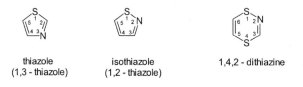

thiazole   isothiazole   1,4,2 - dithiazine
(1,3 - thiazole)   (1,2 - thiazole)

Although in the first example the systematic name is 1,3-thiazole, the locants are generally omitted because, except for isothiazole (1,2-thiazole), no other structural isomers exist. Similar rules apply to oxazole (1,3-oxazole) and isoxazole (1,2-oxazole).

- *Identical systems connected by a single bond*

Such compounds are defined by the prefixes bi-, tert-, quater-, etc., according to the number of systems, and the bonding is indicated as follows:

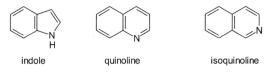

2,2' - bipyridine   2,2' : 4',3" - terthiophene

- *Bicyclic systems with one benzene ring*

Systems in which at least two neighbouring atoms are common to two or more rings are known as fused systems. For several bicyclic benzo-fused heterocycles, trivial names are permitted, e.g.:

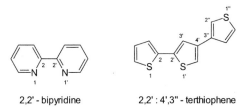

indole   quinoline   isoquinoline

If this is not the case, and only the heterocycle has a trivial name, then the systematic name is formulated from the prefix benzo- and the trivial name of the heterocyclic component as follows:

benzo [b] furan   furan

The system is dissected into its components. The heterocyclic component is regarded as the *base component*. The bonds between the ring atoms are denoted according to the successive numbers of the ring atoms by the letters *a*, *b*, *c*, etc. The letter *b* in brackets between benzo and the name of the base component denotes the atoms of the base component which are common to both rings. The letter must be as early as possible alphabetically and hence benzo[*d*]furan is incorrect.

It is generally accepted that the numbering of the whole system in the case of bi- and also polycyclic systems should be done independently of the numbering of the components, and as follows:

The ring system is projected onto rectangular coordinates in such a way that
— as many rings as possible lie in a horizontal row
— a maximum number of rings are in the upper right quadrant.
The system thus oriented is then numbered in a clockwise direction commencing with that atom which is not engaged in the ring fusion and is furthest to the left
— in the uppermost ring or
— in the ring furthest to the right in the upper row.
C-Atoms which belong to more than one ring are omitted. Heteroatoms in such positions are, however, included. If there are several possible orientations in the coordinate system, the one in which the heteroatoms bear the lowest locants is valid:

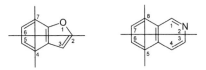

If the base component does not have a trivial name, the entire system is numbered as explained above and the resulting positions of the heteroatoms are placed before the prefix benzo:

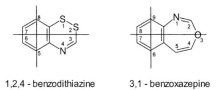

1,2,4 - benzodithiazine   3,1 - benzoxazepine

- *Bi- and polycyclic systems with two or more heterocycles*

First the base component is established. To this end the criteria in the order set out below are applied, one by one, to arrive at a decision. The base component is
— a nitrogen-containing component
— a component with a heteroatom, other than nitrogen, which is as high as possible in Table 1
— a component with as many rings as possible (e.g. bicyclic condensed systems or polycyclic systems which have trivial names)
— the component with the largest ring
— the component with most heteroatoms
— the component with the largest number of heteroatoms of different kinds
— the component with the greatest number of heteroatoms which are highest in Table 1
— the component with heteroatoms which have the lowest locant numbers.

Two isomers are given as an example:

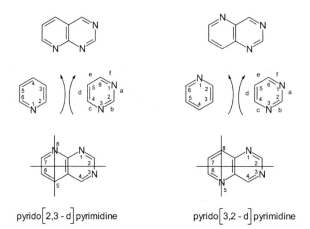

pyrido[2,3-d]pyrimidine   pyrido[3,2-d]pyrimidine

First, the system is dissected into its components. The base component cannot be established until the fifth criterion has been reached: pyrimidine. The bonds between the ring atoms are marked by consecutive lettering according to the serial numbering of the base component. In contrast to the example on p 9, the fused component must also be numbered, always observing the principle of assignment to the lowest possible locants. The name of the fused component, by the replacement of the terminal 'e' with 'o', is put before the name of the base component. The atoms common to both rings are described by numbers and letters in square brackets, wherein the sequence of the numbers must correspond to the direction of the lettering of the base component. Finally the whole system is numbered.

● *Indicated hydrogen*

In some cases, heterocyclic systems occur as one or more structural isomers which differ only in the position of an H-atom. These isomers are designated by indicating the number corresponding to the position of the hydrogen atom in front of the name, followed by an italic capital H. Such a prominent H-atom is called an indicated hydrogen and must be assigned the lowest possible locant.

pyrrole

2H-pyrrole
(not 5H-pyrrole)

3,4-dihydro-2H-pyrrole
(not 4,5-dihydro-3H-pyrrole
or Δ¹ pyrroline)

The name pyrrole implies the 1-position for the H-atom.

Heterocyclic compounds in which a C-atom of the ring is part of a carbonyl group are named with the aid of indicated hydrogen as follows:

phosphinin-2-(1H)-one   pyrazin-2(3H)-one

2.2 Replacement Nomenclature

• *Monocyclic systems*

The type of heteroatom is indicated by a prefix according to Table 1. As all prefixes end with the letter a, replacement nomenclature is also known as 'a' nomenclature. *Position and prefix for each heteroatom are written in front of the name of the corresponding hydrocarbon.* This is derived from the heterocyclic system by replacing every heteroatom by CH_2, CH or C:

silacyclopenta-2,4-diene   cyclopentadiene   1-thia-4-aza-2-silacyclohexane   cyclohexane

Sequence and numbering of the heteroatoms follow the rules given in 2.1. The two compounds chosen as examples could also be named according to the Hantzsch–Widman system: silole, 1,4,2-thiazasilane.

• *Bi- and polycyclic systems*

Again, position and prefix are put in front of the name of the corresponding hydrocarbon, but *the numbering of the hydrocarbon is retained*:

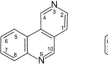

3,9 - diazaphenanthrene   phenanthrene   7 - oxabicyclo 2.2.1 heptane   bicyclo 2.2.1 heptane

The Hantzsch–Widman nomenclature can only be applied to the first example and this then results in different numbering.

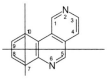

pyrido[4,3 - c]quinoline

2.3 Examples of Systematic Nomenclature

Finally, the systematic nomenclature of heterocyclic compounds will be illustrated by a few complex examples.

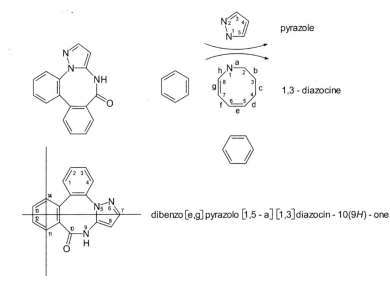

dibenzo [e,g] pyrazolo [1,5 - a] [1,3] diazocin - 10(9H) - one

An analysis of the system reveals two benzene rings, one pyrazole ring and one 1,3-diazocine ring, the latter ring being the base component according to the fourth criterion. The square brackets [1,3] indicate that the position of the two heteroatoms is not the basis for numbering the whole system.

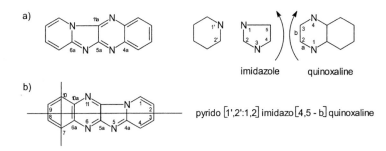

pyrido [1',2':1,2] imidazo [4,5 - b] quinoxaline

According to the third criterion, quinoxaline is the base component. The heterocycle imidazole, which is fused to the base component, is numbered in the usual way; the pyridine ring, however, is denoted by 1', 2', etc., and it is not necessary to mark the double bonds. Pyrido[1',2':1,2]imidazo denotes one ring fusion, imidazo[4,5-b]quinoxaline the other. For numbering polycyclic systems, five-membered rings must be drawn as shown above and not as regular pentagons. For the orientation in a system of coordinates, an additional rule has to be observed, namely that C-atoms common to two or more rings must be given the lowest possible locant. The numbering in (b) is therefore correct, while that in (a) is wrong, because 10a < 11a.

2 - ethoxy - 2,2 - dimethyl - 1,2,3 λ^5 - dioxaphospholane
(the standard bonding number of P is 3)

With ring atoms such as phosphorus, which can be tri- or pentavalent, a non-standard bonding number is indicated as an exponent of the Greek letter λ after the locant. In the example, this is shown by λ^5 (the 1993 Blue Book, p 21).

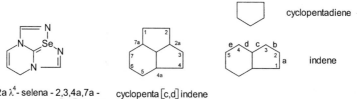

cyclopentadiene

indene

5H - 2a λ^4 - selena - 2,3,4a,7a - tetraazacyclopenta [c,d] indene

cyclopenta [c,d] indene

The name is constructed according to replacement nomenclature. The basic hydrocarbon with the greatest number of noncumulative double bonds is cyclopenta[c,d]indene. Note the retention of the numbering.

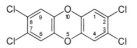

2,3,7,8 - tetrachlordibenzo [1,4] dioxin

In this case, [b,e] is omitted after dibenzo since there is no other possibility for ring fusion. This compound is also known as TCDD or Seveso dioxin.

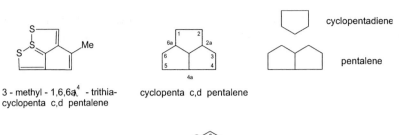

cyclopentadiene

pentalene

3 - methyl - 1,6,6a[4] - trithia-
cyclopenta c,d pentalene

cyclopenta c,d pentalene

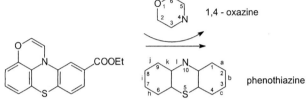

1,4 - oxazine

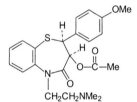

phenothiazine

ethyl [1,4] oxazino [2,3,4 - *kl*] phenothiazine - 6 - carboxylate

(2S, 3S) - 3 - acetoxy - 5 - (2 - dimethylaminoethyl)
2 - (4 - methoxyphenyl) - 2,3,4,5 - tetrahydro - 1,5 -
benzothiazepin - 4 - one

So far in all the examples, the base compound has been the heterocyclic system. If this is not the case, the univalent radical of the heterocyclic system is regarded as a substituent, e.g.:

3 - (4-pyridyl) butyric acid

The names of some univalent heterocyclic substituent groups are to be found in the list of trivial names in the 1993 Blue Book, p 172.

The most important source of information on heterocyclic and isocyclic systems is the *Ring Systems Handbook* of the Chemical Abstracts Service (CAS) published by the American Chemical Society. The 1988 edition is arranged as follows:

Band 1:   Ring Systems File I: RF 1–RF 27595,
Band 2:   Ring Systems File II: RF 27596–RF 52845,
Band 3:   Ring Systems File III: RF 52846–RF 72861,
Band 4:   Ring Formula Index, Ring Name Index.

Since 1991, cumulative supplements have been published annually.

The *Ring Systems File* is a catalogue of structural formulas and data. It lists the systems consecutively with numbering RF 1–RF 72861 on the basis of a ring analysis. The Ring Systems File starts with the following system:

The ring analysis shows:

1 RING: 3
AsPS

1 RING represents a monocycle, 3 denotes the ring size. The ring atoms are listed underneath in alphabetical order followed by

RF 1   88212-44-6
[Ring File (RF) Number]   (CAS Registry Number)
Thiaphospharsirane
AsH_2PS

the systematic name and molecular formula, and furthermore Wiswesser Notation, *Chem.Abstr.* reference (*Chem.Abstr.* volume number, abstract number), structural diagram.

An example from the Ring Systems File 1, p 758, is given below:

3 RINGS: 3,5,5
$C_2N–C_4S–C_5$

RF 15037   113688-14-5
Thieno[3',2':3,4]cyclopent[1,2-*b*]azirin
C_7H_3NS
T B355 CN GSJ
CA 108:112275y

The *Ring Formula Index* is a list of molecular formulas of all ring systems with ring atoms quoted in alphabetical order, H-atoms being omitted, e.g. C_6N_4: 2 RINGS, CN_4-C_6N, 1*H*-Tetrazolo[1,5-*a*]azepine [RF 9225].

With the aid of the *Ring File Number RF 9225*, the structural formula can be found in the Ring Systems File.

The *Ring Name Index* is an alphabetical list of the systematic names of all ring systems, e.g.: Benzo[4,5]indeno[1,2-*c*]pyrrole [RF 40064]. The Ring File Number allows access to the Ring Systems File.

Organization and use of the Ring Systems File, Ring Formula Index and Ring Name Index are, in each case, explained in detail at the beginning of the book.

2.4   Important Heterocyclic Systems

Several possibilities existed for the arrangement of chapters 3–8. For instance, the properties of the compounds could have been emphasized and the heteroarenes dealt with first, followed by the hetero-cycloalkenes and finally the heterocycloalkanes. However, in this book, the reactions, syntheses and synthetic applications of heterocyclic compounds are considered of greatest importance. In many cases, they are characteristic only of a single ring system. For this reason, we have adopted an arrangement for the systems which is similar to that shown on the cover of issues of the *Journal of Heterocyclic Chemistry*. The guiding principle is ring size (see Table 2). Heterocycles of certain ring sizes are further subdivided according to the type of heteroatoms, following the sequence shown in Table 1, starting with one heteroatom, two heteroatoms, etc. The parent compound is covered first, provided it is known or of importance. It is followed by the benzo-fused systems and finally by the partially or fully hydrogenated systems. Moreover, as in Gmélin's *Handbuch der Anorganischen Chemie* and Beilstein's *Handbuch der Organischen Chemie*, the principle of the latest possible classification is applied, i.e. condensed systems of two or more heterocycles are discussed under the parent compound to be found last in the classification. Finally, in view of the fact that there are more than 70,000 known heterocyclic systems, a selection had to be made. We have restricted ourselves to those systems

— which, because of their electronic or spatial structure, provided good examples for a theoretical illustration of molecular structure

— whose reactions afford examples of important reaction mechanisms and whose syntheses illustrate general synthetic principles

— which occur in natural products, drugs or in biologically active or industrially important substances

— which are important as building blocks or auxiliaries for carrying out synthetic transformations.

The description of each heterocyclic system is then arranged as follows:

| A | structure, physical and spectroscopic properties |

| B | chemical properties and reactions |

| C | syntheses |

| D | important derivatives, natural products, drugs, biologically active compounds, industrial intermediates |

| E | use as reagents, building-blocks or auxiliaries in organic synthesis. |

3 Three-Membered Heterocycles

The properties of three-membered heterocycles are mostly a result of the great bond angle strain (BAEYER strain). The resultant ring strain imparts on the compounds high chemical reactivity. Ring opening leading to acyclic products is typical. As set out above, the heterocycles will be treated in decreasing priority, starting with those with one heteroatom. The parent system of the three-membered heterocycles with one oxygen atom is called oxirene. Oxirenes are thermally very labile. They were postulated as intermediates in some reactions. However, oxirane, the saturated three-membered heterocycle with one oxygen, is of great importance.

3.1 Oxirane

A Oxiranes are also known as epoxides. Microwave spectra as well as electron diffraction studies show that the oxirane ring is close to being an equilateral triangle (see Fig. 3.1a).

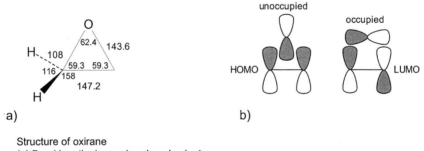

Fig. 3.1 Structure of oxirane
(a) Bond lengths in pm, bond angles in degrees
(b) Model for the bonding MO

The strain enthalpy was found to be 114 kJ mol^{-1}. The ionization potential is 10.5 eV; the electron which is removed derives from a nonbonding electron pair of the O-atom. The dipole moment is 1.88 D with the negative end of the dipole on the O-atom. The UV spectrum of gaseous oxirane has λ_{max} = 171 nm (lg ε = 3.34). The chemical shifts in the NMR spectrum are δ_H = 2.54, δ_C = 39.7. With a rise in the s-orbital component of the relevant C–H bonds, the ^{13}C–H coupling constant increases. The value of 176 Hz for oxirane is much greater than for aliphatic CH$_2$ groups. To explain this fact, one can imagine that the bonding MO of the C–O bonds are formed by interaction of the HOMO of an ethene molecule with an unoccupied AO of an O-atom, and also through interaction of the LUMO of the ethene molecule with an occupied AO of the O-atom (see Fig. 3.1b). As a result, the C–H bonds have more s-character than normal sp^3-hybridized C-atoms.

B Apart from the ring strain, a significant property of oxiranes is their BRÖNSTED and LEWIS basicity, because of the non-bonding electron pairs on the O-atom. Consequently, they react with acids. When handling oxiranes, it should also be borne in mind that many of them are carcinogenic. The most important reactions of oxiranes are described below.

Isomerization to carbonyl compounds

In the presence of catalytic amounts of LEWIS acids, e.g. boron trifluoride, magnesium iodide, or nickel complexes, oxiranes isomerize to give carbonyl compounds. Oxirane itself gives acetaldehyde:

Substituted oxiranes yield mixtures, e.g.:

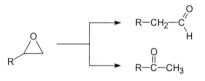

The nickel(II) complex $NiBr_2(PPh_3)_2$ yields aldehydes regioselectively [1].

Ring-opening by nucleophiles

Nucleophiles, e.g. ammonia or amines, cause oxiranes to ring-open to give amino alcohols:

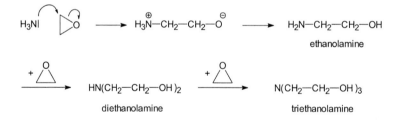

The concerted reaction corresponds to an S_N2 mechanism of a nucleophilic substitution on a saturated C-atom and is stereospecific. For example, from *cis*-2,3-dimethyloxirane, (±)-*threo*-3-aminobutan-2-ol is formed in the following reaction:

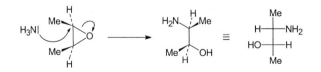

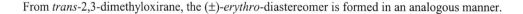

From *trans*-2,3-dimethyloxirane, the (±)-*erythro*-diastereomer is formed in an analogous manner.

Halogens react with oxiranes in the presence of triphenylphosphane or with lithium halides in the presence of acetic acid to give β-halo alcohols (halohydrins) [2], e.g.:

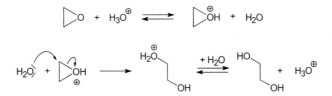

Acid-catalysed hydrolysis to 1,2-diols (glycols)

In this reaction, an acid–base equilibrium precedes the nucleophilic ring-opening of the oxirane ring.

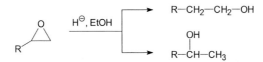

Such an A2 mechanism (A stands for acid, 2 indicates a bimolecular rate-determining step) results in a stereospecific reaction. Thus (±)-butane-2,3-diol is formed from *cis*-2,3-dimethyloxirane and *meso*-butane-2,3-diol from *trans*-2,3-dimethyloxirane. The oxirane obtained by epoxidation of cyclohexene is *trans*-cyclohexane-1,2-diol.

Reduction to alcohols

Oxiranes are reduced by sodium borohydride to give alcohols [3]. This reaction can be viewed as a ring-opening by the nucleophilic hydride ion:

$$R-CH_2-CH_2-OH$$
$$R-\overset{OH}{\underset{|}{CH}}-CH_3$$

Deoxygenation to olefins

A number of reagents deoxygenate oxiranes to give olefins [4]. For instance, a *trans*-oxirane yields a (*Z*)-olefin on treatment with triphenylphosphane at 200°C.

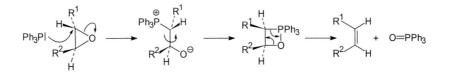

An (*E*)-olefin can therefore be converted into a (*Z*)-olefin via a *trans*-oxirane.

C For the synthesis of oxiranes, four reactions have proved useful. The oxirane syntheses described under (1), (3) and (4) are based on the same principle: an anionic oxygen atom substitutes intramolecularly a leaving group situated on a β-C-atom.

(1)   Cyclodehydrohalogenation of β-halo alcohols

Bases deprotonate β-halo alcohols to give the corresponding conjugate bases. This is followed by an intramolecular displacement of the halogen atom as the rate-determining step.

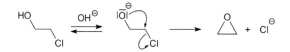

In spite of the ring strain in the product and the considerable activation enthalpy, the reaction occurs rapidly at room temperature owing to favourable entropy. The activation entropy is affected only by the loss of the degree of freedom of the internal rotation in the 2-chloroalkoxide ion because of the monomolecular rate-determining step.

Oxirane was first prepared by WURTZ (1859) by the action of sodium hydroxide on 2-chloroethanol.

(2)   Epoxidation of alkenes

Peroxy acids react with alkenes to give oxiranes. In the PRILESCHAJEW reaction, peroxybenzoic acid, m-chloroperoxybenzoic acid or monoperoxyphthalic acid is used. In weakly polar solvents, the reaction occurs in a concerted manner [5]:

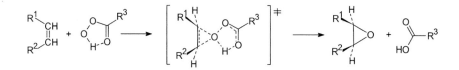

Peroxy acids possess strong intramolecular hydrogen bonds. The concerted progress results in a stereospecific reaction. (Z)-alkenes yield cis-oxiranes, (E)-alkenes trans-oxiranes.

tert-Butyl hydroperoxide is used for the SHARPLESS epoxidation. The epoxidation of allyl alcohol and substituted allyl alcohols with this reagent in the presence of titanium tetraisopropoxide Ti(OCHMe$_2$)$_4$ and (R,R)-(+)- or (S,S)-(−)-diethyl tartrate (DET) occurs enantioselectively (KATSUKI and SHARPLESS 1980) [6]:

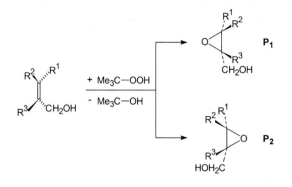

In the presence of (R,R)-(+)-DET, the enantiomer P_1 is formed with ee > 90%, while in the presence of (S,S)-(-)-DET, the enantiomer P_2 is obtained, also with ee > 90%. The SHARPLESS epoxidation is, therefore, an important method for asymmetric synthesis.

(3)   *Darzens reaction (glycidic ester synthesis)*

The reaction of α-halo esters with carbonyl compounds in the presence of sodium ethoxide leads to 2-(ethoxycarbonyl)oxiranes (DARZENS 1904). They are known as glycidic esters. In the first step, the α-halo ester is deprotonated by the base to the corresponding carbanion. This nucleophile adds to the carbonyl compound in a rate-determining step. Finally, the halogen atom is intramolecularly substituted, e.g.:

$$\text{ClCH}_2\text{—COOEt} + \text{EtO}^\ominus \rightleftharpoons \text{ClCH—COOEt} + \text{EtOH}$$

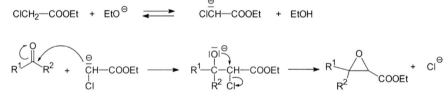

(4)   *Corey synthesis*

In this synthesis, S-ylide nucleophiles derived from trialkylsulfonium or trialkylsulfoxonium halides are made to react with carbonyl compounds (COREY 1962) [7], e.g.:

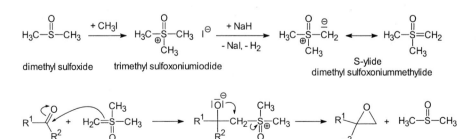

dimethyl sulfoxide   trimethyl sulfoxoniumiodide

S-ylide
dimethyl sulfoxoniummethylide

D **Oxirane** (ethylene oxide), a colourless, water-soluble, extremely poisonous gas of bp 10.5°C, is made on an industrial scale by direct air oxidation of ethene in the presence of a silver catalyst. Oxirane is important as an intermediate in the petrochemical industry. The annual production world-wide is estimated to be seven million tons.

Methyloxirane (propylene oxide) is a colourless, water-miscible liquid, bp 35°C. It is produced commercially from propene and *tert*-butyl hydroperoxide in the presence of molybdenum acetylacetonate [8].

(Chloromethyl)oxirane (epichlorohydrin) is prepared from allyl chloride as follows:

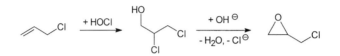

Epichlorohydrin is the starting material for epoxy resins. When used in excess, e.g. with bis-2,2-(4-hydroxyphenyl)propane, the so-called bisphenol A, in the presence of sodium hydroxide, it reacts to give linear polymers with oxirane end-groups.

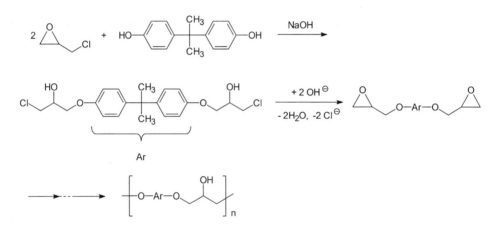

Thus propagation proceeds in two steps which are continuously repeated: opening of the oxirane ring by phenol interaction and closing of the oxirane ring by dehydrogenation. When mixed with diacid anhydrides, diamines or diols, an interaction with the oxirane end-groups of the macromolecules ensues, resulting in cross-linking (hardening). Epoxy resins find use as surface coatings, laminated materials and adhesives.

(Hydroxymethyl)oxirane (glycidol) is produced industrially by the oxidation of allyl alcohol with hydrogen peroxide in the presence of sodium hydrogen tungstate. It serves as a useful starting material in various syntheses [9].

Benzene oxide (7-oxabicyclo[4.1.0]hepta-2,4-diene) was obtained in an equilibrium mixture with the valence isomer oxepine (see p 459) (VOGEL 1967):

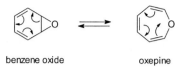

benzene oxide   oxepine

Benzene dioxide and benzene trioxide are also known [10]. Arene oxides are crucial intermediates in the carcinogenic action of benzo[*a*]pyrene and other polycondensed arenes [11]. Oxiranes are found relatively rarely in nature. An example of an oxirane in a natural product is, however, the juvenile hormone of the sphinx moth.

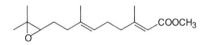

Furthermore, attention must be drawn to the part played by squalene epoxide as an initiator of steroid biosynthesis in eukaryotes. Antibiotics with oxirane rings, e.g. oleandomycine, have also been isolated.

E Oxiranes are of considerable importance as intermediates for multistep stereospecific syntheses of complex target molecules, because closing and opening reactions of the oxirane ring often occur without side reactions. Moreover, they proceed stereospecifically. The first steps in the total syntheses of all 16 stereoisomeric hexoses may serve as an example. These syntheses start from (*E*)-but-2-ene-1,4-diol, **1**, which is obtainable from acetylene and formaldehyde via butyne-1,4-diol [12].

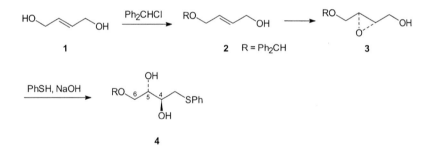

First a hydroxy group is protected by reaction with benzhydryl chloride (**2**). This is followed by a SHARPLESS epoxidation in the presence of (*R,R*)-(+)-DET to give **3**. This reacts with thiophenol and sodium hydroxide to give **4**, in which the C-atoms 4, 5 and 6 of the L-hexoses are already in place. The SHARPLESS epoxidation leads into the D-series with (*S,S*)-(–)-DET. In the course of steps **3** → **4**, two openings and one closure of the oxirane rings are observed:

The presence of the thioether group CH_2SPh in **4** is essential for linking the remaining two C-atoms by a PUMMERER rearrangement and a WITTIG reaction.

3.2   Thiirane

A Thiiranes are also known as episulfides. As a result of the greater atomic radius of the S-atom, the three atoms form an acute-angled triangle (see Fig. 3.2).

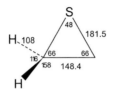

Fig. 3.2   Structure of thiirane
(bond lengths in pm, bond angles in degrees)

The thermochemically determined strain enthalpy of thiirane of 83 kJ mol^{-1} is less than that of oxirane. The ionization potential amounts to 9.05 eV, the dipole moment to 1.66 D. Both values are below those of oxirane. The chemical shifts in the NMR spectra are $\delta_H = 2.27$, $\delta_C = 18.1$.

B The properties of the thiiranes are primarily due to ring strain. In spite of the smaller strain enthalpy, thiirane is thermally less stable than oxirane. Even at room temperature, linear macromolecules are formed because of polymerization of ring-opened products. Substituted thiiranes are thermally more stable. The following reactions are typical for thiiranes [13]:

Ring-opening by nucleophiles

Ammonia, or primary or secondary amines, react with thiiranes to give β-amino thiols:

$$R-NH_2 \ + \ \overset{S}{\triangle} \ \longrightarrow \ R-NH-CH_2-CH_2-SH$$

The mechanism is the same as described on p 18 for oxiranes. However, the yields are lower, due to competing polymerization. Concentrated hydrochloric acid reacts with thiiranes to give β-chlorothiols (protonation on the S-atom and ring-opening by the nucleophilic chloride ion).

Oxidation

Thiiranes are oxidized by sodium periodate or peroxy acids to give thiirane oxides. They decompose at higher temperature into alkenes and sulfur monoxide:

$$\overset{S}{\triangle} \ \xrightarrow{\text{NaIO}_4} \ \overset{\overset{O}{\underset{||}{S}}}{\triangle} \ \xrightarrow{\Delta} \ H_2C{=}CH_2 \ + \ S{=}O$$

Desulfurization to alkenes

Triphenylphosphane, as well as trialkyl phosphites, have proved to be reliable reagents for this purpose. The reaction is stereospecific. *Cis*-thiiranes yield (*Z*)-olefins and *trans*-thiiranes yield (*E*)-olefins. The electrophilic attack of the trivalent phosphorus on the heteroatom is different from that described on p 29.

Metallic reagents, e.g. *n*-butyllithium, also bring about a stereospecific desulfurization of thiiranes.

C   The synthesis of thiiranes starting from *β*-substituted thiols and oxiranes can be achieved as follows.

(1)   *Cyclization of β-substituted thiols*

By analogy with the oxirane synthesis described on p 20, halothiols react with bases to give thiiranes. *β*-Acetoxythiols also yield thiiranes under similar conditions. 2-Sulfanylethanol reacts with phosgene in the presence of pyridine to give 1,3-oxathiolan-2-one, which on heating to 200°C decarboxylates to give thiirane.

(2)   *Ring transformation of oxiranes*

Conversion of one heterocyclic system into another is known as ring transformation. Oxiranes react with aqueous ethanolic potassium rhodate to give thiiranes, probably according to the following mechanism:

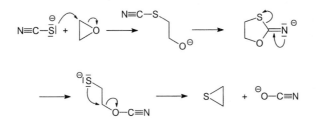

D **Thiirane** (ethylene sulfide) is a colourless liquid, sparingly soluble in water and of bp 55°C.

E A method for C–C coupling, which is based on closing a thiirane ring and opening it by desulfurization, is known as sulfide contraction after ESCHENMOSER, e.g.:

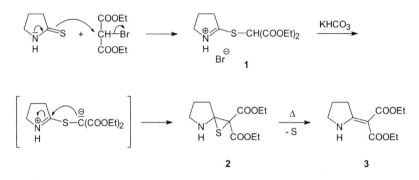

1

2   **3**

Pyrrolidine-2-thione is *S*-alkylated with bromomalonic diethyl ester. On treatment with a solution of $KHCO_3$, the resulting iminium salt **1** yields a thiirane **2** which desulfurizes at 60°C to give **3** [14].

3.3   *2H*-Azirine

A Unlike 1*H*-azirine, 2*H*-azirines can be used preparatively, although the ring strain is substantially greater than that of the saturated three-membered heterocycles. The ring strain enthalpy amounts to approximately 170 kJ mol^{-1}.

B 2*H*-Azirine is thermally unstable and has to be stored at very low temperatures. Substituted 2*H*-azirines are more stable. They are liquids or low melting solids. Their basicity is substantially lower than that of comparable aliphatic compounds. For instance, 2-methyl-3-phenyl-2*H*-azirine is not soluble in hydrochloric acid.

The ring strain endows the C=N double bond with an exceptionally high reactivity. Electrophilic reagents attack the N-atom, nucleophilic reagents the C-atom. For example, methanol added in the presence of a catalytic amount of sodium methoxide produces 2-methoxyaziridines:

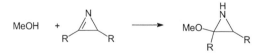

Carboxylic acids also add to the C=N double bond and the products rearrange to more stable compounds with opening of the aziridine ring. A method for peptide synthesis is based on these reactions [15]:

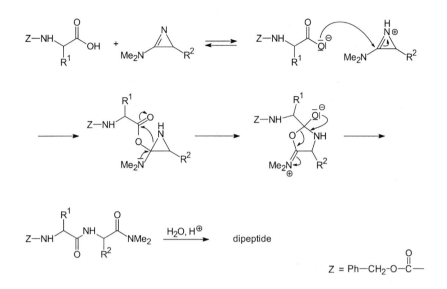

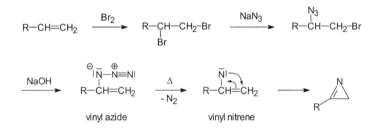

For instance, when the carboxyl group of an *N*-(benzoyloxycarbonyl)amino acid reacts with a 2-substituted or a 2,2-disubstituted 3-(dimethylamino)-2*H*-azirine in diethyl ether at room temperature, the *N,N*-dimethylamide of the dipeptide is obtained quantitatively. Its hydrolysis with 3N hydrochloric acid yields the *N*-(benzyloxycarbonyl)dipeptide.

2*H*-Azirines undergo cycloadditions. They react, for instance, as dienophiles in [4+2] cycloadditions.

| C | 2*H*-Azirines are prepared by thermolysis or photolysis of vinyl azides, which are obtainable from alkenes.

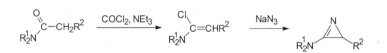

The dediazoniation of the vinyl azide proceeds via vinyl nitrene. A variant of this synthesis enables the preparation of 3-(dialkylamino)-2*H*-azirines starting from *N,N*-disubstituted acid amides [15]:

The methyl iodides of dimethyl hydrazones furnish 2*H*-azirines when treated with bases such as sodium propoxide:

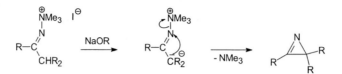

3.4   Aziridine

A   Aziridine was once known as ethylene imine. Bond lengths and bond angles are essentially the same as those in oxirane. The plane in which the N-atom, its nonbonding electron pair and the N–H bond are situated is perpendicular to the plane of the aziridine ring (see Fig. 3.3).

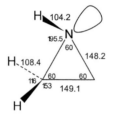

Fig. 3.3   Structure of aziridine
(Bond lengths in pm, bond angles in degrees)

2-Methylaziridine would be expected, therefore, to display diastereoisomerism. Trivalent N-atoms are, however, liable to pyramidal inversion.

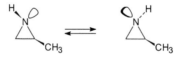

Although the activation enthalpy of this process $\Delta G^{\neq} = 70$ kJ mol^{-1}, i.e. substantially greater than with a secondary aliphatic amine, inversion occurs so rapidly at room temperature that the diastereoisomers are not separable. However, in the case of 1-chloro-2-methylaziridine, where $\Delta G^{\neq} = 112$ kJ mol^{-1}, the mixture of stereoisomers can be separated.

The strain enthalpy of aziridine determined thermochemically amounts to 113 kJ mol^{-1} and is, therefore, almost identical to that of oxirane. The ionization potential was found to be 9.8 eV and the excited electron derives from the nonbonding electron pair of the N-atom. The dipole moment of 1.89 D is almost the same as that of oxirane. The chemical shifts in the NMR spectra are $\delta_H = 1.5$ (CH), 1.0 (NH) and $\delta_C = 18.2$.

B Care is advisable when handling aziridines, because many show considerable toxicity. The following reactions are of importance:

Acid base reactions

Aziridines unsubstituted on the N-atom behave like secondary amines; *N*-substituted aziridines behave like tertiary amines. They react with acids to give aziridinium salts:

The pK_a value of the aziridinium ion is 7.98. Aziridine is thus a weaker base than dimethylamine (pK_a = 10.87), but stronger than aniline (pK_a = 4.62).

The aziridine ring is destabilized by salt formation, and ring-opening by nucleophiles is favoured. Aziridine itself reacts with acids explosively to give polymeric products.

Reactions with electrophilic reagents

Aziridines, like amines, are nucleophiles and react with electrophiles. A nucleophilic substitution on a saturated C-atom and a nucleophilic addition to an alkene bearing an acceptor group serve as examples:

Ring-opening by nucleophiles

Ammonia and primary amines react with azridines to give 1,2-diamines. The mechanism and the stereochemistry of this reaction are similar to the corresponding reactions of the oxiranes.

The ring-opening of the aziridines is catalyzed especially effectively by acids (A2 mechanism, see p 19). The acid-catalysed hydrolysis to give amino alcohols serves as an example:

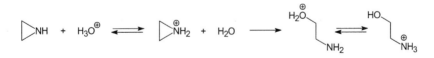

Such reactions can also be interpreted as alkylation of the nucleophile. This explains the cytostatic and antitumour action of aziridines and of bis(2-chloroethyl)amine **1**. An equilibrium exists between **1** and 1-(2-chloroethyl)aziridinium chloride **2**.

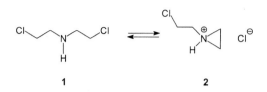

                    1   2

Nucleophilic cell components, e.g. the amino groups of the guanine bases in DNA, are alkylated by the aziridinium ion as a result of a nucleophilic ring-opening. In the case of bis(2-chloroethyl)amine, the reaction can be repeated on a guanine base of the other DNA strand of the double helix. This results in cross-linking of the two DNA strands and consequently blocks replication (see p 31).

Deamination to alkenes

Aziridines with an unsubstituted N-atom are stereospecifically deaminated by nitrosyl chloride via the corresponding *N*-nitroso compound:

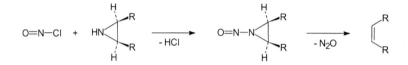

C   The synthesis of aziridines can be achieved from substituted amines or alkenes.

(1)   *Cyclization of β-substituted amines*

β-Amino alcohols, which are conveniently made from oxiranes with ammonia or amines, react with thionyl chloride to give chloroamines, which can be cyclized to aziridines by alkali hydroxide (GABRIEL 1888).

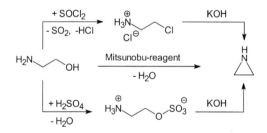

Sulfate esters, obtained from amino alcohols and sulfuric acid, when treated with alkali also form aziridines (WENKER 1935). In both cases, the amine is liberated from the ammonium salts by the base. The leaving group Cl or OSO_3^- is substituted intramolecularly by the amino group on the *β*-C-atom.

     The direct cyclodehydration of *β*-amino alcohols can be effected with the MITSUNOBU reagent (triphenylphosphane/diethyl azodicarboxylate) [16].

(2)   *Thermal or photochemical reaction of azides with alkenes*

Phenyl azide reacts with alkenes to give 4,5-dihydro-1,2,3-triazoles (1,3-dipolar cycloaddition, see p 204), which are thermally or photochemically converted into aziridines through loss of nitrogen:

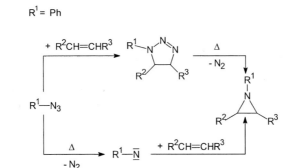

Thermolysis of ethyl azidoformate, however, produces ethoxycarbonylnitrene, which by a [2+1] cycloaddition reacts with alkenes to form aziridines. The mechanism is thus influenced by the azide substituent R^1.

D   **Aziridine**, a colourless, water-soluble, poisonous liquid (bp 57°C) of ammoniacal odour is relatively stable, thermally, but is best stored in a refrigerator over sodium hydroxide.

Some natural products contain an aziridine ring, e.g. mitomycins (**3**: mitomycin C). This is responsible for the cytostatic and antitumour activity of these antibiotics. Many synthetic aziridines have been screened for their antitumour activity. Some have reached the clinic, especially as antileukaemic agents, e.g. **4** and **5**.

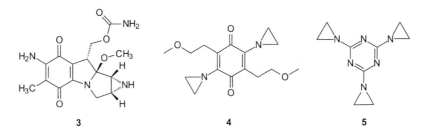

| 3 | 4 | 5 |

E   Aziridines with C_2 symmetry have been used successfully as chiral auxiliaries for alkylations and aldol reactions [17].

3.5 Dioxirane

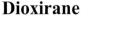

A-C Dioxiranes have been available only since the mid-eighties [18]. They are synthesized by oxidation of ketones with potassium hydrogenperoxysulfate (oxone®), e.g.:

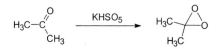

Dimethyldioxirane, together with acetone, is removed from the reaction vessel by distillation. The yellow 0.1-0.2 M solution can be used as an oxidizing agent, e.g. for the epoxidation of olefins [19], for the oxidation of enolates to α-hydroxycarbonyl compounds and for the oxidation of primary amines into nitro compounds:

$$R-NH_2 \ + \ 3\ H_3C \overset{O}{\underset{H_3C}{\triangle}}O \longrightarrow R-NO_2 + 3\ H_3C \overset{O}{\underset{}{\|}} CH_3 + H_2O$$

Boron trifluoride catalyses the isomerization of dimethyldioxiranes to methyl acetate.

Difluorooxirane is formed as a pale-yellow, normally stable gas when an equimolar mixture of FCO_2F and ClF is passed over a CsF catalyst [20].

3.6 Oxaziridine

A,B Oxaziridines are structural isomers of oximes and nitrones. Trialkyl oxaziridines are colourless liquids, sparingly soluble in water. The following reactions are typical for oxaziridine.

Isomerization to nitrones

As a reversal to the photoisomerization of nitrones (see p 33), oxaziridines can be converted into nitrones by thermolysis. The required temperature depends on the type of oxaziridine substituents.

Ring-opening by nucleophiles

On acid-catalysed hydrolysis, 2-alkyl-3-phenyloxaziridines yield benzaldehyde and *N*-alkylhydroxyl-amines, e.g.:

Reduction to imines

Oxaziridines, particularly 2-(phenylsulfonyl)oxaziridines, are used as reagents in a number of oxidation procedures. The oxidation of sulfides to sulfoxides may serve as an example:

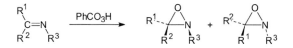

C The synthesis of oxaziridines can be accomplished from imines, nitrones or carbonyl compounds:

(1) *Oxidation of imines with peroxy acids:*

As in the epoxidation of alkenes, (see p 20), a stereospecific *cis*-addition is involved. In the case of 2-substituted oxaziridines ($\Delta G^{\neq}$ = 100-130 kJ mol^{-1}), the activation enthalpy of the pyramidal inversion of the N-atom is so high that the configuration of the N-atom is fixed at room temperature. Thus, the configuration of the starting material is preserved and the racemate of one of the diastereoisomeric oxaziridines is formed. In the case of chiral imines or chiral peroxy acids, the reaction proceeds enantioselectively.

(2) *Photoisomerization of nitrones:*

(3) *Amination of carbonyl compounds:*

In the presence of a base, hydroxylamine-*O*-sulfonic acid or chloramine aminate carbonyl compounds nucleophilically (SCHMITZ 1961), e.g.:

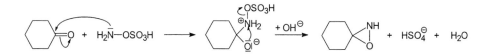

In this reaction, the intramolecular nucleophilic substitution occurs on an N-atom.

E Oxaziridines are oxidizing agents as well as important synthetic intermediates [21]. For instance, *N*-hydroxyaminocarboxylic esters **2** can be prepared from α-aminocarboxylic acid esters with oxaziridines **1** as intermediates as follows:

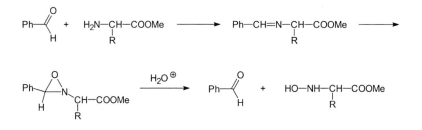

3.7   *3H*-Diazirine

A-D *3H*-Diazirines are structural isomers of diazoalkanes. They are gases or colourless liquids, e.g. 3,3-dimethyldiazirine, bp 21°C. Liquid *3H*-diazirines can decompose explosively. Their basicity is very low. Unlike diazoalkanes, they react with acids only slowly, with the liberation of nitrogen.

The dediazoniation of *3H*-diazirines can be effected thermally or photochemically [22]. In the absence of carbene acceptors, the initially formed carbenes isomerize to give olefins, e.g.:

$$H_3C \text{--} \overset{N}{\underset{N}{\triangle}} \text{--} CH_3 \quad \xrightarrow[-N_2]{\Delta \text{ or } h\nu} \quad H_3C \text{--} \overset{..}{C} \text{--} CH_3 \quad \longrightarrow \quad H_3C \text{--} CH\text{=}CH_2$$

3H-Diazirines are prepared by oxidation of *N*-unsubstituted diaziridines with silver oxide or mercury oxide (PAULSEN 1960, SCHMITZ 1961):

$$R^1 \text{--} \overset{H N}{\underset{R^2}{\triangle}} \text{--} NH \quad + \quad Ag_2O \quad \longrightarrow \quad R^1 \text{--} \overset{N}{\underset{R^2}{\triangle}} \text{--} N \quad + \quad 2\,Ag \quad + \quad H_2O$$

3-Chloro-*3H*-diazirines are formed by oxidation of amidines with sodium hypochlorite:

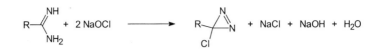

3.8   Diaziridine

A-C Diaziridines are crystalline, weakly basic compounds. As already explained in connection with oxaziridines (see p 33), the N-atoms are configurationally stable so that stereoisomerism is possible.

The acid-catalysed hydrolysis of diaziridines yields ketones and hydrazines:

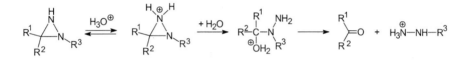

Thus, a synthesis of hydrazines is available starting from imines and hydroxylamine-*O*-sulfonic acid, or from *N*-substituted hydroxylamine-*O*-sulfonic acids.

Diaziridines unsubstituted on the N-atoms can be oxidized to give 3*H*-diazirines.

Diaziridines are prepared by the action of ammonia and chlorine on ketones (PAULSEN, SCHMITZ 1959). Initially, chloramine is formed:

$$2\,NH_3 + Cl_2 \longrightarrow NH_2Cl + NH_4Cl$$

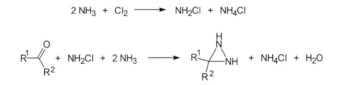

The action of ammonia or primary amines and hydroxylamine-*O*-sulfonic acid upon ketones also yields diaziridines. Likewise, the amination of imines (azomethines) with hydroxylamine-*O*-sulfonic acid yields diaziridines:

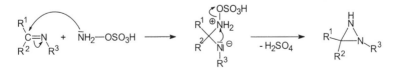

Summary of the general chemistry of three-membered heterocycles

- The reactivity of the compounds is determined mainly by the ring strain, but also by the nature of the heteroatom or heteroatoms.

- A typical reaction of three-membered heterocycles is nucleophilic ring-opening resulting in the formation of 1,2-disubstituted aliphatic compounds.

- A consequence of three-membered heterocycles possessing nonbonding electron pairs is that they behave as BRÖNSTED bases as well as LEWIS bases. Accordingly, they react with BRÖNSTED acids and with electrophiles.

- Some systems isomerize to give aliphatic compounds, namely
  — oxiranes give carbonyl compounds
  — dioxiranes give esters of carboxylic acids
  — oxaziridines give nitrones

- Appropriate reagents remove the heteroatoms to form alkenes (deoxygenation, desulfonation, deamination, dediazoniation).

- The most important synthetic principle is the intramolecular nucleophilic substitution of a β-positioned leaving group
  — by an O-atom (oxiranes)
  — by an S-atom (thiiranes)
  — by an N-atom (aziridines)
  — by an anionic C-atom (2H-azirines)

- Oxygen-containing heterocycles can be synthesized by the action of peroxy compounds on alkenes, ketones or imines.

- Amination of carbonyl compounds or imines yield oxaziridines and diaziridines.

- Azides and alkenes furnish N-heterocycles (aziridines, 2H-azirines)

- Only oxiranes are important in preparative chemistry. In some cases, however, other three-membered heterocycles are useful synthetic intermediates or reagents (2H-azirines, dioxiranes, oxaziridines, diaziridines).

References

[1] A. Miyashita, T. Shimada, A. Sugawara, H. Nohira, *Chem. Lett.* **1986**, 1323.

[2] G. Palumbo, C. Fereri, R. Caputo, *Tetrahedron Lett.* **1983**, *24*, 1307;
J. S. Baywa, R. C. Anderson, *Tetrahedron Lett.* **1991**, *32*, 3021;
C. Bonini, G. Righi, *Tetrahedron* **1992**, *48*, 1531.

[3] A. Ookawa, M. Kitade, K. Soai, *Heterocycles* **1988**, *27*, 213.

[4] H. N. C. Wong, M. Y. Honn, C. W. Tse, Y. C. Yip, J. Tanko, T. Hudlicky, *Heterocycles* **1987**, *26*, 1345.

[5] V. G. Dryuk, *Tetrahedron* **1976**, *32*, 2855.

[6] A. Pfenninger, *Synthesis* **1986**, 89;
E. J. Corey, *J. Org. Chem.* **1990**, *55*, 1693;
P. Besse, H. Veschambre, *Tetrahedron* **1994**, *50*, 8885.

[7] Yu. G. Gololobov, A. N. Nesmeyanov, V. P. Lysenko, I. E. Boldeskal, *Tetrahedron* **1987**, *43*, 2609.

[8] H. Mimoun, *Angew. Chem. Int. Ed. Engl.* **1982**, *21*, 734.

[9] A. Kleemann, R. S. Nygren, R. M. Wagner, *Chem.-Ztg.* **1980**, *104*, 283.

[10] W. Adam, M. Balci, *Tetrahedron* **1980**, *36*, 833;
H.-J. Altenbach, B. Voss, E. Vogel, *Angew. Chem. Int. Ed. Engl.* **1983**, *22*, 410.

[11] R. G. Harvey, *Acc. Chem. Res.* **1981**, *14*, 218;
J. M. Sayer, A. Chadha, S. K. Argawal, H. J. C. Yeh, H. Yagi, D. M. Jerina, *J. Org. Chem.* **1991**, *56*, 20.

[12] S. Y. Ko et al., *Science* **1983**, *220*, 949 ;
Y. Mori, *Chem. Eur. J.* **1997**, *3*, 849.

[13] A. V. Fokin, M. A. Allakhverdiev, A. F. Kolomiets, *Usp. Chim.* **1990**, *59*, 705.

[14] L. F. Tietze, Th. Eicher, *Reactions and Synthesis in the Organic Chemistry Laboratory*, University Science Books: Mill Valley, CA **1989**.

[15] P. Wipf, H. Heimgartner, *Helv. Chim. Acta* **1988**, *71*, 140;
H. Heimgartner, *Angew. Chem. Int. Ed. Engl.* **1991**, *30*, 238;
C. B. Bucher, H. Heimgartner, *Helv. Chim. Acta* **1996**, *79*, 1903;
F. Palacias, A. M. Ochoa de Rentana, E. Martinez de Marigorta, J. M. de los Santos, *Eur. J. Org. Chem.* **2001**, 2391.

[16] J. R. Pfister, *Synthesis* **1984**, 969.

[17] D. Tanner, C. Birgersson, A. Gogoll, *Tetrahedron* **1994**, *50*, 9797;
H. M. I. Osborn, J. Sweny, *Tetrahedron: Asymmetry* **1997**, *8*, 1693.

[18] M. Gilbert, M. Ferrer, F. Sanchez-Baeza, A. Messeguer, *Tetrahedron* **1997**, *53*, 8643;
S. E. Denmark, Z. Wu, *J. Org. Chem.* **1997**, *62*, 8964.

[19] R. W. Murray, D. L. Shiang, *J. Chem. Soc., Perkin Trans. 2*, **1990**, 349.

[20] O. Reiser, *Angew. Chem. Int. Ed. Engl.* **1994**, *33*, 69;
W. Adam, A. K. Smerz, C.-G. Zhao, *J. Prakt. Chem.* **1997**, *339*, 298.

[21] F. A. Davis, A. Kumar, B.-C. Chen, *J. Org. Chem.* **1991**, *56*, 1143;
S. Andreae, E. Schmitz, *Synthesis* **1991**, 327;
J. Aube, *Chem. Soc. Rev.* **1997**, *26*, 269.

[22] M. T. H. Liu, *Chem. Soc. Rev.* **1982**, *11*, 127;
I. R. Likhotvorik, E. L. Tae, C. Ventre, M. S. Platz, *Tetrahedron Lett.* **2000**, *41*, 795.

4   Four-Membered Heterocyles

In four-membered heterocycles, the ring strain is less than in the corresponding three-membered compounds and is approximately equal to that found in cyclobutane. Nevertheless, ring-opening reactions forming acyclic products predominate. At the same time, analogy with the reactivity of the corresponding aliphatic compounds (ethers, thioethers, secondary and tertiary amines, imines) becomes more evident.

4.1   Oxetane

A The oxetane ring represents a slightly distorted square because the bond angle at the O-atom is 92°. The strain enthalpy has been determined thermochemically to be 106.3 kJ mol^{-1} and so only 7.7 kJ mol^{-1} less than that of oxirane, although the bond angles are 30° larger. The reason for this is that the planarity of the oxetane ring causes a considerable PITZER strain due to the eclipsing interactions of the C–H bonds. This strain is reduced by ring-puckering between two nonplanar structures, which simultaneously leads to a reduction in the bond angles.

This results in a compromise between bond angle strain and PITZER strain, which minimizes the total strain energy. The activation energy of the ring inversion amounts to 0.181 kJ mol^{-1}, which is less than the energy of the molecular vibration. Consequently, the process occurs so fast that the molecule should be regarded as planar.

B Oxetanes react like oxiranes with ring-opening at a slower rate and under forcing conditions. Two reactions are of general importance:

Acid-catalysed ring-opening by nucleophiles

Hydrogen halides react with oxetanes to give 3-halo alcohols. The acid-catalyzed hydrolysis yields 1,3-diols.

Cyclooligomerization and polymerization

LEWIS acids, e.g. boron trifluoride, can add to a nonbonding electron pair of the O-atom. Thus, in dichloromethane as solvent, a cyclooligomerization is induced. The main product is the cyclotrimer 1,5,9-trioxacyclododecane [1]:

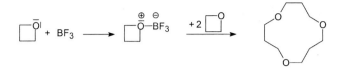

Under different conditions, especially in the presence of water, linear polymers are formed.

C For the synthesis of oxetanes, two methods are useful, namely the cyclization of γ-substituted alcohols and the PATERNO–BÜCHI reaction.

(1)   Cyclization of γ-substituted alcohols

Alcohols with a nucleofuge leaving group in the γ-position can be cyclized to give oxetanes. Thus the cyclodehydrohalogenation of γ-halo alcohols occurs in an analogous way to the oxirane synthesis from β-halogenated alcohols (see p 20). Oxetanes can be prepared from 1,3-diols via monoarene sulfonates:

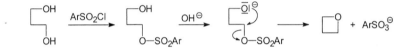

In an alternative synthesis, the 1,3-diol in THF is converted to the lithium alkoxide with n-butyllithium. This is followed by addition of tosyl chloride; cyclization is finally effected with *n*-butyllithium [2].

(2)   Paterno–Büchi reaction

The photochemical [2+2] cycloaddition of carbonyl compounds to alkenes yielding oxetanes is known as the PATERNO–BÜCHI reaction [3]. The carbonyl compound is converted by absorption of a quantum of light into an electronically excited state (n → π* transition), which at first is in the singlet state (in which the spin moments of the electron in the n-MO and the electron in the π*-MO are antiparallel). This is followed by conversion into the lower energy triplet state (in which the spin moments of the two electrons are parallel). The ensuing addition of the alkene should, according to the WOODWARD–HOFFMANN rules, occur in a concerted and, therefore, stereospecific manner. This is indeed observed with alkenes possessing electron-withdrawing groups, e.g.:

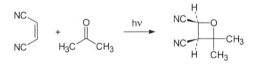

In contrast, alkenes with donor substituents react via radical intermediates, e.g.:

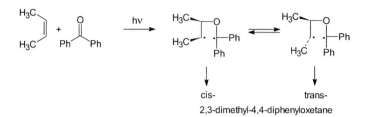

cis-   trans-
2,3-dimethyl-4,4-diphenyloxetane

Even the C–O double bonds of quinones and carboxylic esters can undergo the PATERNO–BÜCHI reaction.

D **Oxetane**, a colourless, water-miscible liquid of bp 48°C, is obtained in 40% yield by heating (3-chloropropyl)acetate with concd KOH solution [1].
Oxetane-2-ones are also β-lactones [4]. They are prepared by cyclodehydration of β-hydroxy-carboxylic acids with phenylsulfonyl chloride in pyridine:

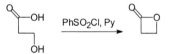

A further method is the [2+2] cycloaddition of aldehydes to ketenes catalysed by LEWIS acids:

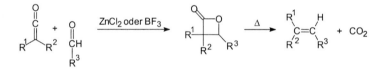

Oxetan-2-ones decarboxylate on heating to give olefins. The synthesis of olefins starting from β-hydroxycarboxylic acids or from ketenes and aldehydes can thus be carried out as an alternative to the WITTIG reaction.

Oxetan-2-ones are more reactive than γ- and δ-lactones because of ring strain. On treatment with sodium hydroxide, salts of β-hydroxycarboxylic acids are formed owing to attack of hydroxide ions on the C-atom of the carbonyl group.

Diketene (4-methyleneoxetane-2-one) is formed by dimerization of ketene which in turn is prepared by pyrolysis of acetone or acetic acid. The compound is an industrial intermediate. It ring-opens with ethanol to give ethyl acetoacetate; the nucleophile attacks the C-atom of the carbonyl group.

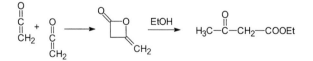

Oxetanes rarely occur in nature. The diterpene alcohol taxol (paclitaxel®) was isolated in 1971 from the bark of the pacific yew tree (*Taxus brevifolia*) native to the northwest USA, and its structure elucidated. The compound, which is now marketed, contains an oxetane ring and displays a strong antitumour and antileukaemia activity. Its total synthesis has been achieved [5].

4.2   Thietane

A The thermochemically determined strain enthalpy of thietane is only 80 kJ mol^{-1}. The activation energy for the ring inversion was found spectroscopically to be 3.28 kJ mol^{-1} and lies above the four lowest vibration levels. The ring is thus not planar:

B The reactivity of thietanes towards nucleophiles is much less than that of thiiranes. For instance, thietane does not react with ammonia or amines at room temperature. Electrophiles attack the S-atom and can thereby cause ring-opening. Thus, addition of acids leads to polymerization. A further example is their ring-opening by haloalkanes:

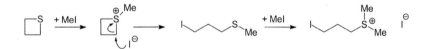

This demonstrates again that a positive charge on the heteroatom destabilizes the ring.
Hydrogen peroxide or peroxy acids oxidize thietanes to 1,1-dioxides (cyclic sulfones) via 1-oxides:

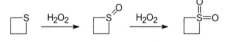

C Thietanes can be prepared from γ-halo thiols or 1,3-dihaloalkanes as follows:

(1)   *Cyclization of γ-halo thiols or their acetyl derivatives by bases:*

$$\text{(structure)} + 2\,OH^{\ominus} \longrightarrow \text{(structure)} + Cl^{\ominus} + CH_3COO^{\ominus} + H_2O$$

(2)   *Action of sodium or potassium sulfide on 1,3-dihaloalkanes:*

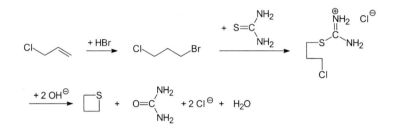

Better yields are obtained from (3-chloropropyl)isothiouronium bromide starting with 1-bromo-3-chloropropane. The latter is available by addition of hydrogen bromide to allyl chloride:

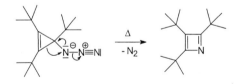

D | **Thietane** is a colourless, water-insoluble liquid, bp 94°C, polymerizing slowly at room temperature, faster on exposure to light.

4.3   Azete

Azete is iso-π-electronic with cyclobutadiene and is therefore the simplest antiaromatic heteroannulene. It would be expected to be thermally unstable and extremely reactive, and as yet the parent compound has not been synthesized. In 1973, tris(dimethylamino)azete was described. In 1986, REGITZ succeeded in synthesizing tri-*tert*-butylazete by thermolysis of 3-azido-1,2,3-tri-*tert*-butylcyclopropene [6]:

Tri-*tert*-butylazete crystallizes as reddish needles, mp 37°C. The space-filling *tert*-butyl groups protect the ring so strongly that polymerization proceeds extremely slowly; $\Delta G^{\neq}$ is high even at elevated temperatures. $\Delta G^{\neq}$ Is also very high for the decomposition of the compound into an alkyne and a nitrile as a consequence of a concerted [2+2] cycloreversion, because the reaction is not allowed according to the WOODWARD–HOFFMANN rules. Tri-*tert*-butyl azete is, therefore, in two ways kinetically stabilized.

4.4   Azetidine

A   Azetidine was previously called trimethyleneimine. The activation energy of the ring inversion is 5.5 kJ mol^{-1} and is therefore only slightly below the value for cyclobutane (6.2 kJ mol^{-1}). The conformer with an equatorial N–H bond is lower in energy:

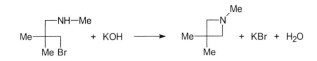

B   Azetidines are thermally stable and less reactive than aziridines. They behave in their reactions almost like secondary alkylamines. The pK_a value of azetidine is 11.29 and so it is more basic than aziridine (pK_a = 7.98) and even dimethylamine (pK_a = 10.73). Azetidines unsubstituted on the N-atom react with alkyl halides to give 1-alkylazetidines which can react further to give quaternary azetidinium salts. With acyl halides, they produce acylazetidines and with nitrous acid, they give 1-nitrosoazetidines.

A positive charge on the N-atom destabilizes the ring, as is the case with the aziridines. Ring-opening by nucleophiles proceeds with acid catalysis. Hydrogen chloride yields γ-chloroamines. 1,1-Dialkylazetidinium chlorides isomerize on heating to give tertiary γ-chloroamines. By contrast, neither bases nor reducing agents open the aziridine ring.

C   The synthesis of azetidines can be accomplished starting with γ-substituted amines or 1,3-dihaloalkanes:

(1)   *Cyclization of γ-substituted amines*

γ-Halogen substituted amines are dehydrohalogenated by bases, e.g.:

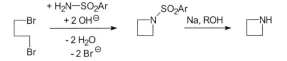

The yields are lower than in the analogous aziridine synthesis. The MITSUNOBU reagent (see p 30) [7] is suitable for the cyclodehydration of γ-amino alcohols.

(2)   *Action of p-toluenesulfonamide and bases on 1,3-dihaloalkanes:*

The tosyl group can be reductively removed from the 1-tosylazetidine.

D **Azetidine** is a water-miscible, colourless liquid of bp 61.5°C. It smells like ammonia and fumes in air.

Azetidin-2-ones are also β-lactams [7a]. The cyclodehydration of β-amino carboxylic acids to give azetidin-2-ones is best carried out with CH_3SO_2Cl and $NaHCO_3$ in acetonitrile [8]. The cyclization of ethyl β-aminopropionate can be carried out with 2,4,6-trimethylphenylmagnesium bromide:

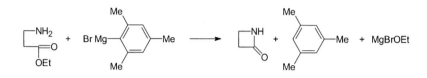

[2+2] Cycloadditions are of great importance for the synthesis of azetidin-2-ones. Three approaches have proved successful.

- *Imines + ketenes*

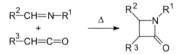

This [2+2] cycloaddition was found by STAUDINGER already in 1907.

- *Imines + activated carboxylic acids:*

Imines react with acyl halides in the presence of triethylamine to give azetidin-2-ones. An intermediate ketene is formed from the acyl chloride and the amine:

$$R^3\text{-}CH_2\text{-}C\overset{O}{\underset{Cl}{\big\langle}} + NEt_3 \longrightarrow R^3\text{-}CH=C=O + \overset{\oplus}{H}NEt_3\ Cl^{\ominus}$$

The activation of the carboxylic acids can be achieved also with the MUKAIYAMA reagent (2-chloro-1-methylpyridinium iodide, see p 308). Carboxylic acids react with this reagent, tri-*n*-propylamine and imines in dichloromethane to give azetidin-2-ones [9].

- *Chlorosulfonyl isocyanate + alkenes*

Chlorosulfonyl isocyanate is prepared from chlorocyanogen and sulfur trioxide. It reacts with alkenes to form 1-chlorosulfonyl azetidin-2-ones, from which the corresponding compounds, unsubstituted in the 1-position, can be obtained by the action of thiophenol. The cycloaddition occurs stereospecifically; *cis*-azetidin-2-one is formed from a *Z*-alkene:

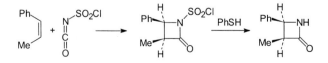

Azetidin-2-ones are more reactive than γ- and δ-lactams because of ring strain. This is true for the alkaline fission to give salts of β-amino carboxylic acids, as well as for the acid-catalysed hydrolysis to β-carboxyethylammonium salts. Starting from alkenes and chlorosulfonyl isocyanate, a stereocontrolled synthesis of β-amino carboxylic acids can be realized. Ammonia and amines react with azetidin-2-ones, also with ring-opening, to produce β-amino carboxylic amides. Hence they are acylated by azetidin-2-ones:

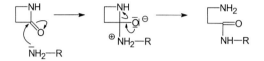

Azetidin-2-ones are reduced chemoselectively by diisobutylaluminium hydride or by chloroaluminium and dichloroaluminium hydrides in THF to form azetidines [10].

The azetidin-2-one system is present in penicillins (see p 159) and cephalosporines (see p 389). These natural products are known as β-lactam antibiotics. They block the biosynthesis of compounds which form the bacterial cell walls. The β-lactam antibiotics are the most prescribed antibiotics today.

(S)-Azetidine-2-carboxylic acid is a cyclic amino acid, not present in proteins, found in agaves and liliaceous plants. It was first isolated from lilies of the valley:

4.5   1,2-Dioxetane

$$\begin{array}{c} \text{O}{-}\text{O} \\ {}^{1}\quad{}^{2} \\ {}_{4}\quad{}_{3} \end{array}$$

A,B

1,2-Dioxetanes are highly endothermic compounds. This is partly due to ring strain, but above all to the low bond energy of the peroxide bond.

The typical reaction of 1,2-dioxetanes is thermal decomposition. On warming tetramethyl-1,2-dioxetane in benzene or other solvents, blue light is emitted. Such a phenomenon is known as chemiluminescence [11]. It has been demonstrated that, according to the principle of conservation of orbital symmetry, one mole of acetone in an electronically excited state is formed. In this way, an electronically excited molecule (symbolized by *) is created by a thermal process. With emission of light, the ground state is restored:

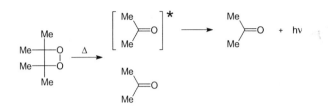

C | Two syntheses are available for the preparation of 1,2-dioxetanes starting from β-halohydro-peroxides or alkenes.

(1) *Dehydrohalogenation of β-halo hydroperoxides*

The electrophilic bromination of alkenes, e.g. with 1,3-dibromo-5,5-dimethylhydantoin, in the presence of concd hydrogen peroxide, leads to β-bromo hydroperoxides. They are cyclized with bases or with silver acetate to give 1,2-dioxetanes (KOPECKI 1973), e.g.:

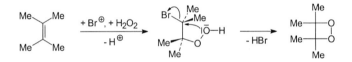

(2) *Photooxygenation of alkenes*

Donor-substituted alkenes in particular react with singlet oxygen to yield 1,2-dioxetanes by a [2+2] cycloaddition. The singlet oxygen is generated in the presence of the alkene by passing oxygen through a solution of the alkene in the presence of a sensitizing dye, e.g. methylene blue under irradiation:

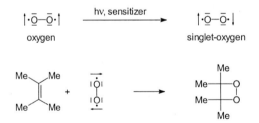

D | **Tetramethyl-1,2-dioxetane**, yellow crystals, mp 76-77°C, emits light a few degrees above its melting point.

1,2-Dioxetan-3-ones are also α-peroxy lactones. They can be prepared in solution at low temperature by cyclodehydration of α-hydroperoxycarboxylic acids with dicyclohexylcarbodiimide. They decompose at room temperature with chemiluminescence:

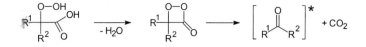

The bioluminescence observed in glowworms and fireflies is due to the decomposition of 1,2-dioxetan-3-ones [12].

The 2,4-dinitrophenyl ester of oxalic acid **1** [13], which reacts with 30% hydrogen peroxide solution to give 1,2-dioxetane-3,4-dione **2**, is an excellent example for the demonstration of chemilumines-cence. It decomposes producing two moles of carbon dioxide, with one mole being generated in an electronically excited state. The light that is emitted on degradation to the ground state lies in the UV region and is made visible by addition of a fluorophore (F), e.g. 9,10-diphenylanthracene:

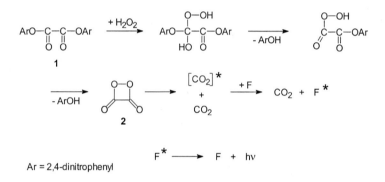

By a similar mechanism, 5-amino-2,3-dihydrophthalazine-1,4-dione **3** (luminol, see p 434) displays an intensely blue chemiluminescence on oxidation with hydrogen peroxide in the presence of complex iron salts, e.g. haemin.

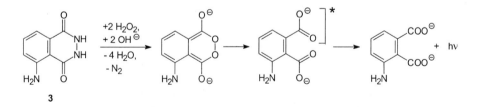

The chemiluminescence of dioxetanes, luminol and other heterocyclic compounds plays an important role in the solution of analytical problems in biochemistry and immunology [14].

4.6   1,2-Dithiete

$$S\underset{4\;\;\;\;3}{\overset{1\;\;\;2}{\rule{1.5em}{0.4pt}}}S$$

A-D   This system is iso-π-electronic with benzene. MO calculations predicted a delocalization energy of 92 kJ mol^{-1}, which overcompensates for the strain enthalpy of 43 kJ mol^{-1} and results in stabilization of the molecule. However, the parent compound has not yet been prepared.
3,4-Bis(trifluoromethyl)-1,2-dithiete, a yellow liquid, bp 95°C, is formed in 80% yield on heating hexafluorobut-2-yne with sulfur.

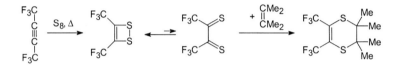

Typical for disubstituted 1,2-dithietes is their valence isomerization, which results in the formation of 1,2-dithiones. The equilibrium favours the 1,2-dithiete with electron-withdrawing substituents such as CF$_3$. The reaction with 2,3-dimethylbut-2-ene to give a hexasubstituted 2,3-dihydro-1,4-dithiine proceeds, however, as a [4+2] cycloaddition via the 1,2-dithione.
3,4-Bis(4-dimethylaminophenyl)-1,2-dithiete exists in solution in equilibrium with the corresponding 1,2-dithione [15]:

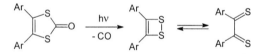

3,4-Di-*tert*-butyl-1,2-dithiete was obtained by heating 2,2,5,5-tetramethylhex-3-yne (di-*tert*-butyl acetylene) with sulfur in benzene in an autoclave at 190°C [16]. It is thermally stable and exists in the dithiete form. Valence isomerization into the dithione form would enhance the steric strain in the molecule.

4.7   1,2-Dihydro-1,2-diazete

$$HN\underset{4\;\;\;\;3}{\overset{1\;\;\;2}{\rule{1.5em}{0.4pt}}}NH$$

Although this system is iso-π-electronic with benzene, only one analogue has been prepared to date, namely 1,2-bis(methoxycarbonyl)-1,2-dihydro-1,2-diazete. Even at room temperature, it undergoes a slow valence isomerism to the corresponding 1,2-diimine [17]:

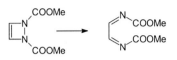

4.8   1,2-Diazetidine

Again, the preparation of the parent compound has, as yet, not been achieved. However, numerous 1,2-diazetidines are known.

The standard synthesis is a [2+2] cycloaddition of electron-rich alkenes such as enol ethers or enamines to azo compounds, e.g.:

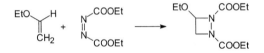

The [2+2] cycloaddition of ketenes to azo compounds yields 1,2-diazetidin-3-ones, e.g.:

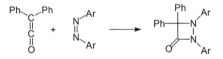

1,2-Diazetidines and 1,2-diazetidin-3-ones are thermally very stable. On strong heating, they decompose either into an azo compound and an alkene or ketene, or into two molecules of imine or an imine and isocyanate.

By analogy with β-lactams, 1,2-diazetidin-3-ones react with nucleophiles with ring-opening, e.g.:

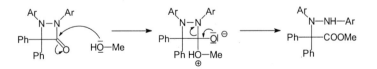

A few 1,2-diazetidin-3-ones are so reactive that ring-opening occurs in moist air.

Summary of the general chemistry of four-membered heterocycles:

- The stability and reactivity of the compounds are determined by the ring strain and the nature of the heteroatom or heteroatoms. While azete, as an antiaromatic system, is extremely reactive, the aromatic systems 1,2-dithiete and 1,2-dihydro-1,2-diazete are hardly any more stable and are very reactive.

- Ring-opening by nucleophiles proceeds more slowly than with three-membered heterocycles and is catalysed by acids.

- Special ring-openings are [2+2] cycloreversions (oxetan-2-ones, 1,2-dioxetanes, 1,2-dioxetan-3-ones, 1,2-diazetidines, 1,2-diazetidin-3-ones) and valence isomerizations (1,2-dithiete, 1,2-dihydro-1,2-diazete).

- Oxetan-2-ones, azetidin-2-ones and 1,2-diazetidin-3-ones are more reactive than five- and six-membered homologues. They are attacked by nucleophiles on the C-atom of the carbonyl group. Ring-opening occurs to give γ-substituted carboxylic acids or carboxylic acid derivatives.

- An important synthetic principle is the intramolecular nucleophilic substitution of a γ-substituted leaving group
  — by an O-atom (oxetanes, oxetan-2-ones, 1,2-dioxetanes, 1,2-dioxetan-2-ones)
  — by an S-atom (thietanes)
  — by an N-atom (azetidines, azetidin-2-ones)
  The rate of reaction is greater than that of the three-membered heterocycles because of the smaller ring strain of the products. At the same time, however, the entropy gain is smaller, because two degrees of freedom of inner rotation are lost en route to the activated complex.

- [2+2] Cycloadditions are of great importance for synthetic purposes
  — carbonyl compounds + alkenes → oxetanes
  — aldehyde + ketenes → oxetan-2-ones
  — imine + ketenes → azetidin-2-ones
  — isocyanates + alkenes → azetidin-2-ones
  — singlet oxygen + alkenes → 1,2-dioxetanes
  — alkenes + azo compounds → 1,2-diazetidines
  — ketene + azo compounds → 1,2-diazetidin-3-ones

- The importance of four-membered heterocycles for organic synthesis is limited. Examples are the alkene synthesis involving oxetan-2-ones and the β-aminocarboxylic acid synthesis involving azetidin-2-ones.

References

[1] J. Dale, *Tetrahedron* **1993**, *49*, 8707.

[2] P. Picard, D. Leclercq, J.-P. Bats, J. Moulines, *Synthesis* **1981**, 550.

[3] M. Braun, *Nachr. Chem. Techn. Lab.* **1985**, *33*, 213.

[4] A. Pomier, J.-M. Pons, *Synthesis* **1993**, 441; **1995**, 729.

[5] K. C. Nicolaou, W.-M. Dai, R. K. Guy, *Angew. Chem. Int. Ed. Engl.* **1994**, *33*, 15; R. A. Holton et al., *J. Am. Chem. Soc.* **1994**, *116*, 1597, 1599.

[6] U.-J. Vogelbacher, M. Regitz, R. Mynott, *Angew. Chem. Int. Ed. Engl.* **1986**, *25*, 842; M. Regitz, *Nachr. Chem. Techn. Lab.* **1991**, *39*, 9.

[7] P. G. Sammes, S. Smith, *J. Chem. Soc., Chem. Commun.*, **1983**, 682.

[7a] M. J. Miller, *Tetrahedron* **2000**, *56*, 5553.

[8] M. F. Loewe, R. J. Cvetovich, G. G. Hazen, *Tetrahedron Lett.* **1991**, *32*, 2299.

[9] G. I. Georg, P. M. Mashava, X. Guan, *Tetrahedron Lett.* **1991**, *32*, 581.

[10] I. Ojima, M. Zhao, T. Yamato, K. Nakahashi, *J. Org. Chem.* **1991**, *56*, 5263.

[11] W. Adam, G. Cilento, *Angew.Chem. Int. Ed. Engl.* **1983**, *22*, 529; W. Adam, W. J. Baader, *Angew. Chem. Int. Ed. Engl.* **1984**, *23*, 166.

[12] W. Adam, *J. Chem. Educ.* **1975**, *52*, 138.

[13] A. G. Mohan, N. J. Turro, *J. Chem. Educ.* **1974**, *51*, 528; B. Z. Shakhashiri, L. G. Williams, G. E. Direen, A. Francis, *J. Chem. Educ.* **1981**, *58*, 70.

[14] S. Albrecht, H. Brandl, W. Adam, *Nachr. Chem. Tech. Lab.* **1992**, *40*, 547; A. Mayer, St. Neuenhofer, *Angew. Chem. Int. Ed. Engl.* **1994**, *33*, 1044.

[15] W. Kusters, P. De Mayo, *J. Am. Chem. Soc.* **1973**, *95*, 2383; S. B. Nielsen, A. Senning, *Sulfur Reports* **1995**, *16*, 371; J. Nakayama, A. Mizumura, Y. Yokomori, A. Krebs, *Tetrahedron Lett.* **1995**, *36*, 8583.

[16] J. Nakayama, K. S. Choi, I. Akiyama, M. Hoshino, *Tetrahedron Lett.* **1993**, *34*, 115; K. S. Choi, I. Akiyama, M. Hoshino, J. Nakayama, *Bull. Chem. Soc. Jpn.* **1993**, *66*, 623.

[17] E. E. Nunn, R. N. Warrener, *J. Chem. Soc., Chem. Commun.*, **1972**, 818.

5   Five-Membered Heterocycles

With this large group of heterocycles, ring strain is of little or no importance. Ring-opening reactions are, therefore, rarer than in three- and four-membered heterocycles. The crucial consideration is rather whether a compound can be regarded as a heteroarene or whether it has to be classified as a heterocycloalkane or heterocycloalkene (see p 2). Various aromaticity criteria apply to heteroarenes, and as a consequence, different opinions have been expressed on this matter [1]. As will be shown by means of examples of the various systems, the nature and number of heteroatoms are the critical factors. The parent compound of the five-membered heterocycles with one oxygen atom is furan.

5.1   Furan

A   Formerly, the positions next to the heteroatom were indicated as α and α'. The univalent residue is known as furyl. All ring atoms of furan lie in a plane and form a slightly distorted pentagon (see Fig. 5.1).

Fig. 5.1   Structure of furan
                              (bond lengths in pm, bond angles in degrees)

The structural representation follows from the fact that the bond length between C-3 and C-4 is greater than that between C-2 and C-3 and between C-4 and C-5. The ionization potential is 8.89 eV, the electron being removed from the third π-MO (see Fig. 5.2b). The dipole moment is 0.71 D, with the negative end situated on the O-atom. In contrast, the dipole moment of tetrahydrofuran is 1.75 D. The small dipole moment of furan confirms that one electron pair of the O-atom is included in the conjugated system and therefore delocalized. Furan has the following UV and NMR data:

UV (ethanol) λ (nm) (ε)	^{1}H-NMR (DMSO-d$_6$) δ (ppm)	^{13}C-NMR (DMSO-d$_6$) δ (ppm)
208 (3.99)	H-2/5: 7.46	C-2/5: 143.6
	H-3/4: 6.36	C-3/4: 110.4

The signals in the region typical for benzenoid compounds indicate that a diamagnetic ring current is induced in the furan molecule. Thus, furan fulfils an important experimental criterion for aromaticity in cyclic conjugated systems.

A description of the electronic structure of the furan molecule is based on the assumption that all ring atoms are sp^2-hybridized (see Fig. 5.2a). The overlap of the five 2p$_z$ atomic orbitals yields delocalized π-MOs, three of which are bonding and two antibonding.

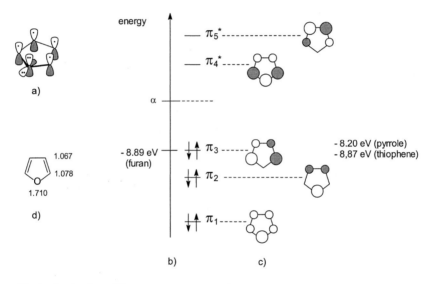

Fig. 5.2   Electronic structure of furan
(a) sp^2-hybridization of the ring atoms
(b) energy level scheme of the π-MO (qualitative) and occupation of electrons
(c) π-MO (the O-atom is situated at the lowermost corner of the pentagon)
(d) π-electron densities calculated by ab initio MO methods [2]

MOs π_2 and π_3, as well as π_4^* und π_5^*, are energetically inequivalent (see Fig. 5.2b and c), contrary to the fact that the carbocycles benzene and the cyclopentadienyl anion are iso-π-electronic with furan. Because the nodal plane of π_3 passes through the heteroatom, in contrast to that of π_2, the degeneracy is lost. Every C-atom contributes one electron and the O-atom two electrons to the cyclic conjugated structure. The six electrons occupy the three bonding π-MOs in pairs. As there are six electrons distributed over five atoms, the π-electron density on each ring atom is greater than one (see Fig. 5.2). Furan is thus a *π-electron excessive heterocycle*.

Resonance energy [1] has been used for many years as the criterion for quantifying aromaticity in cyclic conjugated systems. It is defined as the deficiency in the energy content of a system when compared with nonconjugated or aliphatic reference structures and, therefore implies greater stability. Resonance energy can be theoretically calculated or experimentally determined. The latter is known as *empirical resonance energy*. The values of the empirical resonance energy for furan are reported to lie between 62.3 und 96.2 kJ mol⁻¹, the value varying from laboratory to laboratory using different methods. The energy of the furan π-system is less than the sum of the π-electron energies of the localized fragments $C(sp^2)–C(sp^2)$ and $O(sp^2)–C(sp^2)$ multiplied by two in each case. This difference may be viewed as the energy liberated by delocalizing the π-electrons in the furan molecule. The empirical resonance energy of benzene was calculated to be 150.2 kJ mol⁻¹. As the value for furan is 80 kJ mol⁻¹, its aromaticity is less than that of benzene. The so-called DEWAR resonance energy is based on a corresponding aliphatic polyene which is the hexa-1,3,5-triene in the case of benzene, and the divinyl ether in the case of furan [3]. The value found for benzene was 94.6 kJ mol⁻¹ and that for furan 18.0 kJ mol⁻¹. These values also show that furan is less aromatic than benzene.

An aromaticity index for heteroarenes was proposed based on experimentally determined bond lengths [4].

B   By analogy with benzene, furan undergoes reactions with electrophilic reagents, often with substitution. However, it can also react by addition and/or ring-opening depending on reagent and reaction conditions.

Electrophilic substitution reactions

Furan undergoes electrophilic substitution about 10^{11} times faster than benzene under similar conditions. The reasons for this are:
- The resonance energy of furan is less than that of benzene
- The furan ring has a π-electron excess, while in benzene, the π-electron density is one on each ring atom

The electrophilic substitution reactions of furan, like those of benzene, take place by an addition–elimination mechanism.

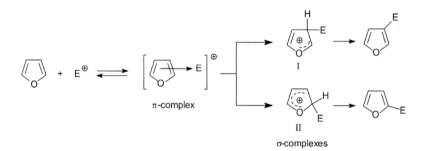

The substitution is regioselective to the α-position; when these positions are occupied, the β-position is substituted. There are two reasons for this:

- The delocalization of the positive charge in the σ-complex II is more efficient, as it is not impaired by the heteroatom.
- The HOMO coefficient is greater on the α-C-atoms than on the β-C-atoms (see Fig. 5.2c).

The importance of the resonance energy for the reactivity of the substrate clearly emerges from the mechanism. The π-system of furan contains more energy than that of benzene. Less energy is, therefore, required in the case of furan to interrupt its cyclic conjugation by way of π-complex → transition state → σ-complex.

Chlorination of furan at -40°C yields 2-chlorofuran and 2,5-dichlorofuran. Bromination with the dioxane–Br_2 complex at -5°C gives 2-bromofuran. Nitration is best carried out with fuming nitric acid in acetic anhydride at -10° to -20°C, and yields 2-nitrofuran. Pyridine–SO_3 or dioxane–SO_3 complex converts furan into furan-2-sulfonic acid and then further into furan-2,5-disulfonic acid. Alkylation and acylation are also possible. The action of mercury(II) chloride and sodium acetate in aqueous ethanol brings about mercuration of furan:

$$\text{furan} \xrightarrow{\text{HgCl}_2,\ \text{CH}_3\text{COONa}} \text{2-(HgCl)furan}$$

Metalation

n-Butyllithium in hexane metalates furan in the 2-position, while excess of reagent at higher temperature produces 2,5-dilithiofuran. This is principally an acid–base reaction, as furan is deprotonated by the strong base butylate:

$$\text{furan} + \text{n-BuLi} \longrightarrow \text{2-Li-furan} + \text{n-BuH}$$

Addition reactions

Furans yield the corresponding tetrahydrofurans by catalytic hydrogenation.

In some addition reactions, furans behave as 1,3-dienes. For example, furan reacts with bromine in methanol in the presence of potassium acetate to give 2,5-dimethoxy-2,5-dihydrofuran by a 1,4-addition:

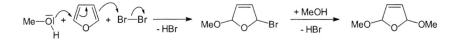

The analogy of the reactivity of furan with that of butadiene is further underlined by the fact that furan undergoes a DIELS-ALDER reaction with dienophiles such as maleic anhydride:

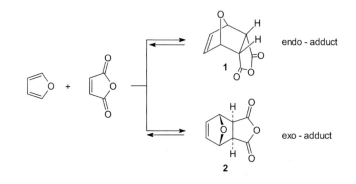

endo - adduct

1

exo - adduct

2

As with butadiene, a 'normal' DIELS-ALDER reaction occurs, i.e. the HOMO of furan (see Fig. 5.2c) interacts with the LUMO of maleic anhydride. The reaction is diastereoselective. ALDER's *endo*-rule applies to the stereochemistry of the cycloadducts **1/2**; thus in acetonitrile at 40°C, the *endo*-adduct **1** is formed 500 times faster than the *exo*-adduct **2** owing to kinetic control. However, with a sufficiently long reaction time, product formation becomes subject to thermodynamic control; the initially formed *endo*-compound is completely converted via the educts into the *exo*-compound, which is more stable by 8 kJ mol[-1].

The DIELS-ALDER reaction of furans has been studied in detail [4a]. With acetylenic dienophiles, e.g. acetylenedicarboxylic ester, adducts are formed (e.g. **3**) which are isomerized by acids to phenols. The selective hydrogenation of **3** to **4** followed by a [4+2] cycloreversion yields the 3,4-disubstituted furan **5**:

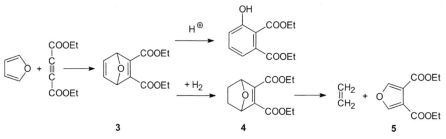

3   4   5

There are even reactions which involve one olefinic π-bond of furan. For instance, furan reacts with ketones under conditions of the PATERNO-BÜCHI reaction (see p 39) to give 2a,5a-dihydro-2*H*-oxeto[2,3-*b*]furans **6**:

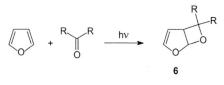

6

Ring-opening reactions

Furans are protonated in the 2-position, and not on the O-atom, by BRÖNSTED acids:

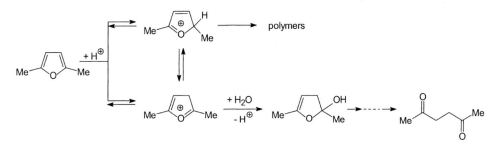

Concentrated sulfuric or perchloric acids induce polymerization of the cations, while dilute acids, e.g. perchloric acid in aqueous DMSO, cause hydrolysis to 1,4-dicarbonyl compounds. Nucleophilic attack by water probably takes place at the 2-position of protonated furan. Finally, in a reversion of the PAAL-KNORR synthesis (see p 58), 2-hydroxy-2,3-dihydrofuran gives rise to the 1,4-dicarbonyl compound hexane-2,5-dione as shown.

C Several methods exist for the preparation of furan. In this book, we shall use the method of retrosynthetic analysis (COREY and WARREN [5]) as a guiding principle for the heterocyclic synthesis of the most important target systems. This procedure leads to 'logical' starting materials and methods for constructing the required heterocycles which can then be compared with existing and preparatively important methods.

When furan is considered in the light of a retrosynthetic analysis, it can be seen to derive from a double enol ether and can, therefore, be dissected retroanalytically in two ways (I, II) according to the following scheme:

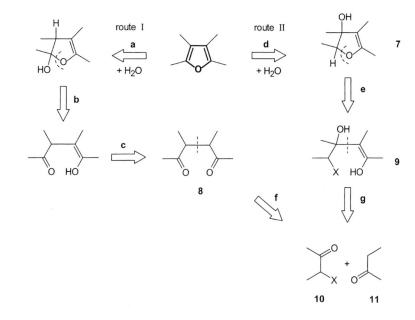

Fig. 5.3 Retrosynthesis of furan

If the retrosynthesis follows route I (addition of water to the furan C-2/C-3 bond followed by bond opening O/C-2, i.e. an enol ether hydrolysis according to steps **a–c**), then the 1,4-dicarbonyl system **8** is obtained as the first suggested educt. Starting from **8**, the furan system should be formed by a cyclic dehydration. Further retroanalysis of **8** leads via **f** to the α-halocarbonyl compound **10** and to the enolate of the carbonyl compound **11**. The latter should be convertible into the 1,4-dicarbonyl system **8** by alkylation with **10**.

Following route II, it can be seen that the primary H$_2$O addition to the furan C-2/C-3 bond can also occur in an opposite direction to **a**, i.e. according to the retrostep **d**. This leads to the intermediate **7** for which the bond cleavage O/C-2 (**e**) is retroanalytically rationalized and leads to the γ-halo-β-hydroxycarbonyl system **9**. A retroaldol operation (**g**) provides the same starting materials **10** and **11** as the retrosynthesis I; route II, however, suggests as the first step aldol addition of **10** and **11** to give **9** followed by intramolecular S$_N$ cyclization of the enolate **9** to the dihydrofuran **7** and conversion into furan by dehydration.

As is shown by the following syntheses, the results of the retrosynthesis can be applied in practice.

(1)   1,4-Dicarbonyl compounds, especially 1,4-diketones, undergo cyclodehydration when treated with concd H$_2$SO$_4$, polyphosphoric acid, SnCl$_2$ or DMSO, providing 2,5-disubstituted furans **12** *(Paal-Knorr synthesis)* :

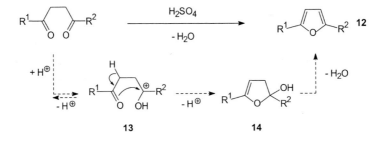

In an acid–base equilibrium, the BRÖNSTED or LEWIS acid adds to one of the carbonyl groups in the 1,4-dicarbonyl system (**13**) enabling a nucleophilic intramolecular attack by the second carbonyl group, to form **14**; finally, the β-elimination, which is also acid-catalysed, occurs (**14 → 12**) [6].

(2)   α-Halocarbonyl compounds react with β-keto carboxylic esters to yield derivatives of 3-furoic acid **15** by cyclocondensation* *(Feist-Benary synthesis)*:

---

* Reactions which yield cyclic products from open-chain starting materials by the elimination of water, hydrogen halides or other low molecular mass compounds are called *cyclocondensations*. They are important methods for synthesizing five-, six- und larger-membered heterocycles.

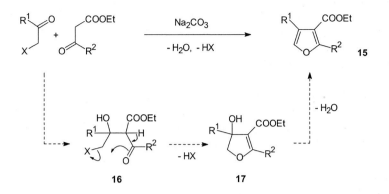

The FEIST-BENARY synthesis requires the presence of a base, e.g. aqueous Na_2CO_3. It proceeds as a multistep reaction involving at least two intermediates (**16/17**) of which 3-hydroxy-2,3-dihydrofurans **17** can be isolated in some cases. The formation of **16** results from an aldol addition, that of **17** from an intramolecular nucleophilic substitution. Cyclic 1,3-diketones also react with α-halocarbonyl compounds according to the FEIST-BENARY synthesis to produce furans, e.g.:

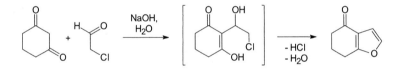

In the reaction of β-keto esters with α-halo ketones, the possible competition between C-alkylation (followed by reaction of the PAAL-KNORR type) and aldol addition (followed by reaction of the FEIST-BENARY type) can result in mixtures of isomeric furans. Regioselectivity can, however, sometimes be controlled by the reaction conditions, as for instance in the interaction of chloroacetone with an acetoacetic acid ester, leading to the furan-3-carboxylates **18/19**:

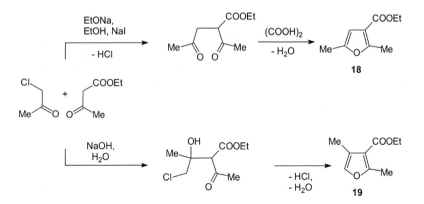

(3)   A furan synthesis which cannot easily be deduced from retrosynthetic considerations is the ring transformation of oxazoles by a DIELS-ALDER reaction with activated alkynes. For example, 4-methyloxazole **20** reacts with dimethyl acetylenedicarboxylate to provide furan-3,4-dicarboxylic ester **22** [6] via a nonisolable adduct **21**:

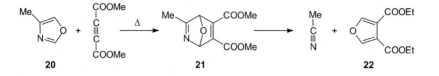

The first step is a [4+2] cycloadditon and the second a [4+2] cycloreversion. The fact that cycloreversion is not a retroaction of the first step is due to the formation of thermodynamically very stable acetonitrile, as well as furan [7].

D **Furan** is produced by the catalytic decarbonylation of 2-furaldehyde or by decarboxylation of 2-furoic acid with copper powder in quinoline. Furan is a colourless, water-insoluble liquid of pleasant odour, bp 32°C. Addition of hydroquinone or other phenols inhibits polymerization, which occurs slowly at room temperature.

2-Furaldehyde (furfural; from Latin for bran) is obtained industrially from plant residues which are rich in pentoses, e.g. bran, by treatment with dilute sulfuric acid followed by steam distillation:

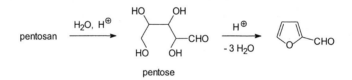

2-Furaldehyde is a colourless, poisonous, water-soluble liquid, bp 162°C, which slowly turns brown in air. Like benzaldehyde, it undergoes the CANNIZZARO reaction, the PERKIN reaction, the KNOEVENAGEL condensation and the acyloin condensation. The catalytic hydrogenation of 2-furaldehyde yields 2-(hydroxymethyl)oxolane **23** (tetrahydro-2-furfuryl alcohol). This compound undergoes a nucleophilic 1,2-rearrangement to give 3,4-dihydro-2H-pyran **24** by the action of acid catalysts:

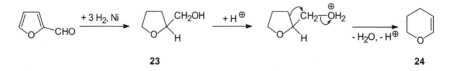

The preparation of the coloured salt **25** of 5-anilino-1(phenylimino)penta-2,4-dien-2-ol by the action of aniline and hydrochloric acid illustrates a ring-opening of 2-furaldehyde:

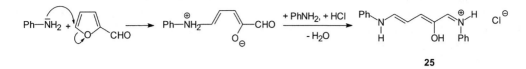

25

2-Furaldehyde is used as a solvent in the manufacture of polymers and as a starting material for syntheses. 2-Furaldehyde could attain the importance of a raw material for the chemical industry if the trend towards the use of 'regenerative raw materials' increases further.

5-Nitro-2-furaldehyde (5-nitrofurfural), mp 36°C, is prepared by the nitration of 2-furaldehyde. Some of its derivatives are bacteriostatic and bacteriocidal, e.g. the semicarbazone (Nitrofural), and are used to combat infectious diseases.

5-(Hydroxymethyl)-2-furaldehyde is formed by the dehydration of saccharose, e.g. with iodine in DMSO.

Furoic acid (furan-2-carboxylic acid) is obtained by dry distillation of D-galactaric acid (mucic acid):

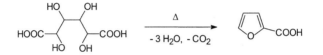

Furoic acid forms colourless crystals, mp 134°C. With a pK_a value of 3.2, it is a stronger acid than benzoic acid (pK_a 4.2).

Furans are occasionally found in plants and microorganisms, e.g. carlina oxide **26**, which is obtained by steam distillation from the roots of the carline thistle (*Carlina acaulis*):

26   **27**

Some natural products containing a furan ring have an intense odour, e.g. 2-furylmethanethiol **27**, a component of coffee aroma, rose furan **28**, a component of rose oil, and menthofuran **29**, which occurs in peppermint oil. The compounds **28** and **29** are reminiscent of terpenoid structures:

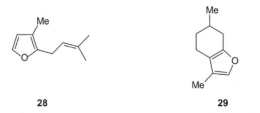

28   **29**

Natural products belonging to the group of furocoumarins are described on p 251.

E Furans are of considerable importance as synthetic intermediates [8]. Their acid-catalysed hydrolysis, as well as their ability to undergo cycloadditions, can be utilized. For instance, the furan **30**, which is accessible from methylfuran and (*Z*)-1-bromohex-3-ene by metallation and alkylation, is cleaved by aqueous acid to give the 1,4-diketone **31**. This is then converted by a base-catalysed intramolecular aldol condensation into (*Z*)-jasmone **32**, a naturally occurring fragrance substance:

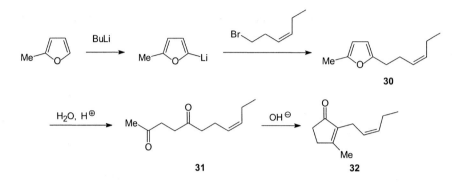

An alternative method of ring-opening is the cleavage of 2-furylmethanols (e.g. **33**) with HCl in alcohols to give esters of 4-oxopentanoic acid (e.g. **34**).

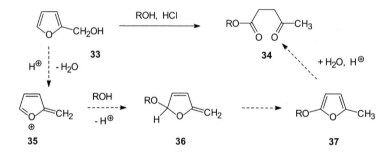

Under similar conditions, 2-furyl vinylcarbonyl systems (ketones, esters, acids) also produce keto esters. For example, 3-(2-furyl)acrylic acid **38**, which is easily accessible from 2-furaldehyde by a decarboxylating KNOEVENAGEL reaction with malonic acid, yields the 3-oxoheptanedicarboxylic ester **39** (MARCKWALD cleavage) [9]:

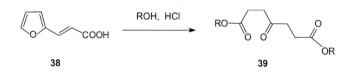

This transformation, which is formally an internal redox process, can be explained by an ionic mechanism involving ROH addition, isomerization and hydrolytic ring-opening (**35** → **36** → **37** → **34**).

5.2   Benzo[*b*]furan

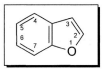

A Benzo[*b*]furan, often called benzofuran, shows the following UV absorption bands and NMR signals:

UV (ethanol)	^{1}H-NMR (acetone-d$_6$)		^{13}C-NMR (CS$_2$)	
λ (nm) (ε)	δ (ppm)		δ (ppm)	
244 (4.03)	H-2: 7.79	H-6: 7.30	C-2: 141.5	C-6: 124.6
274 (3.39)	H-3: 6.77	H-7: 7.52	C-3: 106.9	C-7: 111.8
281 (3.42)	H-4: 7.64		C-4: 121.6	C-3a: 127.9
	H-5: 7.23		C-5: 123.2	C-7a: 155.5

The signals for the furan protons again are found in the region of benzenoid protons. The C-2/C-3 bond, in contrast, behaves chemically rather like a localized olefinic double bond, i.e. it undergoes addition reactions.

B Benzo[*b*]furan, when treated with nitric and acetic acids, yields 2-nitrobenzo[*b*]furan. When the VILSMEIER formylation is used, 2-benzo[*b*]furaldehyde is obtained, and 2-lithiobenzo[*b*]furans are formed with butyllithium.

With addition of bromine, benzo[*b*]furan reacts to give *trans*-2,3-dibromo-2,3-dihydrobenzo[*b*]-furan. On dehydrobromination, this compound yields a mixture of 2-bromo- and 3-bromobenzo[*b*]-furan.

In contrast to furan, the benzene ring in benzo[*b*]furan (because of its large resonance energy) is dominant to such an extent that [4+2] cycloadditions are not possible. On the other hand, photochemical [2+2] cycloaddition occurs readily on the C-2/C-3 double bond leading, e.g. with an acetylene dicarboxylic ester, to the cyclobutene derivative **1**:

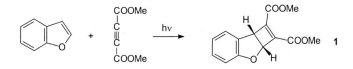

Photooxygenation of 2,3-dimethylbenzo[*b*]furan at -78°C produces a dioxetane which isomerizes at room temperature to give 2-acetoxyacetophenone [10]:

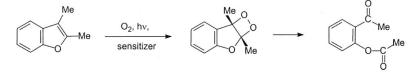

The reactivity of the C-2/C-3 double bond in benzo[*b*]furan corresponds approximately to that of a vinyl ether.

C Benzo[*b*]furan was first prepared from coumarin as follows:

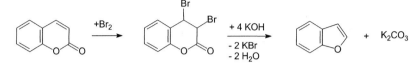

The name coumarone, which was previously used for benzo[*b*]furan, originates from this synthesis. The reaction of the intermediate 3,4-dibromo-3,4-dihydrocoumarin with KOH leading to benzofuran is known as PERKIN rearrangement.

Benzo[*b*]furans are also accessible by reaction of phenolates with halo ketones followed by cyclodehydration with H_2SO_4, polyphosphoric acid or zeolites [11]:

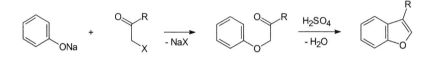

In contrast, the thermal cyclodehydration of 2-alkylphenols leads to 2-alkylbenzo[*b*]furans:

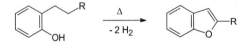

(*o*-Hydroxyaryl)acetylenes **2** can be subjected to transition-metal mediated cyclization by means of a PdI_2 / thiourea / CBr_4 cocatalysis system followed by carbonylation of the Pd-intermediate (simplified as **3**) in methanol to give methyl benzo[*b*]furan-3-carboxylates **4** [11a]:

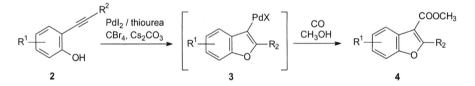

D **Benzo[*b*]furan**, a colourless, oily, water-insoluble liquid, bp 173°C, occurs in coal tar. It is probably formed during coal carbonization by cyclodehydration of 2-ethylphenol. Copolymerization of indene in the presence of BRÖNSTED or LEWIS acids yields the so-called coumarone resins, which find use in industry (adhesives, paints, binders).

Several natural products and pharmaceuticals are derived from benzo[*b*]furan, e.g. the bacteriocidal 2-(4-nitrophenyl)benzo[*b*]furan **5**. Amiodarone **6**, a substituted 3-benzoyl-2-butylbenzo[*b*]furan, is used in the treatment of cardiac arrhythmia:

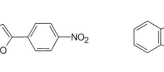

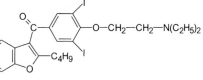

5   6

5.3   Isobenzofuran

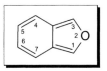

A-D   The name isobenzofuran is retained for benzo[c]furan. It is apparent from its structure that the six-membered ring does not possess a π-electron sextet, but that the compound is of an orthoquinonoid type. The resonance energy of isobenzofuran is thus much less than that of benzo[b]furan. As a consequence, isobenzofuran has so far not been isolated as a pure substance [12]. It is formed by flash vacuum pyrolysis of benzo[b]-7-oxabicyclo[2.2.1]hept-2-ene **1**, and it polymerizes rapidly even at low temperature.

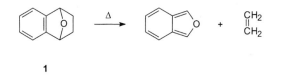

1

In contrast, 1,3-diphenylisobenzofuran **3** can be prepared from 3-phenyl-1,3-dihydroisobenzofuran-1-one **2** as follows:

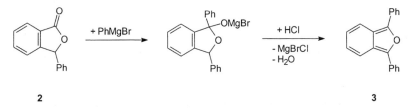

2   3

The intensely yellow compound melts at 127°C. Its solution exhibits a blue-green fluorescence. 1,3-Diphenylisobenzofuran proves to be an extremely reactive diene [12] in [4+2] cycloadditions. It is used to trap unstable alkenes or acetylenes, e.g. 1,2-dehydrobenzene (benzyne) with formation of the adduct **4**:

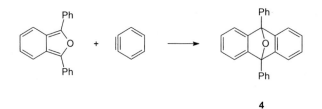

4

The photooxygenation occurs at -50°C as a [4+2] cycloaddition in an analogous way, but differently to that of benzo[*b*]furan. The product **5** is reduced by potassium iodide in acetic acid to 1,2-dibenzoylbenzene:

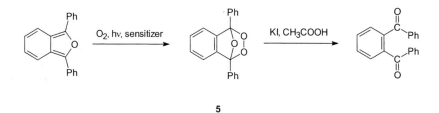

5

The reactions of 1,3-diphenylisobenzofuran are characterized by the fact that the orthoquinonoid structure can transform into a benzenoid structure with a π-electron sextet, thereby liberating resonance energy.

5.4   Dibenzofuran

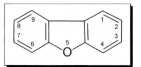

| A-D | The designation [*b,d*] can be omitted, as no other ring fusion can exist. Dibenzofuran was previously known as diphenylene oxide. |

Dibenzofurans behave like *o,o'*-disubstituted diphenyl ethers, i.e. they undergo substitution reactions typical for benzenoid compounds. Halogenation, sulfonation and acylation occur in the 2-position, and subsequently 2,8-disubstituted compounds are formed. Nitration yields 3-nitro- and then 3,8-dinitro compounds. Lithiation and mercuration lead to 4-substituted and finally to 4,6-disubstituted products [13].

The action of lithium in boiling dioxan causes 'ether fission' of dibenzofuran. On hydrolysis, the product yields 2-hydroxybiphenyl:

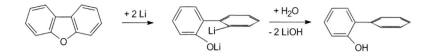

Dibenzofuran is conveniently prepared by the acid-catalysed dehydration of 2,2'-dihydroxybiphenyl:

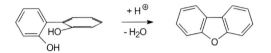

Dibenzofuran, colourless, fluorescent crystals, mp 86°C, bp 287°C, occurs in coal tar.

2,3,7,8-Tetrachlorodibenzofuran and other polychlorodibenzofurans (abbreviated PCDFs) are extremely toxic and, like 2,3,7,8-tetrachlorodibenzo[1,4]dioxin (see p 371), belong to a group of compounds known as supertoxins [14]. The lethal dose for monkeys lies in the region of 0.07 mg/kg body weight. PCDFs are formed in traces during the industrial production of polychlorobenzenes, polychlorophenols and polychlorobiphenyls, as well as during the combustion or thermal decomposition of products which contain such compounds. Among these are pesticides, wood preserved with polychlorophenols, as well as transformer oils, e.g.:

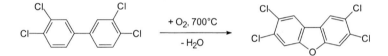

5.5   Tetrahydrofuran

A   The C–O and C–C bond lengths in tetrahydrofuran (oxolan) are 142.8 and 153.5 pm, respectively. These values are close to the corresponding bond lengths in dialkyl ethers. The ring is virtually strain-free, but not planar. There are 10 twist and 10 envelope conformations which interconvert rapidly through pseudorotation (activation energy 0.7 kJ mol^{-1}). This results in a molecule which has almost free conformational mobility (see p 68).

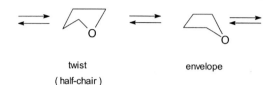

twist   envelope

(half-chair)

Three ring atoms are coplanar in the twist conformation and four in the envelope conformation.
The chemical shifts of the H- and C-atoms of tetrahydrofuran in the NMR spectra are as follows:

^{1}H-NMR (CCl$_4$) ^{13}C-NMR (D$_2$O)
δ (ppm) δ (ppm)

H-2/H-5: 3.61   C-2/C-5: 68.60
H-3/H-4: 1.79   C-3/C-4: 26.20

These values are typical for dialkyl ethers and are significantly different from those of furan (see p 53).

B By analogy with oxetanes, tetrahydrofurans are ring-opened by nucleophiles. For example, 4-
chlorobutan-1-ol is obtained on heating tetrahydrofuran with hydrochloric acid:

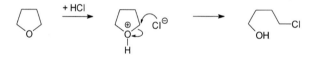

Lithium alkyls, e.g. *n*-butyllithium, also effect ring-opening. First, 2-lithiotetrahydrofuran is formed,
which at room temperature slowly decomposes through a [3+2] cycloreversion into ethene and the lithium enolate of acetaldehyde:

Tetrahydrofurans thus behave essentially like dialkyl ethers.

C The simplest method for obtaining tetrahydrofurans is by cyclodehydration of 1,4-diols:

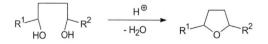

Tetrahydrofuran is produced industrially by catalytic cyclodehydration of butane-1,4-diol.

A further method for the synthesis of tetrahydrofurans is the cyclization of appropriately substituted γ,δ-unsaturated alcohols **1** [15]:

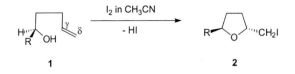

1   **2**

This reaction differs from a simple cyclization because of the introduction of the iodomethyl group in addition to ring formation. The reaction proceeds diastereoselectively producing a *trans*-2,5-disubstituted tetrahydrofuran **2**.

| **D** | **Tetrahydrofuran**, often abbreviated as THF, is a colourless, water-soluble liquid of pleasant odour, bp 64.5°C. Inhalation of its vapours leads to severe poisoning. When stored in air, tetra- |

hydrofuran is converted to a hydroperoxide by autoxidation:

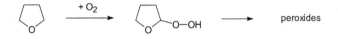

This can easily be demonstrated by the liberation of iodine upon the addition of potassium iodide. The hydroperoxide transforms to highly explosive peroxides.

Tetrahydrofuran is used as a solvent, especially for GRIGNARD reactions, and at low temperature also for the preparation of organolithium compounds.

Many natural products are derived from tetrahydrofuran. Of great biological importance are those carbohydrates (pentoses, and to a lesser extent hexoses and glycosides derived therefrom), which contain a tetrahydrofuran ring and which are therefore known as 'furanoses'.

Muscarine **5**, one of the active ingredients in the toadstool fly agaric (*Amanita muscaria*), contains a 2,3,5-trisubstituted tetrahydrofuran structure. A number of stereocontrolled syntheses have been worked out for muscarine. One of these starts with methyl vinyl ketone [16] and cyclizes stereoselectively the appropriately substituted γ,δ-unsaturated 2,6-dichlorobenzyl ether **3** with iodine:

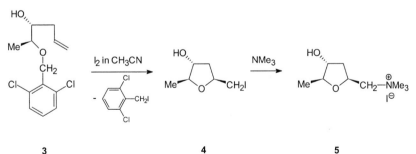

3   **4**   **5**

Differently to the alcohol **1** (see p 69), the diastereomer **4** is formed, whose 2,5-substitutents are in the *cis* configuration; on reaction with triethylamine **4** is transformed to (+)-muscarine **5**.

Cantharidin **7** is the poison of the cantharis beetles, among them the Spanish fly, native to Southern Europe. Cantharidin is a skin irritant and vesicant.

The direct synthesis of cantharidin by a DIELS-ALDER reaction of furan with dimethyl maleic anhydride **6** is not possible. However, cycloaddition can be carried out with 2,5-dihydrothiophene-3,4-dicarboxylic acid anhydride **8** as dienophile. The cycloadduct **9** (*exo/endo* mixture 85:15) is subsequently hydrogenated and the tetrahydrothiophene ring in **10** is cleaved by a reductive desulfurization [17]:

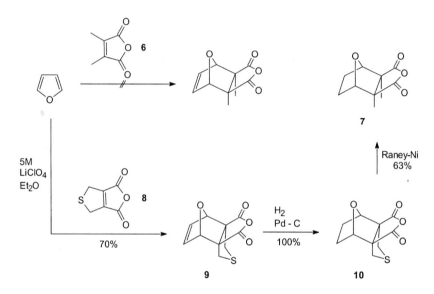

Polyether antibiotics contain tetrahydrofuran rings. In monensin **11**, three tetrahydrofuran rings are linearly connected. The molecule contains 17 asymmetric centres. Stereoselective syntheses for monensin have been elaborated [18]. In nonactin **12**, the rings are interconnected in an α,α'-orientation via ester groupings. Nonactin is therefore classed as a macrolide antibiotic. Polyethers of the type **11/12** are capable of facilitating ion transport across biological membranes; they are, therefore, also known as ionophores.

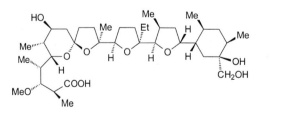

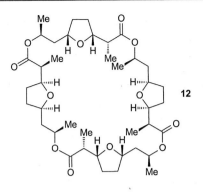

Organic compounds have essential information about their reactivity encoded in their structural formu-
lae from which their reactivity can be deduced. Furan, benzo[*b*]furan, isobenzofuran, dibenzofuran and
tetrahydrofuran are good examples of this.

5.6   Thiophene

A The univalent radical from thiophene is called thienyl. The ring atoms of thiophene are coplanar, as in furan (see Fig. 5.4). The greater atomic radius of sulfur causes the bond between the
heteroatom and one of the α-C-atoms to be longer by 35.2 pm than in furan.

Fig. 5.4   Structure of thiophene
(bond lengths in pm, bond angles in degrees)

The ionization potential of thiophene is 8.87 eV which is, according to KOOPMAN's theorem, equal to
the negative orbital energy of π_3 (see Fig. 5.2c, p 53). Because of the lower electronegativity of sulfur
compared with oxygen, the dipole moment of 0.52 D is even smaller than that of furan. The chemical
shifts in the NMR spectrum lie in the typical aromatic regions, as is the case with furan:

UV (95% ethanol)	^{1}H-NMR (CS$_2$)	^{13}C-NMR (acetone-d$_6$)
λ (nm) (ε)	δ (ppm)	δ (ppm)
215 (3.80)	H-2/H-5: 7.18	C-2/C-5: 125.6
231 (3.87)	H-3/H-4: 6.99	C-3/C-4: 127.3

Thiophene is aromatic. Its electronic structure follows from Fig. 5.2 (see p 53). Thiophene is a π-excessive heterocycle, i.e. the electron densitiy on each ring atom is greater than one. The value of the empirical resonance energy of thiophene is approximately 120 kJ mol^{-1}; the DEWAR resonance energy is quoted as 27.2 kJ mol^{-1}. The aromaticity of thiophene is thus less than that of benzene but greater than that of furan. There are two possible explanations to account for the difference between thiophene and furan:

- Because of the lower electronegativity of sulfur compared with oxygen, the electron pair on sulfur is more effectively incorporated into the conjugated system, i.e. its delocalization produces more energy. This assumption is consistent with the dipole moments.
- Sulfur, as an element of the second short period, is capable of expanding its octet. Thus, its 3d-orbitals can take part in the conjugated system. This can be demonstrated by its resonance structures:

Spectroscopic measurements and MO calculations have shown, however, that the involvement of the 3d-orbitals in the aromatic system is negligible, certainly in the electronic ground state of the thiophene molecule.

B Thiophene prefers reactions with electrophilic reagents. Additions and ring-opening reactions are less important than with furan, and substitution reactions are dominant. Some additional re-actions, such as oxidation and desulfurization, are due to the presence of sulfur and are thus confined to thiophenes.

Electrophilic substitutions

Thiophene reacts more slowly than furan but faster than benzene. The S$_E$Ar reactivity of thiophene corresponds approximately to that of anisole. The reaction mechanism is the same as described for fu-ran (see p 54). Substitution is regioselective in the 2- or in the 2,5-position.

Thiophene is chlorinated by Cl$_2$ or SO$_2$Cl$_2$. Bromination occurs by Br$_2$ in acetic acid or with N-bromosuccinimide. Nitration is effected by concentrated nitric acid in acetic acid at 10°C. Further sub-stitution predominantly yields 2,4-dinitrothiophene. Sulfonation with 96% H$_2$SO$_4$ occurs at 30°C within minutes. Benzene reacts extremely slowly under these conditions. This provides the basis for a method to remove thiophene from coal tar benzene (see p 77). Alkylation of thiophenes often gives only poor yields. However, more efficient procedures are the VILSMEIER-HAACK formylation, which yields thiophene-2-carbaldehyde, and acylation with acyl chlorides in the presence of tin tetrachloride, to give 2-acylthiophenes. Like furan, thiophene is mercurated with mercury(II) chloride.

Metalation

Thiophene is metalated by butyllithium in the 2-position. 2-Alkylthiophenes are prepared from 2-lithiothiophenes by alkylation with haloalkanes:

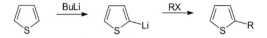

Addition reactions

The palladium-catalysed hydrogenation of thiophene yields thiolanes (tetrahydrothiophenes).

Thiophenes undergo DIELS-ALDER reactions, but their diene reactivity is lower than that of furans. The [4+2] cycloaddition occurs, therefore, only with very reactive dienophiles (arynes and alkynes with acceptor substituents) or under high pressure (example 3). 1,2-Disubstituted benzene derivatives 2 are formed with alkynes, as the primary DIELS-ALDER adduct 1 eliminates sulfur:

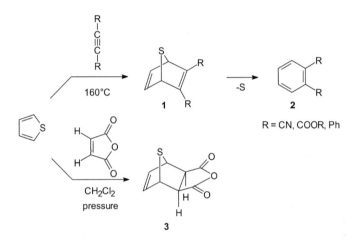

R = CN, COOR, Ph

Thiophenes undergo [2+1] cycloadditions with carbenes across the C–2/C–3 bond, e.g.

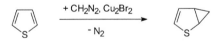

and [2+2] cycloadditions with active alkynes. For example, tetramethylthiophene reacts with dicyanoacetylene under AlCl$_3$ catalysis as shown:

With 3-aminothiophenes (e.g. 4), which are potential enamines, the [2+2] cycloaddition is considerably facilitated and occurs even below 0°C. The cycloadducts (e.g. 5) are converted thermally into 3-amino-phthalic acid derivatives (e.g. **7**). This occurs with electrocyclic ring-opening of the cyclobutene ring via a thiepine (e.g. **6**) and the extrusion of sulfur.

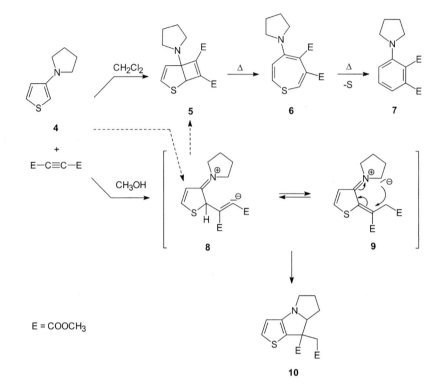

The solvent dependence of this cycloaddition is remarkable. The formation of **5** requires an aprotic medium, whereas in a protic medium (e.g. CH₃OH), the thieno[2,3-*b*]-5,6,7,7a-tetrahydro-1*H*-pyrrolizine **10** is formed. As the rate of product formation is the same in both media, it is plausible that the same primary product **8** of a dipolar [2+2] cycloaddition is formed first. The reaction then proceeds to give **5** in the nonpolar medium and **10** via the ylide **9** in the polar medium [19].

Ring-opening reactions

Thiophenes are neither polymerized nor hydrolysed by moderately concentrated BRÖNSTED acids. Ring-opening demands special reagents, e.g. phenylmagnesium bromide in the presence of di-chlorobis(triphenylphosphane)nickel(II):

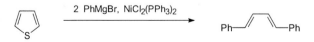

Another ring-opening reaction of thiophene is the reductive desulfurization with RANEY nickel in ethanol to give alkanes:

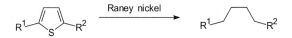

RANEY nickel adsorbs hydrogen during its preparation, which effects the reduction.

In an alternative reaction, hydrogen is used in the presence of molybdenum or tungsten as a catalyst. This hydrodesulfurization is of great industrial importance for the removal of thiophenes and other sulfur compounds from petroleum [20].

Oxidation

Thiophenes are oxidized by peroxy acids to give thiophene 1-oxides **11**, which can proceed further to thiophene 1,1-dioxides **12** [21]:

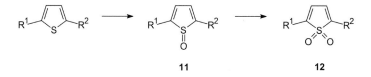

These compounds have a much stronger tendency than thiophene to undergo addition reactions. Thus, in the oxidation of thiophene with an excess of 3-chloroperoxybenzoic acid, the product **13** is produced. It results from a [4+2] cycloaddition of thiophene 1-oxide to the C–2/C–3 bond of thiophene 1,1-dioxide.

Photoisomerization

On irradiation of 2-phenylthiophene, 3-phenylthiophene is formed:

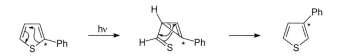

By using ^{14}C-labelled compounds, it can be shown that the phenyl substituent remains bonded to the C-atom of the thiophene ring. Several mechanisms have been put forward to explain the photoisomerization of thiophene [22]. One suggestion has postulated cyclopropene-3-thiocarbaldehyde as intermediate as shown above.

Photoisomerizations of numerous substituted thiophenes, as well as of furans and pyrroles, are known.

C The retrosynthesis of thiophene can be worked out in principle by analogy to that of furan (see p 57). A number of syntheses for thiophene can be deduced.

(1) The simplest method is 'sulfurization' followed by cyclizing dehydration of 1,4-dicarbonyl compounds **14** analogous to the PAAL-KNORR synthesis of furans. This cyclocondensation is carried out with P_4S_{10} or H_2S, and furnishes 2,5-disubstituted thiophenes **15** *(Paal synthesis)*:

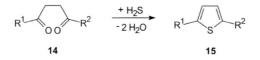

(2) Butane and higher alkanes, as well as corresponding alkenes and 1,3-dienes, undergo a cyclodehydrogenation with sulfur in the gas phase, with the formation of thiophenes, e.g.:

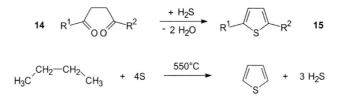

Acetylene, as well as 1,3-diynes, also yield thiophenes with H_2S under similar conditions:

$$2 \ HC{\equiv}CH \quad + \quad H_2S \quad \xrightarrow{400°C} \quad \text{\Large \char"2394} \quad + \quad H_2$$

(3) 1,3-Dicarbonyl compounds or β-chlorovinyl aldehydes react with thioglycolates or other thiols possessing a reactive methylene group to give thiophene-2-carboxylic esters **18** in the presence of pyridine *(Fiesselmann synthesis)*:

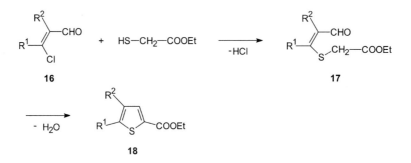

Initially, a formal vinyl substitution of the chlorine atom by MICHAEL addition takes place, followed by loss of HCl to give the intermediate **17**. Finally, cyclization occurs by an intramolecular aldol condensation.

β-Chlorovinyl aldehydes **16** result from α-methylene ketones by the action of DMF/$POCl_3$ (the VILSMEIER-HAACK-ARNOLD reaction):

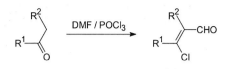

(4)   α-Methylene carbonyl compounds undergo cyclocondensation with cyanoacetic ester or malononitrile and sulfur in ethanol in the presence of morpholine to give 2-aminothiophenes **20** *(Gewald synthesis)* [22a] :

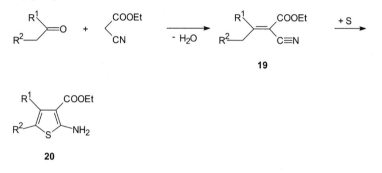

Initially, a KNOEVENAGEL condensation between the carbonyl compound and the reactive methylene group takes place. The α,β-unsaturated nitriles **19** that are formed are cyclized by the sulfur, probably via sulfanyl derivatives of **19** as intermediates.

(5)   1,2-Dicarbonyl compounds can be made to cyclize with esters of 3-thiapentanedioic acid **21** under base catalysis *(Hinsberg synthesis)*. This widely applicable and high-yield synthesis leads to substituted thiophene dicarboxylic esters **22** via a double aldol condensation, with the two CH_2 groups of **21**. Hydrolysis and decarboxylation of the esters yield 3,4-disubstituted thiophenes **23** [23]:

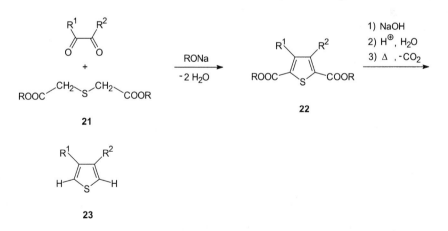

D   **Thiophene**, a colourless, water-insoluble liquid, with an odour like benzene, mp -38°C, bp 84°C, occurs in coal tar. It remains in the benzene fraction (benzene: bp 80°C) on distillation of coal tar, and can be removed by extraction with cold concentrated sulphuric acid (see p 72). A solution of isatin (indole-2,3-dione) in concentrated sulphuric acid turns blue in the presence of thiophene (the indophenine reaction). This test is used to detect thiophene as an impurity in benzene. It was observed that when benzene is made by decarboxylation of benzoic acid, the indophenine reaction, which was originally thought to be caused by benzene, does not take place. This led to the discovery of thiophene by V. MEYER (1882) [24].

Thiophenes occur in fungi and some higher plants, e.g. junipal **24** in the fungus *Daedelia juniperina*. A number of thiophenes have been isolated from compositae, e.g. the 2,2'-bithienyl derivative **25** from the roots of *Echinops spaerocephalus*. Such compounds are nematocidal.

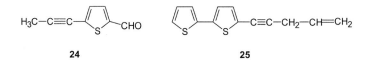

| 24 | 25 |

Many drugs are derived from thiophene [25]. Examples are the antihistamine methaphenilene **26** ([2-(dimethylamino)ethyl]phenyl(2-thenyl)amine) and the antiinflammatory tiaprofenic acid **27** (2-(5-benzoyl-2-thienyl)propionic acid):

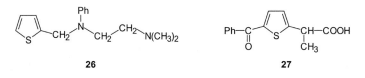

| 26 | 27 |

It is often observed that the pharmacological effect of a thienyl or thenyl substituent is the same or similar to that of a phenyl or benzyl substituent. This phenomenon is known as bioisosterism [26]. Poly(thiophenes) can be used as electric conducting polymers [27].

E Thiophenes are not very important as preparative intermediates. Chiefly, they are used in the synthesis of a saturated C_4 chain by reductive desulfurization. The synthesis of (±)-muscone from 3-methylthiophene serves as an example [28].

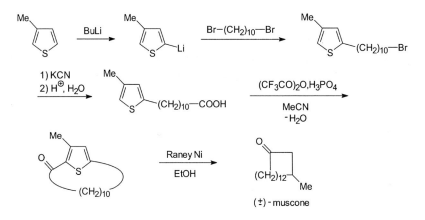

Partial reductive desulfurizations are also successful with 3-methoxythiophenes. This reaction was utilized for the synthesis of pheromones [29].

The systems **28** and **29**, which are derived from thiophene 1,1-dioxide

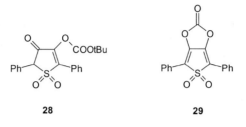

28   **29**

are used as reagents for the transfer of *N*-protective groups (**28**) and for the activation of carboxyl groups (**29**) in peptide synthesis [30]. Their use is illustrated by the synthesis of the protected dipeptide Boc-L-Phe-L-Val-OCH₃ (**31**), which proceeds without racemization. The dipeptide **31** is prepared from Boc-protected L-phenylalanine and L-valine methyl ester:

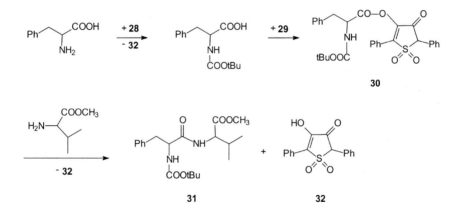

The carboxyl activation proceeds via the enol ester **30**; in this case, as well as in the transfer of the *N*-protecting group, the easily accessible heterocycle **32** (from which the two reagents **28/29** are prepared [31]) is reformed.

5.7 Benzo[*b*]thiophene

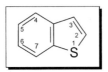

A   Benzo[*b*]thiophene was previously named thionaphthene. It has the following spectroscopic data:

UV (ethanol) λ (nm) (ε)		^{1}H-NMR (CCl$_4$) δ (ppm)		^{13}C-NMR (CDCl$_3$) δ (ppm)	
227 (4.40)	289 (3.22)	H-2: 7.33	H-6: (7.23)	C-2: 126.2	C-6: 124.2
249 (3.83)	296 (3.50)	H-3: 7.23	H-7: (7.29)	C-3: 123.8	C-7: 12
258 (3.83)		H-4: 7.72		C-4: 123.6	C-3a: 139.6
265 (3.63)		H-5: 7.25		C-5: 124.1	C-7a: 139.7

The NMR chemical shift values hardly differ from those of benzo[*b*]furan.

B   Benzo[*b*]thiophene is somewhat less reactive in electrophilic substitutions than thiophene, and also less reactive than benzo[*b*]furan. Moreover, regioselectivity is poor, giving rise to mixtures. Frequently, the 3-position is attacked preferentially over the 2-position, e.g. in halogenation, nitration and acylation. Only the reaction with n-butyllithium is regioselective, giving 2-lithiated benzo[*b*]-thiophene.

Benzo[*b*]thiophenes undergo photochemical [2+2] cycloaddition, for instance with 1,2-dichloro-ethene in the presence of benzophenone as sensitizer:

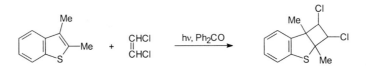

Oxidation of benzo[*b*]thiophene with peroxy acids leads to the 1,1-dioxide.

C   By analogy with benzo[*b*]furans, benzo[*b*]thiophenes can be obtained from thiophenolates and α-halo ketones:

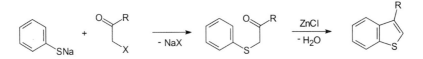

D | **Benzo[b]thiophene**, colourless crystals, mp 32°C, bp 221°C, has a smell similar to that of naphthalene. It is present in the naphthalene fraction of coal tar and can be prepared from sodium thiophenolate and bromoacetaldehyde diethyl acetal. Benzo[b]thiophene occurs in roasted coffee beans.

Various pharmaceuticals and biocides are derived from benzo[b]thiophene, and it is found to be bioisosteric with naphthalene and indole. Mobam **1** [4-(*N*-methylcarbamoyl)benzo[b]thiophene] is an insecticide which is as effective as carbaryl **2**. Both compounds inhibit the enzyme acetylcholinesterase.

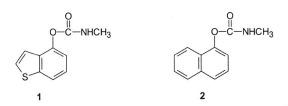

Benzo[b]-3-thienylacetic acid **3** promotes plant growth, as does the corresponding indole compound. 3-(2-Aminoethyl)benzo[b]thiophene **4** has an even stronger action on the central nervous system than tryptamin.

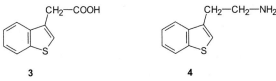

Several compounds which are known as thioindigo dyes are derived fromm benzo[b]thiophene. Thioindigo **6** itself is made from 2-sulfanylbenzoic acid via benzo[b]thiophen-3(2*H*)-one **5** (thioindoxyl) by the following steps (FRIEDLANDER 1905):

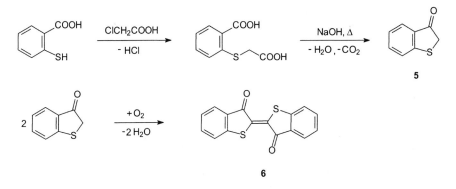

Thioindigo, a vat dye, forms red needles and its vat is bright yellow. As a dye, it has a blue-red colour. Thioindigo dyes were produced in large quantities until the fifties, but since then they have become less important.

5.8   Benzo[c]thiophene

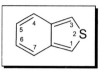

A-D This system lacks a benzene ring and has a *o*-quinonoid structure. Unlike benzo[*c*]furan, benzo[*c*]thiophene has been isolated. Its synthesis proceeded by the following route, starting from 1,2-bis(bromomethyl)benzene:

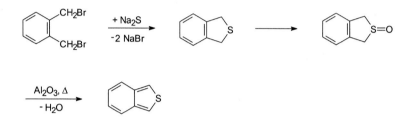

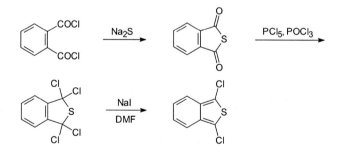

1,3-Dihydrobenzo[*b*]thiophene-2-oxide is heated in a sublimation apparatus with aluminium oxide in vacuo, resulting in the sublimation of benzo[*c*]thiophene. The compound is isolated as colourless crystals of mp 53-55°C. It it thermally unstable and decomposes, even at -30°C, under nitrogen within a few days. Its stability is increased by substituents in the 1,3-positions as in benzo[*c*]furan.

1,3-Diphenylbenzo[*c*]thiophene, yellow needles, mp 118°C, is thermally stable. Its solution displays a green fluorescence. The compound can be prepared by ring transformation of 1,3-diphenyl-isobenzofuran with P_4S_{10} in CS_2.

1,3-Dichlorobenzo[*c*]thiophene, bright yellow crystals, mp 54°C, was made from phthaloyl dichloride [32]:

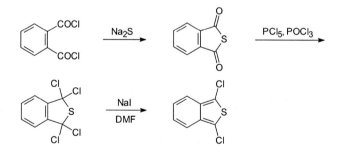

[4+2] Cycloadditions are typical for benzo[*c*]thiophenes, e.g.:

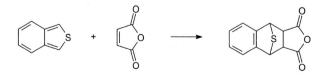

Mixtures of *endo*- and *exo*-diastereomers are usually formed. The reactivity of 1,3-diphenylbenzo[*c*]-thiophene in [4+2] cycloadditions is distinctly lower. Acetylene dienophiles yield adducts which, on heating, are converted into the corresponding substituted naphthalenes with loss of sulfur:

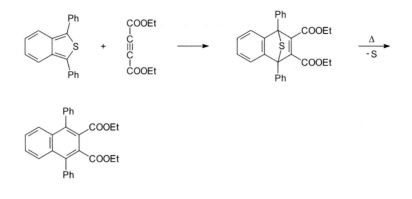

5.9   2,5-Dihydrothiophene

The names Δ^3-thiole and 3-thiole have previously been used for 2,5-dihydrothiophene. A general synthesis for 2,5-dihydrothiophenes with different substituents is based on the MICHAEL addition of α-sulfanylcarbonyl compounds to vinyl phosphonium salts, followed by an intramolecular WITTIG reaction:

2,5-Dihydrothiophenes are oxidized to 1,1-dioxides by *m*-chloroperoxybenzoic acid. These compounds are also accessible by [4+1] cycloaddition of 1,3-dienes and sulfur dioxide. For instance, butadiene reacts even at room temperature with liquid sulfur dioxide to give an adduct 2,5-dihydrothiophene 1,1-dioxide, commonly known by the trivial name 3-sulfolene:

Following WOODWARD, such conversions are called cheletropic reactions. The LUMO of the 1,3-diene surrounds the nonbonding electron pair of sulfur like the claws of a crab (Greek: chele).
3-Sulfolenes are masked 1,3-dienes [33]. At about 150°C they undergo [4+1] cycloreversion (cyclo-elimination) into 1,3-dienes and sulfur dioxide. As this is a thermal concerted reaction, it proceeds in a

disrotatory manner according to the WOODWARD-HOFFMANN rules. Thus a *cis*-2,5-disubstituted 3-sulfolene stereospecifically yields an (*E,E*)-1,3-diene, while the *trans*-diastereomer gives an (*E,Z*)-1,3-diene:

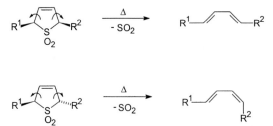

For multistep syntheses that include a DIELS-ALDER reaction, corresponding 1,3-dienes are necessary. They can be prepared by the method described. In many cases, the 1,3-diene is not isolated, but the DIELS-ALDER reaction is carried out with 3-sulfolene and dienophile in boiling xylene.

5.10 Thiolane

A   The bond lengths in thiolane (tetrahydrothiophene) are the same as those in dialkyl sulfides. As in tetrahydrofuran (see p 67), the ring is nonplanar and conformationally flexible. The twist conformation is, however, preferred because of the larger heteroatom. The activation energy for pseudorotation is greater than that for tetrahydrofuran. The chemical shifts in the NMR spectrum correspond to those observed for cycloalkanes and dialkyl sulfides.

B   Thiolanes behave like dialkyl sulfides. With haloalkanes or alcohols in the presence of BRÖNSTED acids, tertiary sulfonium salts are formed:

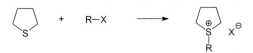

These compounds are good alkylation agents. Ring fission occurs with compounds possessing CH-acidity, e.g.:

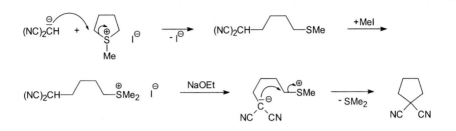

Further methylation, followed by the action of sodium ethoxide in ethanol, leads to substituted cyclopentanes.

Thiolanes can be oxidized to sulfoxides and subsequently to sulfones.

C The reaction of 1,4-dibromo- or 1,4-diiodoalkanes with sodium or potassium sulfide provides thiolanes in good yield:

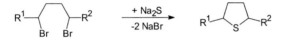

The ring transformation of tetrahydrofurans into thiolanes proceeds with hydrogen sulfide in the presence of aluminium oxide at 400°C:

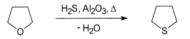

D Thiolane is a colourless, water-insoluble liquid of bp 121°C. It has a distinctive odour similar to town gas. In combination with other compounds, it is responsible for the typical odour of urine after consumption of asparagus.

Thiolane 1,1-dioxide, known by the trivial name sulfolane, is obtained industrially by catalytic hydrogenation of 3-sulfolene. Sulfolane, colourless crystals, mp 27.5°C, bp 285°C, is water soluble. Sulfolane is a polar aprotic solvent and is used for the extraction of sulfur compounds from industrial gases and for the extraction of aromatic substances from pyrolysis fractions. It also serves as a solvent for cellulose acetate, polyvinyl chloride, polystyrene, and polyacrylonitrile.

5.11 Selenophene

A-D The selenophene molecule is planar. The chemical shifts in the NMR spectra are found in the region typical for aromatic compounds.

The synthesis of selenophenes uses a modified FIESSELMANN synthesis for thiophenes and proceeds as follows:

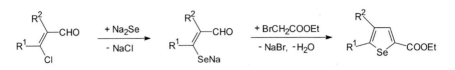

Acetylene reacts at 350-370°C with selenium to produce selenophene. Ring transformation of furan into selenophene occurs with hydrogen selenide over aluminium at 400°C.

Selenophene, a colourless liquid, which, in contrast to thiophene, has an unpleasant smell, has mp -38°C and bp 110°C and is insoluble in water. Selenophene undergoes electrophilic substitution reactions typical of furan and thiophene. It reacts faster than thiophene but much more slowly than furan. Substitution occurs regioselectively in the 2- or 2,5-positions.

Tellurophenes have not been examined in great detail. Their synthesis is analogous to that described for selenophenes, but uses Na_2Te instead of Na_2Se. Tellurophenes are more sensitive towards acids than selenophenes or even thiophenes.

Benzo[b]selenophenes and -tellurophenes, as well as the corresponding dibenzo-condensed systems, are known.

Natural products derived from selenophene or tellurophene have so far not been discovered, which is in contrast to furan and thiophene.

Selenophene proves to be bioisosteric with benzene, thiophene and pyrrole, as in the case of 2-amino-3-(benzo[b]selenophen-3-yl)propionic acid and the proteinogenic amino acid tryptophan:

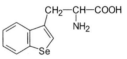

5.12   Pyrrole

A   The univalent radical derived from pyrrole is known as pyrrolyl. All atoms of the pyrrole molecule lie in a plane and the ring forms an almost regular pentagon (see Fig. 5.5).

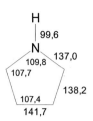

Fig. 5.5   Structure of pyrrole
                 (bond lengths in pm, bond angles in degrees)

The ionization potential was found to be 8.23 eV. The electron is derived from the HOMO, π_3. The dipole moment is 1.58 D. In contrast to furan and thiophene, the heteroatom represents the positive end of the dipole. This could be due to the fact that the heteroatom in pyrrole possesses only one nonbonding electron pair, whereas in furan and thiophene, there are two. As for furan and thiophene, the chemical shifts in the NMR spectrum are to be found in the region typical for aromatic compounds:

UV (ethanol)	^{1}H-NMR (CDCl$_3$)	^{13}C-NMR (CH$_2$Cl$_2$)
λ (nm) (ε)	δ (ppm)	δ (ppm)
210 (4.20)	H-2/H-5: 6.68	C-2/C-5: 118.2
	H-3/H-4: 6.22	C-3/C-4: 109.2

The chemical shift for the NH proton depends on the solvent used.

Pyrrole is aromatic (see Fig. 5.2, p 53). Like furan and thiophene, it belongs to the π-electron excessive heterocycles because the electron density on each ring atom is greater than one:

An acceptable mean value for the empirical resonance energy of pyrrole would be 100 kJ mol^{-1}. The aromaticity of pyrrole is thus greater than that of furan but less than that of thiophene. The value of 22.2 kJ mol^{-1} for the DEWAR resonance energy also fits this picture. If the extent of delocalization of the nonbonding electron pair is decisive for the aromaticity, then the grading of aromaticity, i.e. furan < pyrrole < thiophene < benzene, is correctly reflected by PAULING's electronegativity values for oxygen (3.5), nitrogen (3.0) and sulfur (2.5).

B Pyrroles undergo many reactions, the most important of which are described below.

Acid–base reactions [34]

The pyrrole molecule possesses the NH group typical of secondary amines. The basicity of pyrrole, pK_a = -3.8 for the conjugated acid is, however, much less than that of dimethylamine (pK_a = 10.87). This large difference is due to the incorporation of the nonbonding electron pair of the N-atom into the cyclic conjugated system of the pyrrole molecule. The protonation, moreover, does not occur on the N-atom, but to the extent of 80% on C-2 and of 20% on C-3.

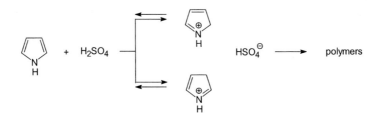

A consequence of the loss of the cyclic 6π-system is rapid polymerization of the generated cations.

By analogy with secondary amines, pyrrole proves to be an NH acid, pK_a = 17.51. For this reason, pyrrole reacts with sodium, sodium hydride or potassium in inert solvents, and with sodium amide in liquid ammonia, to give saltlike compounds:

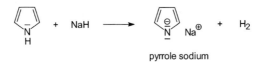

pyrrole sodium

The presence of 'active hydrogen' in pyrrole can also be detected with methylmagnesium iodide according to ZEREWITINOFF:

n-Butyllithium reacts in an analogous manner:

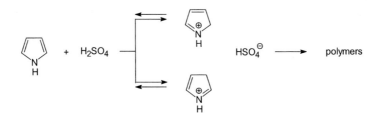

Electrophilic substitution reactions on carbon

Pyrrole reacts in electrophilic substitution reactions about 10^5 faster than furan under similar conditions. This is in spite of the fact that its resonance energy is greater than that of furan; it should, therefore, react more slowly. The discrepancy can be explained by considering the mechanism described on p 54. This postulates that in the case of pyrrole, the σ-complex is especially stablilized by a carbenium–iminium mesomerism:

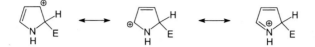

As a consequence, $\Delta H^{\neq}$ for the rate-determinig step could become lower than that for furan (see Fig. 5.6). On the other hand, differences in the stability of the π-complexes could also influence $\Delta H^{\neq}$.

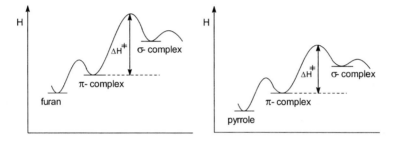

Fig. 5.6   Energy profile of the formation of the π- and σ-complex in the electrophilic substitution of furan and pyrrole

In most electrophilic substitution reactions, pyrrole is preferentially attacked at the α-position. This regioselectivity, however, also depends on whether the reactions are carried out in solution or in the gas phase [35].

Pyrrole reacts with N-chlorosuccinimide to give 2-chloropyrrole. However, with SO_2Cl_2 or aq. NaOCl, one obtains 2,3,4,5-tetrachloropyrrole. N-Bromosuccinimide forms 2-bromopyrrole, and bromine forms 2,3,4,5-tetrabromopyrrole. Pyrroles are nitrated with HNO_3 in acetic anhydride at -10°C to yield 2-nitropyrroles. Concentrated sulfuric acid causes polymerization of pyrroles, but at 100°C, pyridine–SO_3 complex provides the corresponding pyrrole-2-sulfonic acids [35a].

Alkylation of pyrroles proves to be problematic because the usual LEWIS acid catalysts initiate polymerization. The VILSMEIER-HAACK formylation, however, leads to the formation of pyrrole-2-carbaldehyde in good yield. The HOUBEN-HOESCH acylation (reaction with nitriles in the presence of hydrogen chloride) provides 2-acylpyrroles:

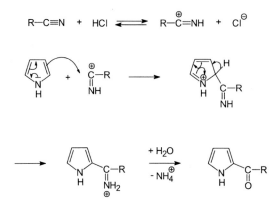

The mechanism of this reaction is illustrated in order to show the low-energy iminium structure of the σ-complex. 2-Acylpyrrole is only formed after the ketiminium salt is hydrolysed with water.

The unusually high reactivity of pyrrole towards electrophiles is demonstrated by two further reactions which are not observed in furan or in thiophene:

- Pyrroles react with arene diazonium salts to give azo compounds, e.g.:

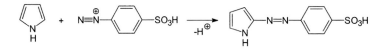

Pyrrole couples even faster than *N,N*-dimethylaniline. With 2,5-disubstituted pyrroles, coupling occurs in the 3-position.

- Pyrroles undergo hydroxymethylation in the 2-position with carbonyl compounds in the presence of acid. The products react further to give dipyrrolylmethanes (see p 484):

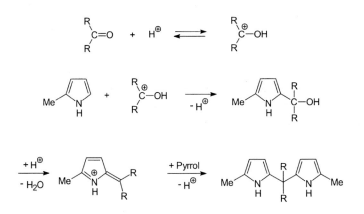

In the case of aldehydes, iron(III) chloride oxidizes the dipyrrolylmethanes to give coloured pyrrolyl(pyrrol-2-ylidene)methanes, which are converted by acids into symmetrically delocalized protonated salts:

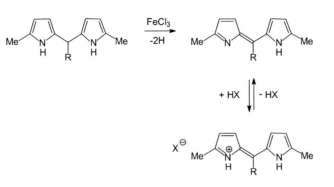

With EHRLICH's reagent, i.e. a solution of 4-(dimethylamino)benzaldehyde in hydrochloric acid, the reaction proceeds to give only the purple-coloured azafulvenium salt:

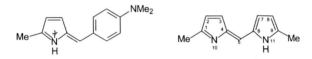

From 2-methylpyrrole and formaldehyde, orange-coloured pyrrolyl(pyrrol-2-ylidene)methane is formed, which is unsubstituted in the 5-position. In acid-catalysed reactions with carbonyl compounds, pyrroles behave similarly to phenols, which give diphenylmethanes via hydroxymethyl compounds.

Electrophilic substitution reactions on nitrogen

Pyrrole sodium and pyrrole potassium yield 1-substituted pyrroles with haloalkanes, acyl halides, sulfonyl halides as well as with chlorotrimethylsilane. On the other hand, 2-methylpyrrole is obtained from pyrrol-1-ylmagnesium iodide and methyl iodide.

1-Acylpyrroles add lithiumorganic reagents to the carbonyl group [36].

1-Phenylsulfonylpyrrole undergoes FRIEDEL-CRAFTS acylation with substitution in the 3-position. 3-Acylpyrroles are obtained from pyrrole as follows:

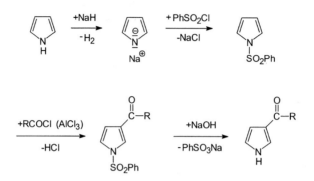

As already mentioned, *n*-butyllithium effects the lithiation of pyrroles in the 1-position. If this position is blocked by a substituent, then 2-lithiopyrroles are formed regioselectively. They can be used for the synthesis of substituted pyrroles, e.g.:

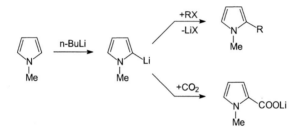

Addition reactions

Hydrogenation of pyrroles to pyrrolidines by RANEY nickel proceeds only under pressure and at high temperatures. Autoxidation of pyrroles, as well as oxidation with hydrogen peroxide, can be considered addition reactions. Attack of O_2 or H_2O_2 occurs first at the 2-position and then at the 5-position, resulting finally in the formation of maleimide or *N*-substitued maleimide:

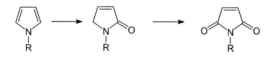

The great reactivity of pyrrole towards electrophiles is the reason why the reaction with maleic anhydride does not result in a DIELS-ALDER addition but in an electrophilic substitution.

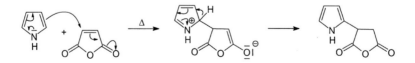

This reaction can also be regarded as a MICHAEL addition of pyrrole to maleic anhydride. Some substituted pyrroles, however, undergo a [4+2] cycloaddition with acetylene dienophiles, e.g. 1-(ethoxycarbonyl)pyrrole with acetylenedicarboxylic ester [37].

Among the [2+2] cycloadditions, the PATERNO-BÜCHI reaction (see p 39) with pyrroles has been investigated. The oxetanes isomerize to give 3-(hydroxyalkyl)pyrroles under the reaction conditions:

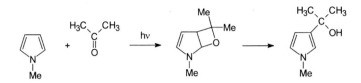

A well-established cycloaddition of pyrroles is the [2+2] cycloaddition with dichlorocarbene. This is in competition with the REIMER-TIEMANN formylation:

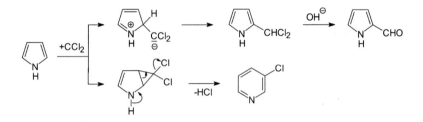

Under strongly basic conditions (generation of dichlorocarbene from chloroform and potassium hydroxide), electrophilic substitution of pyrrole by dichlorocarbene dominates, leading eventually to pyrrole-2-carbaldehyde. In a weakly basic medium (generation of dichlorocarbene by heating sodium trichloroacetate), the [2+1] cycloaddition prevails. The primary product eliminates hydrogen chloride to give 3-chloropyridine.

Ring-opening reactions

The opening of the pyrrole ring leads to clean reactions in only a few cases, because BRÖNSTED as well as LEWIS acids initiate polymerization, and strong bases cause only salt formation.

Hydroxylamine hydrochloride and sodium carbonate in ethanol react with pyrroles to dioximes of 1,4-dicarbonyl compounds [38]:

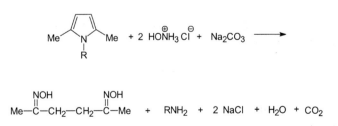

Pyrrole itself yields the dioxime of butandial and ammonia.

Concerning its *retrosynthesis* pyrrole exhibits the function of a double enamine and can, therefore, be dissected retroanalytically in two ways (I/II), analogously to furan. Route I (after retrosynthetic operations of an enamine hydrolysis **a** - **c**) yields 1,4-dicarbonyl compounds as potential starting materials, which should produce pyrroles by cyclocondensation with NH$_3$. When the intermediate **1** i.e. *γ*-keto enamine is treated according to step **d**, a bond cleavage different from that leading to **2** is possible. This reversal of an enamine alkylation gives rise to *α*-halocarbonyls **3** and enamines **4** and thereby suggests alternative starting materials:

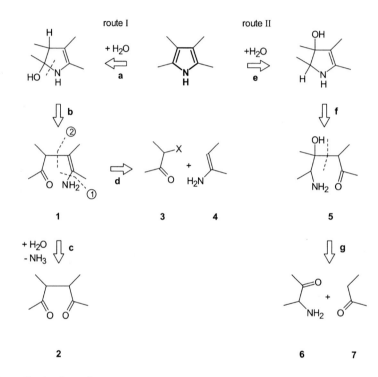

Fig. 5.7 Retrosynthesis of pyrrole

After H$_2$O addition and enamine hydrolysis (**e** / **f**), route II leads to the *γ*-amino aldol intermediate **5**. Aldol fission follows (retroanalysis step **g**) resulting in the formation of *α*-amino carbonyl compounds **6** and methylene ketones **7**. These are possible starting materials for the synthesis of pyrroles.

(1) The *Paal-Knorr synthesis*, in which 1,4-dicarbonyl compounds are treated with NH$_3$ or primary amines (or with ammonium or alkylammonium salts) in ethanol or acetic acid, leads to 2,5-disubstituted pyrroles, and is universally applicable. For instance, hexane-2,5-dione **8** reacts with NH$_3$ to yield 2,5-dimethylpyrrole **9**:

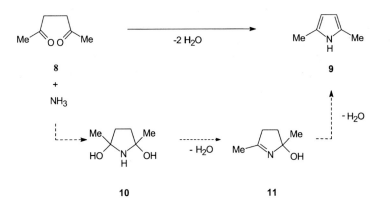

8

+

NH₃

10

11

The primary step leads to the double hemiaminal **10** which, by stepwise H_2O elimination, furnishes the pyrrole system **9** via the imine **11** [39].

(2) α-Halocarbonyl compounds react with β-keto esters or β-diketones and ammonia or primary amines to give 3-alkoxycarbonyl- or 3-acyl-substituted pyrrole derivatives, respectively *(Hantzsch synthesis)*:

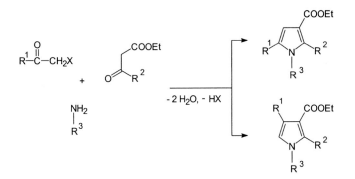

The regioselectivity depends on the substituents in the starting material but gives mainly the 1,2,3,5-tetrasubstituted pyrrole. Exhaustive investigations show that β-keto esters react with ammonia or amine to give a β-aminoacrylic ester **(12)** as a primary step. *C*-Alkylation of the enamine function in **12** by the haloketone produces the 1,2,3,5-substituted pyrrole **13**, while *N*-alkylation leads to a 1,2,3,4-substituted pyrrole **14**:

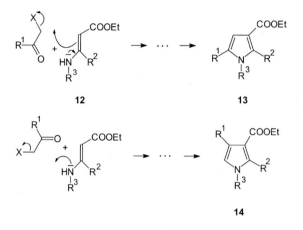

12   **13**

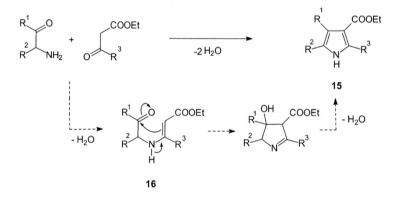

14

(3)   α-Amino ketones undergo cyclocondensation with β-keto esters or β-diketones to give 3-alkoxy-carbonyl- or 3-acyl-substituted pyrroles **15** *(Knorr synthesis)*:

15

16

The KNORR synthesis also proceeds via β-enaminone intermediates **16** [40]. Frequently, the α-amino ketones are not employed as such but generated in situ by reduction of α-oximino ketones. The latter are obtained by nitrosation of ketones with alkyl nitrites in the presence of sodium methoxide:

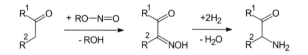

For both the HANTZSCH and KNORR syntheses of pyrroles, several variations have been elaborated [41].

(4)   3-Substituted pyrrole-2-carboxylic esters **19** are synthesized from *N*-tolylsulfonyl glycine ester **17** and vinyl ketones *(Kenner synthesis)* [42]. By MICHAEL addition and intramolecular aldol addition, they first yield pyrrolidine-2-carboxylic esters **18**. These are converted into pyrroles by succesive H_2O and sulfinic acid eliminations.

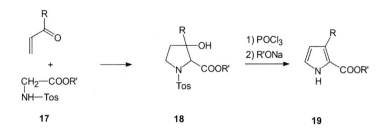

17   **18**   **19**

(5)   Cyclocondensation of nitroalkenes with CH-acidic isocyanides in the presence of bases leads to the formation of trisubstituted pyrroles **20** (*Barton-Zard synthesis*) [43]:

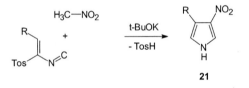

R^3 = COOEt, Tos

20

The first step of this reaction is a MICHAEL addition of isocyanide to the nitroalkene. Cyclization and elimination of HNO_2 follow. On the other hand, α,β-unsaturated isocyanides and nitromethane yield 3-nitropyrroles **21**.

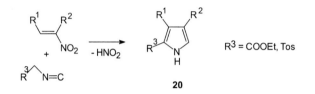

21

D   **Pyrrole** was first isolated from bone oil, but also occurs in coal tar. It can be prepared by the dry distillation of the ammonium salt of D-galactaric acid (mucic acid; see p 61). Pyrrole is a colourless liquid with a characteristic odour reminiscent of chloroform, of mp -24°C and bp 131°C. It is slightly soluble in water and turns brown quickly in air.

The pyrrole ring, although not very common in nature, occurs in some very important natural products. A few antibiotics contain a pyrrole ring, one of the simplest is pyrrolnitrin **22**:

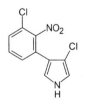

22

The biologically important tetrapyrroles contain four pyrrole rings, which are linked by CH_2 or CH bridges. One differentiates between linear tetrapyrroles (bilirubinoids) and cyclic tetrapyrroles (porphyrins and corrins).

Bilirubinoids are coloured compounds occurring in vertebrates, in some invertebrates and even in algae. They are formed by the biological oxidation of porphyrins. The most important representative is the orange-coloured bilirubin. It occurs in the bile and in gallstones and is excreted in the faeces and in urine. Bilirubin was first isolated by STAEDELER (1864) and can be purified via its crystalline ammonium salt. It is oxidized to blue-green biliverdine by iron(III) chloride:

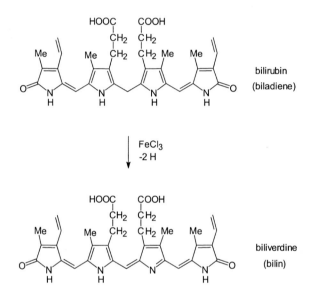

bilirubin
(biladiene)

biliverdine
(bilin)

The corresponding unsubstituted compounds are known as biladiene and bilin [44]. Porphyrins and corrins are discussed in chapter 8.3.

A number of pharmaceuticals are derived from pyrrole, e.g. the analgesic and anti-inflammatory zomepirac **23** [5-(4-chlorobenzoyl)-1,4-dimethylpyrrol-2-ylacetic acid]:

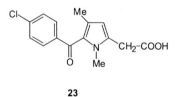

23

Polymers and copolymers of pyrrole are used as organic conductors for special purposes, e.g. in photovoltaic cells [45].

5.13 Indole

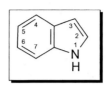

A Benzo[*b*]pyrrole is known by the trivial name indole. The univalent radical is called indolyl. The UV and NMR data of indole are given in the table below:

UV (ethanol) λ (nm) (ε)		^{1}H-NMR (acetone-d$_6$) δ (ppm)		^{13}C-NMR (CDCl$_3$) δ (ppm)	
216 (4.54)	276 (3.76)	H-1: 10.12	H-5: 7.00	C-2: 123.7	C-6: 119.0
266 (3.76)	278 (3.76)	H-2: 7.27	H-6: 7.08	C-3: 101.8	C-7: 110.4
270 (3.77)	287 (3.68)	H-3: 6.45	H-7: 7.40	C-4: 119.9	C-3a: 127.0
		H-4: 7.55		C-5: 121.1	C-7a: 134.8

B Indoles are less reactive than pyrroles. The following reactions are characteristic for the chemical behaviour of indole.

Acid–base reactions [46]

The basicity of indole, pK_a = -3.50, corresponds approximately to that of pyrrole. Protonation occurs mainly on C-3 with formation of the 3*H*-indolium ion which reacts further to give oligomers:

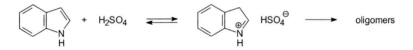

With a pK_a value of 16.97, indole possesses an NH-acidity similar to that of pyrrole. Indole reacts, therefore, with sodamide in liquid ammonia, with sodium hydride in organic solvents, with GRIGNARD reagents and with *n*-butyllithium to give 1-metalated indoles.

Electrophilic substitution reactions on carbon

In most electrophilic substitution reactions, indole reacts more slowly than pyrrole but faster than benzo[*b*]furan. In contrast to pyrrole substitution of the H-atom occurs preferably in the 3-position. There are two reasons for this:
- Attack of the electrophile on the 3-position leads to formation of a low energy iminium structure of the σ-complex. Attack on the 2-position, however, results in a high-energy orthoquinonoid iminium structure:

- With pyrrole, the coefficient of the HOMO is greatest in the 2- and 5-positions, whilst with indole this applies to the 3-position.

If the 3-position already carries a substituent, attack usually occurs first on the 2- and subsequently on the 6-position.

Indole is chlorinated with $SOCl_2$ or aqueous NaOCl to give 3-chloroindole, and 3-bromoindole is formed with N-bromosuccinimide. Action of HNO_3 on indole causes oxidation of the pyrrole ring followed by polymerization. Indoles substituted in the 2-position react with HNO_3 in acetic acid to give 3,6-dinitro compounds. Sulfonation of indole with pyridine–SO_3 complex leads to the formation of indole-3-sulfonic acid.

As is the case with pyrroles, C-alkylation of indoles results in mixtures of products. Formylation and acylation, however, occur more readily. The VILSMEIER-HAACK reaction furnishes indole-3-carbaldehyde; heating with acetic anhydride produces 3-acetylindole. In the HOUBEN-HOESCH acylation, substitution takes place in the 3-position.

Although indoles are less reactive than pyrroles, they react with arenediazonium salts to give 3-(aryl-azo)indoles. The reaction with carbonyl compounds proceeds in an analogous way in the presence of acids. For this reason, indoles unsubstituted in the 3-position give a positive colour test with EHRLICH's reagent. Indole and acetaldehyde react via an azafulvenium salt to give 3-vinylindole, which on further reaction with indole forms 1,1-di(indol-3-yl)ethane:

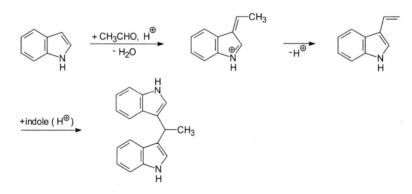

Another electrophilic substitution which indole undergoes with ease is aminoalkylation (MANNICH reaction). Indole, formaldehyde and dimethylamine in acetic acid interact with formation of the natural product gramine [(3-dimethylaminomethyl)indole], which has been isolated from grasses (earlier gramineae; now poaceae):

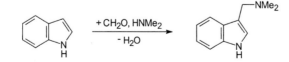

Electrophilic substitutions on nitrogen

The saltlike alkali metal compounds of indole react with electrophiles such as haloalkanes, acyl halides, sulfonyl halides and trimethylchlorosilane to form the corresponding 1-substituted indoles. 1-Benzylindole isomerizes to 2-benzylindole when heated in polyphosphoric acid [47]. 1-Phenylsulfonylindole is lithiated in the 2-position by *n*-butyllithium. Subsequent alkylation with haloalkanes and cleavage of the phenylsulfonyl residue with sodium hydroxide yields 2-alkylindole.

The interaction of (indol-1-yl)magnesium halides and electrophilic reagents results mainly in 3-substituted indoles.

Addition reactions

Catalytic hydrogenation of indoles under pressure and at elevated temperature leads to 2,3-dihydroindoles (indolines). These compounds are also formed by the action of reducing agents (zinc and phosphoric acid or tin and hydrochloric acid) on indoles.

Indoles, like pyrroles, are easily oxidized. During autoxidation, the 3-position is attacked by oxygen leading to a hydroperoxide which gives rise to indol-3(2*H*)-one (indoxyl):

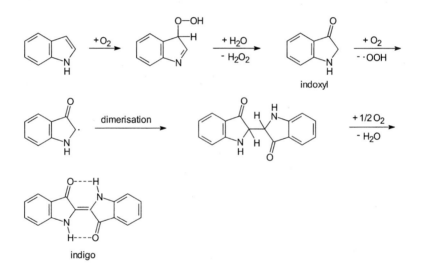

Indoxyl reacts further by radical coupling followed by oxidation to give indigo. Other oxidizing agents, as well as air, cause such reactions. On oxidation, indoles with substituents in the 3-position are converted into indol-2(3*H*)-ones (oxindoles):

Indoles show little inclination to undergo cycloadditions. The [2+1] cycloaddition with dichlorocarbene leads to mixtures of indole-3-carbaldehyde and 3-chloroquinoline (see p 334).

C For the *retrosynthesis* of indole (see Fig. 5.8), two routes (I/II) are proposed, as for pyrrole (see p 94). Route I suggests *o*-aminobenzyl ketone **1** or *o*-alkyl-*N*-acylaniline **2** as starting material on the basis of operations **a** - **c**. Their retroanalysis (**d,e**) in turn leads to 2-alkylaniline **5** and carboxylic acid derivative **6**. Construction of the indole system should thus occur by *N*- or *C*-acylation of **5** (utilizing the *o*-nitrotoluene derivative **4**) followed by cyclodehydration of **1/2**. The alternative route II, based on retrosynthetic analysis **g-i**, leads to aniline via the α-(*N*-phenylamino)ketones **3** and to α-halo ketones **7** as possible precursors for the indole synthesis.

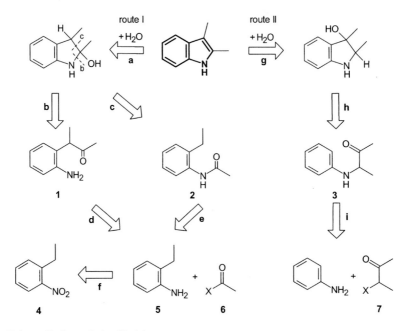

Fig. 5.8   Retrosynthetic analysis of indole

As predicted by retrosynthetic considerations, most indole syntheses start with aniline or 2-alkyl-anilines, upon which the heterocyclic part is constructed according to various methods [48].

(1)   The reductive cyclization of *o*-nitrobenzylcarbonyl compounds as shown for **8**

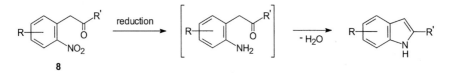

serves above all for the preparation of 2-substituted indoles (*Reissert synthesis*). In the classical version of this method, *o*-nitrophenylpyruvic acid **9**, obtained by a CLAISEN condensation of *o*-nitrotoluene with oxalic ester, is subjected to catalytic hydrogenation (H_2,Pd-C). Reduction of the nitro group to an amino group is followed by a spontaneous cyclodehydration to give the indole-2-carboxylic ester **10**:

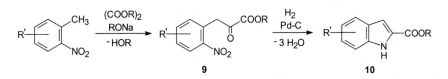

o-Aminobenzyl carbonyl compounds, which are essential for the cyclizing step of the REISSERT synthesis, are also formed from aniline via anilinosulfonium salts **12** by a remarkable reaction sequence. The latter are obtained from N-chloroanilines **11** and α-(methylsulfanyl)ketones; these react with bases to give 3-(methylsulfanyl)indoles **15**, which after a reductive cleavage of the SCH₃ group, furnish indoles **16** by hydrogenolysis (*Gassmann synthesis*):

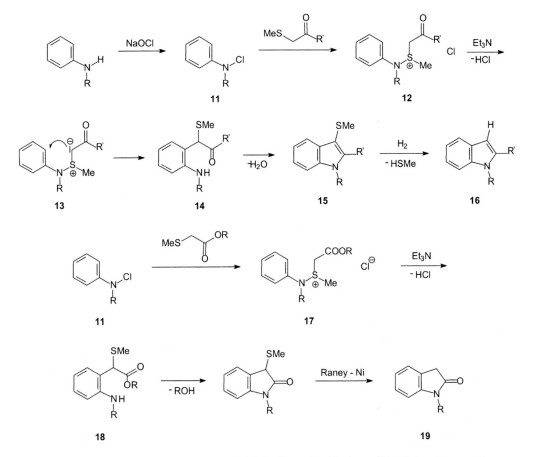

A SOMMELET-HAUSER rearrangement of the initially formed sulfonium ylide **13** is a key reaction step. This brings about the required *ortho*-linkage with the arene, which results in the formation of the *o*-aminobenzyl carbonyl compound **14**.

Anilinosulfonium compounds of type **17** can also be used to synthesize indol-2(3*H*)-ones **19** starting from N-chloroanilines and (methylsulfanyl)acetic ester. The (*o*-aminophenyl)acetic esters **18** obtained via the S-ylide lead to indol-2(3*H*)-ones **19** after cyclization and reductive desulfurization.

(2) 1-Dimethylamino-2-(*o*-nitrophenyl)ethenes **20**, obtained from *o*-nitrotoluene and *N,N*-dimethyl-formamide dimethyl acetal, yield indoles **21** on reductive cyclization of the corresponding *o*-amino-phenyl derivatives:

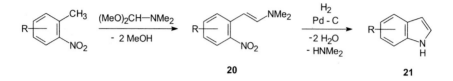

20   **21**

This method (*Batcho-Leimgruber synthesis*) is particularly suitable for the synthesis of indoles substituted on the benzene ring but unsubstituted on the pyrrole moiety [49].

(3) *N*-Acyl-*o*-toluidines **22** can be cyclodehydrated by means of sodamide or n-butyllithium (*Madelung synthesis*). However, this method is essentially confined to the preparation of 2-alkylindoles because of the vigorous reaction conditions:

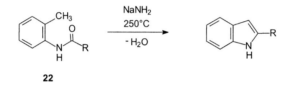

22

The cyclization necessitates deprotonation of the CH_3 group and its coupling with the acyl C-atom, but the mechanism has so far not been elucidated.

The indole formation from *o*-tolylisocyanide **23**, which is brought about by metallation with lithium dialkylamides, is related to the MADELUNG synthesis. The lithium compound **24** can undergo cyclization to form indole (via *N*-lithioindole) or, after alkylation and renewed metallation, to produce 2-substituted indoles:

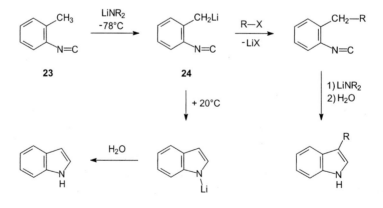

(4) α-Arylamino ketones **25** are easily accessible from arylamines and α-halo ketones. They undergo an intramolecular $S_E Ar$ reaction under acid catalysis, which is followed by H_2O elimination to produce indoles (*Bischler synthesis*):

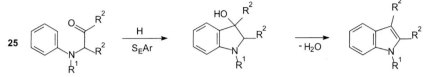

The preparative value of this simple indole synthesis is limited to systems with the same C-2/C-3 substituents.

(5) (*o*-Alkynyl)arylamines and their *N*-acyl or *N*-sulfonyl derivatives **27** are cyclized to the corresponding indoles **28** by treatment with (a) TBAF, (b) Pd (and Cu) complexes [50]:

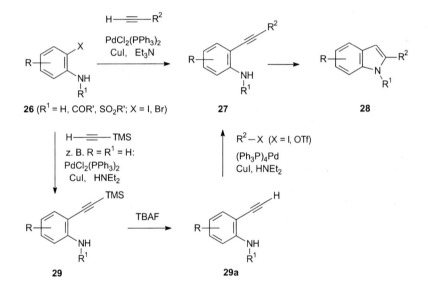

In (b), the organopalladium intermediate (cf. p 64) can be trapped either by protonation to give 2-substitued indoles **28** or by alkylation with R^3-X leading to introduction of an additional substituend R^3 in the 3-position of the indoles **28**.

The (*o*-aminoaryl)acetylenes **27** are easily accessible by Pd-mediated SONOGASHIRA cross coupling reactions of (*o*-halogeno)aniline derivatives **26** (X preferably I) with terminal acetylenes. Alternatively, **27** is prepared by SONOGASHIRA coupling of **26** with TMS-acetylene (to give **29**), desilylation of **29** and base-induced alkylation at the free acetylene terminus (**29 → 29a → 27**).

The potentiality of this method for indole synthesis is demonstrated by solid-phase versions [50a] utilizing polymer-bound (*o*-halogeno)anilines and their transformation according to the sequence **26 → 27 → 28**.

(6)   *N*-Arylhydrazones, derived from enolizable aldehydes or ketones **30**, are converted into indoles **31** under LEWIS (ZnCl$_2$, BF$_3$) or BRÖNSTED acid (H$_2$SO$_4$, polyphosphoric acid, CH$_3$COOH, HCl in ethanol) catalysis with loss of ammonia (*Fischer synthesis*, E. FISCHER 1883):

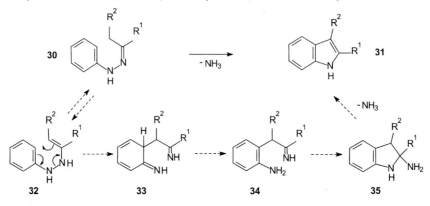

The mechanism of this indole synthesis, which has wide application and is of great general importance, has been the subject of intensive studies [51]. It was shown that the hydrazone **30** first tautomerizes to an enehydrazine **32** which, by a [3,3] sigmatropic rearrangement, establishes a C–C bond in the *ortho*-position of the arene (diaza-COPE rearrangement). The resulting bisimine **33** is converted into 2-amino-2,3-dihydroindole **35** via the intermediate **34**, and finally cyclizes to give indole **31** with elimination of NH$_3$. Experiments with ^{15}N isotopes prove that the nitrogen attached to the arene of **30** is retained in the indole.

The phenylhydrazones required for the FISCHER synthesis are prepared from carbonyl compounds and phenylhydrazine. Alternatively, they can be obtained by a JAPP-KLINGEMANN reaction from CH-acidic compounds (β-diketones, β-keto esters, etc.) or from enamines by interaction with aryldia-zonium salts. More complex indoles are thus accessible by a simple route. This is demonstrated by the synthesis of 3-(indol-3-yl)propionic acid **40**, the starting material of WOODWARD's lysergic acid synthesis [52].

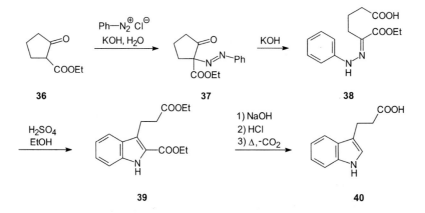

The 2-oxocyclopentane-1-carboxylic ester **36** is coupled with benzenediazonium chloride in a basic medium with simultaneous acid cleavage of the product **37**. The resulting phenylhydrazone of the α-oxo adipic ester **38** is cyclized by H_2SO_4 in ethanol to give the indole derivative **39**. Saponification and selective decarboxylation yield the indole **40**.

(7) Only a few indole syntheses make use of building blocks in which the N-atom is not directly bonded to an arene. The *Nenitzescu synthesis* is of this type. In this synthesis, 1,4-quinones are condensed with 3-aminoacrylic esters to give 5-hydroxyindole-3-carboxylic esters **44**. The mechanism of this synthesis has not been completely clarified. It includes a MICHAEL addition (→ **41**) and a cyclodehydration (**42** → **43**) as well as a redox transformation (**41** → **42** and **43** → **44**) [53]:

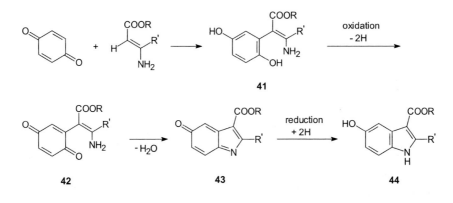

41

42   43   44

D **Indole** is a colourless solid forming leaflet crystals, moderately soluble in water, of mp 52°C and bp 253°C. It occurs in coal tar and jasmine oil. Indole has an unpleasant faecallike odour, but in high dilution has a pleasantly flowery aroma. It was found recently that indole is one of the compounds responsible for the smell of flowering rape fields.

Indole-3(2*H*)-one (indoxyl), bright yellow crystals, mp 85°C, is manufactured by a process developed by HEUMANN and PFLEGER (1890, 1898). Aniline and chloroacetic acid yield (phenylamino)acetic acid. On melting with $KOH/NaOH/NaNH_2$, its potassium salt cyclizes to indoxyl:

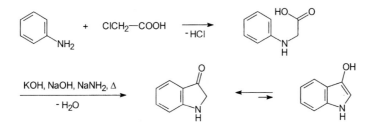

This reaction is the basis of the manufacture of indigo dye, which is produced by aerial oxidation of indoxyl (see p 110). Indoxyl is present predominantly in its keto form.

Indole-2(3*H*)-one (oxindole), colourless needle-like crystals, mp 127°C, is made from aniline and chloroacetyl chloride as follows [54]:

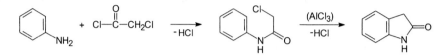

Indole-2,3-dione (isatin), red crystals, mp 204°C, is formed by the oxidation of indigo with nitric acid. It can be synthesized from *o*-nitrobenzoyl chloride via isatinic acid **45**:

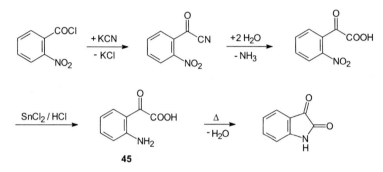

A considerable number of natural products are derived from indole. Of greatest importance is the essential amino acid tryptophan **46**, a constituent of many proteins:

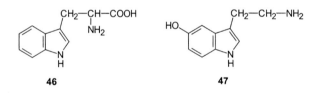

Enzymatic conversion of tryptophan in living organisms produces additional natural products, e.g. serotonin (5-hydroxytryptamine) **47** by hydroxylation and decarboxylation. It occurs in the serum of warm-blooded animals as a vasoconstrictor, it is one of the agents responsible for maintaining vascular tone. Moreover, it acts as a neurotransmitter, i.e. it is essential for conducting impulses between nerve cells.

Bufotenin **48**, a poison occurring in the skin of toads, causes a rise in blood pressure and paralyses the spinal and cerebral motor centres. Psilocin **49**, the psychoactive substance in the Mexican mushroom Teonanacatl, increases excitability and causes hallucinations:

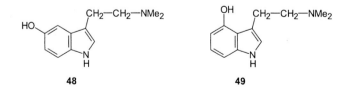

Melanin, the black and brown hair and skin pigment in humans and animals, also contains indole units, in particular 5,6-dihydroxyindole. It is derived from the amino acid tyrosine via DOPA (3,4-dihydroxy-phenylalanine):

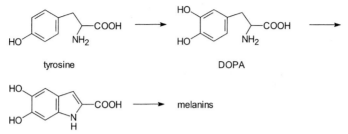

tyrosine   DOPA

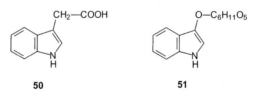

melanins

Melanins are produced by specialized cells, the melanocytes. An accumulation of these cells causes moles. If these cells become malignant, a melanoma develops, a type of skin cancer which requires immediate surgery.

The class of indole alkaloids is so large that only a few of those that are physiologically active and pharmacologically important natural products can be mentioned here, e.g. strychnine, brucine (see p 462), yohimbine, reserpine, vincamine, ergotamine and lysergic acid. Some antibiotics are also derived from indole.

Indole-3-acetic acid **50**, also known as heteroauxin, is a phytohormone. It is mainly formed in buds, seeds and in young blossoms and is a plant growth regulator.

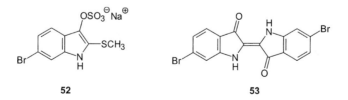

50   **51**

Indican **51**, the β-glucoside of the enol form of indoxyl, occurs in indigo plants (*Indigofera tinctoria*) and in woad (*Isatis tinctoria*). During the extraction of the crushed plant with water, indican is hydrolysed to give indoxyl and glucose by the action of the enzyme indoxylase, which is also present in these plants. From antiquity until about 1890, the extraction of indigo from plants was based on this reaction and subsequent aerial oxidation. Thereafter, synthetically produced indigo came to dominate the market.

Among indole compounds occurring in sea organisms, tyrindolsulfate **52** must be mentioned [55]. It occurs in molluscs of the type murex, purpura and dicathais, which are mainly found in the Mediterranean. The tyrian purple of antiquity (6,6'-dibromoindigo **53**) was extracted from these animals:

52   **53**

The indole ring system forms the basis of several pharmaceuticals, for instance the anti-inflammatory indomethacin **54** and the antidepressant iprindole **55**:

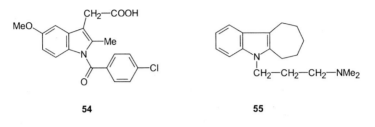

54   55

Indigo and other dyes (see p 81) which contain the chromophore **56**

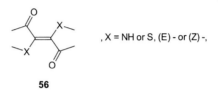

, X = NH or S, (E) - or (Z) -,

56

are known as indigoid dyes, and are vat dyes. These water-insoluble compounds are reduced by sodium dithionite and sodium hydroxide and applied as vat dyes, whereby they become soluble as disodium dihydro compounds, e.g.:

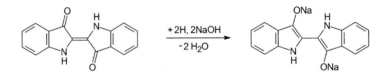

Before the availability of sodium dithionite (1871), the reduction of indigo was brought about by bacteria with reducing properties (fermentation vat). The vat of indigo has a brown-yellow colour. The cloth is dipped in the vat and then exposed to air to allow reoxidation to indigo which is precipitated and finely distributed onto the fibre. A consequence of this process is the low rubbing fastness of the dyes. It causes the faded appearance of indigo-dyed jeans and makes possible the manufacture of 'faded jeans'.

Since the seventies, indole compounds have lost their importance as textile dyes. They have, however, found application in other fields, e.g. in polaroid photography.

5.14 Isoindole

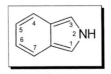

 The name isoindole is admissible for benzo[c]pyrrole. Although isoindole has an *ortho*-quinonoid structure, it can be isolated; it is characterized spectroscopically and chemically [56]:

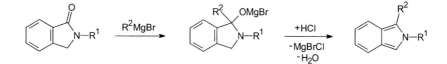

2-(Methoxycarbonyloxy)-1,3-dihydroisoindole (obtained from 2-hydroxy-1,3-dihydroisoindole and methyl (4-nitrophenyl)carbonate) was pyrolysed in vacuo at 500°C and isoindole condensed on a cold-finger condenser filled with liquid nitrogen. Isoindole crystallizes in colourless needles, darkens at room temperature and polymerizes. Solutions in dichloromethane kept under nitrogen are more stable. They give a red colour with EHRLICH's reagent and form the corresponding DIELS-ALDER adducts (endo:exo = 2:3; see p 56) with N-phenylmaleinimide. Isoindoles substituted on the pyrrole ring are thermally more stable and can, for instance, be prepared from 2-substituted 1,3-dihydroisoindol-1-ones by the following route:

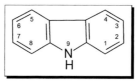

A tautomeric equilibrium exists between isoindoles unsubstituted in the 2-position and the corresponding 1H-isoindoles:

Isomerizations in which an H-atom changes its position in a heterocyclic system are known as *annular tautomerisms*. This is a special case of prototropy (prototropic rearrangement) [57]. For isoindole, the position of the equilibrium depends to a large extent on the type of the substituent in the 1- and/or 3-position [58]. The 1-phenyl compound exists in the 2H-form and couples with benzenediazonium chloride in the 3-position.

5.15 Carbazole

A,B   The numbering of carbazole (benzo[b]indole) predates the introduction of systematic nomenclature and is retained for historical reasons.

Carbazoles behave like *o,o'*-disubstituted diphenylamines. However, the basicity of carbazole, pK_a = -4.94, is much lower than that of diphenylamine (pK_a = 0.78), and also lower than that of indole and pyrrole. As a consequence, carbazole is insoluble in dilute acids but only soluble in concd H_2SO_4 with protonation of the N-atom. On pouring the solution into water, carbazole precipitates without polymerization.

The NH-acidity of carbazole pK_a = 17.06 corresponds approximately to that of indoles and pyrroles. For this reason, carbazole is convertible into *N*-metallated compounds which can be subjected to electrophilic substitution on nitrogen.

Carbazole reacts with electrophiles faster than benzene. Substitution occurs regioselectively in the 3-position, e.g. in the VILSMEIER-HAACK formylation.

There are very few addition and ring-opening reactions of carbazoles.

C | *ortho*-Substituted biphenyl derivatives are a starting point for the synthesis of carbazoles, e.g.:

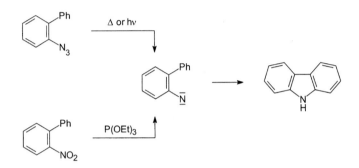

Thermolysis or photolysis of 2-azidobiphenyl produces a nitrene, as does deoxygenation of 2-nitrobiphenyl with triethyl phosphite. The nitrene cyclizes immediately to give carbazole. Carbazoles can also be made by cyclodehydrogenation of diphenylamines.

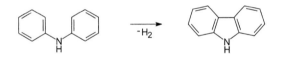

This reaction can be carried out photochemically or with palladium(II) acetate in acetic acid.

D | **Carbazole** forms water-insoluble crystals of mp 245°C, bp 355°C, and occurs in the anthracite fraction of coal tar.

A few alkaloids are derived from carbazole, e.g. murrayanine **1** (1-methoxycarbazole-3-carbaldehyde) and ellipticine **2**, which is a carbazole system with a fused pyridine ring. Ellipticine is one of the substances which can intercalate into human DNA. The molecule is inserted between two paired bases. Ellipticine derivatives are approved in some countries for use as cytostatic agents [59].

Pharmaceuticals containing a carbazole system are rare, but one example of these is the beta-blocker carazolol **3** [1-(carbazol-4-yloxy)-3-(isopropylamino)propan-2-ol]:

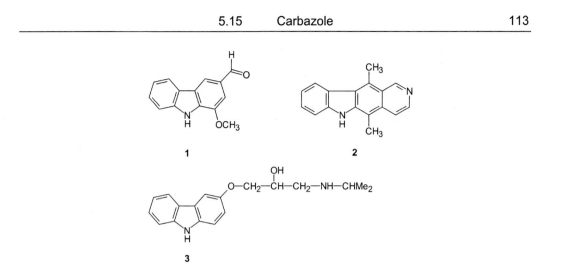

1

2

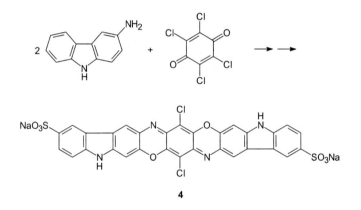

3

The textile dye Sirius Light Blue **4** is produced by a double cyclocondensation of 3-aminocarbazole with chloranil followed by sulfonation:

4

Carbazole is vinylated in the 9-position by acetylene. The resulting polymer **5** (poly-*N*-vinylcarbazole) proves to be a photoconductor:

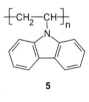

5

Photoconductors are substances which, upon irradiation, increase their electric conductivity [60]. They find use in electrophotography as well as in copying processes.

5.16 Pyrrolidine

A   The pyrrolidine molecule is practically without strain, nonplanar and conformationally mobile. As in the case with tetrahydrofuran (see p 67), the twist and envelope conformations are preferred. The activation energy for pseudorotation is 1.3 kJ mol⁻¹. The chemical shifts in the NMR spectra lie in the region characteristic for cycloalkanes and dialkylamines.

^{1}H-NMR (CDCl₃) ^{13}C-NMR (CDCl₃)
δ (ppm) δ (ppm)

H-2/H-5: 2.75   C-2/C-5: 47.1
H-3/H-4: 1.53   C-3/C-4: 25.7

B   Pyrrolidines and *N*-substituted pyrrolidines undergo reactions typical of secondary or tertiary alkylamines. They can be alkylated, quaternized, acylated and nitrosated. The basicity and nucleophilicity of pyrrolidines are, however, greater than those of diethylamine (pyrrolidine pK_a = 11.27, diethylamine pK_a = 10.49). Because of these properties, pyrrolidine is very suitable for the conversion of carbonyl compounds into enamines:

R¹—CH₂—C(=O)—R² + HN⟨pyrrolidine⟩ →(H⊕) / −H₂O→ R¹—CH=C(N⟨pyrrolidine⟩)—R²

C   Pyrrolidine and *N*-substituted pyrrolidines are produced commercially by ring transformation of tetrahydrofuran with ammonia or primary amines at 300°C on aluminium oxide catalysts.
N-Substituted pyrrolidines are also accessible by photodehydrohalogenation of *N*-alkyl-*N*-chloroamines (HOFMANN-LÖFFLER reaction):

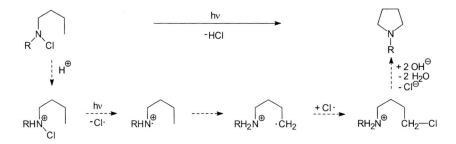

The reaction is first carried out in acid solution. The photolysis of the *N*-chloroammonium ions produces an aminium radical ion which, by abstraction of an H-atom from the methyl group, is converted into an alkyl radical. The latter initiates a chain reaction by abstracting a Cl-atom form a new *N*-

chloroammonium ion. After addition of a base, the cyclodehydrohalogenation occurs via the corresponding δ-chloroalkylamine involving an intramolecular nucleophilic substitution.

D | **Pyrrolidine** is a colourless, water-soluble liquid of a penetrating aminelike odour, of bp 89°C. It fumes in air owing to salt formation with carbon dioxide.

Pyrrolidin-2-one, often called pyrrolidone, is the lactam of 4-aminobutyric acid. Pyrrolidone is prepared from butano-4-lactone and ammonia at 250°C. It is colourless, water-soluble, of mp 25°C and bp 250°C (decomp). Pyrrolidone is vinylated by acetylene. Poly-(N-vinylpyrrolidone) has proved useful as a plasma substitute in blood transfusion.

1-Methylpyrrolidin-2-one is a colourless, water-soluble liquid, bp 206°C, made from butano-4-lactone and methylamine. It is used as a solvent, for instance for the industrial extraction of acetylene from gas mixtures.

Proline (pyrrolidine-2-carboxylic acid), is one of the 20 essential amino acids.

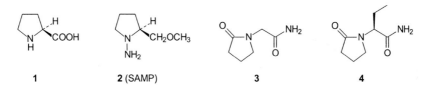

| 1 | 2 (SAMP) | 3 | 4 |

(S)-(-)-Proline **1**, colourless crystals, mp 220°C, $[\alpha]^{20}_D$ -80 (water), occurs abundantly in collagen. It is produced by acid-catalysed hydrolysis from gelatine.

   Piracetam **3** [60a] is medicinally used as anticonvulsant and antiepileptic. Levetiracetam **4** is a chiral second-generation analogue of **3** providing a useful alternative for adjunctive therapy to conventionally anticonvulsants, e.g. benzodiazepines.

The designation 'chiral pool' was introduced to denote an available source of enantiomerically pure natural products. These include the (S)-amino acids, as well as (S)-lactic acid, (S)-malic acid, (R,R)-tartaric acid and β-D-glucose. How the knowledge of their chirality can be utilized for asymmetric syntheses is demonstrated by an example of the chiral auxiliaries (S)- and (R)-1-amino-2-(methoxymethyl)pyrrolidine developed by ENDERS and abbreviated as SAMP (**2**) and RAMP [61]. They are synthesized from (S)- or (R)-proline in several steps [62]. The enantioselective synthesis of the insect pheromone (S)-4-methylheptan-3-one **8** by alkylation of pentan-3-one **1** serves as an example for the use of these chiral auxiliaries:

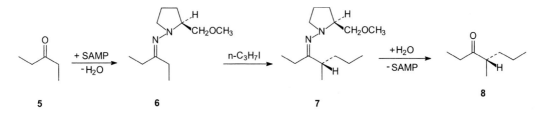

| 5 | 6 | 7 | 8 |

The ketone **5** reacts with SAMP to give hydrazone **6**. The possibility of internal asymmetric induction exists in **6**, so that the following alkylation (action of lithium diisopropylamide in diethyl ether, followed by 1-iodopropane at -110°C) occurs diastereoselectively. In the final step, the auxiliary SAMP is hydrolytically removed from **7**. Thereby, the α-alkylated ketone **8** is formed with ee = 99.5%. The use of RAMP as auxiliary would thus produce the (R)-enantiomer of **8**.

Although it appears to be an enantioselective synthesis, it is in fact a multistep synthesis in which steps **6 → 7** represent a diastereoselective reaction. Nowadays chiral auxiliaries are at our disposal for the approach to the most varied synthetic problems.

4-Hydroxyproline **9** is a proteinogenic amino acid, occurring mainly in collagen. It can be separated from the hydrolysis products of gelatine.

Several alkaloids are derived from pyrrolidine, e.g. hygrin **10**, a minor alkaloid of the coca plant, as well as nicotine (see p 305).

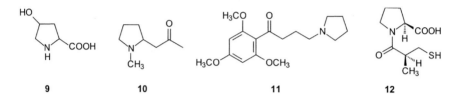

| 9 | 10 | 11 | 12 |

The vasodilator buflomedil **11** and the antihypertensive captopril **12** are drugs containing a pyrrolidine ring.

5.17 Phosphole

Phosphole polymerizes rapidly. 1-Substituted phospholes are thermally more stable. X-Ray studies of 1-benzylphosphole show that the molecule is not planar and that phosphorus retains its pyramidal structure [63].

Consistent with this finding, the NMR spectra show that phosphole has a lower aromaticity than furan.

Phospholes are weak bases and react with strong acids to give phospholium salts. The cleavage of the exocyclic P–C bond by lithium in boiling THF is an interesting reaction of 1-phenyl- and 1-benzyl-phosphole.

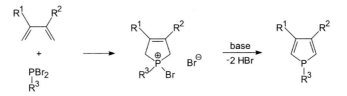

The phosphole anion is planar, aromatic and iso-π-electronic with furan and thiophene.

Phospholes are accessible by [4+1] cycloaddition of buta-1,3-dienes with alkyl- or aryldibromophosphanes followed by dehydrobromination.

Summary of the general chemistry of five-membered heterocycles with one heteroatom:

- The parent compounds of the monocyclic five-membered heterocycles with one heteroatom are aromatic. When considering the three most important systems only, it appears that the aromaticity increases as follows: furan < pyrrole < thiophene (< benzene). This sequence also applies to the respective benzo[*b*] condensed systems.

- Because of their aromaticity, these compounds undergo electrophilic substitution, with the reactivity decreasing in the following order: pyrrole > furan > thiophene (>> benzene). Substitution occurs regioselectively in the 2-position. The corresponding benzo[*b*] condensed systems react more slowly. Substitution occurs on the five-membered ring, but in this case, the regioselectivity is reduced.

- The reactivity as 1,3-dienes in [4+2] cycloadditions is greatest for furan which is also the system most liable to undergo ring-opening.

- The benzo[*b*] condensed systems do not react as 1,3-dienes but undergo [2+2] cycloadditions.

- The benzo[*c*] condensed systems have *o*-quinonoid structures. Their resonance energy is, therefore, lower than that of the corresponding benzo[*b*] condensed systems. The compounds prove to be reactive 1,3-dienes.

- Some heterocycles are accessible by cyclodehydration, namely:
  - furans from 1,4-dicarbonyl compounds
  - benzo[*b*]furans from α-phenoxycarbonyl compounds
  - benzo[*b*]thiophenes from α-(phenylsulfanyl)carbonyl compounds
  - indoles from *N*-acyl-*o*-toluidines or α-arylamino ketones
  - dibenzofurans from 2,2'-dihydroxybiphenyls
  - tetrahydrofurans from 1,4-diols

- Cyclocondensation is an important synthetic method starting from:
  — 1,4-dicarbonyl compounds (thiophenes, pyrroles)
  — α-halocarbonyl compounds and β-keto carboxylic esters (furans, pyrroles)
  — α-amino ketones and β-keto carboxylic esters (pyrroles)
  — 1,3-dicarbonyl compounds or β-chlorovinylcarbonyl compounds (thiophenes, selenophenes, tellurophenes)
  — carbonyl compounds and CH-acidic nitriles (aminothiophenes)

- Indoles are obtained by specific cyclizations (FISCHER synthesis, REISSERT synthesis, BATCHO-LEIMGRUBER synthesis, NENITZESCU synthesis), as are carbazoles from biphenylene or diphenyl-amines and pyrrolidines (HOFMANN-LÖFFLER reaction).

- Ring transformations make possible the preparation of furans, thiolane, pyrrolidine and pyrrolidin-2-one.

- The importance of five-membered heterocycles with one heteroatom, of the benzo and dibenzo con-densed systems and of the partially or completely reduced compounds as natural products, pharma-ceuticals, and starting materials or auxiliaries for syntheses is much greater than for three- or four-membered heterocycles, apart from oxirane.

5.18 1,3-Dioxolane

A-D 1,3-Dioxolanes can be viewed as cyclic acetals or ketals. The ring is nonplanar and confor-mationally mobile.

1,3-Dioxolanes are prepared by cyclocondensation of aldehydes or ketones with 1,2-diols in benzene and with p-toluenesulfonic acid as catalyst:

Quantitative yields are realized by removing the resulting water by azeotropic distillation.

1,3-Dioxolanes are stable to bases. They are hydrolysed by dilute acids even at room temperature in a reversal of their formation. The conversion of aldehydes and ketones into 1,3-dioxolanes is one of the most important methods for the protection of the carbonyl function in multistep syntheses. It is also the standard method for blocking two vicinal cis-positioned hydroxy groups in a carbohydrate, via reaction with acetone, e.g.:

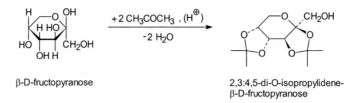

β-D-fructopyranose   2,3:4,5-di-O-isopropylidene-
                                                β-D-fructopyranose

Halogens react with 1,3-dioxolanes via 1,3-dioxolan-2-ylium salts to give β-haloalkyl formates:

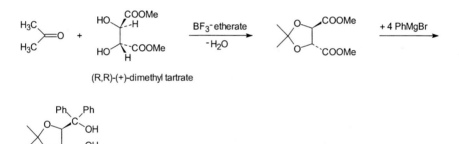

1,3-Dioxolane, a colourless liquid, bp 78°C, is soluble in water.

(4R,5R)- and (4S,5S)-2,2-Dimethyl-α,α,α',α'-tetraphenyl-1,3-dioxolane-4,5-dimethanol were prepared from enantiomeric methyl tartrates as follows:

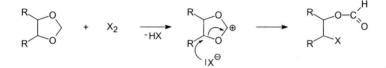

(R,R)-(+)-dimethyl tartrate

(4R,5R)-(-)-enantiomer

The compounds form chiral metal complexes as well as chiral clathrates and can be used as auxiliaries for asymmetric syntheses [64].

5.19 1,2-Dithiole

1,2-Dithiole is not a cyclic conjugated system. However, the 1,2-dithiolylium ions derived from it are aromatic. For instance, the corresponding salts are obtained from 1,3-dicarbonyl compounds and disulfane in acid solution:

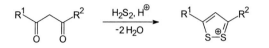

5.20 1,2-Dithiolane

A-D   1,2-Dithiolanes can be viewed as cyclic disulfanes. A C–S–S–C dihedral angle of 26.6° results from the elctronic interaction between the S-atoms. For this reason, a considerable ring strain is caused by the widening of the bond angles, as can be deduced from the Newman projection formula. The strain enthalpy was found to be 67 kJ mol⁻¹.

1,2-Dithiolanes are prepared by oxidation of 1,3-dithioles:

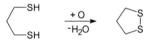

1,2-Dithiolanes are very reactive because of ring strain. 1,2-Dithiolane polymerizes at room temperature. Nevertheless, about 20 1,2-dithiolanes are found in nature. Among them, α-lipoic acid (6,8-disulfanyloctanoic acid) is of great biological importance. The crystalline compound was isolated for the first time in 1951 from the liver of cattle.

(R)-(+)-α-lipoic acid

α-Lipoic acid, bound to proteins, acts as a coenzyme in the oxidative decarboxylation of pyruvic acid and 2-oxopentandioic acid. By taking up hydrogen, it is converted into the corresponding 1,3-dithiole, which is subsequently reoxidized to α-lipoic acid by flavine–adenine dinucleotide.

  Flies are killed upon contact with the marine worm *Lumbriconereis*. Nereistoxin **1** was identified as the active substance in this worm, and its constitution determined as 4-dimethylamino-1,2-dithiolane.

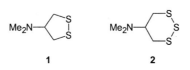

1   2

Consequently, structural analogues were synthesized and introduced as insecticides, e.g. thiocyclam **2**.

5.21 1,3-Dithiole

A-C   1,3-Dithiole, like its 1,2-isomer, is not a cyclic conjugated system. However, the aromatic 1,3-dithiolylium system is derived from it. For instance, salts with this cation can be obtained from 2-oxoalkyl dithiocarbamates as follows:

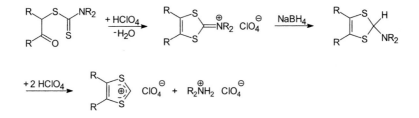

1,3-Dithiolylium salts react with tertiary amines to give tetrathiafulvalenes:

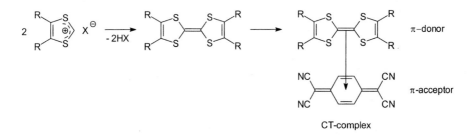

CT-complex

Tetrathiafulvalenes form crystalline CT complexes (charge-transfer complexes, π-complexes) with e.g. 7,7,8,8-tetracyanoquinodimethane, which display high electric conductivity (organic metals) [65].

5.22 1,3-Dithiolane

A–C 1,3-Dithiolanes can also be regarded as cyclic dithioacetals or ketals. They are accordingly obtained by cyclocondensation of aldehydes or ketones with 1,2-dithiols in acetic acid with *p*-toluenesulfonic acid as catalyst:

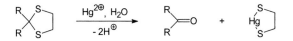

1,3-Dithiolanes are stable towards bases. They are hydrolysed by dilute acids only slowly, in a reversal of their formation, faster in the presence of mercury(II) salts:

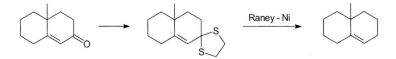

A better method for regenerating carbonyl compounds from 1,3-dithiolanes is the reaction with selenium dioxide in acetic acid [66].

1,3-Dithiolanes are subject to reductive desulfurization by the action of RANEY nickel in ethanol. A method for the chemoselective reduction of α,β-unsaturated ketones is based on this reaction, e.g.:

5.23 Oxazole

A Oxazole (1,3-Oxazole) contains an O-atom bonded as in furan, and also a pyridine-like N-atom (see p 3). The univalent radical is known as oxazolyl. The oxazole molecule is planar and its structure can be represented by a distorted pentagon (see Fig. 5.9):

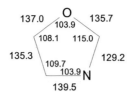

Fig. 5.9   Structure of oxazole
(bond lengths in pm, bond angles in degrees)

The differences in the bond lengths, especially between the bonds N/C-2 and N/C-4 indicate that the delocalization of the π-electrons is affected by the heteroatoms. As in the case of furan, the structural formula with two π-bonds is a good representation of the electronic structure of the molecule. The ionization energy of oxazole is 9.83 eV and its dipole moment is 1.5 D.

The UV absorption and chemical shifts in the NMR spectra are as follows:

UV (methanol)	^{1}H-NMR (CCl$_4$)	^{13}C-NMR (CDCl$_3$)
λ (nm) (ε)	δ (ppm)	δ (ppm)
205 (3.59)	H-2: 7.95	C-2: 150.6
	H-4: 7.09	C-4: 125.4
	H-5: 7.69	C-5: 138.1

Oxazole is thus diatropic and aromatic. All ring atoms are sp^2-hybridized. Consequently, there are two nonbonding electron pairs, one on the O-atom and the other on the N-atom. The π-electron densities were calculated by various SCF/MO methods, e.g.:

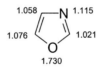

Oxazole, therefore, belongs to the family of π-electron excessive heterocycles. The electronegativity of the pyridine-like N-atom, however, causes the π-electron density to be low, especially on the C-2 atom. Electrophilic substitution should occur in the 5- or 4-position, while reaction with nucleophiles should occur in the 2-position.

B The reactions of five-membered heterocycles with two or more heteroatoms cannot be classified as easily as those systems possessing only one heteroatom. For this reason, they are described according to the system and importance of their reactions. For oxazoles, the following reactions are characteristic [67].

Salt formation

Oxazoles are weak bases. The pK_a value of oxazole is 0.8. They are protonated by strong acids on the N-atom, e.g.:

Oxazolium salts react faster with nucleophiles than oxazole.

Metallation

Oxazoles unsubstituted on the 2-position react with *n*-butyllithium in THF already at -75°C to give 2-lithiooxazoles [68]:

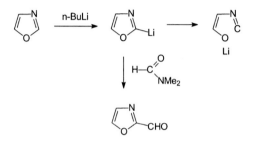

2-Lithiooxazole is subject to slow ring-opening at room temperature to form the lithium enolate of 2-oxoethyl isocyanide. With DMF it reacts to give oxazole-2-carbaldehyde.

Reactions with electrophilic reagents

Oxazoles are quaternized by haloalkanes:

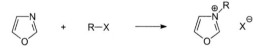

Although electrophilic substitution reactions are possible with oxazoles, they are frequently accompanied by addition reactions, as in furan. The bromination of 4-methyl-2-phenyloxazole with bromine or *N*-bromosuccinimide yields 5-bromo-4-methyl-2-phenyloxazole, and that of 2-methyl-5-phenyloxazole gives 4-bromo-2-methyl-5-phenyloxazole. Mercury(II) acetate in acetic acid acetoxymercurates 4-substituted oxazoles in the 5-position, 5-substituted oxazoles in the 4-position, and 4,5-disubstituted oxazoles in the 2-position. The acetoxymercury group can be substituted by electrophiles, e.g.:

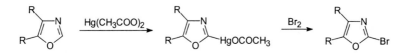

In general, it can be stated that the pyridine-like N-atom in the oxazole molecule impedes electrophilic substitution reactions. This is particularly evident in the nitration of phenyloxazoles. The substitution occurs on the benzene ring, e.g.:

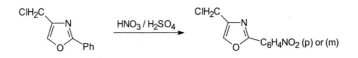

Reactions with nucleophilic reagents

Oxazoles are attacked in the 2-position by nucleophiles, even if a substituent is already present. Ring-opening occurs and, depending on the reagent, is followed by ring-closure, e.g. with ammonia:

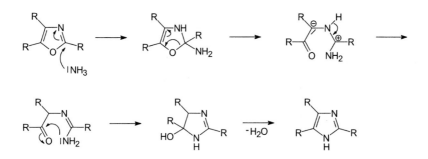

On heating with ammonia, formamide or primary amines, oxazoles thereby undergo a ring transformation into imidazoles.

With nucleophiles, oxazolium salts react much faster. For instance, the acid-catalysed hydrolysis of 2,4,5-triphenyloxazole yields benzaldehyde, benzoic acid and ammonium chloride analogously to the mechanism described above:

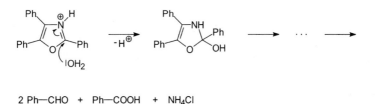

2 Ph—CHO + Ph—COOH + NH₄Cl

As a consequence of the low π-electron density on C-2, nucleophilic substitution reactions of 2-halo-oxazoles proceed rapidly, e.g.:

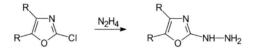

Cycloadditions

Oxazoles can react as 1,3-dienes, and with dienophiles such as maleic acid derivatives undergo DIELS-ALDER reactions (see p 131). With acetylene dienophiles, furans are formed via the corresponding DIELS-ALDER adducts (see p 60).

The photooxygenation of oxazoles also occurs as a [4+2] cycloaddition. The primary products de-compose in various ways depending on the solvent. 2,4,5-Triphenyloxazole in methanol yields *N,N*-dibenzoylbenzamide:

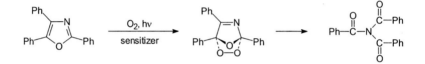

Cornforth rearrangement

Oxazoles in which the C-4 atom is bonded to a carbonyl group isomerize on heating. Two substituents change places in this process, e.g.:

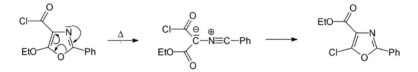

Results of experiments carried out with isotopically labelled compounds indicate an intermediate which is derived from the ring-opened starting material. It has the structure of a nitrile ylide and is converted into the final product by ring-closure.

To summarize, the reactivity of oxazoles closely resembles that of the furans, especially in their readi-ness to undergo ring-opening reactions and [4+2] cycloadditions. The pyridine-like N-atom impedes electrophilic substitution reactions. On the other hand, it facilitates the attack of nucleophiles onto the C-2 atom. Although oxazole is considered to be a heteroarene, because of its six delocalized ring elec-trons, its aromaticity is low and comparable to that of furan.

C   When considering the *retrosynthesis* (see 5.10), it must be borne in mind that the oxazole structure combines the functionalities of an imido ester (C-2) and of an enediol ether (C-4 and C-5).
Retrosynthetic considerations can then move in two directions (I/II) by analogy to furan–thiophene–
pyrrole.

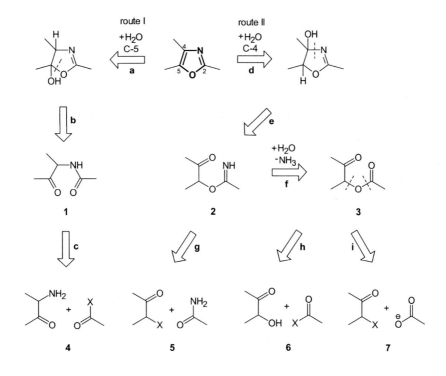

Fig. 5.10   Retrosynthesis of oxazole

According to route I, addition of water after the retrosynthetic steps **a/b** leads to the α-acylamino-
carbonyl system **1** which should be formed according to **c** from the two fragments **4**.

The more productive retroanalytical route is II, in which water is added, according to **d**. After cleavage of the N/C-4 bond (i.e. via **e**), this leads to the intermediate stage **2**. This can split further in several
ways, namely **g**, which leads to an α-halo ketone and an acid amide (fragments **5**), or alternatively **f**,
which yields NH_3 and the α-hydroxy- or α-halocarbonyl compound, i.e. by combination of the fragment pair in **6** or **7** in a reversal of the retrosynthetic operation indicated by **h** and **i**.

This elucidates the principles underlying the important syntheses of oxazoles [69].

(1) α-Acylamino ketones, esters or -amides are cyclodehydrated by H_2SO_4 or polyphosphoric acid to
give oxazoles **8** (*Robinson-Gabriel synthesis*):

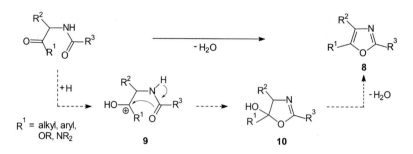

R^1 = alkyl, aryl,
OR, NR_2

9   **10**

α-Acylamino ketones are accessible, for instance, by the DAKIN-WEST reaction (heating acid anhy-drides with α-amino acids in the presence of pyridine). Labelling the α-acylamino ketones with ^{18}O in the keto carbonyl group does not produce ^{18}O-labelled oxazoles. However, if the acyl carbonyl group is labelled with ^{18}O, then all the label is found in the oxazole. These experiments support the suggested mechanism (intermediates **9** and **10**).

(2)   α-Acyloxy ketones **11**, obtainable from α-halo ketones and salts of carboxylic acids, form oxa-zoles on treatment with ammonia:

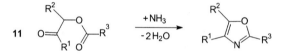

An enamine is formed from **11** which dehydrates to give the heterocycle.

(3)   α-Halo and α-hydroxy ketones condense with acid amides via an *O*-alkylation to give oxazoles (*Blümlein-Lewy synthesis*). Formamide yields oxazoles unsubstituted in the 2-position, urea yields 2-aminooxazoles:

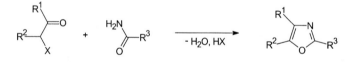

(4)   Oxazole syntheses employing isocyanides as starting materials are of considerable preparative value. In the *van Leusen synthesis*, tosylmethyl isocyanide (TosMIC) reacts with aldehydes under base catalysis, e.g. in the presence of K_2CO_3. The primary products are 4,5-dihydro-1,3-oxazoles **12**, which are converted into oxazoles by elimination of sulfinic acid (see p 172):

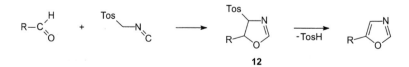

12

In the *Schöllkopf synthesis*, α-metalated isocyanides **13** (from isocyanides and *n*-butyllithium) react with acid chlorides to give 4,5-disubstituted oxazoles **14** via *C*-acylation and electrophilic C–O bond formation.

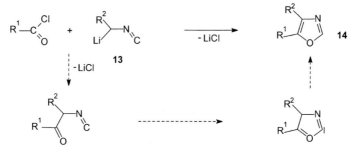

(5) α-Diazocarbonyl compounds undergo addition to nitriles with elimination of N_2 in the presence of LEWIS acids or transition metal compounds [Cu(II), Pd(II), especially Rh(II)] as catalysts to give oxazoles.

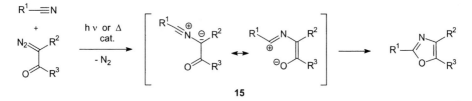

This reaction is likely to proceed via intermediate formation of nitrile ylides **15** and their electrocyclic ring closure as 1,5-dipoles (cf. p 144) yielding the oxazole system [69a].

D **Oxazole** is a colourless liquid with a smell similar to pyridine, of bp 69-70°C, and is soluble in water.

1,3-Oxazolium-5-olates **16** are known by the trivial name of Münchnones. Their reactions have been extensively studied by HUISGEN in Munich. They are prepared by cyclodehydration of *N*-substituted *N*-acyl α-amino acids with acetic anhydride:

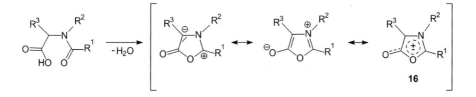

Mesomeric zwitterions like Münchnones are known as *mesoionic* compounds, which also occur in other heterocyclic systems [70]. Münchnones are crystalline compounds and are reactive 1,3-dipoles. For instance, they react with acetylenes via the corresponding cycloadducts **17** to give pyrroles **18**.

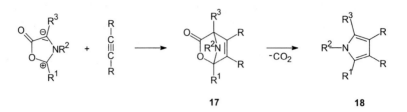

17   18

Oxazoles are rarely found in natural products. Pimprinin **19**, isolated from *Streptomyces pimprina*, provides an example:

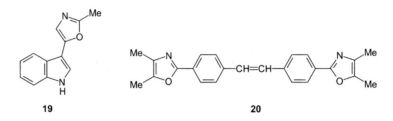

19   20

The oxazole ring is found in some macrocyclic antibiotica and several alkaloids.

Only a few pharmaceuticals derived from oxazole are in use. The anti-inflammatory and analgesic action of 2-diethylamino-4,5-diphenyloxazole is known.

Aryl-substituted oxazoles are strongly fluorescent. In solution, they are therefore suitable as luminous substances for liquid scintillation counters and also as optical brighteners (brightening agents). For instance, 4,4'-bisoxazol-2-ylstilbenes (see **20**) are added to washing agents. During the washing process, they are absorbed by the fibres, so that the clothes appear to be 'whiter than white' as a result of the blue fluorescence.

Polymethine dyes with oxalyl terminal groups are used as photographic sensitizers of silver halogen emulsions. 2,5-Diphenyloxazole is added as antioxidant to hydraulic fluids and high temperature lubricating oils.

| E | Oxazoles are used as auxiliaries for transformations in organic synthesis in various ways. |

(1)   The photooxygenation with singlet oxygen, when applied to 4,5-cycloalkeno-1,3-oxazoles, e.g. **21**, leads via **22** to mononitriles of dicarboxylic acids **23**:

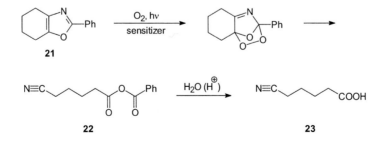

22   23

(2)   Transformations of DIELS-ALDER adducts from oxazoles and activated multiple bond systems are important. In this case, alkynes lead to furan derivatives (see p 60). Alkenes lead to adducts, regioselectively in the case of unsymmetrical alkenes, which eliminate H_2O in acidic media and are thereby converted into pyridine derivatives; in this way, acrylic acid and oxazole **24** furnish pyridine-4-carboxylic acid **26** via the DIELS-ALDER adduct **25**.

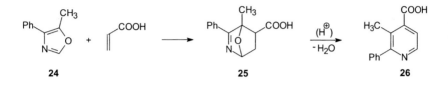

If the oxazole contains cyano or alkoxy groups in the 5-position, or if the dienophile possesses appropriate leaving groups, then these groups can be eliminated and 3-hydroxypyridines are formed. The industrial synthesis of pyridoxine **27** (vitamin B_6, see p 305) is based on these reactions.

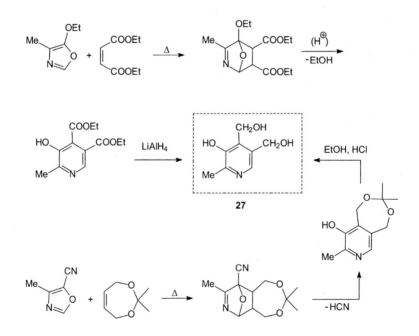

5.24 Benzoxazole

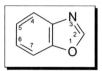

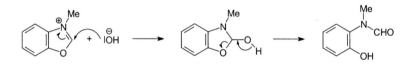

| **A** | The numbering 1,3- is omitted, as there is no other possibility for ring condensation. |

As in the case of benzo[*b*]furan, the signal for the proton on C-2 appears in the region of benzenoid protons in the ¹H-NMR spectrum, with a small downfield shift.

| UV (ethanol) | ¹H-NMR (methanol-d₄) | ¹³C-NMR (CDCl₃) | |
λ (nm) (ε)	δ (ppm)	δ (ppm)	
231 (3.90)	H-2: 7.46	C-2: 152.6	C-3a: 140.1
263 (3.38)	H-4: 7.67	C-4: 120.5	C-7a: 150.5
270 (3.53)	H-5: 7.80	C-5: 125.4	
276 (3.51)	H-6: 7.79	C-6: 124.4	
	H-7: 7.73	C-7: 110.8	

| **B** | Salt formation and quaternization of benzoxazoles occurs by analogy to oxazoles. In benzoxazoles, the electrophilically accessible atoms C-4 and C-5 of the oxazole ring are blocked. Nitrating acid causes substitution of the benzene ring in the 5- or 6-position. However, nucleophiles attack benzoxazoles, benzoxazolium salts and *N*-alkylbenzoxazolium salts in the 2-position, e.g.: |

Nucleophilic substitution reactions of 2-halobenzoxazoles occur rapidly and *N*-alkyl-2-chlorobenzoxazolium salts are even more reactive. They are efficient dehydrating agents [71] , e.g. for the synthesis of alkynes from aryl ketones:

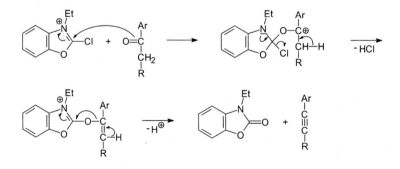

The low electron density on the C-2 atom, which is due to electron withdrawal by the N-atom, is responsible for the CH-acidity of 2-alkylbenzoxazoles. Thus 2-methylbenzoxazole undergoes a CLAISEN condensation, e.g.:

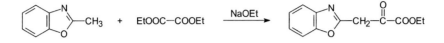

As expected, benzoxazoles do not react as 1,3-dienes, but [2+2] cycloadditions are possible. The photodimerization serves as an example:

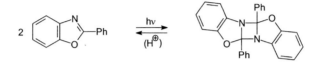

In the dark and in solution, the photodimer undergoes an acid-catalysed fission in an exothermic reaction, $\Delta H° = -116$ kJ mol^{-1}. This energy is only a small part of the liberated resonance energy. The strain enthalpy of the doubly condensed diazetidine system makes the major contribution.

C The standard synthesis for benzoxazoles is the cyclocondensation of o-aminophenol with carboxylic acids or their derivatives [72]:

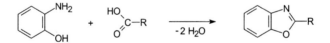

The reaction proceeds via o-(acylamino)phenols, which can be isolated and subsequently subjected to cyclodehydration.

Benzoxazoles unsubstituted in the 2-position are obtained by cyclocondensation of o-aminophenols with orthoformic trimethyl ester in the presence of concd hydrochloric acid [73], e.g.:

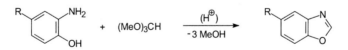

D **Benzoxazole** forms colourless crystals of mp 31°C.
Some benzoxazoles are pharmaceuticals, e.g. 2-amino-5-chlorobenzoxazole is a sedative.

5.25 4,5-Dihydrooxazole

A 4,5-Dihydrooxazole was previously known as Δ^2-oxazoline or 2-oxazoline. It follows from microwave spectra that the ring is planar.

B 4,5-Dihydrooxazoles are weak bases and form salts with strong acids. They undergo a stepwise hydrolysis in aqueous solution to give salts of β-amino alcohols and carboxylic acids. The nucleophile attacks at the 2-position, as in oxazolium salts:

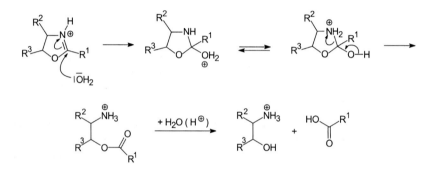

On the basis of this reaction, 4,5-dihydrooxazoles can be regarded as carboxylic acid dervatives, i.e. as cyclic imido esters [74].

C 4,5-Dihydrooxazoles are prepared from β-amino alcohols (from oxiranes and ammonia, see p 18) and carboxylic acids or carboxylic esters [75]. N-(2-Hydroxyalkyl)carboxylic acid amides can be isolated as intermediates and subsequently subjected to thermal cyclodehydration or to the action of H_2SO_4, $SOCl_2$ or to other dehydrating agents:

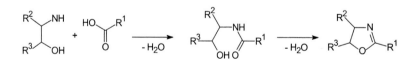

E 4,5-Dihydrooxazoles are widely used as building blocks and auxiliaries in organic synthesis because of their accessibility and the mild hydrolytic conditions for ring-opening. The CH-acidity of 2-alkyl-4,5-dihydrooxazoles, the electrophilic reactivity at C-2 in N-alkyl-4,5-dihydrooxazolium salts and the activation of 2-aryl substituents in the 4,5-dihydrooxazole system in metalation reactions have all been synthetically exploited as shown in the following examples:

(1)   With arylaldehydes in a type of KNOEVENAGEL reaction, 2-alkyl-4,5-dihydrooxazoles **1** yield condensation products **2**, which on acid hydrolysis give 2-alkyl-3-arylpropenoic acid **3**:

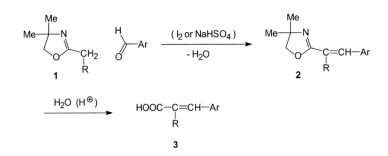

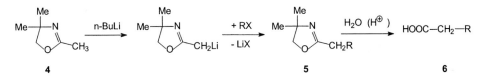

(2)   2-Methyl-4,5-dihydrooxazole **4** can be metalated by *n*-butyllithium and subsequently alkylated with haloalkanes. The hydrolysis of **5** yields carboxylic acid **6** in which the chain of the haloalkane has been lengthened by two C-atoms:

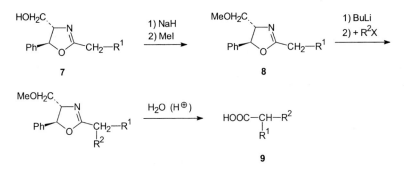

A chiral 2-alkyl-4,5-dihydrooxazole **7** is obtained by the use of (+)-(1*S*,2*S*)-1-phenyl-2-aminopropane-1,3-diol, available from the chiral pool (see p 115). From this, the methyl ether **8** is prepared using sodium hydride and iodomethane. As a result of internal asymmetric induction, the alkylation of its lithium derivative occurs diastereoselectively. In the case of R^1 = Me, R^2 = Et, hydrolysis yields the (+)-(*S*)-enantiomer of 2-methylbutanoic acid **9**, with ee = 67 %, as the main product:

This synthetic application of 4,5-dihydrooxazoles is known as the MEYERS oxazoline method [76].

(3)   The quaternary salt **10**, which is easily accessible from 4,4-dimethyl-4,5-dihydrooxazole and iodomethane, adds arylmagnesium halides to yield the oxazolidine **11**, which hydrolyses with formation of arylaldehydes and exposure of the C-2 functionality. Thereby 4,5-dihydrooxazole brings about the conversion Ar-Hal → Ar-CHO [77]:

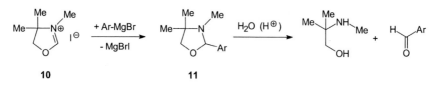

10   **11**

(4)   The *ortho*-positions of the benzene ring in 4,4-dimethyl-2-phenyl-4,5-dihydrooxazole **13**, prepared from benzoyl chloride and the amino alcohol **12**, are activated to such an extent that lithiation is
possible. On reaction with an electrophile, e.g. a haloalkane, and followed by hydrolysis, 2-substituted
or 2,6-disubstituted benzoic acids **14** or **15** are obtained [78]:

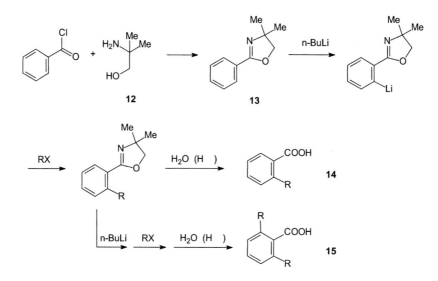

For organic syntheses, oxazol-5(4*H*)-ones **17** are of importance [79]. They are known by the trivial
name of azlactones and are obtained by cyclodehydration of *N*-acylamino acids **16** with acetic anhydride:

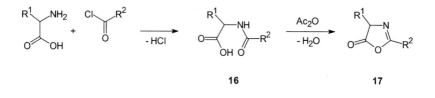

16   **17**

Azlactones are CH-acidic. Thus azlactones **18** derived from glycine condense with benzaldehyde to
give 4-benzylideneoxazol-5(4*H*)-ones **19**:

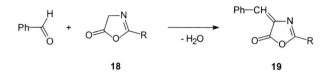

18   **19**

Azlactones **22** are intermediates in the *Erlenmeyer synthesis* (1893) of α-amino acids. They are made by the action of acetic anhydride/sodium acetate upon a mixture of *N*-benzoylglycine (hippuric acid) **21** and benzaldehyde via 2-phenyloxazol-5(4*H*)-one **20**:

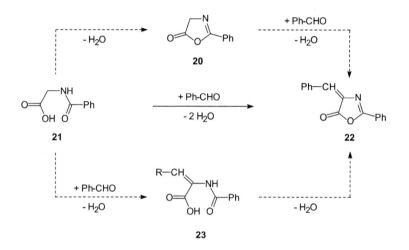

If polyphosphoric acid is used as the dehydrating agent, then by analogy to the PERKIN reaction, benzylidenehippuric acid **23** is formed in the intermediate step. Under these conditions, ketones can also be used in the ERLENMEYER synthesis.

α-Amino acids **25** are obtained by catalytic hydrogenation of 4-alkylideneoxazole-5(4*H*)-ones **24** followed by acid hydrolysis:

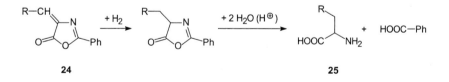

24   **25**

Carbanions **26** are derived from azlactones by the action of bases. They undergo a MICHAEL addition with alkenes possessing electron-withdrawing substituents such as acrylonitrile. On hydrolysis, the adducts **27** yield γ-keto nitriles **28** [80]:

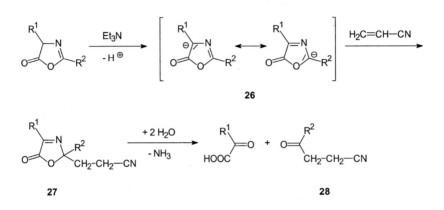

26

27   **28**

If R^2 is a bulky substituent, e.g. a mesityl group, then the addition of the MICHAEL acceptor occurs in the 4-position of the anion **26** [81].

5.26 Isoxazole

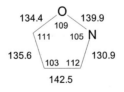

A | Isoxazole (1,2-oxazole) contains a pyridine-like N-atom, but differs from oxazole by the presence of an N–O bond. The bond energy of such a σ-bond amounts only to 200 kJ mol^{-1}, much lower than that of N–C or O–C bonds. The univalent radical is known as isoxazolyl.

The isoxazole molecule is planar (see Fig. 5.11). Again, it is evident that the heteroatoms impair the delocalization of the π-electrons. This is more pronounced than in oxazole, as can be deduced from a comparison of the bond lengths between ring atoms 3 and 4.

```
              O
        134.4    139.9
           109
         111    105  N
   135.6            130.9
        103   112
         142.5
```

Fig. 5.11   Structure of isoxazole
(bond lengths in pm, bond angles in degrees)

The ionization energy of isoxazole is 10.17 eV and its dipole moment of 2.75 D is greater than that of oxazole. Isoxazole, like oxazole, has a very short-wave UV absorption. The chemical shifts in the NMR spectrum are found in the region typical for benzenoid compounds.

UV (H$_2$O)	^{1}H-NMR (CCl$_4$)	^{13}C-NMR (CHCl$_3$)
λ (nm) (ε)	δ (ppm)	δ (ppm)
211 (3.60)	H-3: 8.19	C-3: 149.1
	H-4: 6.32	C-4: 103.7
	H-5: 8.44	C-5: 157.9

B Isoxazole is aromatic like its structural isomer oxazole. It is a π-excessive heterocycle. The following π-electron densities have been calculated:

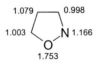

Therefore, electrophilic substitutions occur at the 4-position, whereas nucleophiles prefer the 3-position. The following reactions are typical for isoxazoles.

Salt formation

Isoxazoles are very weak bases. The pK_a of isoxazole is -2.97. Protonation occurs at the N-atom.

Reactions with electrophilic reagents

Isoxazoles are quaternized by iodoalkanes or dialkyl sulfates. Electrophilic substitutions such as halogenation, nitration, sulfonation, VILSMEIER-HAACK formylation and acetoxymercuration, occur at the 4-position, provided this position is unsubstituted. As with oxazole, the pyridine-like N-atom impedes electrophilic substitution, so that isoxazoles are less reactive than furans but more reactive than benzene.

Reactions with nucleophilic reagents

Nucleophiles react with isoxazoles and even faster with isoxazolium salts, but differently in each case and usually with ring cleavage. A special and synthetically useful reaction of isoxazoles unsubstituted in the 3-position is the ring-opening by bases, e.g.:

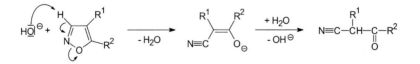

The base does not attack at the C-3 atom but at the H-atom. Via an E2-type mechanism, facilitated by the weak N–O bond, a (Z)-cyano enolate is formed leading to an α-cyano ketone.

Reductive ring opening

On catalytic hydrogenation, isoxazoles give rise to enamino ketones, which can be hydrolysed to produce 1,3-diketones:

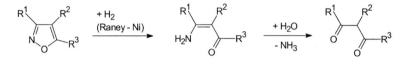

Isoxazoles are reduced by sodium in liquid ammonia in the presence of *tert*-butanol to yield β-amino ketones, which are converted on heating or by acid into α,β-unsaturated ketones:

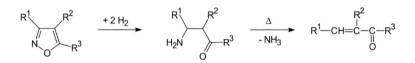

The reactions of isoxazoles differ considerably from those of oxazoles, although both systems are aromatic. The reason for this lies in the relatively weak N–O bond in the isoxazole molecule, which is cleaved in all ring-opening reactions. Moreover isoxazoles, unlike oxazoles, do not react with dienophiles to form DIELS-ALDER adducts [82].

C For the *retrosynthesis* of the isoxazole system (see Fig. 5.12), it is essential that the heterocycle possesses the functionality of an oxime and of an enol ether, and that C-3/C-5 are at the oxidation level of a carbonyl function. Therefore, a logical retrosynthetic route (**a** – **c**) leads by way of the monoxime **2** to the 1,3-diketone and hydroxylamine. If the retrosynthetic operation **a** to **d** is generalized, one arrives at the 4,5-dihydroisoxazole **1**. Its analysis, according to a retroanalytically permitted cycloreversion, leads to an alkene unsubstituted by a leaving group and to a nitrile oxide **3**. These fragments represent the two components of a 1,3-dipolar cycloaddition.

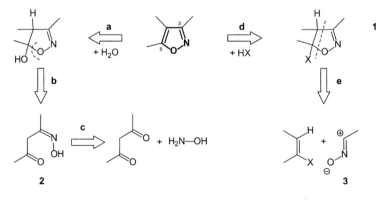

Fig. 5.12   Retrosynthesis of isoxazole

These two synthetic principles proved of advantage in the synthesis of isoxazoles.

(1) β-Diketones yield 3,5-disubstituted isoxazoles **4** with hydroxylamine (*Claisen synthesis*) :

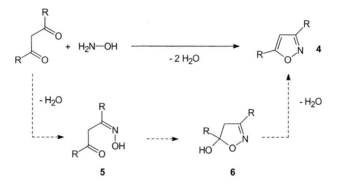

The cyclocondensation proceeds according to the retroanalytical prognosis via the isolable intermediates of a monoxime **5** and a 5-hydroxy-4,5-dihydrooxazole **6**. In the case of unsymmetrically substituted β-diketones, it is still possible to control the regioselectivity by using variable carbonyl electrophilicity and observing strict reaction conditions. α-Hydroxymethylene ketones, the corresponding enol ethers and ethynyl ketones also yield isoxazoles by reaction with hydroxylamine.

(2)   Nitrile oxides **7** react as 1,3-dipoles with alkynes in a [3+2] cycloaddition to give isoxazoles (*Quilico synthesis*). The nitrile oxides are generated in situ. They are obtained, for instance, by dehydrohalogenation of hydroxamic acid chlorides (α-chloraldoximes) with triethylamine:

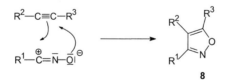

One of the resonance structures of the nitrile oxides is represented by a 1,3-dipole. With alkynes as dipolarophiles, a concerted [3+2] cycloaddition occurs to give isoxazoles **8**:

$$R^2-C\equiv C-R^3$$

$$R^1-C=N-O$$

8

The regioselectivity depends on the nature of the substituents present. Alkynes with a terminal triple bond yield 3,5-disubstituted isoxazoles (**8**, $R^2 = H$).

The 1,3-dipolar cycloaddition of nitrile oxides to alkenes furnishes 2,3-dihydroisoxazoles (see p 144), which can be converted into isoxazoles with suitable oxidizing agents.

 Isoxazole is a colourless liquid of a pyridine-like odour, bp 94.5°C. It is soluble at room temperature in six times its volume of water.

A few oxazoles occur in nature, e.g. muscimol **9**, a CNS depressant in the fly agaric (*Amanita muscaria*) [83]. The compound acts as an antagonist of the neurotransmitter 4-aminobutyric acid.

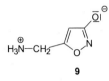

9

Among the isoxazoles, many biologically active compounds are found. Some of them are important as drugs or biocides, e.g. the long-acting sulfonamide sulfamethoxazole **10**, the anti-inflammatory isoxicam **11**, the antiarthritic and antirheumatic leflunomide **12** and the fungicide 3-hydroxy-5-methylisoxazole.

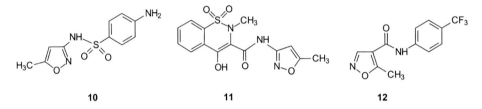

10   **11**   **12**

> **E** | Isoxazoles have a considerable synthetic potential because of their ring-opening reactions as masked 1,3-dicarbonyl systems [84].

(1) The ring-opening, occurring with C-3 deprotonation and O–N fission (see p 139), is utilized for the synthesis of α-cyano ketones, especially with cycloalkanones and in the steroid series. The isoxazole ring behaves as an electrophilic cyanide equivalent:

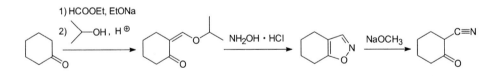

(2) In contrast, C-3 deprotonation of quaternized isoxazolium salts causes fission of the O–N bond with formation of acyl ketenimines:

The application of isoxazolium salts as reagents for activating carbonyl groups in peptide synthesis is based on this reaction [85].

(3)   The 4-(chloromethyl)isoxazole **13**, which is readily accessible from 3,5-dimethyloxazole, serves as a C$_4$-building-block in annulations to cycloalkanones (isoxazole annelation according to STORK). The primary step is alkylation leading to product **14**, a masked triketone. On hydrogenation, the isoxazole ring is reductively opened and cyclization via the enaminone **15** leads to the enamine **16**. On treatment with sodium hydroxide, this is converted into the bicycloenone **17** by hydrolysis, acid fission of the β-dicarbonyl system and an intramolecular aldol condensation (analogous to a ROBINSON annulation):

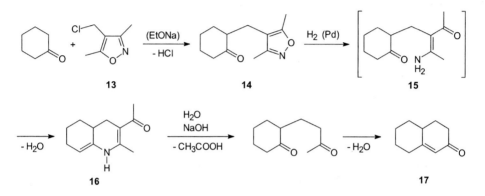

(4)   The conversion of β-ionone **18** into the structural isomer β-damascenone **22** [86] exemplifies a transformation which can only be achieved with difficulty by other routes:

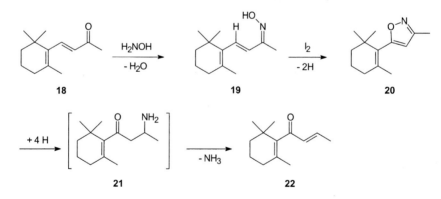

In this reaction, β-ionone is converted into its oxime **19** which is oxidized to the isoxazole **20** by a special method (reaction with I$_2$, KI and NaHCO$_3$ in THF/water). The compound **20** is converted into β-damascenone by use of sodium in liquid ammonia in the presence of *tert*-butanol without isolation of the intermediate β-amino ketone **21**; yield 72% (based on β-ionone). In this reaction, the heterocycle facilitates a 1,3-carbonyl migration within an α,β-unsaturated carbonyl system.

(5)   Numerous synthetically relevant ring transformations of isoxazoles are known. For example the *N*-isoxazolyl thiourea **23** isomerizes thermally to give 1,2,4-thiadiazole **24**:

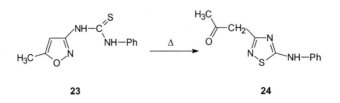

23 **24**

Another rearrangement occurs in the photoisomerization of isoxazoles to oxazoles, which is interpreted in terms of a 1,5-electrocyclization (cf. p 129) involving an isolable 3-acylazirine **25** as an intermediate:

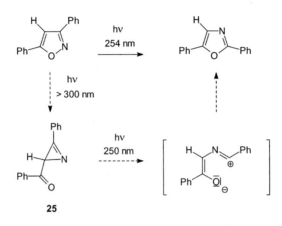

25

5.27 4,5-Dihydroisoxazole

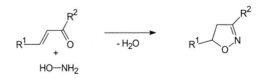

A,C 4,5-Dihydroisoxazole has previously been known as Δ^2-isoxazoline or 2-isoxazoline. 4,5-Dihydroisoxazoles are accessible mainly by two synthetic pathways:

(1) Cyclocondensation of α,β-unsaturated ketones with hydroxylamine. This reaction is analogous to isoxazole formation via the corresponding ketoxime:

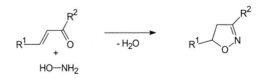

(2) 1,3-Dipolar cycloaddition of nitrile oxides to alkenes. This reaction is the standard synthesis for 4,5-dihydroisoxazoles [87]. For instance, 4-(3-phenyl-4,5-dihydroisoxazole)carboxylic acid **1** is formed regioselectively from benzonitrile oxide and acrylic acid:

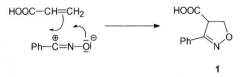

1

Moreover, the reaction proceeds stereospecifically, e.g. (*E*)-methyl cinnamate yields methyl *trans*-4-(3,5-diphenyl-4,5-dihydroisoxazole)carboxylate.

| **B** | 4,5-Dihydroisoxazoles display characteristic and synthetically important reactions. |

Reductive ring-opening

Reductive ring-opening of 4,5-dihydroisoxazole can be carried out by various methods. Hydrogenation with RANEY Ni or Pd/C in the presence of boric acid in methanol/water at room temperature leads to β-hydroxy ketones **3**:

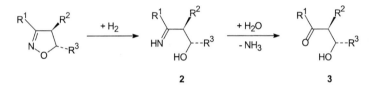

Hydrogenolysis of the N–O bond results first in the formation of a β-hydroxyimine **2**, which under the reaction conditions, is liable to hydrolysis. The configuration at C-atoms 4 and 5 is preserved. A special reagent for ring fission of 4,5-dihydroisoxazoles to give β-hydroxy ketones is molybdenum hexacarbonyl in acetonitrile/water [88].

4,5-Dihydroisoxazoles yield β-amino alcohols **4** when reduced by sodium in ethanol or by NaBH$_4$ and NiCl$_2 \cdot$ (H$_2$O)$_6$ in methanol at -30°C.

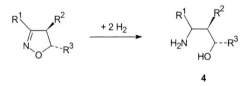

4

In this ring-opening reaction, the configuration of the C-atoms 4 and 5 is again preserved.

Oxidation

N-Bromosuccinimide or KMnO$_4$ in acetone oxidizes 4,5-dihydroisoxazoles to give isoxazoles.

| **D** | The 4,5-dihydroisoxazole structure is found in some pharmacologically interesting natural products, e.g. in the antibiotic cycloserin **5**, which is used in the treatment of tuberculosis. It is |

also found in the antibiotic acivicine **6**, an α-amino acid with antitumour action.

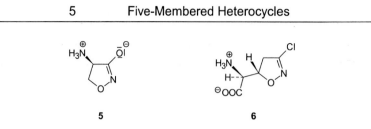

5   6

The following three examples illustrate the numerous synthetic applications of the 'isoxazoline route' [89].

(1) *Synthesis of flavanones* [90]:

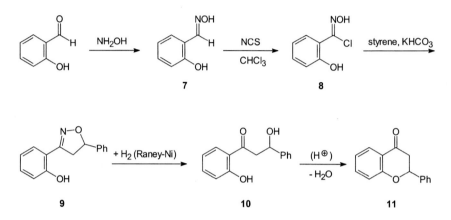

The oxime **7**, prepared from salicyl aldehyde, is converted with *N*-chlorosuccinimide into the hydroxamic acid chloride **8**. From this, the nitrile oxide is obtained with KHCO$_3$, which reacts regioselectively with styrene to give the 3,5-diaryl-4,5-dihydroisoxazole **9**. Catalytic hydrogenation leads to the *β*-hydroxy ketone **10**, which on acid-catalysed cyclodehydration gives the flavanone **11**.

(2) *Synthesis of higher monosaccharides* [91]:

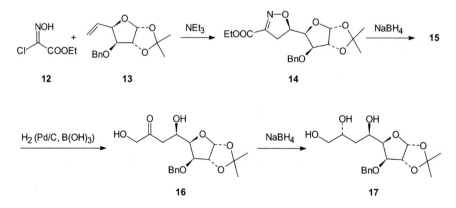

The alkene **13** is produced from D-glucose. In the presence of triethylamine, it is converted into a mixture of two diastereomeric 4,5-dihydroisoxazoles by interaction with the hydroxamic chloride **12**, with the (5*R*)-compound **14** in excess. It is separated by column chromatography and the ethoxycarbonyl group is reduced to give a hydroxymethyl derivative. On catalytic hydrogenation, the intermediate **15** furnishes the β-hydroxy ketone **16**. Its reduction with NaBH$_4$ leads to a mixture of 6-deoxyoctose derivatives with excess of the (7*S*)-epimer **17**.

(3) *Stereocontrolled synthesis of homoallylamines* [92]:

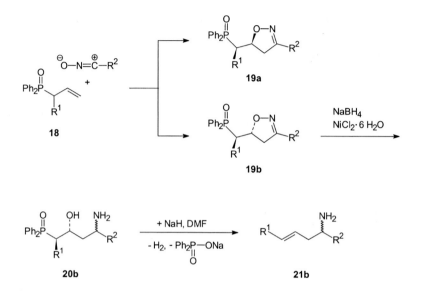

Allyldiphenylphosphane oxides **18**, accessible from allyl alcohols and chlorodiphenylphosphane, react regioselectively with nitrile oxides to give diastereoisomeric 3,5-disubstituted 4,5-dihydroisoxazoles **19a** and **19b**, which are separable by column chromatography. Reduction of **19b** yields the β-amino alcohol **20b**, which is transformed into the (*E*)-homoallylamine **21b** by elimination with sodium hydride in DMF. Starting from **19a**, the (*Z*)-diastereomer is obtained.

5.28 2,3-Dihydroisoxazole

4 3
5 1 2 NH
O

 2,3-Dihydroisoxazole has also been known as Δ^4-isoxazoline or 4-isoxazoline. Its standard synthesis is via the 1,3-dipolar cycloaddition of nitrones to alkynes (*Huisgen synthesis*):

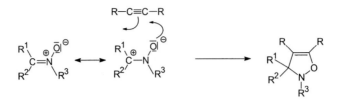

2,3-Dihydroisoxazoles are unstable and very reactive compounds [93]. A typical reaction is the thermal isomerization to 2-acylaziridines:

If the nitrones used for the synthesis of 2,3-dihydrooxazole are *N*-oxides of cyclic imines, then the 2-acylaziridines isomerize further to form betaines, e.g.:

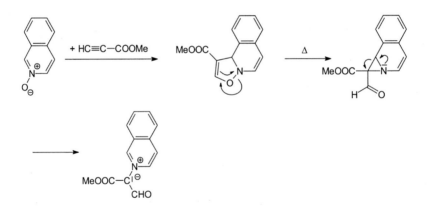

B 2,3-Dihydrooxazoles react with iodomethane in THF with ring fission to form α,β-unsaturated ketones [94]:

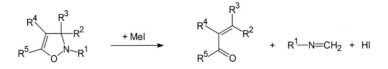

5.29   Thiazole

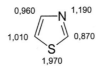

A Thiazole (1,3-thiazole) possesses a pyridine-like N-atom and an S-atom as present in thiophene. The univalent radical is known as thiazolyl. The thiazole molecule is planar and the C–S bond length is 171.3 pm, similar to that in thiophene (see Fig. 5.13):

$$
\begin{array}{c}
\text{S} \\
171{,}3 \quad 89{,}3 \quad 171{,}3 \\
109{,}6 \quad 115{,}2 \\
136{,}7 \quad 115{,}8 \quad 130{,}4 \\
110{,}1 \quad \text{N} \\
137{,}2
\end{array}
$$

Fig. 5.13   Structure of thiazole
(Bond lengths in pm, bond angles in degrees)

A comparison with the bond lengths of oxazole (see Fig. 5.9, p 123) leads to the conclusion that the delocalization of the π-electrons in thiazole is greater. Thus, the aromaticity of thiazole is greater than that of oxazole. The ionization potential is 9.50 eV and its dipole moment 1.61 D. UV and NMR data are listed in the following table:

UV (ethanol) λ (nm) (ε)	^{1}H-NMR (CCl$_4$) δ (ppm)	^{13}C-NMR (CCl$_4$) δ (ppm)
207.5 (3.41)	H-2: 8.77	C-2: 153.6
233.0 (3.57)	H-4: 7.86	C-4: 143.3
	H-5: 7.27	C-5: 119.6

This shows that a diamagnetic ring current is induced in the thiazole molecule during NMR experiments.

Thiazole is aromatic. Four $2p_z$-orbitals and one $3p_z$-orbital form delocalized π-MOs to which the three C-atoms and the N-atom each contribute one electron and the S-atom two electrons. One can also consider the possibility of spd-hybridization for the S-atom. Calculated π-electron densities are as follows:

$$
\begin{array}{c}
0{,}960 \quad \quad 1{,}190 \\
\text{N} \\
1{,}010 \quad \quad 0{,}870 \\
\text{S} \\
1{,}970
\end{array}
$$

B Thiazole is a π-excessive heterocycle, but the π-excess is concentrated mainly on the hetero-atoms. Again the pyridine-like N-atom attracts π-electrons, decreasing the π-electron density on the C-2 atom. Electrophilic substitution should occur at the 5-position, and if the 5-position is blocked, at the 4-position. Nucleophilic attack should take place at the 2-position. The following reactions are typical for thiazoles:

Salt formation

Thiazole is more basic than oxazole but less so than pyridine. The pK_a of thiazole is 2.52. Protonation occurs on the N-atom. Unlike oxazoles, thiazoles form crystalline and stable salts, e.g. picrates.

Metalation

Thiazoles unsubstituted in the 2-position react with alkylmagnesium halides and with organolithium reagents [95], e.g.:

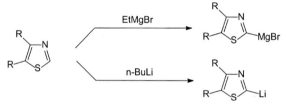

From these products, the corresponding 2-thiazoles can be synthesized by interaction with electrophiles such as haloalkanes, carbon dioxide or carbonyl compounds.

Reactions with electrophilic reagents

The thiazole molecule can be attacked by electrophiles at the S-atom, N-atom or at the C-5 atom. Alkylation with haloalkanes occurs only at the N-atom. The 3-alkylthiazolium salts react rapidly with nucleophiles.

As in the case of oxazoles, the pyridine-like N-atom makes electrophilic substitution reactions more difficult. Thus thiazole does not react with halogens. Donor substituents enhance the reactivity, e.g. 2-methylthiazole reacts with bromine to give 5-bromo-2-methylthiazole. Thiazole cannot be nitrated. 4-Methylthiazole reacts slowly to yield the 5-nitro compound, 5-methylthiazole even more slowly to give the 4-nitro compound, but 2,4-dimethylthiazole reacts fastest producing 2,4-dimethyl-5-nitrothiazole. Sulfonation of thiazole demands the action of oleum at 250°C in the presence of mercury(II) acetate, and occurs at the 5-position. Acetoxymercuration of thiazole with mercury(II) acetate in acetic acid/water proceeds stepwise by way of the 5-acetoxymercury compound and the 4,5-disubstituted product to 2,4,5-tris(acetoxymercury)thiazole.

Reactions with nucleophilic reagents

The nucleophilic substitution of an H-atom succeeds only at the 2-position, and then only with reagents of high nucleophilicity, e.g. sodamide:

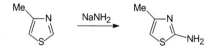

Nucleophilic substitution occurs faster if there is a nucleofuge leaving group, e.g.:

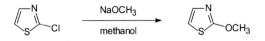

Quaternization strongly enhances the reactivity of the thiazole ring towards nucleophiles. The following reactions have been observed:

- Addition of a nucleophile to the 2-position to give a pseudobase, followed by ring opening, e.g. with sodium hydroxide solution:

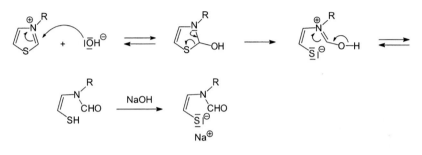

2-Alkyl substituted thiazolium salts react in an analogous way.

- Deprotonation of the 2-position to an N-ylide. The deuterodeprotonation of 3-alkylthiazolium salts by D_2O proceeds via such an N-ylide:

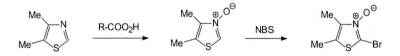

The reaction is catalysed by bases such as triethylamine.

Oxidation

N-Oxides are formed by the action of peracids on thiazoles. They are electrophilically halogenated in the 2-position by N-chloro- or N-bromosuccinimide [96], e.g.:

Reactions of 2-alkylthiazoles

2-Alkylthiazoles are CH-acidic compounds. They are deprotonated by bases on the α-C-atom of the alkyl group [97]. The resulting carbanions, stabilized by conjugation, react, for example, with carbonyl compounds to form alcohols:

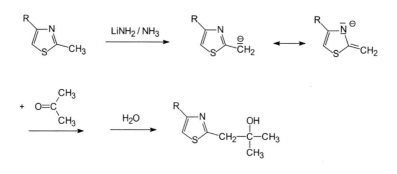

Unlike oxazoles, thiazoles do not react as 1,3-dienes, i.e. they do not undergo a DIELS-ALDER reaction. This is further evidence for the greater aromaticity of thiazole as compared with oxazole.

C   The methods used for the synthesis of thiazoles are the same as those applied in the synthesis of oxazoles (see p 127).

(1)   The cyclocondensation of α-halocarbonyl compounds with thioamides (*Hantzsch synthesis*) offers considerable scope:

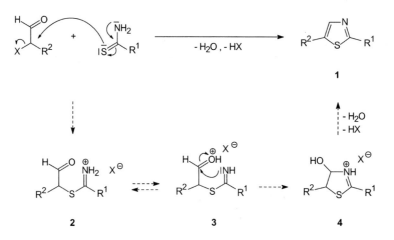

The HANTZSCH synthesis involves three intermediate steps. In the first, the halogen atom of the α-halo aldehyde or α-halo ketone is nucleophilically substituted. The resulting *S*-alkyliminium salt **2** undergoes a proton transfer (**2** → **3**); cyclization produces a salt of a 4-hydroxy-4,5-dihydrothiazole **4** which is converted into a 2,5-disubstituted thiazole **1** in protic solvents by an acid-catalysed elimination of water.

Numerous variations of the HANTZSCH synthesis are known [98], for instance, *N*-substituted thioamides yield 3-substituted thiazolium salts **5**:

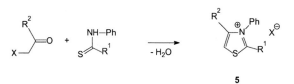

5

Thiourea reacts with α-halocarbonyl compounds to produce 2-aminothiazoles. 2-Amino-4-hydroxythiazoles **6** are formed with α-halocarboxylic acids:

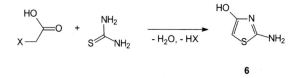

6

Salts or esters of dithiocarbamic acid yield 2-sulfanylthiazoles **7**:

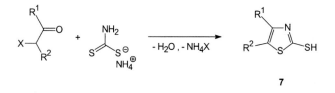

7

(2) α-Aminonitriles react with CS_2, COS, salts or esters of dithiocarboxylic acids and with isothiocyanates, under mild conditions, to give 2,4-disubstituted 5-aminothiazoles **8** (*Cook-Heilbron synthesis*), e.g.:

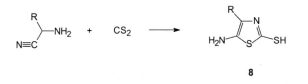

8

(3) α-(Acylamino)ketones react with P_4S_{10} yielding thiazoles **9** (*Gabriel synthesis*):

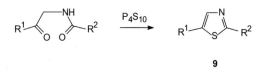

9

D   **Thiazole** is a colourless, water-soluble, putrid-smelling liquid, mp -33°C, bp 118°C.
2-Aminothiazole forms colourless crystals of mp 90°C. It couples with diazonium salts leading to azo compounds, e.g. **10**:

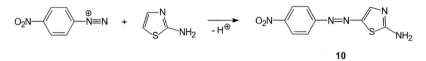

10

2-Aminothiazole can be diazotized. The diazonium salt is reduced to thiazole by hypophosphorous acid. By means of the SANDMEYER reaction, 2-halothiazoles and 2-cyanothiazole are obtained.

Among natural products derived from thiazole, thiamine **11** (aneurine, vitamin B_1) is of great importance:

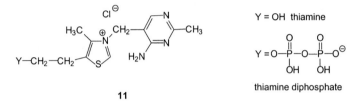

11

Thiamine occurs in yeast, in rice husk and other cereals. Deficiency of vitamin B_1 causes beriberi and damage to the nervous system (polyneuritis). The daily adult intake should be about 1 mg vitamin B_1.

Although thiamine, a thiazolium salt, contains a pyrimine ring, it is the thiazole ring that is responsible for its biological action, thiamine dihosphate being the coenzyme of decarboxylases. The mechanism of the catalytic decarboxylation (e.g. of pyruvic acid to acetaldehyde) was interpreted by BRESLOW in 1958. The active species is the N-ylide **12** formed from thiamine diphosphate and basic cell components:

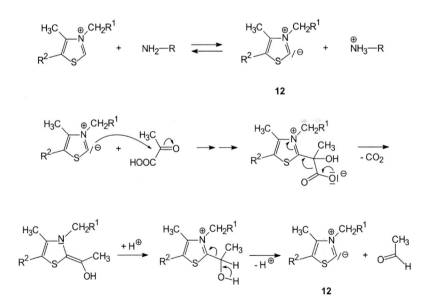

This is an example of nucleophilic catalysis. The adduct from the ylide and pyruvic acid eliminates carbon dioxide to give an enamine which accepts a proton on its exocyclic C-atom. Upon removal of a proton, this intermediate product decomposes to the nucleophilic catalyst and acetaldehyde [99].

Compounds structurally similar to thiamine, e.g. 5-(2-hydroxyethyl)-3,4-dimethylthiazolium iodide **13**, in the presence of triethylamine, catalyse the condensation of aldehydes to acyloins [100]. By analogy to a cyanide ion, the ylide acts as a nucleophilic catalyst.

Thiazoles occur in nature as substances with an aromatic odour, for instance 4-methyl-5-vinylthiazole in the aroma of cocoa beans and passion fruit, 2-isobutylthiazole in tomatoes and 2-acetylthiazole in the aroma of roasted meat.

Many thiazole derivatives are biologically active compounds. For instance, 2-(4-chlorophenyl)-thiazole-4-acetic acid **14** is a pharmaceutical used as an anti-inflammatory agent and niridazol **15** is applied in the treatment of bilharzia (schistosomiasis).

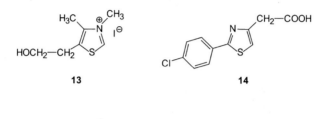

13   **14**

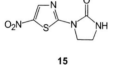

15

5.30 Benzothiazole

A,C   Benzothiazole (1,3-benzothiazole) shows the following UV and NMR spectra:

UV (ethanol) λ (nm) (ε)	^{1}H-NMR (CDCl$_3$) δ (ppm)	^{13}C-NMR (DMSO-d$_6$) δ (ppm)	
217 (4.27)	H-2: 9.23	C-2: 155.2	C-3a: 153.2
251 (3.74)	H-4: 8.23	C-4: 123.1	C-7a: 133.7
285 (3.23)	H-5: 7.55	C-5: 125.9	
295 (3.13)	H-6: 7.55	C-6: 125.2	
	H-7: 8.12	C-7: 122.1	

By analogy to the synthesis of benzoxazoles, benzothiazoles are obtained by cyclocondensation of *o*-aminothiophenols or their salts with carboxylic acids, their derivatives or with aldehydes [101]:

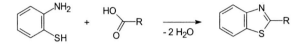

This reaction proceeds by way of the isolable *o*-(acylamino)thiophenols as intermediates. *N*-Arylthioamides can be cyclized oxidatively to give benzothiazoles:

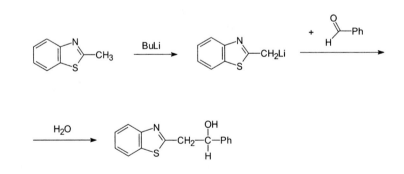

B Benzothiazole (pK_a = 1.2) is a weaker base than thiazole (pK_a = 2.52). Butyllithium metallates in the 2-position and haloalkanes produce the quaternary 3-alkylbenzothiazolium salts. Electrophilic substitutions occur only on the benzene ring. For instance, nitration with nitrating acid at room temperature yields a mixture of 4-, 5-, 6-, and 7-nitrobenzothiazole.

All reactions with nucleophilic reagents mentioned on p 150/151 for thiazole can also be achieved with benzothiazole. Only slight differences are observed. For instance, 2-chlorobenzothiazole reacts about 400 times faster with sodium methoxide or with sodium thiophenolate than 2-chlorothiazole, to give the corresponding substitution products 2-methoxy- and 2-(phenylsulfanyl)benzothiazole.

2-Alkylbenzothiazoles, like 2-alkylthiazoles, are CH-acidic. They are deprotonated by *n*-butyllithium in THF at -78°C. The lithium compounds react with aldehydes or ketones giving alcohols [102], e.g.:

The CH-acidity of the corresponding 3-alkylbenzothiazolium salts is even more pronounced. For instance, 3-ethyl-2-methylbenzothiazolium iodide, on heating with triethylamine in pyridine, reacts to give a trimethincyanine:

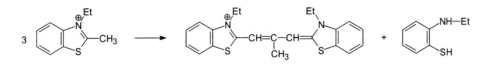

The mechanism of this interesting reaction is complex. First a carbanion stabilized by conjugation is formed, which, as a strong nucleophile, interacts with a second benzothiazolium ion:

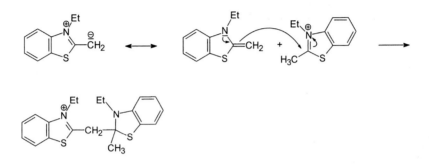

By inclusion of the third benzothiazolium ion, further intermediates are formed from the adduct. Fission of a thiazole ring with formation of 2-(ethylamino)thiophenol finally produces the trimethincyanine.

D **Benzothiazole** is a colourless liquid, bp 227°C, sparingly soluble in water. It is an aroma constituent of cocoa beans, coconuts, walnuts and beer.

Luciferin **1**, which occurs in fireflies and glowworms, upon enzymatic oxidation causes bioluminescence in these insects [103].

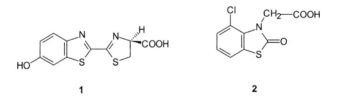

1   2

The herbicide Benazoline **2** serves as an example of a synthetic benzothiazole derivative with biological activity.

Polymethine dyes derived from benzothiazoles are employed for the spectral sensitization of photographic emulsions.

2-Sulfanylbenzothiazole **3** (2-mercaptobenzothiazole) is produced industrially from 1-chloro-2-nitrobenzene, sodium sulfide and carbon disulfide. In this synthesis, nucleophilic substitution of the Cl-atom is followed by reduction of the nitro group and finally by cyclocondensation:

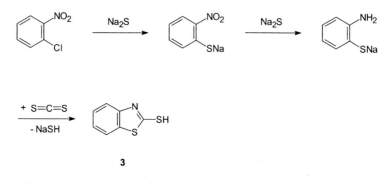

3

2-Sulfanylbenzothiazole is an accelerator in the vulcanization of polybutadiene and polyisoprene, acting as a radical transfer agent.

☐**E** The synthetic applications of benzothiazoles appeared to be limited by the fact that the thiazole ring cannot be hydrolytically cleaved. However, this problem was solved by quaternization followed by reduction with NaBH$_4$ to produce 3-methyl-2,3-dihydrobenzothiazole, as shown in the synthesis of cyclohexene-1-carbaldehyde [104]:

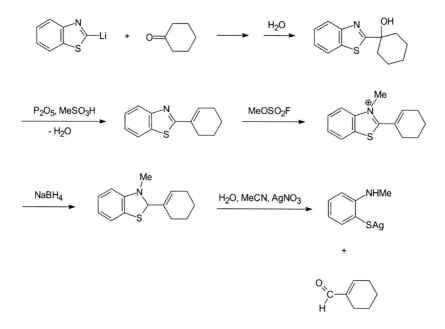

The benzothiazole acts as a carbonyl equivalent. By analogy, 2-lithiomethylbenzothiazole (see p 156) can be considered a synthetic enolate equivalent for the following reason: the product produced by quaternization, reduction and hydrolysis after cleavage of the thiazole ring could be obtained from an aldol reaction [102].

3-Ethylbenzothiazolium bromide is a suitable catalyst for the chain elongation of aldoses with formaldehyde to the next higher ketose [105].

5.31 Penam

A-C   In this bicyclic system, a 1,3-thiazole ring is fused on to an azetidine ring. In most publications, the numbering deviates from the IUPAC rules as shown above. The molecule is chiral.
   In 1929, FLEMING discovered that the mould *Penicillium notatum* inhibits the growth of bacteria. In
1941 FLOREY and CHAIN succeeded in isolating the active agent, known as penicillin, in the form of its
sodium salt. The structural elucidation was achieved by chemical degradation and was confirmed in
1945 by X-ray analysis of penicillin G (benzylpenicillin). The structures were shown to be (3S,5R,6R)-
6-(acylamino)-2,2-dimethyl-7-oxopenam-3-carboxylic acids:

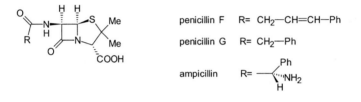

The residue R in the acylamino substituent differs in the various penicillins as shown in the above examples for penicillin F and G. By virtue of the 7-oxo group, penicillins are also β-lactams (see p 44).
The β-lactam ring is essential for the biological acitivity. Its presence enables the penicillins to irre-
versibly acylate the amino groups of those enzymes which are necessary for the synthesis of the peptidoglycans in the bacterial cell wall. However, some bacteria have developed resistant strains to penicillins derived from microorganisms. They synthesize the enzyme β-lactamase that opens the β-lactam
ring hydrolytically. There are two possibilities for suppressing this resistance.
* Application of structurally similar lactams that are bactericidal and that are not hydrolysed by
  β-lactamase [106]. A few semisynthetic penicillins fulfil these requirements. For instance, 6-
  amino-2,2-dimethyl-7-oxopenam-3-carboxylic acid (also known as 6-aminopenicillanic acid) can
  be prepared from a natural penicillin. This is then converted into a penicillin by acylation with another residue R. One of the most effective compounds of this type is ampicillin.
* Use of compounds which inhibit the enzyme β-lactamase, e.g. the synthetically prepared sulbactam
  1 [107].

1

The most successful method is the combination of a semisynthetic penicillin with a β-lactamase inhibitor. There is a seemingly endless conflict between chemists, continually synthesizing new compounds, and bacteria, constantly developing new mechanisms of resistance.

Although total syntheses of the penicillins have been elaborated, they are too expensive relative to the extraction from moulds or to semisynthetic procedures. This also applies to biomimetic syntheses, i.e. to total syntheses based on the biosynthesis of the compounds [108]. The biosynthesis of the penicillins starts from a peptide **2** which is derived from the amino acids cysteine and valine:

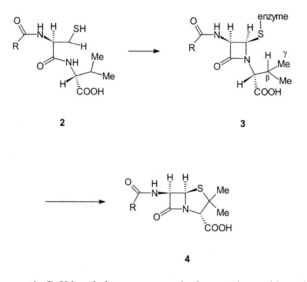

2   **3**

4

In the first step, an enzymatic C–H bond cleavage occurs in the cysteine residue with formation of a C–N bond. From the intermediate **3**, which is bound to the enzyme, penicillin is formed by a C–H bond cleavage in the valine residue and subsequently a C–S linkage (**3** → **4**) is formed.

5.32 Isothiazole

A   In isothiazole (1,2-thiazole), the pyridine-like N-atom is bonded to the S-atom. This σ-bond is also the weakest link in the molecule and is cleaved by ring-opening reactions.

The isothiazole molecule is planar; its ionization energy amounts to 9.42 eV and its dipole moment to 2.4 D. Isothiazole absorbs at longer wavelengths than isoxazole and thiazole, due to a $\pi \to \pi^*$ transition:

UV (ethanol) λ (nm) (ε)	^{1}H-NMR (CCl$_4$) δ (ppm)	^{13}C-NMR (CDCl$_3$) δ (ppm)
244 (3.72)	H-3: 8.54	C-3: 157.0
	H-4: 7.26	C-4: 123.4
	H-5: 8.72	C-5: 147.8

B Isothiazole is aromatic. The NMR spectra confirm a largely undisturbed delocalization of the π-electrons. In consequence, the aromaticity of isothiazole is greater than that of isoxazole, just as the aromaticity of thiophene is greater than that of furan. From the calculated π-electron densities, it follows by analogy to isoxazole (see p 138) that electrophilic substitution should occur at the 4-position, while nucleophiles should attack the 3-position. The most important reactions of isothiazoles can be summarized as follows:

Salt formation

Isothiazoles are weak bases with a pK_a value of -0.51. Protonation occurs on the N-atom. Liquid isothiazoles can be characterized by their crystalline perchlorates, e.g.:

Metalation

Isothiazoles unsubstituted in the 5-position are metalated by *n*-butyllithium [95]. 5-Lithioisothiazoles react with electrophilic reagents, e.g. with haloalkanes to give 5-alkylisothiazoles.

Reactions with electrophilic reagents

Isothiazoles are quaternized by iodoalkanes, dialkyl sulfate, trialkyloxonium tetrafluoroborate or diazomethane.

Electrophilic substitutions, e.g. halogenation, nitration and sulfonation, take place regioselectively at the 4-position. The pyridine-like N-atom again impairs electrophilic substitution. For this reason, isothiazole reacts more slowly than thiophene, but faster than benzene.

Reactions with nucleophilic reagents

Isothiazoles react more slowly with nucleophiles than isoxazoles. They are not affected by alkali hydroxides or alkoxides. 2-Alkylisothiazolium salts are more reactive. By the action of aqueous alkali hydroxide, ring-opening occurs with formation of polymeric products. Carbanions cause ring-opening by nucleophilic attack on the S-atom [109], e.g.:

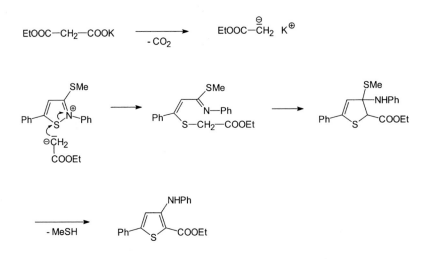

The (ethyloxycarbonyl)methanide ion is produced in the reaction mixture from the potassium salt of the monoethyl ester of the malonic acid. Ring opening by cleavage of the N–S bond is followed by cyclization and β-elimination. This results in a ring transformation to give substituted thiophenes.

Oxidation

Trisubstituted isothiazoles are oxidized by peroxy acids to 1-oxides and further to 1,1-dioxides. Isothiazoles, unsubstituted in the 3-position, yield isothiazol-3(2H)-one-1,1-dioxide with H_2O_2 in acetic acid at 80°C [110].

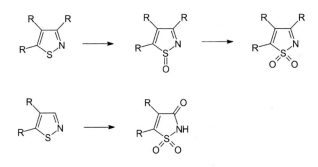

C   The synthesis of isothiazoles is usually carried out by one of the two methods:

(1)   The oxidation of β-imino thiones with iodine or hydrogen peroxide gives 3,5-disubstituted isothiazoles **1**:

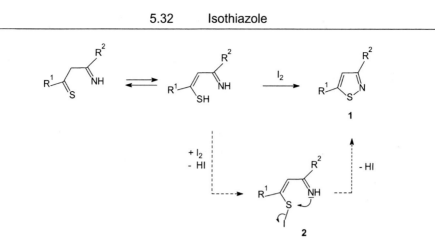

The β-imino thione is isomeric with its thiol form. Cyclization occurs by a nucleophilic substitution on the S-atom via **2**. The reaction can tolerate wide variations in R^1 and R^2. For instance, β-imino thioamides ($R^1 = NH_2$) yield 5-aminoisothiazoles.

(2)   The cyclocondensation of β-chlorovinyl aldehydes (see p 76) with two equivalents of ammonium thiocyanate produces 4,5-disubstituted isothiazoles **3** [110]:

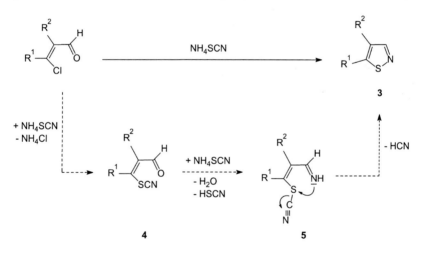

First, an isolable 3-thiocyanatopropenal **4** is formed. It reacts with ammonium thiocyanate to an imine **5**, which forms the isothiazole **3** by nucleophilic substitution at the S-atom.

D   **Isothiazole** is a colourless liquid with a pyridine-like odour, bp 113°C, and sparingly soluble in water.

   The isothiazole moiety is very rarely found in natural products. The fungicidal Brassilexin **6** was isolated from the leaves of the cruciferous plant *Brassica juncea*. This compound is a derivative of isothiazoloindole [111]:

6

Many synthetic isothiazoles are biologically active. For instance, the 5-acetylisothiazolothiosemicarbazone **7** has a virostatic action and 2-octylisothiazol-3(2*H*)-one **8** is fungicidal and algicidal.

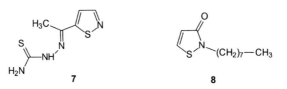

7   **8**

Saccharin **9**, the earliest synthetic sweetening agent (1879), is derived from 1,2-benzothiazole. Saccharin is produced from 2-methylbenzenesulfonyl chloride as follows:

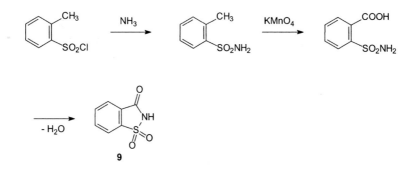

9

Oxidation of 2-methylbenzenesulfonamide yields the corresponding carboxylic acid, which is transformed into saccharin by cyclodehydration.

Saccharin is a crystalline, virtually water-insoluble compound of mp 244°C. As its water-soluble sodium salt, it is used as a sweetening agent [112]. It is 300 to 500 times as sweet as saccharose, but has a bitter-metallic aftertaste.

E Like *N*-bromosuccinimide, *N*-bromosaccharin can be used as a brominating or oxidizing agent.
N-Acylsaccharins serve as acylating agents for tertiary alcohols and convert amino alcohols selectively into *N*-acyl derivatives.

5.33 Imidazole

A Imidazole contains one pyrrole- and one pyridine-like N-atom, located in the 1- and 3-positions, respectively. Its systematic name is 1,3-diazole. The univalent radical is known as imidazolyl. The imidazole molecule is planar and an almost regular pentagon (see Fig. 5.14).

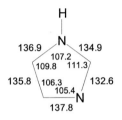

Fig. 5.14   Structure of imidazole
(bond lengths in pm, bond angles in degrees)

The ionization energy of imidazole was calculated to be 8.78 eV. The removed electron is derived from the HOMO π_3. From a comparison with the value of 8.23 eV for pyrrole, it follows that the pyridine-like N-atom reduces the HOMO energy and thereby stabilizes the π-system. This also applies to the furan–oxazole and thiophene–thiazole system.

The dipole moment of imidazole is 3.70 D in the gas phase. In solution, the values depend on the concentration, because of strong intermolecular hydrogen bonds (see p 172). The NMR spectra show that a tautomeric equilibrium is rapidly achieved at room temperature.

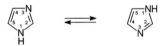

In this annular tautomerism, one averaged signal appears for 4-H and 5-H, as well as for C-4 and C-5.

UV (ethanol)	^{1}H-NMR (CDCl$_3$)	^{13}C-NMR (CDCl$_3$)
λ (nm) (ε)	δ (ppm)	δ (ppm)
207-208 (3.70)	H-2: 7.73	C-2: 135.4
	H-4: 7.14	C-4: 121.9
	H-5: 7.14	C-5: 121.9

Imidazole is aromatic. The pyrrole-like N-atom contributes two electrons to the π-electronic sextet. The pyridine-like N-atom and the C-atoms each donate one electron. As can be deduced from the NMR spectra, the π-electrons are largely delocalized, whereas the nonbonding electron pair is localized onto the pyridine-like N-atom. The following electron densities have been calculated by SCF/MO methods:

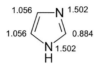

Therefore, imidazole is a π-excessive heterocycle, with six electrons distributed over five atoms, but mainly concentrated on the N-atoms. Electrophilic substitutions should be possible in the 4- or 5-position. In the 2-position, i.e. between the two N-atoms, the π-electron density is less than one. Nucleophilic attack would thus be expected to occur onto this position.

B The most important reactions of imidazoles can be summarized as follows.

Acid–base reactions

Imidazoles are moderately strong bases. The pK_a value of the conjugate acid of imidazole is 7.00. They form salts with many acids, e.g. chlorides, nitrates, oxalates and picrates:

It follows from the ^{1}H-NMR spectra of the salts that the imidazolyl cation possesses the symmetrical structure shown above. It reacts more slowly than imidazole with electrophiles, but faster with nucleophiles.

Imidazoles unsubstituted in the 1-position are weak acids. The pK_a value of imidazole is 14.52. Hence its acidity is greater than that of pyrrole and of ethanol. The sodium salt of imidazole is formed with sodium ethoxide in ethanol, and the sparingly soluble silver salt can be obtained with aqueous silver nitrate solution.

The imidazolyl anion has a symmetrical structure and is a nucleophile which can react with a variety of electrophiles.

Imidazole thus behaves amphoterically, i.e. like a combination of pyridine and pyrrole. Substituents can alter the basicity and acidity of imidazole within certain limits.

Annular tautomerism

A direct consequence of the amphoteric character of 1,3-unsubstituted imidazoles is the rearrangement of 4-substituted imidazoles to the corresponding 5-substituted isomers and vice versa, owing to a rapid proton transfer from position 1 to 3:

This special case of prototropy is known as annular tautomerism (see p 111). In solution, equilibria are established so rapidly that the separate tautomers cannot be isolated. However, their presence can be demonstrated by spectroscopic methods. In this case, e.g. R = CH$_3$, the compound is known as 4(5)-methylimidazole. With certain substituents R, the equilibrium lies predominantly to one side, for instance, in the case of the nitro compound (4-nitroimidazole) or with the methoxy compound (5-methoxyimidazole). Annular tautomerism has also been demonstrated for 4,5-disubstituted imidazoles:

Formation of metal complexes

Imidazoles form complexes with many metal ions in which the pyridine-like N-atom provides the donor, e.g., dichlorodiimidazole cobalt(II):

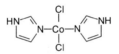

Haemoglobin is an iron(II) complex of haem with the imidazole ring of the amino acid histidine contained in the protein globin.

Metalation

Imidazoles substituted in the 1-position, e.g. 1-methylimidazole, react with *n*-butyllithium in diethyl ether to form the corresponding 2-lithioimidazoles [113]:

Many 1,2-disubstituted imidazoles are accessible by the action of electrophiles on these compounds. For instance iodomethane produces 1,2-dimethylimidazole, and chlorotrimethylsilane produces 1-methyl-2-trimethylsilylimidazole. Imidazoles with substituents in the 2-position can also be synthesized provided there is a hydrolysable protecting group such as an ethoxymethyl group in the 1-position.

   Lithiation with 1,2-disubstituted imidazoles occurs in the 5-position. Metalated imidazoles are prepared by metal–halogen exchange from haloimidazoles which can be made to react with electrophiles [114].

Deprotonation of 1,3-diadamant-1-ylimidazolium chloride with sodium hydride in THF in the presence of DMSO provides the crystalline 1,3-diadamant-1-ylimidazol-2-ylidene of mp 240-241°C. This reaction represents the first isolation of a carbene that is stable at room temperature [115]:

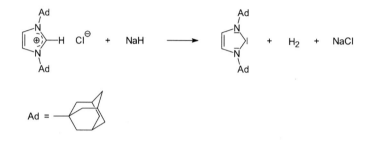

This carbene is stabilized both thermodynamically, by incorporating the bivalent C-atom into an electron-rich π-system, and kinetically, due to steric shielding of the bivalent C-atom by the space-filling adamantyl substituents.

Reactions with electrophilic reagents

Alkylation, acylation, sulfonation and silylation occur on the N-atoms of imidazoles [116]. Other reagents substitute onto the C-atoms 4 and 5, which are equivalent as a result of annular tautomerism.

Imidazoles react with haloalkanes in the absence of strong bases because the pyridine-like N-atom effects a nucleophilic substitution of halogen. The quaternary salts that are formed initially usually undergo rapid deprotonation to 1-alkylimidazoles. These can react with a second mole of haloalkane to give 1,3-dialkylimidazolium salts:

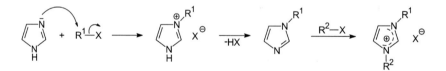

However, in the presence of strong bases, the imidazolyl anion reacts with the haloalkane:

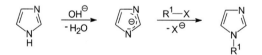

High yields are obtained by preparing the sodium imidazolide from imidazole and sodium hydroxide, which subsequently reacts with the haloalkane or a dialkyl sulfate in dichloromethane, acetonitrile or methanol [116]. Because of the ambient character of the imidazolyl anion, 4- or 5-substituted imidazoles yield mixtures of 1,4- and 1,5-disubstituted imidazoles, e.g.:

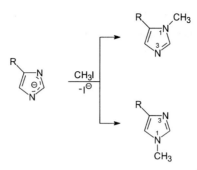

The sodium salt of imidazole reacts with acyl chlorides, sulfonyl chlorides or trimethylsilyl chloride in dichloromethane to give the corresponding 1-substituted imidazoles. As for acylation, 4- or 5-substituted imidazoles yield mixtures of products.

When imidazole undergoes halogenation and azo-coupling in electrophilic substitution reactions at carbon, it displays high reactivity similar to pyrrole. Accordingly, chlorination with sulfuryl chloride yields 4,5-dichloroimidazole, bromination with bromine in aqueous solution produces 2,4,5-tribromoimidazole and iodination with iodine in aqueous alkaline solutions gives 2,4,5-triiodoimidazole. Azo coupling is carried out in aqueous alkaline solution and it is the imidazolyl anion which reacts with the electrophile. Substitution occurs at the 2-position, as the negative charge is delocalized over positions 1 to 3:

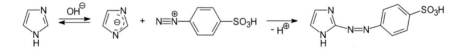

Azo coupling occurs on the imidazole ring of the amino acid histidine in the colour test for proteins with PAULY's reagent.

In contrast to halogenation and azo coupling, nitration and sulfonation proceed very slowly because the reactions take place in an acidic medium with formation of imidazolyl cations, for instance in the nitration with nitric acid:

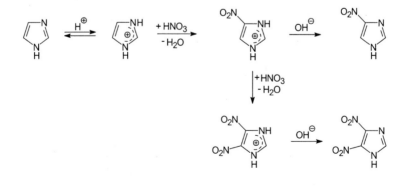

4-Nitroimidazole and, under more drastic conditions, 4,5-dinitroimidazole are obtained. Sulfonation of imidazole with oleum at 160°C yields 4-imidazolesulfonic acid.

Reactions with nucleophilic reagents

Reactions of imidazoles with nucleophiles occur slowly as with thiazoles and demand vigorous conditions. Attack of the nucleophile occurs in the 2-position. For instance, 1-methyl-4,5-diphenylimidazole reacts with potassium hydroxide at 300°C to give imidazol-2(3*H*)-one:

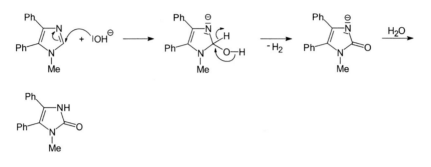

The nucleophilic substitution reactions of 2-haloimidazoles are only possible under drastic conditions, e.g.:

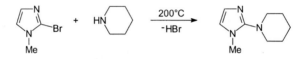

A nitro group in the 4- or 5-position enhances the reactivity. 1,3-Dialkyl- or 1,3-diacylimidazolium salts react faster with nucleophiles, usually with ring opening.

In short, it can be stated that the dual nature of the N-atoms in the imidazole molecule makes possible an extraordinary variety of reactions and this is the principal reason for the great biological importance of the amino acid histidine.

C   For the *retrosynthesis* of the imidazole system (see Fig. 5.15), it is essential that the heterocycle possesses the functionality of an amidine on C-2 and that of a 1,2-enediamine on C-4/C-5. Retroanalysis should, therefore, consider two approaches:

Route I leads via the retrosynthetic operations **a** and **b** to the α-acylamino carbonyl system **2**. This suggests NH_3 or a primary amine as starting materials. According to route II, addition of H_2O (**c**) leads to intermediate **1** which can be further dissected in line with the usual retroanalytical scheme. Route **d** points to an α-halo- or α-hydroxycarbonyl compound and amidine, whereas route **e** suggests an α-amino ketone and acid amide or nitrile as starting materials.

The most important principles underlying the syntheses of imidazole thus become apparent.

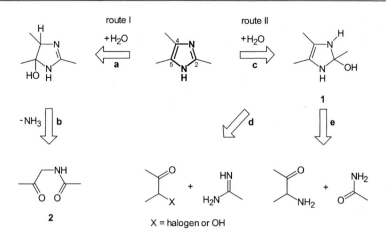

Fig. 5.15   Retrosynthesis of imidazole

(1)   1,2-Dicarbonyl compounds undergo cyclocondensation with ammonia and aldehydes to form imidazole derivatives **3** [117]:

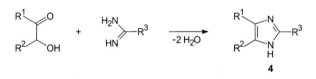

By this route, imidazole was first synthesized from glyoxal, ammonia and formaldehyde, and was known as glyoxaline.

(2)   Imidazoles **4**, with a variable substitution pattern, are obtained from α-halo or α-hydroxy ketones and amidines:

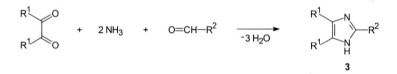

2-Aminoimidazoles are obtained from guanidine, whereas urea or thiourea yield imidazole-2(3*H*)-one or -thione, respectively. Imidazoles **5** unsubstituted in the 2-position are obtained from α-hydroxy ketones and formamide (*Bredereck synthesis*):

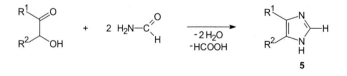

(3)   α-Amino ketones can be condensed with cyanamide to form 2-aminoimidazoles **6**:

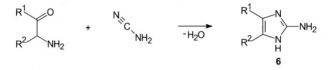

This synthetic principle (*Marckwald synthesis*) can be applied in various ways; cyanates yield imidazole-2(3*H*)-ones, thiocyanates imidazole-2(3*H*)-thiones and alkylisothiocyanates 1-alkylimidazole-2(3*H*)-thiones.

(4)   Aldimines react in the presence of K_2CO_3 with tosylmethylisocyanide (TosMIC) to form 1,5-disubstituted imidazoles **8** [118]:

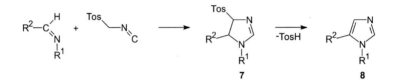

By analogy to the VAN LEUSEN synthesis of oxazoles (see p 128), the carbanion formed from tosylmethylisocyanide adds to the aldimine. The addition product cyclizes to give a 4,5-dihydroaimidazole **7**, which eliminates *p*-toluenesulfinic acid to give the imidazole **8**.

> **D**   **Imidazole**, colourless crystals, mp 90°C, bp 256°C, is soluble in water and other protic solvents. Imidazoles have high melting and boiling points when compared to pyrrole, oxazole and thiazole because the imidazole molecule is a donor as well as an acceptor of hydrogen bonds, and only intermolecular hydrogen bridges can be formed.

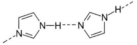

In the solid state, imidazole forms chainlike associations with angled structures imparting a fibrous texture to the crystals. When dissolved in water, the N–H---N bonds are replaced by N–H---O and N---H–O bonds. In contrast, 1-methylimidazole is a liquid, mp -6°C, bp 198°C. It is only slightly soluble in water.

Imidazoles are extremely thermally stable. Imidazole itself decomposes above 500°C.

The most important natural product derived from imidazole is the proteinogenic amino acid histidine **9**:

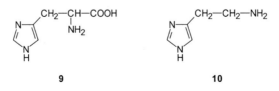

        9   **10**

The imidazole ring, with a physiological pH value of 7.4, exists in the histidine building blocks of proteins as a free base and as a conjugate acid ($pK_a = 7.00$, see p 166) because of a regulating acid–base equilibrium. Especially in enzymes, the ring can act as a BRÖNSTED base or as a BRÖNSTED acid as the occasion demands, i.e. it acts as a buffer. It is also able to form complexes with metal ions. Such properties are not found in any other proteinogenic amino acids [119].

Histamine **10** is formed by enzymatic decarboxylation of histidine. It is a vasodilator and thus lowers the blood pressure. It also contracts smooth muscles and regulates gastric acid secretion. Too high a histamine level in the blood causes allergic reactions, e.g. hayfever. Such reactions can be suppressed by administering antihistamines which act as antagonists of histamine mainly by blocking the allergy-causing histamine receptors (H_1 receptor).

Cimetidine **11** is used in the treatment of duodenal and gastric ulcers. It reduces gastric acid secretion by blocking the histamine receptor which stimulates gastric acid secretion (H_2 receptor), but does not affect the H_1 receptor.

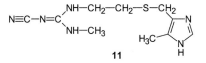

11

Further examples of synthetically produced imidazole derivatives are the drugs **12** and **13**:

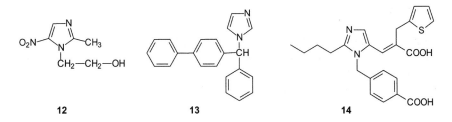

12   **13**   **14**

Metronidazole **12** is used in the treatment of trichomoniasis and amoebal infections. It is incompatible with ethanol. Bifonazole **13** shows antifungal activity. Eprosartan **14** is an angiotensin II inhibitor and used as antihypertensive agent.

The biologically important ring system purine, in which an imidazole ring is fused to a pyrimidine ring, will be discussed in chapter 6.31.

E Among the synthetic applications of imidazoles, the use of 1-acylimidazoles as acylating agents is of great importance. By analogy with amides, 1-acylimidazoles **15** are often named imidazolides. However, in contrast to amides, no nitrogen lone pair is available for an amide mesomerism [120]:

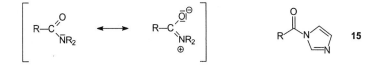

15

For this reason, acylimidazoles are much more reactive than *N,N*-dialkylamides and, like acid anhydrides or acid chlorides, are able to transfer the acyl group to water, alcohols, phenols and amines:

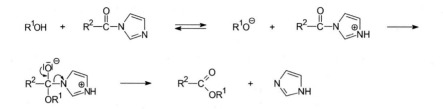

Crystalline 1,1'-carbonyldiimidazole **16**, prepared from phosgene and imidazole, is particularly reactive. Even at room temperature, it reacts vigorously with water to give imidazole and carbon dioxide. With carboxylic acids in aprotic solvents, 1-acyloxycarbonylimidazole **17** is formed, which is capable of an intermolecular transacylation:

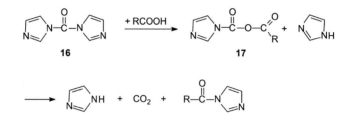

The resulting 1-acylimidazoles can be used as acylating agents as described above.

Imidazole itself acts as a nucleophilic catalyst for the hydrolysis of carboxylic acid derivatives, e.g.:

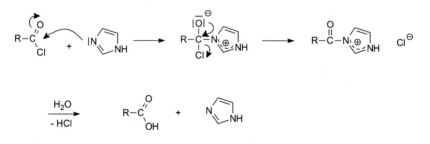

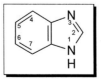

5.34 Benzimidazole

A The UV and ^{13}C-NMR spectra of benzimidazole show the following characteristics:

UV (ethanol)	^{13}C-NMR (methanol-d$_4$)
λ (nm) (ε)	δ (ppm)

244 (3.74), 248 (3.73), 266 (3.69),   C-2: 141.5, C-4: 115.4, C-5: 122.9, C-6: 122.9

272 (3.71), 279 (3.73)   C-7: 115.4, C-3a: 137.9, C-7a: 137.9

The short wavelength bands at 244 and 248 nm are due to electronic excitation of the imidazole ring, the others to electronic transitions in the benzene ring.

There is no detailed analysis of the A_2B_2 system of the two benzene-type protons in the ^{1}H-NMR spectrum. The H-2 signal is in the region of $\delta = 7.59 \pm 0.58$ (CDCl$_3$), depending on the substituents in the benzene ring.

B   Benzimidazole, $pK_a = 5.68$, is less basic than imidazole, but with $pK_a = 12.75$ is more strongly NH-acidic [121]. Like imidazoles, benzimidazoles display annular tautomerism in solution, e.g.:

Benzimidazoles substituted in position 1, e.g. 1-methylbenzimidazole, react with *n*-butyllithium at low temperature to give 2-lithio compounds. At room temperature, the reaction proceeds as follows:

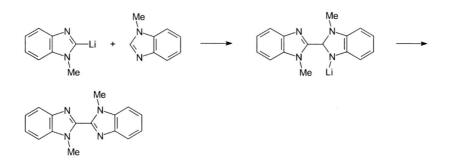

By analogy to imidazoles, alkylation by haloalkanes of benzimidazoles on the N-atom occurs in neutral and in basic media. 1-Alkylbenzimidazoles are obtained from benzimidazole, sodium hydroxide and bromoalkanes. Benzimidazoles unsubstituted in the 1-position undergo the MANNICH reaction:

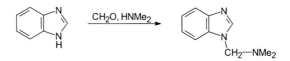

1-(2-Hydroxyethyl)benzimidazole is formed with oxirane.

Electrophilic substitutions on carbon take place first in the 5-position and then in the 7- or 6-position.

1-Methylbenzimidazole yields a mixture of 5-bromo- and 5,7-dibromo-1-methylbenzimidazole with bromine. Mixtures of 5-nitro-, 5,6-dinitro- and 5,7-dinitro compounds are produced with nitrating acid.

Nucleophiles react faster with benzimidazoles than with imidazoles, the attack occurring at the 2-position. For instance, on treatment with sodium amide in xylene, 1-alkylbenzimidazoles give the corresponding 2-amino compounds (*Chichibabin reaction*, see p 278).

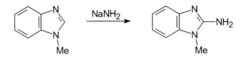

The halogen in 2-halobenzimidazoles can be substituted by nucleophiles, e.g. alkoxides, thiolates or amines. However, the reactions proceed more slowly than with 2-halobenzoxazoles and 2-halobenzothiazoles.

2-Alkylbenzimidazoles possess an acidic CH-group. For instance, 2-methylbenzimidazole reacts with 2 mol of *n*-butyllithium in THF/hexane at 0°C to give dilithio compound **1** which, on treatment with aldehydes, yields secondary alcohols **2**:

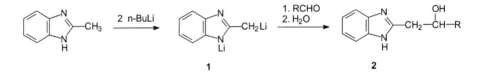

C   The standard synthesis for benzimidazoles is the cyclocondensation of *o*-phenylenediamine or substituted *o*-phenylenediamines with carboxylic acids or their derivatives.

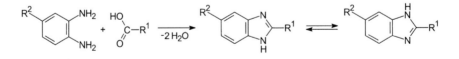

o-Phenylendiamine reacts with formic acid at 100°C to give benzimidazole in a yield of over 80%. *N*-Monosubstituted *o*-phenylenediamines react with other carboxylic acids more slowly, necessitating the addition of hydrochloric or phosphoric acid. A mixture of trifluoromethanesulfonic acid anhydride and triphenylphosphane oxide in dichloromethane is a very efficient dehydrating agent [122].

In contrast, *o*-phenylenediamine reacts with cyclohexanone under mild conditions (in hot water) to give 1,3-dihydro-2*H*-benzimidazole-2-spirocyclohexane **3**, which can be oxidized with active manganese dioxide to yield 2*H*-benzimidazole-2-spirocyclohexane **4** [123].

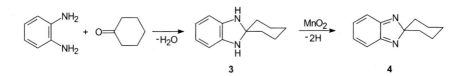

3   4

D **Benzimidazole**, colourless crystals, mp 171°C, bp 360°C, is moderately soluble in water and very soluble in ethanol.

The presence of the benzimidazole system in a natural product is most striking in the case of vitamin B$_{12}$ (cyanocobalamin). It was isolated from liver extracts and from the fungus *Streptomyces griseus*. It is an antipernicious anaemia factor. Elucidation of its complex structure was achieved by X-ray analysis (CRAWFOOT-HODGKIN 1957). 5,6-Dimethylbenzimidazole is bonded via the atom N-1 to D-ribose as an *N*-glycoside; N-3 is linked to a cobalt ion which is situated in the centre of a corrin system (see p 489).

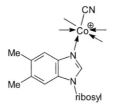

Examples of synthetically produced and biologically active benzimidazoles are the fungicides methyl 1-butylcarbamoyl-2-benzimidazole carbamate (benomyl) **5** and 2-(4'-thiazolyl)benzimidazole (thiabendazole) **6**. The latter is used extensively as a preservative for fruit (E 233) and as an anthelmintic in veterinary medicine:

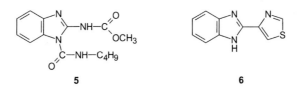

5   6

The 2*H*-benzimidazole-2-spirocyclohexane **4** can be used as a protected *o*-phenylenediamine for the synthesis of compounds not easily accessible by other routes, e.g.:

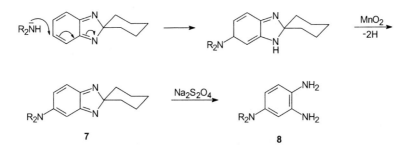

7   8

Nucleophiles, e.g. secondary amines, react with **4** in the presence of activated manganese dioxide to give **7** from which a 5-substituted *o*-phenylenediamine **8** can be obtained by reduction with sodium dithionite. The conversion of *o*-phenylenediamine into **4** can be regarded as an 'umpolung', since the former reacts only with electrophiles [123].

5.35 Imidazolidine

A-D Imidazolidines can also be regarded as cyclic aminals. Some interesting oxo compounds are derived from them.

Imidazolidin-2-one **1** is formed in 75% yield by heating ethylenediamine with urea:

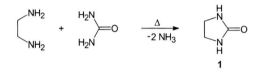

The analogous cyclocondensation of the alkaloid (-)-ephedrine with urea gives (4*R*,5*S*)-1,5-dimethyl-4-phenylimidazolidin-2-one **2**. This compound can be used as a chiral auxiliary for asymmetric syntheses [124]:

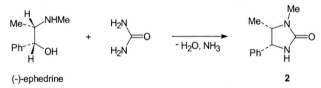

(-)-ephedrine   **2**

(+)-Biotin **3** (vitamin H) occurs in egg yolk and in yeast. It promotes the growth of microorganisms. In biotin, an imidazolidine ring is condensed with a thiolane ring. Several syntheses have been elaborated [125]. Some start from the proteinogenic amino acid (*R*)-cysteine:

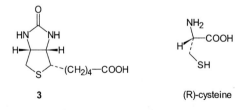

3   (*R*)-cysteine

The vitamin action is due to the coenzymic action of (+)-biotin towards carboxylase.

Imidazolidin-2,4-diones **4** have the trivial name hydantoins. They are made by a two-step synthesis from α-amino acids and potassium cyanate:

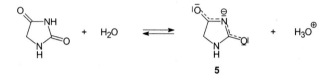

4

As in the WÖHLER urea synthesis, addition takes places to the C–N double bond of the isocyanic acid which is in hydrolytic equilibrium with potassium cyanate. The cyclodehydration is brought about by heating with hydrochloric acid.

Hydantoins unsubstituted on N-3 are NH-acids. The pK_a value of hydantoin (R = H) is 9.12. This is due to the delocalization of electrons in the conjugate base **5**:

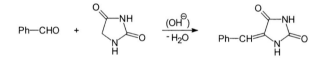

5

Hydantoin has an acidic C–H group and can, therefore, undergo an aldol condensation, e.g.:

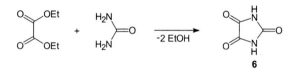

Imidazolidine-2,4,5-trione **6** has the trivial name parabanic acid and is an even stronger NH-acid, pK_a = 5.43. It reacts as a dibasic NH-acid. Parabanic acid is considered to be the ureide of oxalic acid. It is prepared from diethyl oxalate and urea:

6

5.36 Pyrazole

A Pyrazole, like its structural isomer imidazole, contains a pyrrole-like and a pyridine-like N-atom, but in the 1- and 2-positions (1,2-diazole).

The pyrazole molecule is planar. Bond lengths and bond angles have been calculated from microwave spectra (see Fig. 5.16). Consistent with the structural formula, the bond between atoms 3 and 4 is the longest.

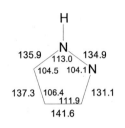

Fig. 5.16   Structure of pyrazole
              (Bond lengths in pm, bond angles in degrees)

The ionization energy of pyrazole is 9.15 eV. It follows from a comparison with pyrrole (8.23 eV) that the pyridine-like N-atom reduces the energy of the HOMO, indeed even more so than in the case of imidazole (8.78 eV).

The dipole moment of pyrazole in benzene was calculated to be 1.92 D. This value depends on the concentration, because cyclic dimers form at higher concentrations (see p 184). The dipole moment is directed from the centre of the molecule to the bond between atoms 2 and 3.

Pyrazole shows the following UV and NMR data:

UV (ethanol)	^{1}H-NMR (CCl$_4$)		^{13}C-NMR (CH$_2$Cl$_2$)	
λ (nm) (ε)	δ (ppm)		δ (ppm)	
201 (3.53) $\pi \rightarrow \pi^{*}$	H-1: 12.64	H-4: 6.31	C-3: 134.6	C-5: 134.6
	H-3: 7.61	H-5: 7.61	C-4: 105.8	

Pyrazole, like its structural isomer imidazole, is aromatic [126]. The following π-electron densities and π-bond orders were calculated by LCAO/MO methods:

Thus, pyrazole is a π-excessive heterocycle. Electrophilic substitution occurs preferably at the 4-position and nucleophilic attack at the 3- or 5-position. Delocalization of π-electrons follows from the bond order of the π-bonds. At the same time, it becomes evident that the N–N bond is the weakest link in the pyrazole ring.

B In most reactions of pyrazoles, an analogy with imidazoles is apparent, and comparisons are possible.

Acid–base reactions

Pyrazoles are much weaker bases than imidazoles, but can be precipitated as picrates. The conjugate acid of pyrazole has a pK_a value of 2.52. The difference is due to the fact that the positive charge in the pyrazolium ion is less delocalized than in the imidazolium ion (see p 166):

The gas-phase basicity (intrinsic basicity) for pyrazoles and imidazoles has been determined, as have their thermodynamic and kinetic basicities and proton affinities [127].

Pyrazoles unsubstituted in the 1-position show NH-acidity. The pK_a value of pyrazole is 14.21 and equals that of imidazole. Pyrazole reacts with sodium to give the sodium salt. The sparingly soluble silver salt is formed with aqueous silver nitrate solution.

Annular tautomerism

Pyrazoles unsubstituted in the 1,2-position undergo tautomerism [128]:

In solution, equilibrium is attained so rapidly that the existence of tautomers can only be demonstrated by means of ^{13}C and ^{15}N NMR spectroscopy. Other than for R = CH$_3$, the equilibrium lies to the left i.e. the 3-substituted isomer predominates.

Formation of metal complexes

Many metal complexes with pyrazole as ligand have been prepared with the pyridine-like N-atom acting as donor, for instance in dichlorotetrapyrazole nickel(II). In another series of metal complexes, the pyrazole anion is the ligand, e.g. in the following gold(I) complex:

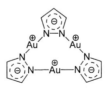

Metalation

Pyrazoles substituted in the 1-position are metalated with *n*-butyllithium in the 5-position. 5-Substituted pyrazoles can be synthesized if the 1-substituent is a removable protecting group, e.g. dimethylsulfamoyl [129]:

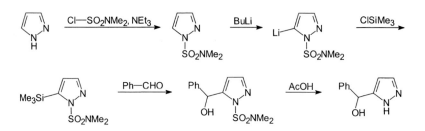

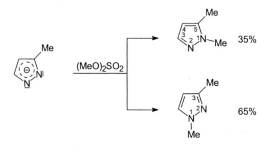

Reactions with electrophilic reagents

The best procedure for methylation of pyrazole is via the sodium salt which reacts with iodomethane or dimethyl sulfate:

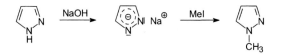

Benzylation, acetylation, benzoylation, methylsulfonation, methoxycarbonylation and trimethylsilylation of pyrazole are effected by analogous methods [116].

Mixtures of 1,3- and 1,5-disubstituted pyrazoles are formed from 3- and 5-substituted pyrazoles because of the ambident nature of the pyrazolyl anion, e.g.:

Electrophilic substitution on the C-atoms of pyrazole proceeds more slowly than for pyrrole and at about the same rate as for benzene. The pyrazole anion reacts faster and the pyrazolium ion much more slowly.

The corresponding 4-halopyrazoles are produced by the action of chlorine or bromine in acetic acid. Nitrating acid yields 4-nitropyrazoles and, dependent on the substituents in the pyrazole ring, reaction takes place either with pyrazole itself or the pyrazolium ion. Sulfonation involves the pyrazolium ion. For this reason, heating in oleum is necessary, which leads to pyrazole-4-sulfonic acid. Pyrazoles with substituents in the 1-position yield pyrazole-4-carbaldehyde in the VILSMEIER-HAACK formylation and are amenable to FRIEDEL-CRAFTS acylation. 4- and 5-aminopyrazoles can be diazotized.

Reactions with nucleophilic reagents

Pyrazoles either do not react with nucleophiles, or react with them only very slowly. For instance, pyrazoles unsubstituted in the 3-position undergo ring opening on heating with alkali hydroxides. Nucleophilic substitution of a halogen in halopyrazoles is also difficult.

Photoisomerism

On exposure to light, pyrazoles rearrange to imidazoles and undergo ring-opening to form 3-aminopropenenitriles, e.g. 1-methylpyrazole **1**:

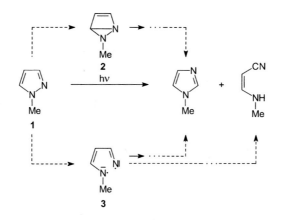

The reaction proceeds according to various mechanisms [130]. The first step is either electrocyclization to form a 1,5-diazabicyclo[2.1.0]pentene **2** or homolysis of the N–N bond to give a diradical **3**.

C   There are many syntheses of pyrazoles [131]. Two of them are especially versatile and widely applicable.

(1)   Hydrazine, and alkyl- or arylhydrazines undergo cyclocondensation with 1,3-dicarbonyl compounds to give pyrazoles **4**.

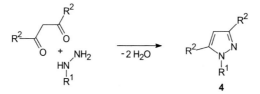

Unsymmetrical 1,3-diketones give mixtures of structural isomers. The mechanism of the reaction depends greatly on the nature of the substituent R as well as on the pH of the medium. In another variant of this synthesis, acetylenic ketones are used as bifunctional components [132], e.g.:

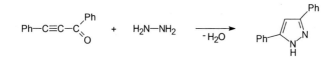

(2)   The 1,3-dipolar cycloaddition of diazoalkanes to alkynes leads to pyrazoles, e.g.:

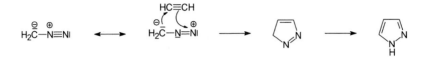

Diazomethane reacts with acetylene in a concerted [3+2] cycloaddition to give 3*H*-pyrazole which rapidly isomerizes to pyrazole. The 1,3-dipolar cycloaddition of diazoalkanes to alkenes yields dihydropyrazoles.

D **Pyrazole**, colourless needles, mp 70°C, bp 188°C, is soluble in water. Pyrazole exists in the solid state and in concentrated solution as a dimer with two intermolecular hydrogen bonds:

For this reason, the melting point and boiling point of pyrazole are higher than those of pyrrole and pyridine, but lower than those of imidazole (see p 172).

Natural products containing pyrazole rings are rare. It seems that the evolution of organisms has produced few enzymes which cause formation of an N–N bond. However, many synthetically produced pyrazoles are biologically active. Some are used as pharmaceuticals, e.g. the analgesic, anti-inflammatory and antipyretic difenamizole **5**. Betazole **6** is bioisosteric with histamine and selectively blocks the H_2 receptor. Celecoxib **7** is a powerful COX-2 inhibitor and exhibits analgesic and antiarthritic effects. The pyrzolopyrimidine zaleplon **8** belongs to a new generation of hypnotics structurally not based on the benzodiazepin system.

Other biologically active pyrazoles are the herbicide difenzoquat **9** and the insecticide dimetilan **10**:

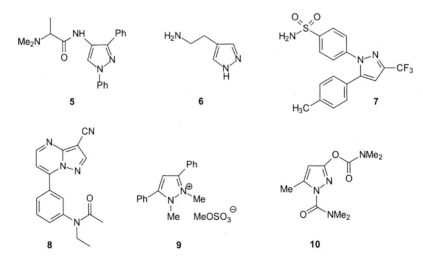

5.37 Indazole

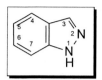

Indazole (benzo[d]pyrazole) has the following UV and NMR spectral data:

UV (H_2O (pH 4)) λ (nm) (ε)	^{1}H-NMR (DMSO-d_6) δ (ppm)		^{13}C-NMR (DMSO-d_6) δ (ppm)	
250 (3.65)	H-3: 8.08	H-6: 7.35	C-3: 133.4	C-6: 125.8
284 (3.63)	H-4: 7.77	H-7: 7.55	C-4: 120.4	C-7: 110.0
296 (3.52)	H-5: 7.11		C-5: 120.1	

B Indazole, pK_a = 1.25, is less basic than pyrazole but a stronger N–H acid, pK_a = 13.86 [121]. The tautomerism of indazole is a special case inasmuch as the 2H-indazole possesses an o-quinonoid structure:

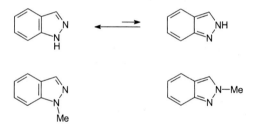

For this reason, the equilibrium lies to the left. However, the energy difference between the tautomers is very small. In the gas phase, for instance, 2H-indazole is higher in energy by only 15 kJ mol^{-1} [133]. This is borne out by the fact that the difference in the chemical shifts in the NMR spectra of 1-methylindazole and 2-methylindazole is not large. Furthermore, the UV spectra are similar in the long wavelength region.

1-Methylindazole reacts with n-butyllithium to give 1-lithiomethylindazole, while 2-methylindazole produces 3-lithio-2-methylindazole.

Alkylation of indazole in the presence of bases proceeds via the ambident indazolyl anion and gives a mixture of 1- and 2-alkylindazoles.

The N-arylation of indazoles is possible, as for pyrazoles, imidazoles and benzimidazoles, with aryllead triacetate in the presence of copper(II)acetate [134].

Halogenation of indazole occurs preferentially in the 5-position. Nitration with fuming nitric acid gives 5-nitroindazole. Sulfonation with oleum, however, yields indazole-7-sulfonic acid. Indazole couples with diazonium salts in the 3-position.

C Most syntheses of indazole start from *o*-substituted anilines, e.g.:

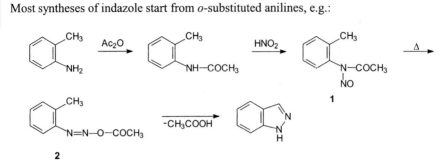

1

2

o-Toluidine is acetylated followed by nitrosation. The *N*-nitroso compound **1** rearranges in benzene at 45-50°C giving the acetoxy compound **2** which cyclizes to indazole [135].

D **Indazole**, colourless crystals, mp 145-149°C, is soluble in hot water.
Indazoles, like pyrazoles, are analgesic, anti-inflammatory and antipyretic, e.g. benzydamine **3**:

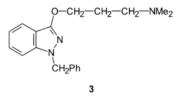

3

5.38 4,5-Dihydropyrazole

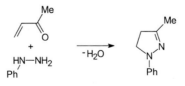

A-D 4,5-Dihydropyrazole has often been referred to as Δ^2-pyrazoline or 2-pyrazoline.
The standard synthesis of 4,5-dihydropyrazole is the cyclocondensation of hydrazine, or alkyl- or arylhydrazines with α,β-unsaturated carbonyl compounds, e.g.:

The oxo compounds derived from 4,5-dihydropyrazole are more important, they are often referred to as pyrazolones. They display tautomerism in the ring as well as in the side chain (ring–chain tautomerism) [136]:

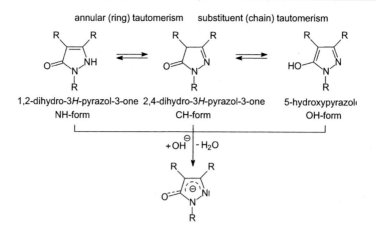

annular (ring) tautomerism   substituent (chain) tautomerism

1,2-dihydro-3*H*-pyrazol-3-one   2,4-dihydro-3*H*-pyrazol-3-one   5-hydroxypyrazole
NH-form   CH-form   OH-form

$+OH^{\ominus}$ | $-H_2O$

The equilibrium position depends on the type of substituent and on the solvent used. In the gas phase and in apolar solvents, the 2,4-dihydro-3*H*-pyrazol-3-one dominates. Substituent tautomerism in heterocycles is observed in oxo compounds as well as in the corresponding thiones and imines.

The systematic nomenclature for the CH-form and the NH-form can be appreciated on the basis of the following diagram:

3*H*-pyrazol-3-one

2,4-Dihydro-3*H*-pyrazol-3-ones are also CH-acids. The pK_a value of the unsubstituted compound is 7.94. They react with bases to give ambient anions which are stabilized by conjugation and can be attacked by electrophiles at C-4, at N-1, or at the O-atom. Thus 2,4-dihydro-3*H*-pyrazol-3-ones undergo aldol condensations and react with nitrous acid to give isonitroso compounds, e.g.:

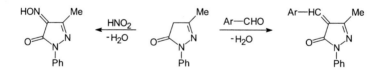

2,4-Dihydro-3*H*-pyrazol-3-ones couple with arenediazonium salts in alkaline solution to give azo compounds with the dihydropyrazole ring in the OH-form, e.g.:

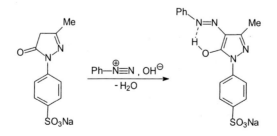

The compound is used as a textile dye known as flavazin L or acid yellow.

2,4-Dihydro-3*H*-pyrazol-3-ones can be alkylated on the C-, N- or O-atom depending on the substituents, reagent and reaction conditions. 5-Methyl-2-phenyl-2,4-dihydro-3*H*-pyrazol-3-one is methylated with iodomethane or dimethyl sulfate to give the 1,5-dimethyl derivative, which was one of the first synthetic drugs (1884, phenazone or antipyrine). It has antipyretic and antirheumatic action:

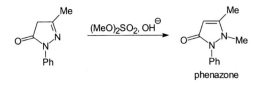

The standard synthesis for 2,4-dihydro-3*H*-pyrazol-3-ones is the cyclocondensation of hydrazine, alkyl- or arylhydrazines with β-ketocarboxylic esters (*Knorr synthesis*, 1883), e.g.:

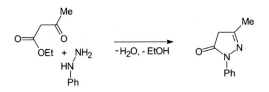

In this reaction, the phenylhydrazone of ethyl acetoacetate was isolated as an intermediate and converted into the product. Further intermediates were detected by means of ^{13}C-NMR spectroscopy [137]. Acetylenecarboxylic esters also react with hydrazines to give pyrazolones.

There are many biologically active compounds among the 4,5-dihydropyrazoles. For instance the (-)-(4*S*)-enantiomer of 4,5-dihydropyrazole **1** is an insecticide [138]:

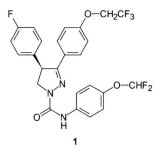

1

Among the pyrazolones used as pharmaceuticals, phenazone has already been mentioned. By its nitrosation, reduction of the nitroso group to an amino group, followed by methylation, the 4-dimethyl-amino-1,5-dimethyl-2-phenyl-1,2-dihydro-3*H*-pyrazol-3-one **2** is produced (STOLZ 1896). Up to the eighties, this compound (amidopyrine or pyramidon) was used worldwide as an analgesic and antipyretic, but has now been withdrawn in several countries. Other examples of pharmaceuticals derived from pyrazolone are the analgesic metamizol **3** [139] and the diuretic and antihypertensive drug muzolimin **4**:

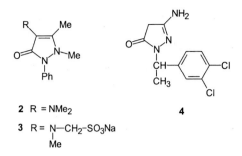

2 R = NMe$_2$

3 R = N—CH$_2$–SO$_3$Na
       |
       Me

4

Flavazin L has already been described as a textile dye derived from pyrazolone. Tartrazine **5** is one of the few synthetic dyes which are approved for colouring foodstuffs and cosmetics. The yellow compound, which is the trisodium salt, is produced by coupling 2,4-dihydro-3*H*-pyrazol-3-ones with diazotized sulfanilic acid. The required pyrazole is accessible from diethyl 2-oxobutanedioate and 4-hydrazinobenzene-1-sulfonic acid.

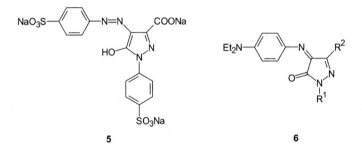

5   **6**

2,4-Dihydro-3*H*-pyrazol-3-ones are also used as couplers in colour photography. During the chromogenic development of the exposed colour film, they react with the developer, e.g. *N,N*-diethyl-*p*-phenylenediamine, to give a purple dye **6**.

1,3-Diaryl-4,5-dihydropyrazoles are used as optical brighteners, e.g. the 1-[4-(aminosulfonyl)-phenyl]-3-(4-chlorophenyl)-4,5-dihydropyrazole [140].

5.39 Pyrazolidine

Several oxo compounds are derived from pyrazolidine, a cyclic *N,N*-disubstituted hydrazine. Pyrazolidin-3-ones, as well as pyrazolidin-4-ones are known [141]. The anti-inflammatory phenylbutazone **1** is a substituted pyrazolidine-3,5-dione. It is made by *C*-alkylation of 1,2-diphenylpyrazolidin-3,5-dione. For the industrial synthesis, cyclocondensation of 1,2-diphenylhydrazine with diethyl 2-butyl-malonate is preferred:

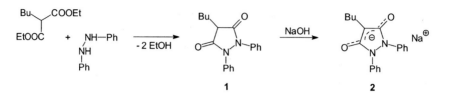

Phenylbutazone, $pK_a = 4.5$, is soluble in sodium hydroxide and forms the salt **2**.

Summary of the general chemistry of five-membered heterocycles with two heteroatoms:

- The parent compounds are aromatic. Increasing aromaticity in the series furan < pyrrole < thiophene is essentially maintained (see Fig. 5.17):

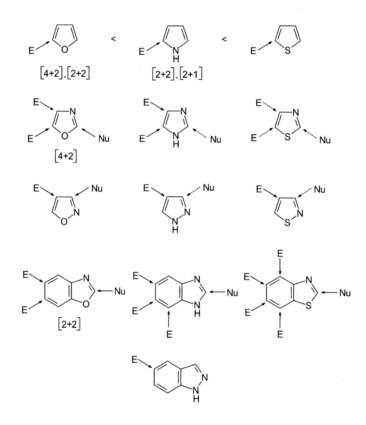

Fig. 5.17   Reactivity and selectivity of the reactions of five-membered heteroarenes with electrophiles E and nucleophiles Nu (1,2-benzoxazoles and 1,2-benzothiazoles are also known but are not discussed in this book).

- Compared to furan, pyrrole and thiophene, the pyridine-like N-atom stabilizes the π-systems. Several indicators support this. The most obvious is the high thermal stability of the compounds. For instance, 1-alkyl and 1-arylimidazoles are some of the most thermally stable organic substances. In the absence of air, they decompose above 600°C.

- The pyridine-like N-atom is also responsible for the following properties and reactions:
    — basicity, is greatest for imidazole, followed by benzimidazole;
    — greater NH-acidity of imidazoles, pyrazoles, benzimidadzoles compared to pyrrole or indole;
    — annular tautomerism in imidazoles, pyrazoles, benzimidazoles and indazoles;
    — quaternization;
    — retardation of electrophilic substitution, the position of substitution is indicated in Fig. 5.17 (exception: azo-coupling of imidazole);
    — nucleophilic substitution, always occurs in the position α to the pyridine-like N-atom.

The statements concerning the reactivity and selectivity of the reactions of azoles with electrophiles and nucleophiles can be further defined on the basis of the donor–acceptor concept [142].

- In benzo-condensed systems, electrophilic substitution occurs on the benzene ring (exception: azo-coupling of indazole).

- Oxazoles, imidazoles and thiazoles are lithiated in the 2-position, and pyrazoles and isothiazoles in the 5-position. With isoxazoles, n-butyllithium causes ring-opening.

- Under normal reaction conditions, only oxazoles and benzoxazoles undergo cycloadditions.

- Methyl groups in the 2-position of the 1,3-diheterocycles are CH-acidic, which is especially marked in the benzo-condensed systems.

- The most important principle of synthesis is cyclocondensation. Oxazoles, benzoxazoles, isoxazoles, 4,5-dihydroisoxazoles, thiazoles, benzothiazoles, isothiazoles, imidazoles, pyrazoles and 4,5-dihydropyrazoles are synthesized by this method.

- Isoxazoles, 2-isoxazoles, 4-isoxazoles and pyrazoles are accessible by 1,3-dipolar cycloaddition.

- Among the parent compounds, oxazoles, isoxazoles, imidazoles and benzothiazoles are important for synthetic transformations. The partially or completely hydrogenated systems, especially 1,3-dioxolanes, 1,3-dithiolanes, 4,5-dihydrooxazoles, oxazol-5(4H)ones, 4,5-dihydroisoxazoles, imidazolidine-2,4-diones and 2,4-dihydro-3H-pyrazol-3-ones are of wider application.

5.40 1,2,3-Oxadiazole

There are eight structurally isomeric oxadiazoles and thiadiazoles:

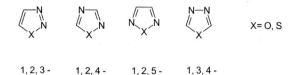

1, 2, 3 -   1, 2, 4 -   1, 2, 5 -   1, 3, 4 -

A few selected systems derived from these parent compounds will be discussed.

1,2,3-Oxadiazoles are not known. Although they are formed in some reactions, they isomerize imme-diately to α-diazoketones.

Sydnones, like münchnones (see p 129), are mesoionic compounds. The synthesis of the first example, namely 3-phenylsydnone **1**, was achieved by EARL and MACKNEY at the University of Sydney. It was carried out by cyclodehydration of *N*-nitroso-*N*-phenylglycine with acetic anhydride:

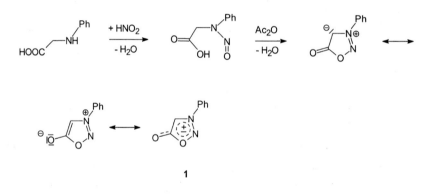

1

Sydnones are crystalline compounds, stable in acid solution and ring-opened in alkaline solution. They react as a 1,3-dipole in cycloadditions, e.g. with acetylenedicarboxylate, to give the pyrazole derivative **2**:

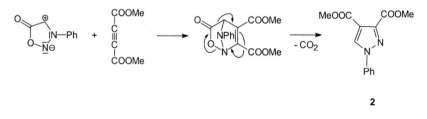

2

Sydnoneimines 3 are obtained from α-aminonitriles by nitrosation followed by cyclization with hy-drogen chloride:

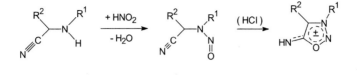

3

5.41 1,2,5-Oxadiazole

| **A** | The trivial name furazan is still in wide usage for 1,2,5-oxadiazole. The 1,2,5-oxadiazole mole-

cule is planar and exists as a regular pentagon. The ionization energy is 11.79 eV and the dipole moment 3.38 D. Both values are greater than those of isoxazole. The chemical shift (δ) in the ^{1}H-NMR spectrum is 8.19, and in the ^{13}C-NMR spectrum 139.4 (both in CDCl$_3$).

1,2,5-Oxadiazole is aromatic. The following π-bond orders were calculated by the HMO method:

A considerable delocalization of the π-electrons is evident from these figures, and again the values for the bonds between the heteroatoms are the lowest. Strictly, 1,2,5-oxadiazole should be regarded as a π-excessive heterocycle with six electrons distributed over five atoms. However, the π-electron density on the heteroatoms is so great that the values for the C-atoms are smaller than one, where π-deficiency prevails, thereby influencing the reactivity [143].

| **B** | 1,2,5-Oxadiazole, with pK$_a$ $\approx$ -5, is less basic than isoxazole (pK$_a$ = -2.97). 1,2,5-Oxadiazoles

do not react with electrophiles, or react slowly. Thus quaternization with dimethyl sulfate in sulfolane proceeds much more slowly than that of isoxazoles with iodomethane. Halogenation and nitration of 3-phenyl-1,2,5-oxadiazole and of 1,2,3-benzoxadiazole occur exclusively on the benzene ring. The 1,2,5-oxadiazole ring shows little reactivity towards oxidizing agents, as can be expected from the high ionization energy. 3,4-Dimethyl-1,2,5-oxadiazole, for instance, is oxidized by potassium permanganate to give 1,2,5-oxadiazole-3,4-dicarboxylic acid.

In spite of the low π-electron density on the C-atoms, 1,2,5-oxadiazoles do not react at all or only slowly with nucleophiles. Nucleophiles which are also strong bases, e.g. sodium hydroxide in methanol, bring about ring-opening to form sodium salts of α-oximinonitriles. The mechanism corresponds to that occurring in the analogous reaction of isoxazoles (see p 139).

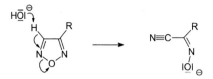

3-Methyl-1,2,5-oxadiazoles are lithiated on the methyl group by *n*-butyllithium yielding the corresponding carboxylic acid with carbon dioxide:

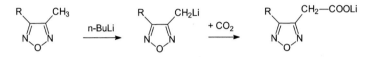

|C| Cyclodehydration of dioximes of 1,2-dicarbonyl compounds has proved useful as a standard synthesis for 1,2,5-oxadiazoles [143]:

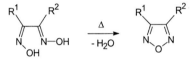

1,2,5-Oxadiazole has been prepared by heating glyoxal dioxime with succinic acid anhydride at 150-170°C. For the synthesis of substituted compounds, mere heating is sufficient in some cases, whereas the action of thionyl chloride in 1,2-dichloroethane gives good results in others.

1,2,5-Oxadiazoles are also accessible by deoxygenation of 1,2,5-oxadiazole-2-oxides with triphenylphosphane.

|D| **1,2,5-Oxadiazole** is a colourless, water soluble liquid, mp -28°C and bp 98°C, stable indefinitely at room temperature.

1,2,5-Oxadiazole-2-oxides (furoxans) are made by oxidation of the dioximes of 1,2-dicarbonyl compounds. Sodium hypochlorite, lead(IV) acetate or dinitrogen tetroxide have all proved useful in this reaction. Electrochemical oxidation is also possible [144].

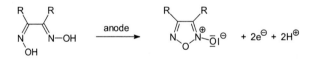

1,2,5-Oxadiazole-2-oxides are also formed by dimerization of nitrile oxides:

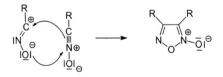

In this [3+2] cycloaddition, nitrile oxide reacts simultaneously as a 1,3-dipole and a dipolarophile (see p 144). This is possible because nitrile oxides possess a high energy HOMO and a low energy LUMO.

1,2,3-Benzoxadiazole-1-oxide is obtained by heating *o*-nitrophenylazide in acetic acid or by oxidation of *o*-nitroaniline with sodium hypochlorite:

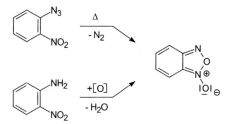

By the second method, the compound can be produced cheaply on a large scale.

1,2,5-Oxadiazole-2-oxides with various substituents isomerize on heating, probably via 1,2-dinitrosoolefins as intermediates [144]:

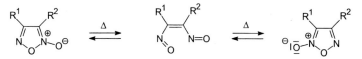

At temperatures above 150°C, a [3+2] cycloreversion takes place to give nitrile oxides.

1,2-Dinitrosobenzene, which was detected spectroscopically, is formed by photoisomerization of 1,2,3-benzoxadiazole-1-oxide [145]:

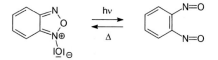

1,2,5-Oxadiazoles, their *N*-oxides, as well as their benzo-fused systems, are biologically active compounds. Some of their derivatives are important because of their anthelmintic, fungicidal, bactericidal and herbicidal action. They have also been found to have antitumour activity.

E | 1,2,5-Oxadiazole-2-oxides are often used for synthetic transformations. Three examples are given below.

(1) *Synthesis of diisocyanates* **1**:

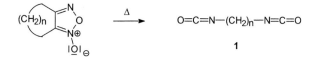

Thermolysis yields a bis-nitrile oxide which, in the absence of dipolarophiles, isomerizes to a diisocyanate.

(2)   *Synthesis of o-quinone dioximes* **2**:

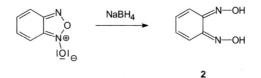

2

(3)   *Synthesis of quinoxaline-1,4-dioxide* **4** (*Beirut reaction*, see p 436):

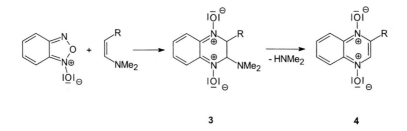

3   **4**

In this interesting reaction, the dihydro compounds **3** are formed from 1,2,3-benzoxadiazole-1-oxide and enamines. β-Elimination leads to the products **4**. Enolates react in an analogous way [146].

5.42 1,2,3-Thiadiazole

A   1,2,3-Thiadiazoles, in contrast to 1,2,3-oxadiazoles, can be isolated and are thermally relatively stable. NMR spectra indicate that the molecules are diatropic.

1,2,3-Thiadiazole is aromatic. Calculation of the π-electron densities by the HMO method gives the following values:

B   This is again a π-excessive heterocycle with some π-deficiency on the C-atoms. Therefore, electrophilic reagents should attack on the heteroatoms. Electrophilic substitution on the C-atoms would be difficult and nucleophiles should prefer the 5-position.

1,2,3-Thiadiazoles are weak bases. On quaternization, e.g. with dimethyl sulfate, mixtures of 2- and 3-methyl-1,2,3-thiadiazoles are formed. Electrophilic substitution of the C-atoms could not be achieved. With 1,2,3-benzothiazoles, substitution occurs on the benzene ring. For instance, with nitric acid the 4- and 7-nitro-1,2,3-benzothiadiazoles are obtained. Nucleophiles bring about ring-opening, e.g.:

Thermolysis and photolysis of 1,2,3-thiazoles and 1,2,3-benzothiazoles lead to elimination of nitrogen. Depending on the substituents in the 4- and 5-positions, the fragments attain stability to different degrees, forming mainly thioketenes and/or thiirenes [147]:

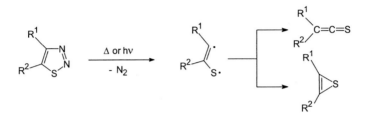

C   1,2,3-Thiadiazoles are prepared by cyclocondensation of tosylhydrazones derived from α-methylene ketones with thionyl chloride or sulfur dichloride (*Hurd-Mori synthesis*) [148]:

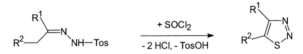

1,2,3-Benzothiadiazoles are obtained by the reaction of 2-aminobenzenethiol with sodium nitrite in acetic acid:

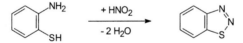

D   **1,2,3-Thiadiazole** is a yellow liquid, bp 157°C, soluble in water.

E   So far 1,2,3-thiadiazoles have found little application in organic synthesis. Flash pyrolysis to give thioketenes and ring opening to produce acetylenes can occur. The latter are preferably prepared by thermolysis of 1,2,3-selenadiazoles:

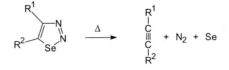

5.43 1,2,4-Thiadiazole

A 1,2,4-Thiadiazole, like its structural isomer 1,2,3-thiadiazole, is aromatic and is to be regarded as a π-excessive heterocycle, with relatively π-deficient C-atoms. The π-electron density calculated by the HMO method is lowest on C-5 (0.7888). Therefore, nucleophiles attack in this position.

B 1,2,4-Thiadiazoles are weak bases. Methylation with iodomethane occurs at N-4, with trimethyloxonium tetrafluoroborate onto both N-atoms:

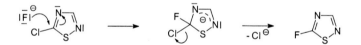

Electrophilic substitution at the C-atoms could not be achieved.

1,2,4-Thiadiazole reacts rapidly with alkali hydroxides, undergoing ring-opening. With hydrochloric acid, hydrolytic ring-opening occurs via the 1,2,4-thiadiazolium ion. Nucleophiles attack at the 5- and/or 3-position. 3,5-Diphenyl-1,2,4-thiadiazole is much less sensitive to alkali hydroxides and mineral acids.

5-Chloro-1,2,4-thiadiazoles are reactive compounds and undergo numerous nucleophilic substitutions, e.g. with silver fluoride:

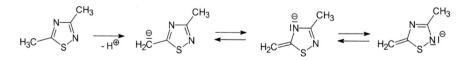

In the case of 3-chloro-1,2,4-thiadiazoles, delocalization of the negative charge in the intermediate with involvement of both N-atoms is not possible. For this reason they do not react, or react only slowly with nucleophiles.

3,5-Dimethyl-1,2,4-thiadiazole is selectively metalated by *n*-butyllithium on the 5-methyl group, which indicates that its CH-acidity is greater than that of the 3-methyl group. With regard to the methyl group, the 1,2,4-thiadiazol-5-yl substituent possesses distinct acceptor properties because of the two pyridine-like N-atoms which stabilize the corresponding carbanion:

Again, an extensive delocalization of the negative charge in the deprotonation of the 3-methyl group is not possible.

5-Amino-1,2,4-thiadiazoles can be diazotized with sodium nitrite in acetic acid because of the strong electron attraction of the 1,2,4-thiadiazol-5-yl moiety and the resulting diazonium ions are highly electrophilic. They can even couple with the hydrocarbon mesitylene:

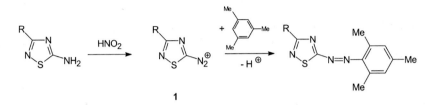

1

C 1,2,4-Thiadiazoles **1**, with identical substituents in the 3- and in the 5-positions, are obtained from thioamides by oxidation with hydrogen peroxide or by the action of $SOCl_2$, SO_2Cl_2 or PCl_5. The mechanism has not been completely elucidated:

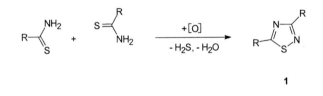

1

In a second type of synthesis, amidines are the starting materials. They are, for instance, thioacylated by means of thiocarboxylic esters and subsequentely oxidatively cyclized to give the thiadiazole system **2**:

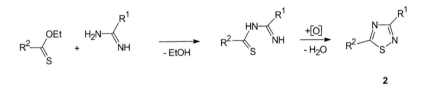

2

5-Amino-1,2,4-thiadiazoles **3** are accessible from amidines and potassium thiocyanate by oxidation with sodium hypochlorite:

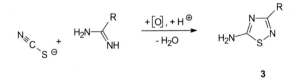

3

The cyclocondensation of amidines with trichloromethylsulfenyl chloride furnishes 5-chloro-1,2,4-thiadiazoles **4**:

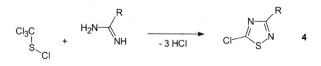

D 1,2,4-**Thiadiazole** is a colourless liquid, mp -34°C, bp 121°C, soluble in water.
Although the 1,2,4-thiazole moiety does not occur in natural products, many synthetic derivatives are biologically active and are used as insecticides, fungicides, herbicides and antibacterials. For
instance, 5-ethoxy-3-trichloromethyl-1,2,4-thiadiazole (etridiazole) is used to combat or prevent fungal
infestation of plants, fruit, cotton, seeds and soil [149].

Azo dyes derived from diazotized 5-amino-1,2,4-thiadiazoles are used as dyestuffs for polyester and
polyacrylonitrile fibres.

5.44 1,2,3-Triazole

A This five-membered heterocycle contains one pyrrole-like and two pyridine-like N-atoms in the
1,2,3-positions. It was known as v-triazole (v meaning vicinal). Since all ring atoms are sp²-
hybridized, its six available electrons are in delocalized π-MOs. Thus, 1,2,3-triazole is aromatic. Its
ionization energy, measured by photoelectron spectroscopy, amounts to 10.06 eV, which is greater than
that of imidazole (8.78 eV) and pyrazole (9.15 eV), i.e. the HOMO of 1,2,3-triazole is lower [150]. The
dipole moment in benzene is 1.82 D and is similar to that of pyrazole.
1,2,3-Triazole **1** and 1-methyl-1,2,3-triazole **2** have the following UV and NMR data:

UV (ethanol) λ (nm) (ε)	¹H-NMR (DMSO) δ (ppm)	¹³C-NMR (DMSO) δ (ppm)
1: 210 (3.64) $\pi \rightarrow \pi^*$	H-2: 13.50	
	H-4: 7.91	C-4: 130.3
	H-5: 7.91	C-5: 130.3
2: 213 (3.64)	H-4: 7.72	C-4: 134.3
	H-5: 8.08	C-5: 125.5

The UV spectra are similar to those of pyrazole and pyrrole. The δ-values for the protons and C-atoms
in the 4- and 5-positions of **1** are in each case identical, because the compound exists as a 2*H*-tautomer
in solution (see p 201) and equilibrium is established so rapidly that at room temperature, only one

signal appears in the NMR. However, if, as in **2**, tautomerism is blocked by a substituent in position 1, then the values differ. In 2-methyl-1,2,3-triazole, they are again identical (^{1}H: $\delta = 7.77$, ^{13}C: $\delta = 133.2$; in DMSO).

The π-electron densities calculated by various MO methods again show the highest values for the heteroatoms. However, the π-deficiency on the C-atoms is not as pronounced as in the case of the oxadiazoles. The situation is more in keeping with that found in the pyrazole molecule.

 The following reactions are typical for 1,2,3-triazoles.

Acid–base reactions

1,2,3-Triazoles are weak bases. 1,2,3-Triazole, with $pK_a = 1.17$, is less basic than pyrazole.

Triazoles unsubstituted on the N-atom are NH-acidic. The acidity of 1,2,3-triazole with a pK_a value of 9.3 is greater than that of pyrazole and is similar to that of HCN. This is mainly due to the continued delocalization of the negative charge in the conjugate base:

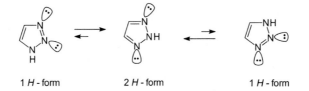

The silver salt of 1,2,3-triazole is water insoluble.

Annular tautomerism

Unsubstituted 1,2,3-triazole has three tautomeric forms, only two of which are identical. This is in contrast to imidazole and pyrazole.

1 H - form   2 H - form   1 H - form

The 2*H*-form predominates in most solvents. The equilibrium constant in water is K_T = [2*H*-form] / [1*H*-form] $\approx$ 2 [151]. A destabilization of the 1*H*-form, owing to repulsive forces between the non-bonding electron pairs in the 2- and 3-positions, is considered to be responsible. Therefore, for a C-monosubstituted 1,2,3-triazole, three structural isomers are conceivable, i.e. 4-methyl-1,2,3-triazole, 4-methyl-2*H*-1,2,3-triazole und 5-methyl-1,2,3-triazole.

Metalation

N-Substituted 1,2,3-triazoles are metalated by *n*-butyllithium at low temperature [152], e.g.:

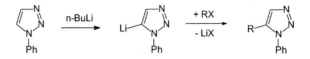

Reactions with electrophilic reagents

The action of dimethyl sulfate on sodium 1,2,3-triazole in dichloromethane produces an 88% yield of a mixture of 1-methyl- and 2-methyl-1,2,3-triazole in a proportion of 1.9 : 1 [116]. Only the 1-methyl compound is quaternized by iodomethane:

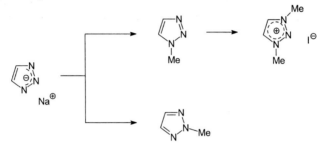

Diazomethane reacts with 1,2,3-triazole giving 2-methyl-1,2,3-triazole, while chlorotrimethylsilane reacts to give 2-trimethylsilyl-1,2,3-triazole.

Acetylation and tosylation of 1,2,3-triazole with the corresponding acid halides usually lead to mixtures of 1- and 2-acetyl or tosyl compounds.

Among the electrophilic substitution reactions at carbon, as with pyrrole and imidazole, halogenation is the fastest. Bromine reacts with 1,2,3-triazole to give 4,5-dibromo-1,2,3-triazole with an NH-acidity of pK_a = 5.37. This is greater than that of 1,2,3-triazole. The high reactivity of 1,2,3-triazole towards halogenation is apparently due to the pyrrole-like N-atom, because 2-methyl-1,2,3-triazole reacts much more slowly. 2-Phenyl-1,2,3-triazole undergoes nitration first on the benzene ring and then on the 1,2,3-triazole ring:

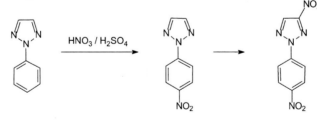

Dimroth rearrangement

With nucleophiles, 1,2,3-triazoles, like pyrazoles, do not react at all or react only slowly with ring-opening. When 1,2,3-triazoles are heated in suitable solvents, ring fission often occurs to give intermediates which recyclize to yield a structural isomer of the starting material. Such isomerizations, which have also been observed with other heterocycles possessing several N-atoms, are known as DIMROTH rearrangements. The rearrangement of 5-amino-1-phenyl-1,2,3-triazole **3** to give, in boiling pyridine, 5-phenylamino-1,2,3-triazole **6**, is an example:

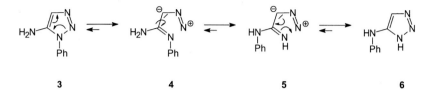

| **3** | **4** | **5** | **6** |

Ring opening occurs by cleavage of the N–N bond in **3** to the diazoimine **4**. This rearranges to another diazoimine **5**, which cyclizes to the product **6**. Many DIMROTH rearrangements are reversible. In the chosen example, the rearrangement occurs in the direction shown (**3 → 6**) because the NH-acidity of the product is greater than that of the starting material and salt formation occurs with the basic solvent.

Dediazoniation

1,2,3-Triazoles are subject to ring cleavage by pyrolysis or photolysis with loss of nitrogen. The fragments arising from 1-substituted 1,2,3-triazoles have a diradical or an iminocarbene structure **7**. They cyclize to give 1*H*-azirines **8** which isomerize in the gas phase or in inert solvents to give mainly 2*H*-azirines **9**:

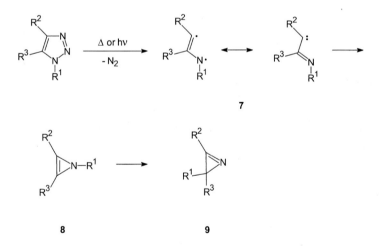

Pyrolysis or photolysis of 1-unsubstituted 1,2,3-triazoles yields nitriles:

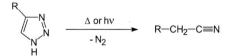

The thermal dediazoniation of 4,5-dihydro-1,2,3-triazoles occurs faster and is a method for synthesizing aziridines (see p 31).

| C | The following methods are particularly applicable to the synthesis of 1,2,3-triazoles. |

(1) Hydrazoic acid or azides react with alkynes in a 1,3-dipolar cycloaddition:

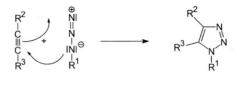

10

If terminal alkynes are used, 1,4-substituted 1,2,3-triazoles are formed with high regioselectivity (**10**, R^3 = H). For instance, trimethylsilylacetylene reacts quantitatively with aryl azides to give 1-aryl-4-trimethylsilyl-1,2,3-triazoles [153].

The 1,3-dipolar addition of azides to alkenes proceeds more slowly and yields 4,5-dihydro-1,2,3-triazoles (see p 31).

(2) The cycloaddition of azides to CH-acidic compounds in the presence of sodium methoxide leads to 5-amino-1,2,3-triazoles **11**:

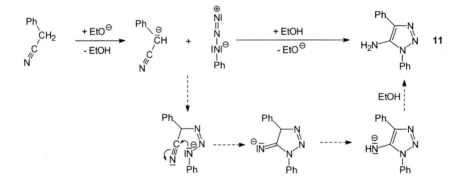

In contrast to a 1,3-dipolar cycloaddition, a multistep reaction occurs which ensures 100% regioselectivity.

(3) The oxidation of the bishydrazones of 1,2-dicarbonyl compounds yields 1-amino-1,2,3-triazoles **12**:

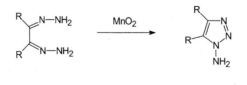

12

However, 1,2-bisphenylhydrazones (osazones) on heating or on oxidation yield 2-phenyl-1,2,3-triazoles **13** (osotriazoles):

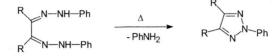

| **D** | **1,2,3-Triazole** forms colourless, sweet-tasting, hygroscopic crystals, mp 24°C, bp 209°C, which are soluble in water. |

Natural products derived from 1,2,3-triazole are not known. The 1,2,3-triazole structure is contained in a number of pharmaceuticals (see p 473). It is used, for example, to modify the pharmacokinetic properties of the clinically employed β-lactam antibiotic cefatrizin **14**:

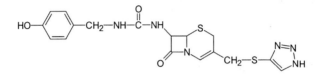

14

The synthesis of 2-aryl-1,2,3-triazoles, e.g. **15**, which are in use as optical brighteners, is the subject of numerous patents [140]:

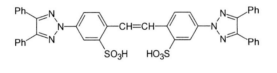

15

Among the synthetic applications of 1,2,3-triazoles and 4,5-dihydro-1,2,3-triazoles, the dediazoniation to form 2*H*-azirines or aziridines is of importance.

5.45 Benzotriazole

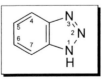

| **A** | The UV spectra of simple benzotriazoles show the following peaks (nm, log ε): |

Benzotriazole   259 (3.75), 275 (3.71)
1-Methylbenzotriazole   255 (3.81), 283 (3.68)
2-Methylbenzotriazole   274 (3.96), 280 (3.98), 285 (3.97)

Benzotriazole is an extremely weak base, but with a pK_a = 8.2, it is a stronger NH-acid than indazole, benzimidazole or 1,2,3-triazole. The fused benzene ring imparts additional stabilization to the conjugate base:

Benzotriazole forms complexes with numerous metals, acting frequently as a bridging ligand.

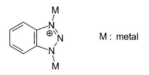

B   By analogy to 1,2,3-triazole, benzotriazole also has three tautomers, namely two 1*H*-forms and one 2H-form (see p 201). In solution, the equilibria lie almost entirely on the side of the 1*H*-forms [154]. Although the 2*H*-tautomer possesses an orthoquinonoidal structure, NMR spectra of 2-methylbenzotriazole confirm that the π-electrons are largely delocalized, as in the case of 2-methyl-indazole (see p 185).

Alkylation of benzotriazoles yields mixtures of 1- and 2-alkylbenzotriazoles, the proportion of which depends on the alkylating agent, as for 1,2,3-triazole. However, acylation and sulfonation occur at N-1. 1-Phenylcarbamoylbenzotriazole is formed with phenylisocyanate:

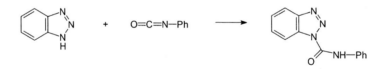

For electrophilic substitution reactions, only the benzene ring carbon atoms are available. Thus chlorination with a mixture of concentrated hydrochloric and nitric acid yields 4,5,6,7-tetrachlorobenzo-triazole, whereas nitration gives mainly 4-nitrobenzotriazole. Action of KMnO$_4$ oxidatively opens the benzene ring and produces 1,2,3-triazole-4,5-dicarboxylic acid.

Dediazoniation of 1-phenylbenzotriazole leads via a diradical intermediate almost quantitatively to the formation of carbazole (GRAEBE-ULLMANN reaction):

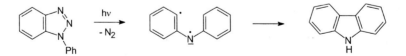

C   The standard synthesis of benzotriazoles is the cyclocondensation of o-phenylenediamines with sodium nitrite in acetic acid:

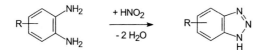

D   **Benzotriazole**, colourless crystals of mp 99°C, cannot be distilled at normal pressure.
Benzotriazoles do not occur in nature but nevertheless have found several uses. For instance, the dopamine antagonist alizaprid **1** is an antiemetic:

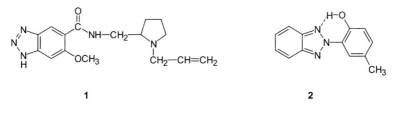

1   2

Benzotriazole is a useful corrosion inhibitor, especially for copper. It renders the metal passive by forming a layer of insoluble copper(I) benzotriazole. Benzotriazole stabilizes photographic emulsions and delays fogging.

Compounds such as 2-(2-hydroxy-5-methylphenyl)benzotriazole **2** (tinuvin P) absorb UV light in the region of 300 - 400 nm and are used as radiation protection agents, e.g. for the prevention of sunburn, and also as photostabilizers for plastics, rubber and chemical fibres [155]. Tests have been carried out as to the suitability of polynitro-1-phenylbenzotriazoles as explosives [156].

E   Most synthetic applications of benzotriazoles are based on the acceptor action of the benzotria-
zol-1-yl moiety i.e. on the stability of the benzotriazolyl anion. For instance 1-chlorobenzotriazole is used as a chlorinating agent. It transfers a Cl$^+$ ion to the substrate. An example of this reaction is the chlorination of carbazole in dichloromethane to give 3-chlorocarbazole.

1-Hydroxybenzotriazole is obtained by condensation of 1-chloro-2-nitrobenzene with hydrazine:

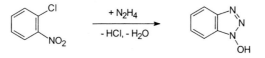

Use of this compound as a coreagent in the DCC method of peptide coupling reduces racemization and prevents the formation of N-acylureas [157].

The reactive intermediate 1,2-dehydrobenzene (benzyne) is generated by oxidation of 1-aminobenzotriazole with lead(IV) acetate. It can be trapped with anthracene to form triptycene:

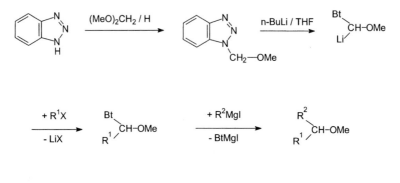

The preparation of a variety of organic compounds can be carried out with benzotriazole as starting material [158]. For instance, the synthesis of methyl ethers is achieved as follows:

Bt : benzotriazol-1-yl

By a similar method, syntheses of diarylmethanes, dieneamines, *N,N*-disubstituted thioureas and 4-substituted pyrylium salts (see p 225) were achieved by the use of benzotriazole (benzotriazole-mediated methodology, KATRITZKY 1990).

5.46 1,2,4-Triazole

A This system was known as s-triazole (s for symmetric). It is aromatic, like its structural isomer 1,2,3-triazole. Its ionization potential is 10.00 eV. Therefore, its HOMO is as low as that of 1,2,3-triazole. The dipole moment in the gas phase is 2.72 D and in dioxane 3.27 D. The absorption maximum in the UV (measured in dioxane) is 205 nm (log ε = 0.2). The ^{1}H-NMR spectrum of 1,2,4-triazole (in HMPT) shows a CH-signal at δ = 8.17 and an NH-signal at δ = 15.1. Only one signal at δ 147.4 is observed in the ^{13}C-NMR spectrum (methanol-d$_4$) because of time-averaging due to tautomerism. At -34°C, however, the CH-signal is split into two peaks, namely δ = 7.92 for 3-H and 8.85 for 5-H.

Every C-atom in 1,2,4-triazole is linked to two N-atoms. As a consequence, the C-atoms are π-deficient. MO calculations yield values up to 0.744 for the π-electron density in positions 3 and 5. The electron density on the N-atom is correspondingly high.

| **B** | 1,2,4-Triazoles are noted for the following reactions. |

Acid–base reactions

1,2,4-Triazole is a weak base, $pK_a = 2.19$. Protonation occurs at N-atom 4. 1,2,4-Triazoles unsubstituted on nitrogen are NH-acidic and their copper and silver salts are sparingly soluble. The parent compound has a pK_a of 10.26. Hence 1,2,4- and 1,2,3-triazole differ very little in basicity and acidity.

Annular tautomerism

Three tautomers are possible with unsubstituted 1,2,4-triazole, namely two 1*H*-forms and one 4*H*-form:

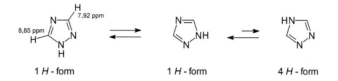

Microwave spectra show that only the 1*H*-forms exist in the gas phase. They are also dominant in solution, as can be seen from the ¹H-NMR spectra at low temperature. 4-Methyl-1,2,4-triazole shows only one CH signal at $\delta = 8.34$, apart from the CH$_3$ signal at $\delta = 3.64$, even at room temperature.

Reactions with electrophilic reagents

The N-atoms of 1,2,4-triazole and of the 1,2,4-triazolyl anion are preferentially attacked by electrophiles. The reaction of dimethyl sulfate with sodium 1,2,4-triazolide in acetonitrile gives a 100% yield of a mixture of 1-methyl- and 4-methyl-1,2,4-triazole in a proportion of 8.8 : 1. Similarly, benzylation, methoxycarbonylation and trimethylsilylation can be carried out, again leading mainly to 1-substituted compounds [159]. Quaternization of 1-methyl-1,2,4-triazole with trimethyloxonium tetrafluoroborate proceeds stepwise:

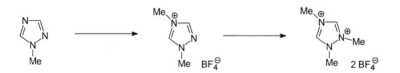

Acylation occurs preferentially at the 1-position.

Electrophilic substitution at carbon, e.g. nitration and sulfonation, proceed very slowly. The action of chlorine or bromine leads to (3)5-chloro- or 3(5)-bromo-1,2,4-triazole by way of 1-halo intermediates:

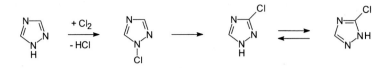

3(5)-Halo-1,2,4-triazoles can be nucleophilically substituted. The reactivity is increased by salt forma-tion with acids or by quaternization.

Dediazoniation can only be achieved under extreme conditions because a convenient N=N grouping is not available for a thermolytic nitrogen extrusion.

C Most syntheses of 1,2,4-triazoles start from hydrazine or substituted hydrazines [160].

(1) Hydrazines condense with diacylamines to give 1,2,4-triazoles **2** (*Einhorn-Brunner synthesis*), e.g.:

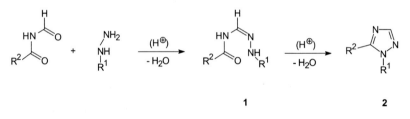

With an *N*-formyl acid amide, the reaction proceeds regioselectively by way of the acylamidrazone of the formic acid **1** to give a 1,5-disubstituted 1,2,4-triazole.

(2) Acid hydrazides cyclize with acid amides or thioamides to form 1,2,4-triazoles of type **3** (*Pelliz-zari synthesis*):

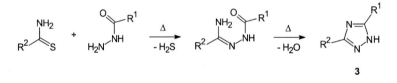

The reaction requires higher temperatures and also proceeds via acylamidrazones.

(3) 1,2-Diacylhydrazines undergo cyclocondensation with ammonia:

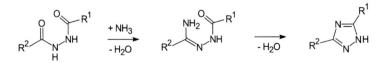

Therefore, acylamidrazones are intermediates in all three reactions. For this reason, 1,2,4-triazoles can also be prepared from amidrazones and acylating agents such as acid esters or acid chlorides.

D | **1,2,4-Triazole** forms colourless crystals, mp 121°C, bp 260°C. It is soluble in water.
3-Amino-1,2,4-triazole, colourless crystals of mp 152°C, is made by heating aminoguanidinium formate to 120°C.

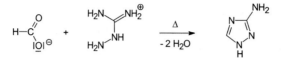

3-Amino-1,2,4-triazole can be diazotized, but when treated with sodium nitrite and hydrochloric acid, 3-chloro-1,2,4-triazole is formed due to dediazoniation:

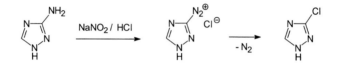

3-Amino-1,2,4-triazole reacts with isoamyl nitrite and *N*,*N*-dialkylanilines in methanol/water under a CO_2 pressure of 53.5 bar to give a quantitative yield of azo dyes [161]. Presumably the reaction proceeds via a diazonium hydrogen carbonate:

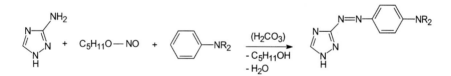

1,2,4-Triazoles have not yet been found in nature.

Semisynthetic β-lactam antibiotics with a 1,2,4-triazolyl substituent have been prepared and are in clinical use. However, the most important application of 1,2,4-triazoles is as biocides. 3-Amino-1,2,4-triazole (amitrole, amizol) is an unselective herbicide. Triadimenol **4** is one of the most effective fungicides. Its configuration (1*S*, 2*R*) is essential for its action [162]. The 1,2,4-triazole fungicides inhibit the biosynthesis of ergosterol in fungi.

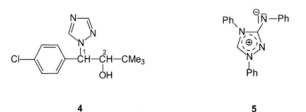

4   5

E 1,2,4-Triazoles are important as reagents. 1,2,4-Triazole is useful as a catalyst for transacylation reactions, e.g. for the synthesis of peptides from *N*-protected amino-4-nitrophenyl esters and amino acids without racemization. Nitron **5**, a mesoionic 1,2,4-triazole, forms an almost insoluble nitrate and is used for the detection and gravimetric determination of nitrate ions. Nitron forms a covalent compound with poly(4-chloromethylstyrene) which removes nitrate from drinking water.

Five-membered heterocycles with four heteroatoms are also known. Apart from tetrazole, there are four structurally isomeric oxatriazoles and thiatriazoles:

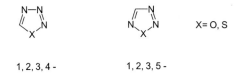

1, 2, 3, 4 -   1, 2, 3, 5 -

Only tetrazole will be discussed here [163].

5.46 Tetrazole

A Tetrazole contains three pyridine- and one pyrrole-like N-atom. It is aromatic because it contains six delocalized π-electrons. Tetrazole has the highest ionization potential of all azoles, i.e. 11.3 eV. The tetrazole ring resists even strong oxidizing agents because of its low HOMO. The dipole moment in the gas phase is 2.19 D and in dioxane 5.15 D. From this great difference, it can be concluded that the 2*H*-form predominates in the gas phase and the 1*H*-form in solution. Its associated nature in solution could contribute to this difference. The tetrazole ring gives rise only to a weak absorption in the region of 200-220 nm in the UV spectrum. In the ^{1}H-NMR spectrum of tetrazole (in D$_2$O), the 5-H signal appears at $\delta = 9.5$; the ^{13}C signal (in DMF) occurs at $\delta = 143.9$. Compared with the other five-membered heterocycles, the π-deficiency at the C-atom in tetrazole should be most marked.

B The following reactions are characteristic of tetrazoles.

Acid–base reactions

Tetrazole is a very weak base, pK_a = -3.0. Protonation occurs in the 4-position.

Of all azoles, tetrazole has the strongest NH-acidity, pK_a = 4.89. Therefore, the acidity of tetrazole is comparable to that of acetic acid (pK_a = 4.76). 5-Substituted tetrazoles can be regarded as nitrogen analogues of carboxylic acids:

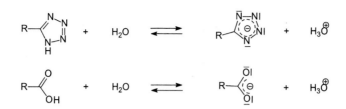

Since -CN$_4$H and -CO$_2$H have a similar spatial demand, bioisosterism is observed. Tetrazoles form poorly soluble, explosive copper and silver salts.

Annular tautomerism

The NMR spectra and the dipole moments confirm that the 1*H*-form predominates over the 2*H*-form in solution. In contrast to 1,2,3-triazole, only two structurally isomeric dimethyltetrazoles **1** and **2** exist:

Ring–chain tautomerism

1,5-Disubstituted tetrazoles can isomerize to give azidoimines [164]:

The equilibrium position depends on the nature of the substituents R^1 and R^2. Ring–chain tautomerism is frequently observed in bi- and polycyclic systems, e.g. imidazole[2,3-*e*]tetrazole **3**. An equilibrium with 2-azidoimidazole **4** is established by the action of bases.

Metalation

Although 1-methyltetrazole is lithiated by *n*-buyllithium in THF at -60°C in the 5-position, the product suffers dediazoniation above -50°C to give the lithium salt of methyl cyanamide:

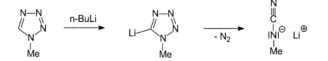

5-Alkyltetrazoles substituted in the 1-position undergo metallation at the alkyl substituent as a result of the acceptor action of the tetrazolyl moiety. The products are stable at room temperature and can be further modified by reaction with electrophiles, e.g.:

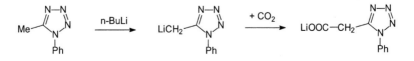

Reactions with electrophilic reagents

Sodium tetrazolide reacts with iodomethane in methanol to yield a mixture of 1- and 2-methyltetrazole in the proportion 1.9 : 1. Both 1- and 2-substituted tetrazoles can be quaternized, e.g.:

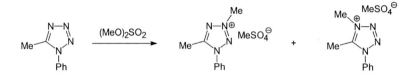

Acylation with acyl halides occurs in the 2-position, but the products are not stable.

   Among electrophilic substitution reactions on the tetrazole C-atom, only bromination and acetoxy-mercuration of 1- and 2-phenyltetrazole are known. Nitrating acid causes substitution in the benzene ring.

Reactions with nucleophilic reagents

5-Halotetrazoles react with nucleophiles leading to substitution, e.g. with phenols and K_2CO_3 in acetone:

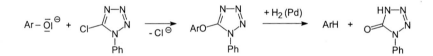

On catalytic hydrogenation, the resulting phenol ethers produce arenes. A method for the deoxygenation of phenols is based on these reactions.

   Substitution of the halogen in 5-halotetrazoles occurs with nitrogen nucleophiles.

Dediazoniation

By analogy with 1,2,3-tetrazoles, dediazoniation of 2,5-disubstituted tetrazoles can be accomplished thermally or photochemically. However, nitrilimines are produced at the same time [165]:

$$R^2 \overset{N=N}{\underset{N-R^1}{\diagdown}} \xrightarrow[-N_2]{\Delta \text{ or } h\nu} R^2-C\equiv\overset{\oplus}{N}-\overset{\ominus}{N}-R^1$$

At room temperature, crystalline tetrazoles are liable to decompose explosively with dediazoniation.

   2-Acyltetrazoles in solution are subject to dediazoniation at room temperature. However, the acyl nitrilimines cyclize immediately to give 1,3,4-oxadiazoles (HUISGEN reaction):

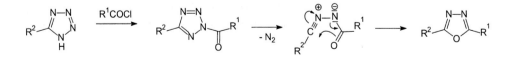

C The following reactions are important for the synthesis of tetrazoles:

(1) Azide ions (e.g. sodium azide in DMF) react with nitriles in a [3+2] cycloaddition with formation of 5-substituted tetrazoles **5** [166]:

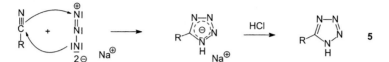

Cycloaddition of alkyl-, aryl- or trimethylsilyl azide with nitriles or isonitriles yields 1,5- and/or 2,5-disubstituted tetrazoles.

(2) Imidoyl halides react with sodium azides with formation of 1,5-disubstituted tetrazoles **7**:

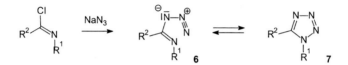

The imidoyl azides **6** are formed first and cyclize in the course of a ring–chain tautomerism to give 1,5-disubstituted tetrazoles.

(3) Tetrazolium salts **9** are produced by the oxidative cyclization of formazans **8**:

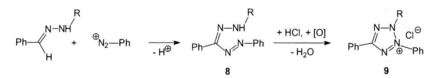

Formazans are accessible by interaction of hydrazones with arenediazonium salts. When R = Ph, the colourless 2,3,5-triphenyltetrazolium chloride is formed. It is reconverted into the red formazan by reducing agents, e.g. by reducing enzymes. By this method, the part of a cell in which biological reductions occur can be stained.

If R is a good leaving group, e.g. a tosyl residue, then the formazan cyclizes to give a 2,5-diphenyltetrazole.

D **Tetrazole**, colourless crystals, mp 156°C, is soluble in water. The existence of intermolecular hydrogen bonds has been demonstrated spectroscopically.

5-Aminotetrazole, colourless crystals, mp 203°C, is prepared either by cyclocondensation of aminoguanidine with nitrous acid or by cycloaddition of hydrogen azide (hydrazoic acid) with cyanamide:

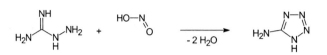

5-Aminotetrazole is diazotized with sodium nitrite and hydrochloric acid. The highly explosive diazonium salts can only be handled in solution. Hypophosphorous acid brings about reduction to tetrazole. Warming the solution leads to dediazoniation forming monoatomic carbon:

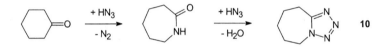

Many tetrazoles are biologically active. 1,5-Pentamethylenetetrazole **10** (pentetrazol, Cardiazol) is the most widely known of such compounds. It is a CNS-active cardiac and respiratory stimulant and narcotic antagonist and is injected in cases of barbiturate poisoning. It is prepared via hexano-6-lactam by addition of sodium azide and sulfuric acid to cyclohexanone (SCHMIDT reaction):

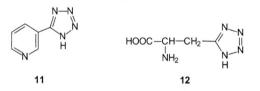

Some synthetically prepared β-lactam antibiotics contain a 1-methyltetrazol-5-yl substituent.

For every biologically active carboxylic acid, there is a bioisosteric compound in which the carboxyl group is replaced by a tetrazol-5-yl moiety. The tetrazole compounds are degraded biologically more slowly than carboxylic acids. For instance, the tetrazole analogue **11** of nicotinic acid reduces the fatty acid and cholesterol levels in blood.

11   12

Moreover, peptides have been produced which contain a tetrazole analogue **12** of aspartic acid.

5-Sulfanyltetrazoles stabilize photographic emulsions and retard fogging. Tetrazoles are also used as additives in special explosives and rocket fuels.

| E | Synthetic applications of tetrazoles are deoxygenation of phenols, preparation of nitrilimines and preparation of 1,3,4-oxadiazoles by the HUISGEN reaction. All these have been previously described. |

Summary of the general chemistry of five-membered heterocycles with three and four hetero-atoms:

- The parent compounds are aromatic. Pyridine-like N-atoms stabilize the π-systems, which is most pronounced in tetrazole.

- The basicity of the pyridine-like N-atoms is low. The weakest bases are 1,2,5-oxadiazole, benzotriazole and tetrazole.

- The NH-acidity increases with the number of pyridine-like N-atoms, because the anions (conjugate bases) become more stable. The pK_a values of tetrazoles thus correspond approximately to those of carboxylic acids.

- Systems with pyridine- and pyrrole-like N-atoms display tautomerism. This applies to 1,2,3-triazole, benzotriazole, 1,2,4-triazole and tetrazole.

- Metalation of the ring carbon is possible with 1,2,3-triazole and tetrazole. Metalation of alkyl substituents in the α-position of the ring succeeds with 1,2,5-oxadiazole, 1,2,4-thiadiazole, benzotriazole and tetrazole.

- Systems with a pyrrole-like N-atom can be alkylated, acylated, sulfonated and silylated. Subsequent conversion into quaternary compounds is possible.

- Electrophilic substitution on the C-atoms of the ring proceeds slowly or not at all.

- Reactivity towards nucleophilic reagents is not very high. 1,2,5-Oxadiazoles, and 1,2,3- and 1,2,4-thiadiazoles are subject to ring cleavage. Halogenated compounds such as 5-chloro-1,2,4-thiadiazole, 3(5)-chloro-1,2,4-triazole and 5-chlorotetrazole undergo nucleophilic substitution reactions.

- Systems possessing the structural fragment –N=N– are subject to thermal and photochemical dediazoniation.

- 5-Amino-1,2,4-thiadiazole, 3-amino-1,2,4-triazole und 5-aminotetrazole can be diazotized.

- Methods for the synthesis of these systems include the following: cyclocondensation (1,2,3- and 1,2,4-thiadiazoles, benzotriazoles, 1,2,4-triazoles), cycloaddition (1,2,3-triazoles, tetrazoles), cyclodehydration (1,2,5-oxadiazoles) and cyclization (1,2,5-oxadiazole-2-oxides, 1,2,4-thiadiazoles, 1,2,3-triazoles, tetrazoles).

- 1,2,5-Oxadiazole-2-oxides, 1,2,3-triazoles, benzotriazoles, 1,2,4-triazoles and tetrazoles all particularly find application in organic syntheses.

At this point, the question must be raised as to whether the C-atom in tetrazole can be replaced by a pyridine-like N-atom, i.e. whether pentazoles are capable of existing or not. So far, only arylpentazoles have been isolated from arenediazonium salts and sodium azide solutions at temperatures below -30°C.

4-Dimethylaminophenylpentazole has proved to be the most stable derivative. However, even this compound is subject to dediazoniation above 50°C. The decomposition can occur explosively [167].

References

[1]   P. v. Rague Schleyer: Aromaticity. *Chem. Rev.*
      2001, *101*, 1115.
[2]   I. G. John, L. Random, *J. Am. Chem. Soc.* **1978**,
      100, 3981.
[3]   R. S. Hosmane, J. F. Liebman, *Tetrahedron Lett.*
      1992, *33*, 2303.
[4]   C. W. Bird, *Tetrahedron* **1997**, *53*, 13111, 17195;
      B. S. Jursic, *J. Heterocycl. Chem.* **1997**, *34*, 1387.
[4a] C. O. Kappe, S. S. Murphree, A. Padwa, *Tetrahe-
      dron* **1997**, *53*, 14179.
[5]   E. J. Corey, X.-M. Cheng, *The Logic of Chemical
      Synthesis*, Wiley, New York **1989**;
      S. Warren, *Organic Synthesis: The Disconnection
      Approach*, 5th ed., Wiley, New York **1991**.
[6]   V. Amarnath, K. Amarnath, *J. Org. Chem.* **1995**,
      60, 301.
[7]   Xue Long Hou et al, *Tetrahedron* **1998**, *54*, 1955;
      G. Hajos, Z. Riedl, G. Kollenz, *Eur. J. Org.
      Chem.* **2001**, 3495..
[8]   M. E. Maier, *Nachr. Chem. Tech. Lab.* **1993**, *41*,
      696;
      G. Piancatelli, M. D'Auria, F. D'Onofrio, *Synthe-
      sis* **1994**, 867.
[9]   L. F. Tietze, Th. Eicher, *Reactions and Synthesis
      in the Organic Chemistry Laboratory*, p 382, Uni-
      versity Science Books, Mill Valley, CA **1989**.
[10] M. Sauter, W. Adam, *Acc. Chem. Res.* **1995**, *28*,
      289.
[11] Z. Chen, X. Wang, W. Lu, J. Yu, *Synlett* **1991**,
      121.
[11a] Y. Nan, H. Miao, Z. Yang, *Org. Lett.* **2000**, *2*,
      297.
[12] R. Rodrigo, *Tetrahedron* **1988**, *44*, 2093.
[13] T. Keumi, N. Tomioka, K. Hamanaka, H. Kakiha-
      ra, M. Fukushima, T. Morita, H. Kitayima *J. Org.
      Chem.* **1991**, *56*, 4671.
[14] O. Hutzinger, M. Fink, H. Thoma, *Chemie in
      unserer Zeit* **1986**, *20*, 165.
[15] J.-C. Harmange, B. Figadere, *Tetrahedron: Asym-
      metry* **1993**, *4*, 1711.
[16] R. Amouroux, B. Gerin, M. Chastrette, *Tetrahe-
      dron Lett.* **1982**, 4341.
[17] W. G. Dauben, C. R. Kessel, K. H. Takemura,
      J. Am. Chem. Soc. **1980**, *102*, 6893;
      P. A. Grieco, J. J. Nunes, M. D. Gaul,
      J. Am. Chem. Soc. **1990**, *112*, 4595.
[18] U. Koert, *Synthesis* **1995**, 115.
[19] D. N. Reinhoudt, J. Greevers, W. P. Trompenaars,
      S. Harkema, G. J. van Hummel, *J. Org. Chem.*
      1981, *46*, 424.
[20] J. R. Angelici, *Acc. Chem. Res.* **1988**, *21*, 387;
      A. Müller, E. Diemann, F.-W. Baumann, *Nachr.
      Chem. Tech. Lab.* **1988**, *36*, 18;

C. Bianchini, A. Meli, *Acc. Chem. Res.* **1998**, *31*,
      109.
[21] J. Nakayama, *Sulfur Reports* **2000**, *22*, 123.
[22] K. Jug, H.-P. Schluff, *J. Org. Chem.* **1991**, *56*,
      129.
[22a] R. W. Sabnis, D. W. Rangnekar, N. D. Sonawane,
      J. Heterocyclic Chem. **1999**, *36*, 333.
[23] Y. Miyahara, *J. Heterocycl. Chem.* **1979**, *16*,
      1147.
[24] G. V. Tormas, K. A. Belmore, M. P. Cava, *J. Am.
      Chem. Soc.* **1993**, *115*, 11512.
[25] J. Engel, *Chem.-Ztg.* **1979**, *103*, 160;
      R. Böhm, G. Zeiger, *Pharmazie* **1980**, *35*, 1.
[26] G. A. Patani, E. J. LaVoie, *Chem. Rev.* **1996**, *96*,
      3147.
[27] J. Roncali, *Chem. Rev.* **1992**, *92*, 711.
[28] G. Catoni, C. Galli, L. Mandolini, *J. Org. Chem.*
      1980, *45*, 1906.
[29] C. R. Noe, M. Knollmüller, K. Dungler, P. Gärt-
      ner, *Monatsh. Chem.* **1991**, *122*, 185.
[30] G. Schorrenberg, W. Steglich, *Angew.Chem. Int.
      Ed. Engl.* **1979**, *18*, 307.
[31] see ref 7, p 532.
[32] Y. Okuda, M. V. Lakshmikantham, M. P. Cava,
      J. Org. Chem. **1991**, *56*, 6024.
[33] T.-S. Chou, H.-H. Tso, *Organic Preparations and
      Procedures Int.* **1989**, *21*, 257;
      T.-S. Chou, S.-Y. Chang, *J. Chem. Soc. Perkin
      Trans. 1, 1992*, 1459.
[34] M. Balon, M. C. Carmona, M. A. Munoz, J. Hi-
      dalgo, *Tetrahedron* **1989**, *45*, 7501.
[35] M. Speranza, *Pure Appl. Chem.* **1991**, *63*, 243.
[35a] Contrary to previous results sulfonation of pyrrole
      and 1-methyl pyrrole with SO₃-pyridine complex
      was recently shown to afford mainly the 3-
      sulfonic acids: A. Mizuno et al., *Tetrahedron Lett.*
      2000, *41*, 6605.
[36] S. Brandange, E. Holmgren, H. Leijonmarck, B.
      Rodriguez, *Acta Chem. Scand.* **1995**, *49*, 922.
[37] R. W. M. Aben et al., *Tetrahedron Lett.* **1994**, *35*,
      1299.
[38] S. P. Findley, *J. Org. Chem.* **1956**, *21*, 644.
[39] V. F. Ferreira et al., *Organic Preparations and
      Procedures Int.* **2001**, *33*, 411.
[40] E. Fabiano, B.T. Golding, *J. Chem. Soc. Perkin
      Trans. 1*, **1991**, 3371.
[41] See ref 7, p 325.
[42] T. D. Lash, M. C. Hoehner, *J. Heterocycl. Chem.*
      1991, *28*, 1671;
      also see ref 7, p 327.
[43] R. ten Have, F. R. Leusink, A. M. van Leusen,
      Synthesis **1996**, 871;
      N. P. Pavri, M. L. Trudell, *J. Org. Chem.* **1997**,
      62, 2649;

A. R. Katritzky, D. Cheng, R. P. Musgrave, *Heterocycles* **1997**, *44*, 67.

[44] G. P. Moss, *Pure Appl. Chem.* **1987**, *59*, 807.

[45] D. Curran, J. Grimshaw, S. D. Perera, *Chem. Soc. Rev.* **1991**, *20*, 391.

[46] M. Munoz, P. Guardado, J. Hidalgo, M. Balcon, *Tetrahedron* **1992**, *48*, 5901.

[47] P. Wiedenau, S. Blechert, *Synth. Commun.* **1997**, *27*, 2033.

[48] J. R. Sundberg: Indoles, Best Synthetic Methods, Academic Press, San Diego **1996**;
G. W. Gribble, *J. Chem. Soc. Perkin Trans. 1*, **2000**, 1045.

[49] R. Cournoyer, D. H. Evans, S. Stroud, R. Boggs, *J. Org. Chem.* **1991**, *56*, 4576.

[50] A. Yasuhara, Y. Kanamori, M. Kaneko, A. Numata, Y. Kondo, T. Sakamoto, *J. Chem. Soc. Perkin Trans 1*, **1999**, 529;
A. Arcadi, S. Cacchi, F. Marinelli, *Tetrahedron Lett.* **1992**, *33*, 3915.

[50a] M. D. Collini, J. W. Ellingboe, *Tetrahedron Lett.* **1997**, *38*, 7963;
H.-Ch. Zhang, H. Ye, A. F. Moretto, K. K. Brumfield, B. E. Maryanoff, *Org. Lett.* **2000**, *2*, 89.

[51] D. L. Hughes, *Organic Preparations and Procedures Int.* **1993**, *25*, 607;
K. Bast, T. Durst, R. Huisgen, K. Lindner, R. Temme, *Tetrahedron* **1998**, *54*, 3745;
Y. Murakami et al, *Tetrahedron* **1998**, *54*, 45;
S. Wagaw, B. H. Yang, St. L. Buchwald, *J. Am. Chem. Soc.* **1999**, *121*, 10251.

[52] E. C. Kornfeld, E. J. Fornefeld, G. B. Kline, M. J. Mann, D. E.Morrison, R. G. Jones, R. B. Woodward, *J. Am. Chem. Soc.* **1956**, *78*, 3087;
see ref 7, p 332.

[53] D. M. Ketscha, L. J. Wilson, D. E. Portlock, *Tetrahedron Lett.* **2000**, *41*, 6254.

[54] G. M. Karp, *Organic Preparations and Procedures Int.* **1993**, *25*, 481.

[55] M. Alvarez, M. Salas, *Heterocycles* **1991**, *32*, 1391.

[56] R. Bonnett, R. F. C. Brown, R. G. Smith, *J.Chem. Soc. Perkin Trans. 1*, **1973**, 1432.

[57] A. R. Katritzky, M. Karelson, P. A. Harris., *Heterocycles* **1991**, *32*, 329;
Y. Kurasawa, A. Takada, *Heterocycles* **1995**, *41*,1805, 2057.

[58] R. P. Kreher, H. Hennige, M. Konrad, J. Uhrig, A. Clemens, *Z. Naturforsch.* **1991**, *46*, 809.

[59] U. Pindur, M. Haber, K. Sattler, *Pharmazie in unserer Zeit* **1992**, *21*, 21.

[60] J. M. Pearson, *Pure Appl. Chem.* **1977**, *49*, 463.

[60a] Th. Eicher, H. J. Roth, Synthese, Gewinnung und Charakterisierung von Arzneistoffen, p 134, Thieme, Stuttgart 1986.

[61] D. Enders, A. S. Demir, B. E. M. Rendenbach, *Chem. Ber.* **1987**, *120*, 1731;

D. Enders, W. Gatzweiler, U. Jegelka, *Synthesis* **1991**, 1137.

[62] D. Enders, P. Fey, H. Kipphardt, *Organic Synthesis* **1987**, *65*, 173;
see ref 7, p 443.443.

[63] F. Mathey, *Phosphorus, Sulfur and Silicon* **1994**, *87*, 139;
G. Keglevich, Z. Böcskei, G. M. Keserü, K. Ujszaszy, L. D. Quin, *J. Am. Chem. Soc.* **1997**, *119*, 5095.

[64] B. Weber, D. Seebach, *Angew. Chem. Int. Ed. Engl.* **1992**, *31*, 84;
Y. N. Ito et al, *Helv. Chim. Acta* **1994**,*77*, 2071.

[65] M. Hanack, G. Pawlowski, *Naturwissenschaften* **1982**, *69*, 266;
J. M. Williams, M. A. Beno, H. H. Wang, P. C. W. Leung, T. D. Emge, U. Geiser, K. D. Carlson *Acc. Chem. Res.* **1985**, *18*, 261.

[66] S. A. Haroutounian, *Synthesis* **1995**, 39.

[67] A. Hassner, B. Fischer, *Heterocycles* **1993**, *35*, 1441.

[68] B. Iddon, *Heterocycles* **1994**, *37*, 1321;
C. M. Shafer, T. F. Molinsky, *Tetrahedron Lett.* **1998**, *39*, 2903.

[69] S. Marcaccini, T. Torroba, *Organic Preparations and Procedures Int.* **1993**, *25*, 141;
W. Huang, J. Pei, B. Chen, W. Pai, X. Ye, *Tetrahedron* **1996**, *52*, 10131;
C. M. Shafer, T: F. Molinski, *Heterocycles* **2000**, *53*, 1167.

[69a] C. J. Moody, K. J. Doyle, *Prog. Hetrocycl. Chem.* **1997**, *9*, 1.

[70] M. H. Osterhout, W. R. Nadler, A. Padwa, *Synthesis* **1994**, 123.

[71] T. Mukaiyama, K. Kawata, A. Sasaki, M. Asami, *Chem. Lett.* **1979**, 1117.

[72] F. Wang, J. R. Hauske, *Tetrahedron Lett.* **1997**, *38*, 6529.

[73] K. R. Kunz, E. W. Taylor, H. M. Mutton, B. J. Blackburn, *Organic Preparations and Procedures Int.* **1990**, *22*, 613.

[74] T. G. Gant., A. I. Meyers, *Tetrahedron* **1994**, *50*, 2297.

[75] P. Zhou, J. E. Bluebaum, C. D. Burns, N. R. Natale, *Tetrahedron Lett.* **1997**, *38*, 7019.

[76] A. I. Meyers, W. Schmidt, M. J. McKennon, *Synthesis* **1993**, 250.

[77] see ref 7, p 384.

[78] M. Reuman, A. I. Meyers, *Tetrahedron* **1985**, *41*, 837.

[79] A. K. Mukerjee, *Heterocycles* **1987**, *26*, 1077.

[80] see ref 7, p 381.

[81] H. Wegmann, W. Steglich, *Chem. Ber.* **1981**, *114*, 2580.

[82] J. Gonzalez, E. C. Taylor, K. N. Houk, *J. Org. Chem.* **1992**, *57*, 3753.

[83] P. Pevarello, M. Varasi, *Synth. Commun.* **1992**,*22*, 1939.

[84] N. R. Natale, Y. R. Mirzaei, *Organic Preparations and Procedures Int.* **1993**, *25*, 515.

[85] R. B. Woodward, R. A. Olofson, *Tetrahedron* **1966**, Suppl. 7, 415.

[86] G. Büchi, J. C. Vederas, *J. Am. Chem. Soc.* **1972**, *94*, 9128.

[87] S. Kanemasa, O. Tsuge, *Heterocycles* **1990**, *30 (Special Issue)*, 719;
C. J. Easton et al, *Tetrahedron Lett.* **1994**, *35*, 3589;
R. H. Wallace, J. Liu, K. K. Zong, A. Eddings, *Tetrahedron Lett.* **1997**, *38*, 6791, 6795.

[88] P. G. Baraldi, A. Barco, S. Benetti, S. Manfredini, P. Simoni, *Synthesis* **1987**, 276.

[89] V. Jäger et al, *Bull. Soc. Chim. Belg.* **1994**, *103*, 491.

[90] K. Gothelf, I. Thomsen, K. B. G. Torssell, *Acta Chem. Scand.* **1992**, *46*, 494.

[91] R. Müller, Th. Leibold, M. Pätzel, V. Jäger, *Angew. Chem. Int. Ed. Engl.* **1994**, *33*, 1295.

[92] S. K. Armstrong, S. Warren, E. W. Collington, A. Naylor, *Tetrahedron Lett.* **1991**, *32*, 4171.

[93] J. P. Freeman, *Chem. Rev.* **1983**, *83*, 241.

[94] U. Chiaccio, A. Lignori, A. Rescifina, G. Romeo, F. Rossano, G. Sindona, N. Uccella, *Tetrahedron* **1992**, *48*, 123.

[95] B. Iddon, *Heterocycles* **1995**, *41*, 533.

[96] M. Begtrup, L. B. L. Hansen, *Acta Chem. Scand.* **1992**, *46*, 372.

[97] P. Kornwall, C. P. Dell, D. W. Knight, *J. Chem. Soc. Perkin Trans. 1*, **1991**, 2417;
A. Dondoni, *Synthesis* **1998**, 1681.

[98] E. Aguilar, A. I. Meyers, *Tetrahedron Lett.* **1994**, *35*, 2473.

[99] M. Lobell, D. H. G. Crout, *J. Am. Chem. Soc.* **1996**, *118*, 1867.

[100] H. Stetter, G. Dämbkes, *Synthesis* **1977**, 403.

[101] S. P. G. Costa et al, *J. Chem. Res.* **1997**, 314.

[102] H. Chikashita, S. Ikegami, T. Okumura, K. Itoh, *Synthesis* **1986**, 375;
M. V. Costa, A. Brembilla, D. Roizard, P. Lochon, *J. Heterocycl. Chem.* **1991**, *28*, 1933.

[103] Y. Toya, M. Takagi, T. Kondo, H. Nakata, M. Isobe, T. Goto, *Bull. Soc. Chem. Jpn.* **1992**, *65*, 2604.

[104] E. J. Corey, D. L. Boger, *Tetrahedron Lett.* **1978**, *19*, 5, 9, 13.

[105] T. Matsumoto, T. Enomoto, T. Kurosaki, *J. Chem. Soc. Chem. Commun.*, **1992**, 610.

[106] R. Kirrstetter, W. Dürckheimer, *Pharmazie* **1989**, *44*, 177.

[107] M. I. Page, A. P. Laws, *J. Chem. Soc. Chem. Commun.* **1998**, 1609.

[108] J. E. Baldwin, R. M. Adlington, R. Bohlmann, *J. Chem. Soc. Chem. Commun.* **1985**, 357.

[109] D. McKinnon, K. A. Duncan, L. M. Millar, *Can. J. Chem.* **1984**, *62*, 1580.

[110] B. Schulze, G. Kirsten, S. Kirrbach, A. Rahm, H. Heimgartner, *Helv. Chim. Acta* **1991**, *74*, 1059.

[111] M. Devys, M. Barbier, *Synthesis* **1990**, 214.

[112] L. Hough, *Chem. Soc. Rev.* **1985**, *14*, 357;
B. Schulze, K. Illgen, *J. Prakt. Chem.* **1997**, *339*, 1;
D. J. Ager, D. P. Pantaleone, S. A. Henderson, A. R. Katritzky, I. Prakash, D. E. Walters, *Angew. Chem. Int. Ed. Engl.* **1998**, *37*, 1802.

[113] B. Iddon, R. I. Ngochindo, *Heterocycles* **1994**, *38*, 2487.

[114] M. P. Groziak, L. Wei, *J. Org. Chem.* **1991**, *56*, 4296;
G. Shapiro, M. Marzi, *Tetrahedron Lett.* **1993**, *34*, 3401.

[115] A. J. Arduengo, III, R. L. Harlow, M. Kline, *J. Am. Chem. Soc.* **1991**, *113*, 361;
A. J. Arduengo, III, *Acc. Chem. Res.* **1999**, *32*, 913;
V. P. W. Böhm, W. A. Herrmann, *Angew. Chem. Int. Ed. Engl.* **2000**, *39*, 4036.

[116] M. Begtrup, P. Larsen, *Acta Chem. Scand.* **1990**, *44*, 1050.

[117] A. A. Gridnev, I. M. Mihaltseva, *Synth. Commun.* **1994**, *24*, 1547.

[118] C. Lamberth, *J. Prakt. Chem.* **1998**, *340*, 483;
B. A. Kulkarni, A. Ganesan, *Tetrahedron Lett.* **1999**, *40*, 5637;
J. Sisko, A. J. Kassick, M. Mellinger, J. J. Filan, A. Allen, M. A. Olsen, *J. Org. Chem.* **2000**, *65*, 1516.

[119] R. Breslow, *Acc. Chem. Res.* **1991**, *24*, 317.

[120] T. H. Fife, *Acc. Chem. Res.* **1993**, *26*, 325.

[121] J. Catalan, R. M. Claramunt, J. Elguero, J. Laynez, M. Menendez, F. Anvia, J. H. Quian, M. Taajepera, R. W. Taft, *J. Am. Chem. Soc.* **1988**, *110*, 4105.

[122] J. B. Hendrickson, M. S. Hussoin, *J. Org. Chem.* **1987**, *52*, 4137.

[123] S. Schwoch, W. Kramer, R. Neidlein, H. Suschitzky, *Helv. Chim. Acta* **1994**, *77*, 2175.

[124] S. E. Drewes, D. G. S. Malissar, G. H. P. Roos, *Chem. Ber.* **1991**, *124*, 2913.

[125] E. Poetsch, M. Casutt, *Chimia* **1987**, *41*, 148;
E. J. Corey, M. M. Mehrotra, *Tetrahedron Lett.* **1988**, *29*, 57.

[126] T. M. Krygowski, R. Anulewicz, M. K. Cyranski, A. Buchala, D. Rasala, *Tetrahedron* **1998**, *54*, 12295.

[127] L.-Z. Chen, R. Flamang, A. Maquestiau, R. W. Taft, J. Catalan, P. Cabildo, R. M. Claramunt, J. Elguero, *J. Org. Chem.* **1991**, *56*, 179.

[128] M. Ramos, I. Alkorta, J. Elguero, *Tetrahedron* **1997**, *53*, 1403.

[129] F. Effenberger, M. Roos, R. Ahmad, A. Krebs, *Chem. Ber.* **1991**, *124*, 1639.

[130] J. W. Pavlik, E. M. Kurzweil, *J. Org. Chem.* **1991**, *56*, 6313;
R. E. Connors, J. W. Pavlik, D. S. Burns, E. M. Kurzweil, *J. Org. Chem.* **1991**, *56*, 6321.

[131] K. Makino, H. S. Kim, Y. Kurasawa, *J. Heterocycl. Chem.* **1998**, *35*, 489; *J. Heterocycl. Chem.* **1999**, *36*, 321.

[132] H. Garia, S. Iborra, M. A. Miranda, I. M. Morera, J. Primo, *Heterocycles* **1991**, *32*, 1745.

[133] J. Catalán, J. L. G. de Paz, J. Elguero, *J. Chem. Soc. Perkin Trans. 2* **1996**, 57.

[134] P. López-Alvarado, C. Avendaño, J. C. Menéndez, *Tetrahedron Lett.* **1992**, 659.

[135] C. Dell'Erba, M. Novi, G. Petrillo, C. Tavani, *Tetrahedron* **1994**, *50*, 3529; S. Caron, E. Vazques, *Synthesis* **1999**, 588.

[136] A. R. Katritzky, M. Karelson, P. A. Harris, *Heterocycles* **1991**, *32*, 329; J. Kurasawa, A. Takada, *Heterocycles* **1995**, *41*, 1805

[137] A. J. Katritzky, P. Barczynski, D. L. Ostercamp, *J. Chem. Soc. Perkin Trans. 2*, **1987**, 969.

[138] A. Bosum-Dybus, H. Neh, *Liebigs Ann. Chem.* **1991**, 823.

[139] Th. Eicher, H. J. Roth, *Synthese, Gewinnung und Charakterisierung von Arzneistoffen,* p 137, Thieme, Stuttgart **1986**.

[140] A. Dorlars, C.-W. Schellhammer, J. Schroeder, *Angew. Chem.* **1975**, *87*, 693.

[141] R. M. Claramunt, J. Elguero, *Organic Preparations and Procedures Int.* **1991**, *23*, 273.

[142] M. Begtrup, *Heterocycles* **1992**, *33*, 1129.

[143] W. Sliva, *Heterocycles* **1984**, *22*, 1571.

[144] W. Sliva, A. Thomas, *Heterocycles* **1985**, *23*, 399; J. Stevens, M. Schweizer, G. Rauhut, *J. Am. Chem. Soc.* **2001**, *123*, 7326.

[145] N. P. Hacker, *J. Org. Chem.* **1991**, *56*, 5216.

[146] M. J. Haddadin, C. H. Issidorides, *Heterocycles* **1993**, *35*, 1503.

[147] B. D. Larsen, H. Eggert, N. Harrit, A. Holm, *Acta Chem. Scand.* **1992**, *46*, 482.

[148] M. Fujita, T. Kobori, T. Hiyama, K. Kondo, *Heterocycles* **1993**, *36*, 33.

[149] L. Zirngibl: Antifungal Azoles Wiley-VCH, **1998**.

[150] K. Nielsen, I. Sötofte, J. Johansen, *Acta Chem. Scand.* **1993**, *47*, 943.

[151] A. Albert, P. J. Taylor, *J. Chem. Soc. Perkin Trans. 2*, **1989**, 1903.

[152] M. R. Grimmet, B. Iddon, *Heterocycles* **1995**, *41*, 1525.

153] P. Zanirato, *J. Chem. Soc. Perkin Trans. 1*, **1991**, 2789.

[154] J. Catalan, P. Perez, J. Elguero, *J. Org. Chem.* **1993**, *58*, 5276.

[155] R. J. Greenwood, M. F. Mackay, J. F. K. Wilshire, *Aust. J. Chem.* **1992**, *45*, 965.

[156] J. L. Flippen-Anderson, R. D. Gilardi, A. M. Pitt, W. S. Wilson, *Aust. J. Chem.* **1992**, *45*, 513.

[157] M. Begtrup, P. Vedsö, *Acta Chem. Scand.* **1996**, *50*, 549.

[158] A. R. Katritzky, X. Lan, J. Z. Yang, O. V. Denisko, *Chem. Rev.* **1998**, *98*, 409; A. R. Katritzky, S. A. Henderson, B. Yang, *J. Heterocyclic Chem.* **1998**, *35*, 1123; A. R. Katritzky, *J. Heterocyclic Chem.* **1999**, *36*, 1501.

[159] M. Balasubramanian, J. G. Keay, E. F. V. Scriven, N. Shobana, *Heterocycles* **1994**, *37*, 1951.

[160] K. Paulvannan, T. Chen, R. Hale, *Tetrahedron* **2000**, *56*, 8071.

[161] R. Raue, A. Brack, K. H. Lange, *Angew. Chem.* **1991**, *103*, 1689.

[162] G. M. Ramos Tombo, D. Bellus, *Angew. Chem.* **1991**, *103*, 1219.

[163] S. J. Wittenberger, *Organic Preparations and Procedures Int.* **1994**, *26*, 499; D. Moderhack, *J. Prakt. Chem.* **1998**, 340, 687.

[164] E. Cubero, M. Orozco, F. J. Luque, *J. Am. Chem. Soc.* **1998**, *120*, 4723.

[165] G. Maier, J. Eckwerth, A. Bothur, H. P. Reisenauer, C. Schmidt, *Liebigs Ann. Chem.* **1996**, 1041.

[166] K. Koguro, T. Oga, S. Mitsui, R. Orita, *Synthesis* **1998**, 910; Z. P. Demko, K. B. Sharpless, *J. Org. Chem.* **2001**, *66*, 7945.

[167] R. N. Butler, S. Collier, A. F. M. Fleming, *J. Chem. Soc. Perkin Trans. 2*, **1996**, 801; R. N. Butler, A. Fox, S. Collier, L. A. Burke, *J. Chem. Soc. Perkin Trans. 2*, **1998**, 2243.

6   Six-membered Heterocycles

As in five-membered heterocycles, ring strain in six-membered heterocycles is of little or no importance. Pyran and thiine (thiopyran), with an oxygen or sulfur atom, respectively, and pyridine, with a nitrogen atom, are the parent compounds of six-membered neutral heterocycles with one heteroatom and the maximum number of noncumulative double bonds. In contrast to pyran and thiine, pyridine exists as a cyclic conjugated system. However, by (formal) abstraction of a hydride ion, both pyran and thiine can be converted into the corresponding cyclic conjugated cations, i.e. the pyrylium and the thiinium ions (thiopyrylium ion).

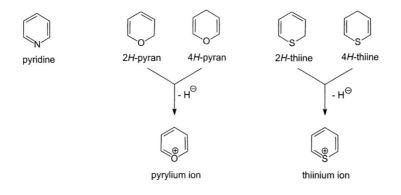

Therefore, the simplest benzene analogue of the heterocycles with one oxygen atom is the pyrylium ion.

6.1   Pyrylium ion

A | The pyrylium ion possesses the structure of a planar, slightly distorted hexagon with C–C and C–O bonds of more or less equal length. This follows from the X-ray analysis of substituted pyrylium salts, e.g. the 3-acetyl-2,4,6-trimethyl system (see Fig. 6.1).

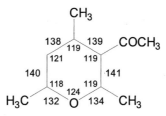

Fig. 6.1   Bond parameters of the pyrylium system in the 3-acetyl-2,4,6-trimethyl pyrylium ion
(bond lengths in pm, bond angles in degrees)

The spectroscopic data, in particular the NMR spectra, indicate that the pyrylium ion is an aromatic system in which the π-electron delocalization is strongly perturbed by the positively charged heteroatom:

UV (ethanol) λ (nm)	^{1}H-NMR (CF$_3$COOD) δ (ppm)		^{13}C-NMR (CD$_3$CN) δ (ppm)	
270	H-2/H-6:	9.22	C-2/C-6:	169.2
	H-3/H-5:	8.08	C-3/C-5:	127.7
	H-4:	8.91	C-4:	161.2
Comparison with pyridinium ion (CDCl$_3$)	H-2/H-6:	9.23	C-2/C-6:	142.5
	H-3/H-5:	8.50	C-3/C-5:	129.0
	H-4:	9.04	C-4:	148.4

The influence of the oxonia-oxygen is especially noticeable in the ring positions C-2, C-4 and C-6; their ^{1}H- and ^{13}C-resonances are shifted further downfield than in the iso-π-electronic pyridinium ion (see p 270). For this reason, the distribution of the π-electron density in the pyrylium ion can be represented by simple mesomeric structures.

B Among the reactions of pyrylium salts, addition of nucleophiles to the ring positions 2/6 and 4, and subsequent reactions, dominate [1]. The point of attack and the nature of the product depend on the steric and electronic properties of the nucleophile, and on those of the ring substituent.

For instance, pyrylium ions **1** react with aqueous alkali giving ene-1,5-diones **4** or their enol tautomers **3**:

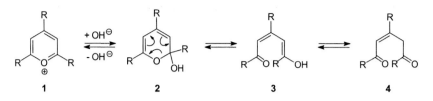

Attack of the nucleophile at C-2 and formation of the 2-hydroxy-2*H*-pyrans **2** occur first, followed by electrocyclic ring-opening, yielding the products **3** and **4**. This ring-opening, which in the case of the unsubstituted pyrylium ion already occurs with H_2O, is reversible, i.e. acids reconstitute the pyrylium system from **3** and **4**.

The 2,4,6-triphenylpyrylium ion undergoes C-2 and C-4 addition. For instance, methoxide produces methoxy-substituted 2*H*- and 4*H*-pyrans **5** and **6**:

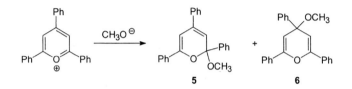

The 2,6-diphenylpyrylium ion **7** undergoes mainly C-4 addition. There is some evidence that the direction of the addition is subject to kinetic or thermodynamic control. Thus 4*H*-pyrans **8** are observed when methoxide is added to **7** at low temperature, whereas the ring-opened products **10**, which derive from a primary C-2 addition to the 2*H*-pyrans **9**, result at higher temperature:

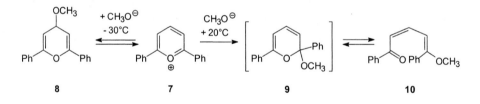

Accordingly, the pyrylium ion **7** reacts with CH-acidic compounds and activated arenes as nucleophiles to give new pyrylium systems with dehydrogenation.

Organometallic compounds, when added to pyrylium ions, predominantly form 2*H*-pyrans, which tautomerize to give dienones. Benzylmagnesium chloride is an exception; for unknown reasons, it leads to 4*H*-pyrans.

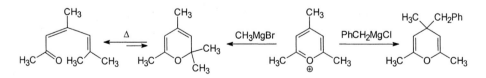

Reactions of pyrylium ions with nucleophiles, which first lead to 2*H*-pyrans **11** followed by ring-cleavage to the dienones **12**, are important in preparative chemistry. On ring closure, which occurs with incorporation of the nucleophilic center, new carbocyclic or heterocyclic systems **13** are produced. This reaction principle has many synthetic applications (see p 228.)

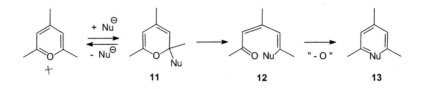

This basic reactivity of pyrylium ions is reversed by donor substituents in the 2-, 4- and 6-positions. For instance, 2,4,6-tris(dialkylamino)pyrylium ions **14** are stable towards nucleophilic attack, but are easily substituted by electrophiles (e.g. HNO_3/H_2SO_4 or $BrCN/AlCl_3$) to give **15** (SCHROTH 1989):

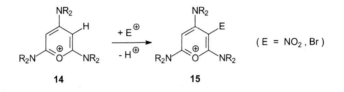

This behaviour may be due to the fact that the donor-substituted pyrylium ion **14** does not possess the structure of cyclic delocalized 6π-systems, but rather that of a localized trimethine cyanine [2].

A formally electrophilic substitution at the 4-position of pyrylium ions with a substitution pattern like **16** can be achieved using benzotriazole(Bt)-mediated methodology (KATRITZKY [3]). Pyrylium salts **16** are readily converted to 4-Bt-substituted 4*H*-pyranes **17** by addition of Bt-Na. Subsequent deprotonation, trapping of the Bt-stabilized anion by alkylation (to give **18**) and acid cleavage of the benzotriazole moiety in **18** affords the 4-substituted pyrylium ions **19** [4]:

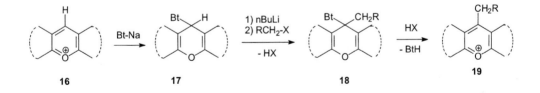

Alkyl groups in positions 2, 4 or 6 of the pyrylium system display marked CH-acidity. The action of bases causes deprotonation on CH groups attached to the ring forming 2- or 4-methylenepyrans **16** or **17**:

16   **17**

The terminal C-atoms of the enol systems **16** and **17** can be attacked by electrophiles, which is useful for carrying out C–C-forming reactions (aldol condensation, CLAISEN condensation, etc.), and exemplifies side-chain-reactivity in heterocyclic systems (see also p 281). The symmetrical conjugated stabilization of the 4-methylene pyran system (**17**) favours electrophilic attack on the 4-alkyl substituent. For example, the 2,4,6-trimethylpyrylium salt **18** undergoes a regioselective aldol condensation with benzaldehyde yielding only the 4-styryl derivative **19**:

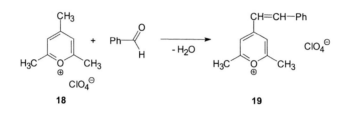

18   **19**

C The *retrosynthesis* of pyrylium ions (Fig. 6.2) is based on the ring-opening reaction by hydroxide ions (see p 224) and leads to 1,5-dicarbonyl compounds **20-22**. From these starting materials the synthesis of pyrylium ions can occur by cyclocondensation, their oxidation level is brought about by removal of a hydride ion or of a suitable leaving group:

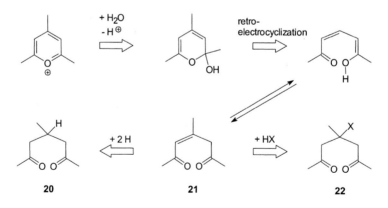

20   **21**   **22**

Fig. 6.2   Retrosynthesis of the pyrylium ion

(1)   1,3-Dicarbonyl compounds and aryl methyl ketones condense in acetic anhydride in the presence of strong acids to give 2,4,6-trisubstituted pyrylium salts **24**:

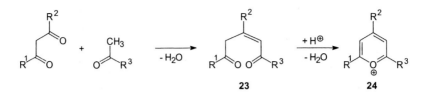

Pent-2-ene-1,5-diones **23** are intermediates in this reaction. Variation of this synthesis is achieved using related bis-electrophiles. For instance, 2,6-disubstituted pyrylium salts **25** are formed from two molecules of aryl methyl ketone and orthoformic ester in the presence of strong acids (HClO$_4$, HBF$_4$):

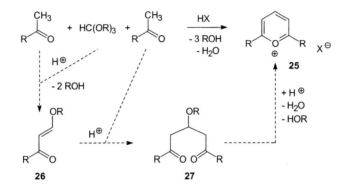

In the first step, the orthoformic ester transforms the methyl ketone, via a carboxonium ion HC$^+$(OR)$_2$, into an alkoxyenone **26**. This adds a second molecule of methyl ketone giving 3-alkoxy-1,5-dione **27**, which cyclizes with elimination of H$_2$O and ROH.

Chlorovinyl ketones or chlorovinyl immonium salts also condense with aryl methyl ketones yielding pyrylium salts [5].

(2) Propene derivatives produce pyrylium salts **28** (*Balaban synthesis*) by a double acylation with acid chlorides or anhydrides in the presence of LEWIS acids, e.g. AlCl$_3$:

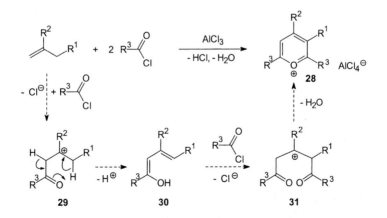

This involves a complex mechanistic reaction sequence. First, the alkene is acylated to produce the cationic intermediate **29**, followed by deprotonation to the enol **30**. Renewed acylation furnishes the cationic 1,5-dicarbonyl system **31** which, by dehydration, is transformed into **28**. Formation of the product can be controlled by the use of LEWIS acid. The use of alcohols or halogen compounds for the *in situ* generation of alkenes is a preparatively important variation of the BALABAN synthesis [6].

(3)   When reacted with aryl methyl ketones in acetic anhydride in the presence of a hydride acceptor, e.g. FeCl$_3$, chalcones **32** yield trisubstituted pyrylium salts **34** (*Dilthey synthesis*):

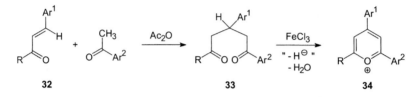

| 32 | 33 | 34 |

Pentane-1,5-diones **33** are formed by MICHAEL addition. In a simpler preparative method, two molecules of methyl ketone condense directly with an aldehyde without an α-hydrogen by means of strong acid and dehydration agents to give symmetrical pyrylium salts with identical substituents in the 2- and 6-positions.

D **Pyrylium perchlorate 38** forms colourless crystals which are prone to hydrolysis and decompose explosively at 275°C. It can be obtained from *N*-acceptor substituted pyridinium ions (see p 272), e.g. by the SO$_3$ complex **35**, on treatment with NaOH followed by addition of HClO$_4$. Initially, the orange-coloured sodium salt of 5-hydroxypenta-2,4-dienal (glutaconic dialdehyde in the enol form) **36** is formed which subsequently cyclizes via the red protonated species **37** [7]:

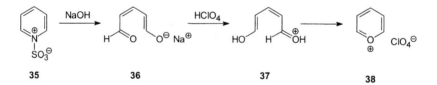

| 35 | 36 | 37 | 38 |

E Pyrylium salts are useful intermediates for the formation of a range of carbocyclic and heterocyclic systems. In these preparations, the sequence nucleophilic C-2 opening/recyclization (see p 225) and side-chain reactivity are used.

(1)   The 2,4,6-triphenylpyrylium ion reacts with nitromethane to give the dienone **39**. This is followed by an intramolecular nitro-aldol condensation forming 2,4,6-triphenylnitrobenzene **40**. By way of the ring-opened phosphorane **41**, action of phosphorylidene produces the 2-substituted 1,3,5-triphenyl-benzene **42**:

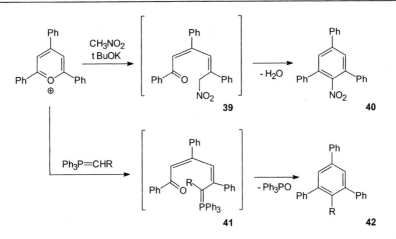

Benzene derivatives can also be obtained from pyrylium salts **43** which are substituted by R–CH$_2$ in the 2- or 6-position. This is effected by C-2 addition of hydroxide ions, leading to a base-catalysed intramolecular aldol condensation. The reaction produces highly substituted phenols **44**, which incorporate the R–CH$_2$ carbon in the benzene ring:

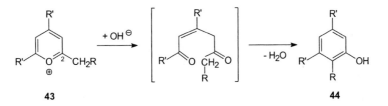

(2) The formation of azulenes by cyclocondensation of cyclopentadienyl anions is a prime example of a synthesis with pyrylium salts (HAFNER 1973 [8], cf. 307). Thus, 4,6,8-trimethylazulene **46** is produced by the reaction between 2,4,6-trimethylpyrylium salts and sodium cyclopentadienide, resulting in ring-opening at C-2 (**45**) followed by ring closure of the cyclopentadienyl system **47** which is a probable intermediate.

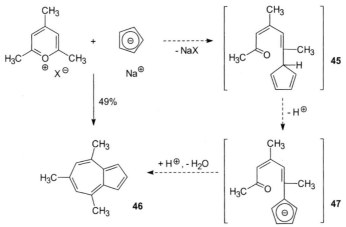

(3)   Pyrylium salts yield pyridines **48** with ammonia, whereas pyridinium salts **49** are obtained from primary amines:

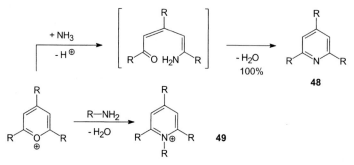

The conversion into the pyridine system occurs smoothly and under mild conditions [9]. Hydroxylamines, acylhydrazines and ureas also yield pyridine derivatives.

However, secondary amines yield aniline derivatives **50** with pyrylium ions in analogy to the example **43 → 44**:

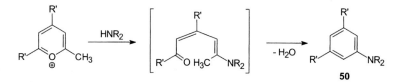

(4)   Other heterocycles can be constructed by means of pyrylium ions. For instance, sodium sulfide cleaves the ring at C-2 resulting in the formation of the blue dienone thiolate **51**, which on acid treatment recyclizes to give the thiinum salt **52**. Phosphanes, e.g. P(SiMe$_3$)$_3$, can replace the pyrylium oxygen nucleophilically, a reaction which was used for the synthesis of phosphabenzenes,e.g. **53** (see p 368):

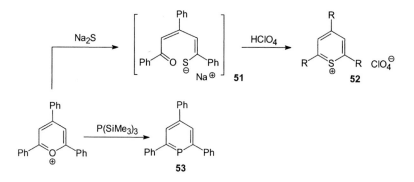

(5)   The addition of nucleophiles to the C-2 position of the pyrylium ion can also lead to ring enlargements to seven-membered heterocycles. Thus 2,4,6-triphenylpyrylium salts react with hydrazine to give 4*H*-1,2-diazepine **54** (see p 471):

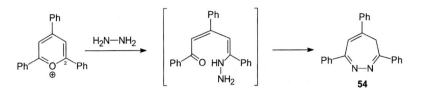

2,3,5,6-Tetraphenylpyrylium salts add azide producing 2-azido-2*H*-pyran **55**, which on photolysis rearranges into the azirine **56** with N_2 elimination. The latter recyclizes thermally to form the oxazepine derivative **57**:

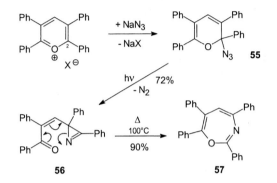

(6)   The side-chain reactivity of the pyrylium ion facilitates a targeted synthesis of cyanine dyes with pyrylium or 4*H*-pyran systems as terminal groups (see p 323). In this way, the dye **60** is produced by VILSMEIER formylation of the 4-methylpyrylium ion **58** to give the formylmethylene-4*H*-pyran **59** followed by aldol condensation with a second molecule of **58** [10]:

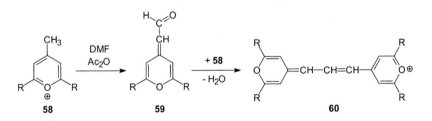

6.2   2*H*-Pyran

| A-C | The parent compound has not yet been isolated. However, 2,2-disubstituted derivatives have been prepared, e.g. **1** and **2**: |

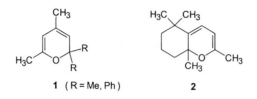

1 (R = Me, Ph)   **2**

2*H*-Pyrans as dienol ethers show typical C=C stretching bands at 1600-1650 cm^{-1} in the IR spectrum. In the ^{1}H-NMR spectrum, chemical shifts of olefinic protons, e.g. **2** at δ = 4.89/5.60 (3-H/4-H; CCl$_4$) are observed.

2*H*-Pyrans behave like oxacyclohexadienes. For instance, the O/C-2 bond is subject to thermal ring-opening with formation of dienones, e.g. **3** to **4**:

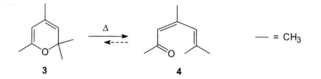

3   **4**

This electrocyclic process is reversible, therefore dienones can be used to prepare 2*H*-pyrans. Thus, the bicyclic structure **2** is obtained by irradiation of (*E*)-β-ionone with the *Z*-isomer as an intermediate (**5 → 6**):

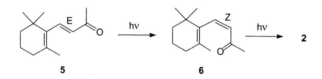

5   **6**

Moreover, 2*H*-pyrans can undergo [4+2] cycloadditions with activated multiple bonds. For instance, compound **3** undergoes a regioselective DIELS-ALDER reaction with propiolic ester to adduct **6** which yields methyl 2,4-dimethylbenzoate by cycloreversion and acetone elimination:

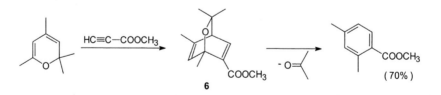

6

2*H*-Pyrans are also formed by addition of C-nucleophiles to pyrylium ions (see p 224).

The most important derivative of 2*H*-pyrans is the corresponding carbonyl system, i.e. 2*H*-pyran-2-one, which will be discussed in chapter 6.3.

6.3   *2H*-Pyran-2-one

A 2*H*-Pyran-2-one **1**, according to microwave spectroscopy, possesses the bond parameters of an enol-lactone system with localized C–C double and single bonds (see Fig. 6.3).

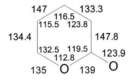

Fig. 6.3   Bond parameters of 2*H*-pyran-2-one
(bond lengths in pm, bond angles in degrees)

2*H*-Pyran-2-one shows no significant diamagnetic ring-current effect in its ^{1}H and ^{13}C-NMR spectra, and only minor charge delocalization, although the electron-attracting ring constituents (O and CO) cause downfield shifts corresponding to values in aromatic systems [11]:

^{1}H-NMR (CDCl$_3$) ^{13}C-NMR (acetone-d$_6$)
δ (ppm)   δ (ppm)

H-3: 6.38   H-5: 6.43   C-2: 162.0   C-4: 144.3   C-6: 153.3
H-4: 7.58   H-6: 7.77   C-3: 116.7   C-5: 106.8

2*H*-Pyran-2-ones possess characteristic IR absorptions at ≈ 1730 cm^{-1}. A comparison with other six-membered ring lactones and with the influence of adjacent C=C double bonds on their C=O absorptions (**2-4**)

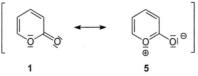

$\nu_{C=O}$ 1730   1735   1775   1710
[cm^{-1}]

led to the conclusion that the betaine structure **5** does not play an important part in the description of the 2*H*-pyran-2-one system **1**:

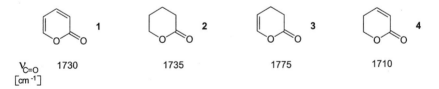

2*H*-Pyran-2-ones show a characteristic fragmentation pattern in the mass spectrum, with furan and cyclopropenylium fragment ions (**6** and **7**) as the main components:

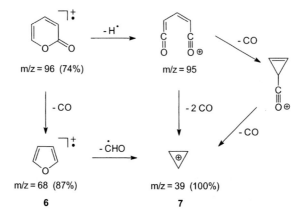

| m/z = 96 (74%) | m/z = 95 |

| m/z = 68 (87%) | m/z = 39 (100%) |
| **6** | **7** |

B   In their reactions, 2*H*-pyran-2-ones behave like 1,3-dienes and also like lactones. This is manifested above all in a series of intra- and intermolecular pericyclic transformations.

Matrix photolysis of 2*H*-pyran-2-one at 8 K leads to electrocyclic ring opening of the bond O/C-2 and formation of the aldoketene **8** which, by addition of CH_3OH, can be intercepted as ester **10**. Photolysis at higher temperatures or in ether leads to electrocyclization of the 1,3-diene system and formation of the β-lactone **9** which, by addition of CH_3OH, yields the enol ether **12**. Further photolysis of the β-lactone **9** causes decarboxylation resulting in cyclobutadiene **11** and its products, e.g. the dimer **14** (CHAPMAN) [12]:

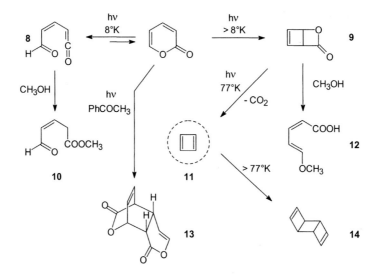

These transformations occur via an excited singlet state of the 2*H*-pyran-2-one. Irradiation of 2*H*-pyran-2-one in the presence of a triplet sensitizer results in dimerization to the tricyclic structure **13** or its regioisomer.

2*H*-Pyran-2-ones undergo DIELS-ALDER reactions with activated alkenes or alkynes. For instance, from 2*H*-pyran-2-one and maleic anhydride, the *endo*-adduct **15** is formed which undergoes thermal decarboxylation to the 1,2-dihydrophthalic anhydride **16**:

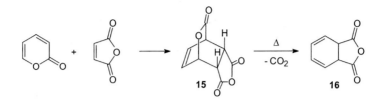

This reaction is useful in the synthetic applications of 2*H*-pyran-2-ones (see p 239).

Nucleophiles frequently attack 2*H*-pyran-2-one on the C-atom of the carbonyl group. For instance, *N*-substituted 2-pyridones **17** are obtained from 2*H*-pyran-2-one and primary amines while Grignard reagents yield 2,2-disubstituted 2*H*-pyrans **18**:

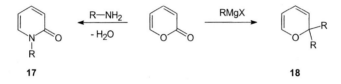

Electrophiles substitute the 2*H*-pyran-2-one in position 3. At higher temperatures, the reaction of 2*H*-pyran-2-one with bromine leads to 3-bromo-2*H*-pyran-2-one **20**. However, at lower temperatures, the addition product *trans*-dibromide **19** is obtained in quantitative yield. At higher temperatures, dibromides **21** or **22** are formed which change into the substitution product **20** with elimination of HBr:

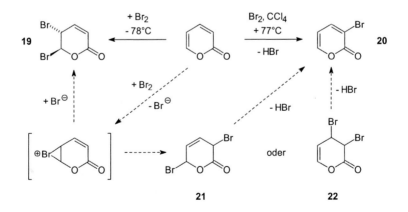

Alkylation of 2*H*-pyran-2-one occurs on the carbonyl oxygen and yields 2-alkoxypyrylium ions, as for instance **23** with trimethyloxonium tetrafluoroborate:

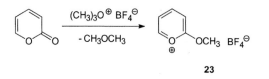

23

C   The synthesis of 2*H*-pyran-2-one follows the route indicated by the *retrosynthesis* (see Fig. 6.4). According to **a** and **b**, the retrosynthetic analysis involves the O/C-2 bond and suggests the pentenoic acids **24** and **25** with the carbonyl function in the δ-position as starting materials.

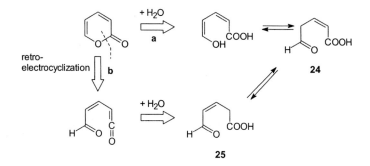

Fig. 6.4   Retrosynthesis of 2*H*-pyran-2-ones

(1) Base-catalysed cyclocondensation of alkynones with malonic esters, or 1,3-diketones with acetylenic acids, yields esters **27** of 2*H*-pyran-2-one-3-carboxylic acid or 5-acyl-2*H*-pyran-2-ones **29**, respectively:

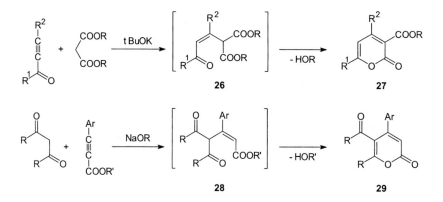

The first steps are MICHAEL additions leading to δ-oxopentenoic ester intermediates **26** and **28** which provide the 2*H*-pyran-2-one system by enolization and lactone formation.

(2) Crotonic esters (unsubstituted or γ-substituted) condense with oxalic esters under base catalysis giving δ-oxopentenoic esters **30**. In acid, they cyclize to 2*H*-pyran-2-one-6-carboxylic esters **31** which, on ester hydrolysis followed by decarboxylation, provide 2*H*-pyran-2-ones **32**:

(3) Pyridinium enol betaines (see p 284) react with cyclopropenones as electrophilic C_3-building blocks [13] to yield 3,4,6-trisubstituted *2H*-pyran-2-ones **35**:

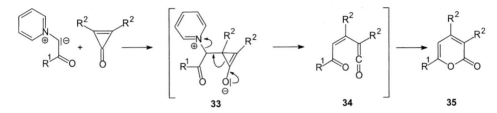

First a MICHAEL addition of the pyridinium ylide to cyclopropenone occurs. This is followed by a reorganization of the intermediate **33** with pyridine elimination giving the keto ketene **34**, which by ring closure yields the *2H*-pyran-2-one system. Sulfonium and phosphonium ylides can be used instead of pyridinium enol betaines.

(4) Pd-catalyzed reaction of allenyl stannanes with β-iodo acrylic acids **36** gives *2H*-pyran-2-ones **37**, this annulation process presumably involves a STILLE coupling sequence [14]:

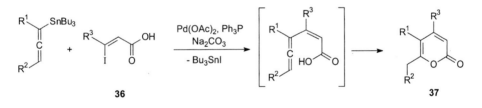

D | **2H-Pyran-2-one**, bp 208°C, is a colourless liquid. *2H*-Pyran-2-one is obtained by decarboxylation of *2H*-pyran-2-one-5-carboxylic acid **39** (coumalic acid). This is formed by H_2SO_4 treatment of malic acid involving self-condensation of formylacetic acid **38** produced as an intermediate:

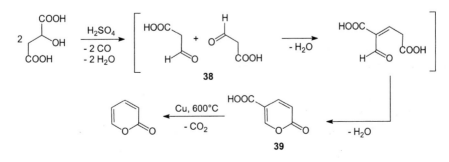

Dehydroacetic acid 40 (3-acetyl-4-hydroxy-6-methyl-2*H*-pyran-2-one) is obtained by self-condensation of acetoacetic ester in the presence of Na$_2$CO$_3$ [15]:

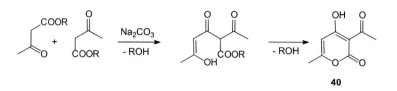

40

Bufadienolides are A/B-*cis* or Δ^4, B/C-*trans*- and C/D-*cis*-steroids. Their characteristic feature is a 2*H*-pyran-2-one substituent at C-17 in the D-ring. Bufalin **41** and bufotalin **42** occur as esters of suberylarginine in the venom of the secretions of poisonous toads (e.g. Bufo bufo). Scillarigenin **43** and hellebrigenin **44** occur as 3-O-glycosides in fresh squill (*Urginea maritima*) and the Christmas rose (*Helleborus niger*) respectively.

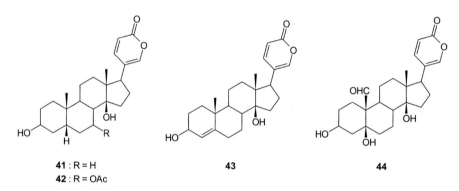

41 : R = H
42 : R = OAc

43

44

The bufadienolides are cardiotonic and have a positive inotropic action (increase in the force of heart muscle contraction).

E   2*H*-Pyran-2-ones can be subjected to (1) base-catalysed ring formations and (2) DIELS-ALDER reactions leading to the preparation of highly substituted benzene derivatives.

(1)   On treatment with KOH in methanol, 6-phenacyl-4-hydroxy-2*H*-pyran-2-one **45** yields 2-benzoylphloroglucinol **47**:

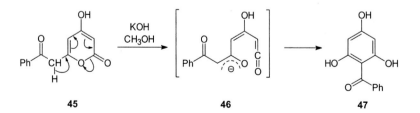

45   **46**   **47**

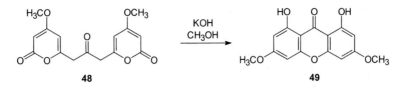

48   **49**

Isomerization of the anion **46**, analogous to a DIMROTH rearrangement (see p 202), is assumed to be the key step in this reaction. An application of this reaction principle is to be found in the synthesis of xanthone **49**, which is effected by a base-catalysed isomerization of the bis-2H-pyran-2-one **48**:

(2)   The DIELS-ALDER reaction of the methyl ester of 2H-pyran-2-one-6-carboxylic acid with acetylenedicarboxylic ester leads to the 1,2,3-benzenetricarboxylic ester **51** (hemimellitic triester), because the cycloadduct **50** undergoes thermal decarboxylation:

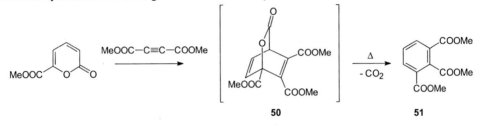

50   **51**

In an intramolecular variation of this reaction, the use of sterically defined alkenes permits a stereoselective construction of stereogenic centers of six-membered heterocycles, as demonstrated in the following example for the key building block **52** of a yohimbine synthesis [16]:

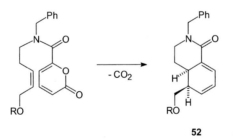

52

6.4   3,4-Dihydro-2*H*-pyran

A   3,4-Dihydro-2H-pyran and 5,6-dihydro-2H-pyran are the oxa-analogues of cyclohexene derived from 2H-pyran.

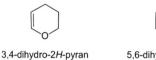

3,4-dihydro-2*H*-pyran   5,6-dihydro-2*H*-pyrar

3,4-Dihydro-2*H*-pyran has the spectroscopic and structural characteristics of a cyclic enolic ether, as is shown by its NMR and IR spectroscopic data:

IR (film) ν (cm^{-1})	^{1}H-NMR (CDCl$_3$) δ (ppm)	^{13}C-NMR (CDCl$_3$) δ (ppm)
1630 C=C stretching band	H-2: 3.97 H-5: 4.65 H-3/H-4: 1.90 H-6: 6.37	C-2: 65.8 C-4: 19.6 C-6: 144.2 C-3: 22.9 C-5: 100.7

Examination of RAMAN spectra shows that at room temperature, 3,4-dihydro-2*H*-pyran exists predominantly in a half-chair conformation with a twist-angle of 23°. According to ^{1}H-NMR spectra the conformers of 2-alkoxy- and 2-aryloxy-3,4-dihydro-2*H*-pyrans are in equilibrium. The conformer with the 2-OR in an axial position predominates due to an anomeric effect (see p 244):

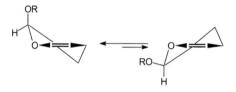

B As expected, 3,4-dihydro-2*H*-pyrans show reactions of electron-rich double-bonded systems. For instance, electrophilic addition of HX and HOX (X = halogen) as well as hydroboration occur regioselectively with unsymmetrical reagents:

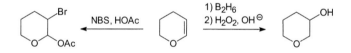

[2+1], [2+2] and [4+2] cycloadditions are also possible. The reaction of 3,4-dihydro-2*H*-pyran with dichlorocarbene to give the adduct **1** and its dehydrohalogenation resulting in ring-opening to give the 2,3-dihydrooxepin system **2** is of preparative interest:

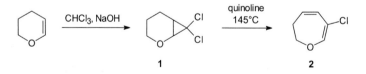

Alcohols, as well as phenols, add to 3,4-dihydro-2*H*-pyran under acid catalysis giving 2-alkyl(aryl)-oxytetrahydropyrans **3** [16a]. These cyclic acetals are stable to base but prone to hydrolysis even with dilute acids. Thus, OH functions can be blocked reversibly (THP protecting group in organic synthesis):

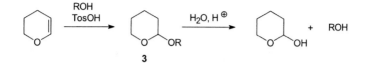

3

3,4-Dihydro-2H-pyran is oxidized by HNO_3 yielding glutaric acid and is dihydroxylated by OsO_4/H_2O_2 to the corresponding 1,2-glycol. Ozonolysis leads to the hydroperoxide **4**, which can be transformed either into the aldehyde **5** by $(n\text{-BuO})_3P$ or to the ester **6** by TosCl / pyridine:

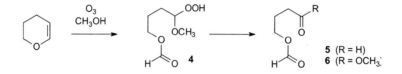

5 (R = H)
6 (R = OCH₃)

C The synthesis of 3,4-dihydro-2H-pyrans can be carried out by the following methods.

(1) α,ß-Unsaturated carbonyl compounds undergo [4+2] cycloadditions with vinyl ethers resulting in the formation of 2-alkoxy-3,4-dihydro-2H-pyrans **7**:

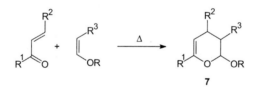

7

This reaction is an example of a hetero-DIELS-ALDER reaction with an inverse electron demand [17]. 2-Alkoxy-3,4-dihydro-2H-pyrans **7** can be employed as masked 1,5-dicarbonyl compounds (see p 299).

(2) 4-Acyloxybutylphosphonium salts **8** produce 6-substituted dihydropyrans **9** by an intramolecular WITTIG reaction:

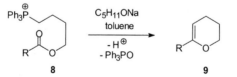

8   **9**

The starting compound **8** is obtained from 1,4-dibromobutane by a succession of S_N reactions with triphenylphosphane and carboxylates. This cyclization is one of the few examples of the conversion of an ester carbonyl group into an olefinic moiety by a phosphorus ylide.

(3) 2,6-Disubstituted 5,6-dihydro-2H-pyranes can be prepared stereoselectively utilizing the LEWIS-acid catalyzed cyclization of vinylsilanes carrying an acetalized glyoxylic ester functionality in β-position [18]. Interestingly, the geometry of the double bond in the educt controls the stereochemistry in the products **10/11**:

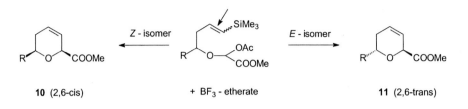

10 (2,6-cis)   + BF₃ - etherate   **11** (2,6-trans)

(4)   An intramolecular hetero-DIELS-ALDER reaction of α,β-unsaturated carbonyl compounds, e.g. **12** [19] derived from citronellal, results in structurally complex 3,4-dihydro-2*H*-pyrans, e.g. the system **13** derived from tetrahydrocannabinol (see p 269):

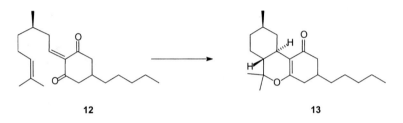

12   **13**

> **D**   **3,4-Dihydro-2*H*-pyran** is a colourless liquid, bp 86°C. It is obtained by catalytic dehydration of tetrahydrofurfuryl alcohol **14**; **14** in turn is obtained by hydrogenation of furfural (see p 60).

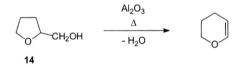

14

Iridoids, a widely occurring class of heterobicyclic monoterpenes, are derived from 2-hydroxy-3,4-dihydro-2*H*-pyran with a 3,4-fused methylpentane ring. Iridodial **15** is the parent system. As an enol lactol, it is in equilibrium with a dialdehyde:

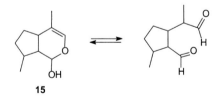

15

In natural products, the 2-hydroxy group is protected either as a glycoside or as an ester. The two bitter principles loganin **16**, found in the buck-bean (*Menyanthes trifoliata*), and gentiopicrosid **16**, occurring in the root of gentian (*Gentiana* species), are derived from iridodial. The sedatives valepotriates obtained from valerian (*Valeriana officinalis*), e.g. valtrat **19**, isovaltrat **20** and didrovaltrat **21**, are triesters of unsaturated iridoid alcohols containing a reactive oxirane ring.

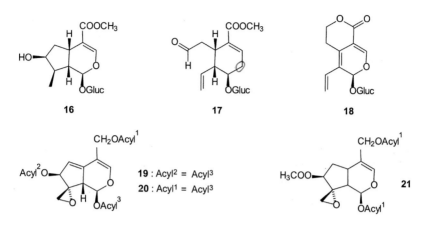

16   17   18

19 : Acyl² = Acyl³
20 : Acyl¹ = Acyl³

21

Secologanin **17** is a key precursor in alkaloidal biogenesis from which over 1000 alkaloids (indol-, cinchona-, ipecacuanha-, pyrroloquinone-alkaloids) are derived. A stereoselective total synthesis of secologanin has been achieved (TIETZE 1983) [20].

6.5   Tetrahydropyran

A The structure of tetrahydropyran in the gaseous state has been determined by electron diffraction and microwave spectroscopy. These show that the saturated oxygen six-membered ring exists in a chair conformation with C_s symmetry (see Fig. 6.5), which is somewhat flattened when compared with cyclohexane:

```
        110.7
        108.3
                  153.1
        111.8
      111.5
        O    142.0
```

Fig. 6.5   Bond parameters of tetrahydropyran
(bond lengths in pm, bond angles in degrees)

The free enthalpy of activation for the ring inversion has been calculated to be 42.3 kJ mol⁻¹ (at 212 K). It is similar to that of cyclohexane (43.0 kJ mol⁻¹), but lower than that of piperidine (46.1 kJ mol⁻¹). The inversion barriers of the saturated six-membered ring heterocycles containing one group-16 atom decrease with the size of the heteroatom; they are 39.3 for the sulfide, 34.3 for the selenide and 30.5 kJ mol⁻¹ for the telluride.

The NMR spectra of tetrahydropyran show the chemical shifts expected for a cyclic ether.

^{1}H-NMR (CDCl$_3$) ^{13}C-NMR (CDCl$_3$)
δ (ppm)   δ (ppm)

H-2/H-6:   3.65   C-2/C-6: 68.6   C-4: 24.3
H-3/H-4/H-5: 1.65   C-3/C-5: 27.4

Electronegative substituents (alkoxy groups, halogens) in the 2(6) position of the tetrahydropyran system prefer to adopt the axial position in the conformational equilibrium.

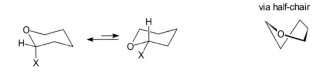

via half-chair

This structural characteristic, called anomeric effect, plays an important part in the chemistry of the carbohydrates, since the pyranose form of sugars corresponds structurally to the 2-substituted tetrahydropyrans as shown above.

C The synthesis of tetrahydropyrans can be carried out by

(1) cyclodehydration of 1,5-diols;

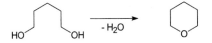

(2) acid-catalyzed cyclization of 4-hydroxybutyloxiranes (e.g. **2**) to give 2-hydroxymethyltetrahydro-pyrans (e.g. **3**);

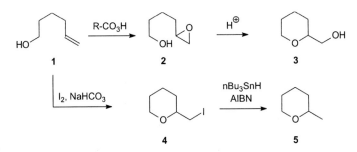

(3) cyclization of hex-5-en-1-ol (**1**, the educt for the oxiranes **2**) with electrophilic halogen reagents to give 2-substituted tetrahydropyrans (e.g. **4/5**).
  If the systems **1** and **2** bear a phenyl substituent in the 3-position, the ring closure reaction proceeds stereoselectively to give *trans*-2,4-disubstituted tetrahydropyrans **3/5** [21].
  The formation of 2-hydroxymethyltetrahydropyrans is of special importance for the stereoselective construction of substituted tetrahydropyran units in polyether antibiotics [22].
  Avermectin **6** is a natural product containing tetrahydropyran moieties. It is a potent acaricide used in plant protection. Its hydrogenated product ivermectin **7** is applied as an antiparasitic agent in

veterinary medicine. The eicosanoid thromboxane A_2 **8** stimulates platelet aggregation and is at the same time converted into inactive thromboxane B_2 **9**.

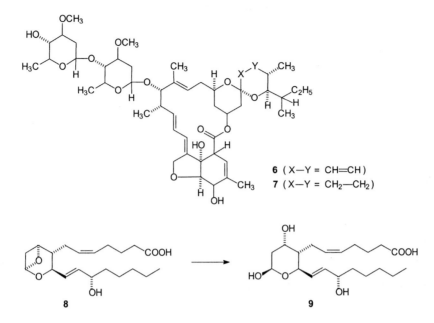

6 (X—Y = CH=CH)
7 (X—Y = CH₂—CH₂)

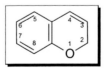

8   **9**

2-Alkoxytetrahydropyrans are formed by addition of alcohols to 3,4-dihydro-2H-pyrans (see p 239). 5-Hydroxycarbonyl compounds are often in equilibrium with 2-hydroxytetrahydropyrans (lactols), e.g.:

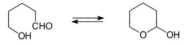

6.6   *2H*-Chromene

A,C The systematic name is 2H-1-benzopyran. 2H-Chromene shows the following shifts in its ¹H-NMR spectrum: δ = 4.53 (2-H), 5.38 (3-H) and 6.20 (4-H) in CDCl₃.

The synthesis of 2H-chromenes generally starts from phenols.

(1) *Cyclization of 2-propynyl aryl ethers*

This thermal isomerization of **1** is initiated by a [3,3] sigmatropic reaction involving the phenyl ring. The allene **3** is produced which rearranges to the *o*-quinomethane **4** by a [1,3] sigmatropic process; finally, electrocyclization of **4** results in formation of 2*H*-chromene **2**:

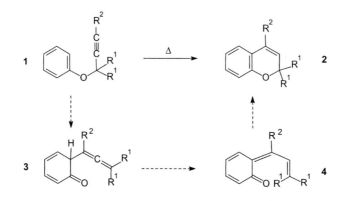

(2) *Cyclocondensation of phenols with α,β-unsaturated carbonyl compounds*

Chromene formation is catalyzed by base, aldehydes are favoured as cyclocondensation components.

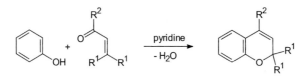

(3) *Oxidative cyclization of (β-aryl)allyl alcohols*

6-Hydroxychromenes **7** result from the oxidative cyclization of (*Z*)-1-hydroxy-3-(3'-hydroxyphenyl)-2-propenes **5** by means of Ph-I(OAc)$_2$ and subsequent reduction of the initial product **6** [23]:

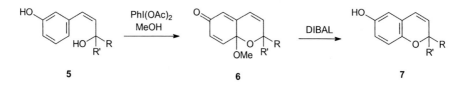

D	The 2*H*-chromene system, especially the 2,2-dimethyl substituted derivative, is a constituent of numerous natural products, such as evodionol **8** and lapachenol **9**, which occur in plants, and

precocene I/II **10**, a juvenile hormone antagonist.

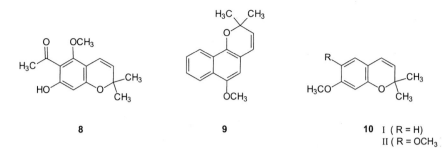

8   **9**   **10** I (R = H)
                                                                 II (R = OCH₃)

Spiro-2*H*-chromenes of type **11** are photochromic: On irradiation, the 2*H*-pyran ring is opened reversibly involving a ring–chain valence isomerization. This leads to the formation of coloured zwitterionic merocyanines **12** which recyclize spontaneously:

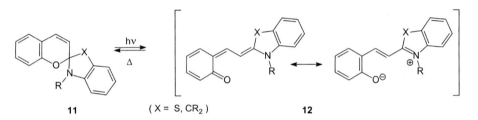

11   (X = S, CR₂)   **12**

Coumarin **13** (2*H*-1-benzopyran-2-one), the 1-benzopyrylium ion **14** and the flavylium ion **15** (2-phenyl-1-benzopyrylium ion) derived from flavane (2-phenyl-2*H*-chromene) are important derivatives of 2*H*-chromene. They are discussed in chapters 6.7 and 6.8.

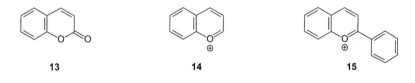

13   **14**   **15**

6.7   2*H*-Chromen-2-one

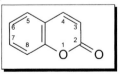

A | 2*H*-Chromen-2-one possesses the trivial name coumarin. X-ray analysis shows coumarin to be nearly planar (see Fig. 6.6):

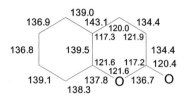

Fig. 6.6   Bond parameters of coumarin
(bond lengths in pm, bond angles in degrees)

Its spectroscopic properties correspond to those of the parent compound 2*H*-pyran-2-one. Coumarin has the character of an enol lactone rather than that of a heteroarene, as is shown by its IR and NMR spectra:

IR (KBr) ν (cm^{-1})	^{1}H-NMR (CDCl$_3$) δ (ppm)		^{13}C-NMR (CDCl$_3$) δ (ppm)	
1710	H-3: 6.43	H-6: 7.22	C-2: 159.6	C-4a: 118.1
(C=O bond)	H-4: 7.80	H-7: 7.45	C-3: 115.7	C-8a: 153.1
	H-5: 7.36	H-8: 7.20	C-4: 142.7	

B Characteristic reactions of coumarin are, accordingly, additions to the C-3/C-4 double bond and nucleophilic opening of the lactone function [24].

Thus, bromine reacts to give the dibromide **1** which eliminates HBr by the action of bases and leads to the formation of 3-bromocoumarin:

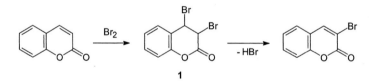

Very reactive electrophiles can attack the carbonyl oxygen of coumarin. For instance MEERWEIN's salt [Et$_3$O]BF$_4$ effects *O*-alkylation to give the benzopyrylium ion **2**:

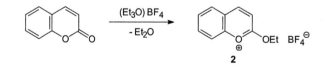

Hydroxide ions open the lactone ring of coumarin forming the dianion of the (*Z*)-*o*-hydroxycinnamic acid **3** (coumarinic acid) which recylizes to coumarin when acid is added. The dianion **3** can be stabilized as the (*Z*)-methoxy ester **5** by methylation with dimethyl sulfate. However, in prolonged reaction, it changes into the *E*-compound **4**:

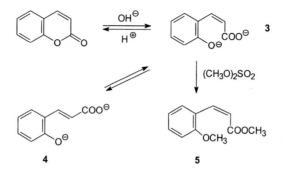

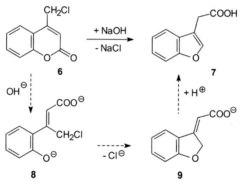

A number of rearrangements are observed in the coumarin system, e.g. the ring contraction of 3,4-dibromo-3,4-dihydrocoumarin **1** with alkali to give benzo[*b*]furan (coumarone, see p 64). The rearrangement of 4-chloromethylcoumarin **6** into coumarone-3-acetic acid **7** by aqueous alkali is mechanistically related to this reaction. This occurs via an S_Ni cyclization of the phenolate **8** into the carboxylate **9** which then isomerizes:

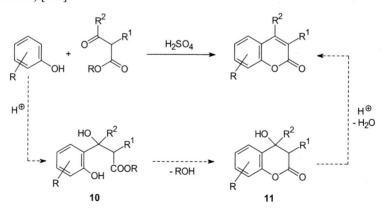

C For the synthesis of coumarins several methods are used.

(1)   Phenols undergo cyclocondensation with β-keto esters under the influence of strong acids (*v. Pechmann synthesis*) [24a]:

The course of this process corresponds to a S_EAr reaction of phenol with the presumably protonated carbonyl group of the β-keto ester **10**. This is followed by lactonization giving **11** and H_2O elimination to produce the coumarin system.

An alternative [25] uses the addition of *o*-metalated phenolic ethers **12** to alkoxymethylene malonic ester as a first step. Treatment with acid is followed by removal of the phenolic protective group, lactoniziation and ROH elimination to give the ester **13**:

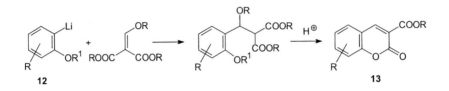

This modification allows a regioselective attack of the 1,3-biselectrophile on substituted phenols to occur, which is often not the case in the V.PECHMANN synthesis.

(2) Cyclocondensation of *o*-hydroxybenzaldehydes with reactive methylene compounds (malonic ester, cyanoacetic ester, malononitrile) in the presence of piperidine and other bases (*Knoevenagel synthesis*):

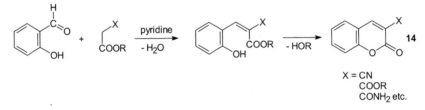

In the course of this reaction, derivatives of coumarin-3-carboxylic acid are produced. This synthesis proceeds under much milder conditions than the condensation based on the Perkin reaction which involves condensation of salicylaldehydes with acetic anhydride/acetate.

(3) Pd(0)-initiated cyclocondensation of alkynoates with donor-substituted phenols leads to coumarins, e.g.:

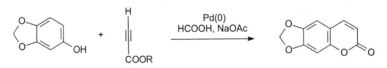

Yields are good even in instances where the V. PECHMANN synthesis is reported to be unsatisfactory [26].

D **Coumarin**, mp 68°C, is a colourless crystalline compound; it is the aromatic principle in woodruff and also occurs in other plants, e.g. lavender and melilot. Coumarin is prepared from salicylaldehyde by a PERKIN reaction with acetic anhydride or by cyclocondensation with 1,1-dimorpholinoethene [27]:

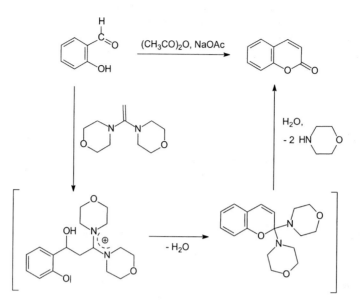

Many natural products contain the coumarin structure [28]. Aesculetin **15** is extracted from horse chestnuts and psoralen **16** from the Indian plant *Psoralea corylifolia*. Furocoumarins such as **16** are photochemically active. On UV irradiation, they induce processes in the cell which lead to an increase in skin pigmentation and inhibition of cell division. This is due to the formation of cyclobutane with the pyrimidine bases of nucleic acids. Psoralens are used in the treatment of psoriasis.

Aflatoxins are highly toxic and carcinogenic coumarin derivatives of a complex structure, e.g. aflatoxin B$_1$ **17**. It is formed as a secondary metabolite by *Aspergillus flavus* which occurs in mouldy food.

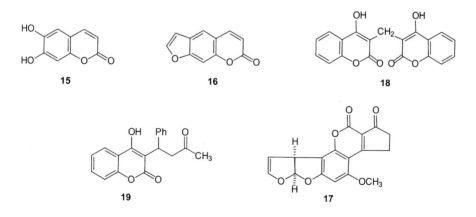

Dicumarol **18** and warfarin **19** are anticoagulants used in the treatment of thrombosis. Warfarin is synthesized by a MICHAEL addition of benzylidene acetone to 4-hydroxycoumarin [29].

6.8   1-Benzopyrylium ion

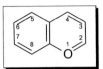

A 1-Benzopyrylium salts (chromylium salts) are coloured and possess longwave UV/VIS maxima at ≈ 385 nm. A bathochromic shift is observed in the presence of a 2-phenyl substituent, i.e. for 2-phenyl-1-benzopyrylium ions (flavylium ions). ^{1}H-NMR data (see pyrylium ion, p 223) show the influence of the positively charged oxygen which reduces the electron density on C-2 and C-4: δ = 9.75 (2-H), 8.40 (3-H) and 8.75 (4-H) (CF$_3$COOD).

This effect is even more pronounced with two benzannulations; thus, the dibenzo[b,e]pyrylium ion (xanthylium ion) shows shifts at δ = 10.18 (9-H) = 165.1 (C-9) (CF$_3$COOD).

B Among the reactions of 1-benzopyrylium salts [30], nucleophilic additions and subsequent reactions are of importance. GRIGNARD compounds add to C-2 and C-4 with formation of 2H- and 4H-benzopyrans. Hydroxide ions attack at C-2 to yield 2H-1-benzopyran-2-ol **1**, which is in equilibrium with the (Z)-o-hydroxycinnamaldehyde **2**. Benzopyranol **1** as a 'pseudobase' regenerates the 1-benzopyrylium ion on treatment with acid:

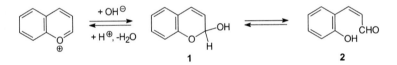

C The synthesis of 1-benzopyrylium ions is accomplished by the following methods.

(1)   Cyclocondensation of o-hydroxybenzaldehydes or -acetophenones with methylene ketones

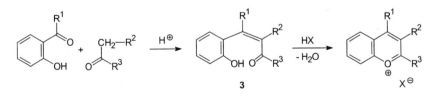

is carried out in acid medium, e.g. acetic anhydride/HClO$_4$. It proceeds via the intermediate **3** of the KNOEVENAGEL condensation product.

(2)   Cyclocondensation of activated phenols and β-diketones in an acid medium gives rise to formation of 1-benzopyrylium ions, for β-ketoesters cf. p 249:

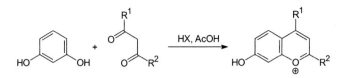

(3) Cyclocondensation of *o*-hydroxyacetophenones, orthoformic esters and arylaldehydes in the presence of a strong acid, usually $HClO_4$, leads to 4-alkoxy-2-aryl-1-benzopyrylium salts **4** [31]:

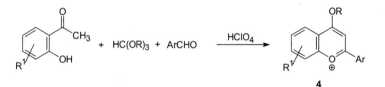

D **2-Phenyl-1-benzopyrylium salts** (flavylium salts) are important natural products (WILLSTÄTTER 1913). Numerous red, blue and violet pigments of blossoms and leaves are derived from the 3,5,7-trihydroxy-2-phenyl-1-benzopyrylium system **5** which has additional OH functions on the 2-phenyl substituent:

4'-OH	:	pelargonidin
3'-,4'-OH	:	cyanidin
3'-,4'-,5'-OH	:	delphinidin

They usually occur as glycosides (anthocyanins). Dilute acid causes hydrolysis to give a sugar and the corresponding aglycon 2-phenyl-1-benzopyrylium salt (anthocyanidin) [32].
Cyanidin chloride 6 occurs in the petals of the cornflower (*Centaurea cyanus*). It forms dark-red needles, soluble in H_2O but insoluble in diethyl ether or hydrocarbons. It is cleaved in molten potassium hydroxide giving phloroglucinol and 3,4-dihydroxybenzoic acid, a reaction which served to establish its structure.

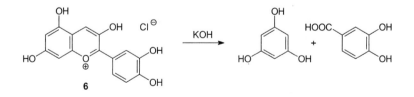

The colour of anthocyanidin is pH dependent. The cyanidin system occurs at pH 3 as a red 2-phenyl-1-benzopyrylium ion **6**; at pH = 8, deprotonation leads to the quinonoidal violet coloured base **7**. In a stronger basic medium (pH = 11), it is converted into the blue anion **8** which forms complexes with Al or Fe ions (E.BAYER 1966). The blue colour of certain blossoms, e.g. the cornflower, is due to the formation of these complexes [33] and not, as originally proposed, to a basic cell environment. Other aromatic compounds contribute to the colour intensity (copigmentation).

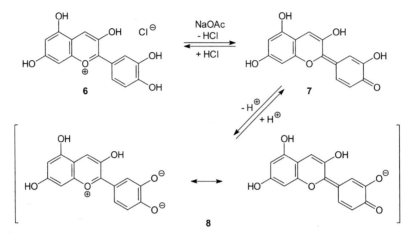

In general, 4'-hydroxy-2-phenyl-1-benzopyrylium ions undergo reversible ring opening of the pyrylium system. For instance, hydroxide ions first deprotonate compound **9** to the *p*-quinomethane **10**, and at higher hydroxide ion concentration cause ring opening to the chalcone **11**. Recyclization is achieved by treatment with acid to give the cation **9**:

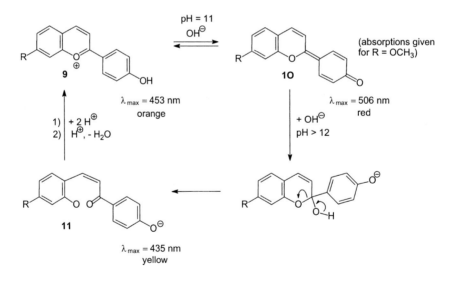

General methods for the synthesis of anthocyanidins have beene mentioned previously (see p 252). For instance, cyanidin chloride **6** is made from 2-benzoyloxy-4,6-dihydroxybenzaldehyde **12** by cyclocondensation with 2-acetoxy-1-(3,4-diacetoxyphenyl)ethanone in the presence of HCl (ROBINSON 1934).

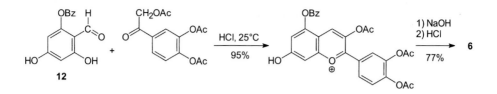

6.9   *4H*-Pyran

A In contrast to 2*H*-pyran, the parent compound 4*H*-pyran is known and spectroscopically characterized. The same applies to the 4,4-disubstituted derivatives. 4*H*-Pyrans have an IR absorption of the C=C bonds at ≈ 1700 and ≈ 1660 cm⁻¹ distinguishing them from 2*H*-pyrans (see p 232). The NMR spectra of 4*H*-pyrans show data expected for a bis-enol ether system:

^{1}H-NMR (CCl₄) ^{13}C-NMR (CCl₄)
δ (ppm)   δ (ppm)

H-2: 6.16   C-2: 141.1
H-3: 4.63   C-3: 101.1
H-4: 2.65

B Among the reactions of 4*H*-pyran, a few ring transformations are of synthetic interest. For instance, the triphenyl-4-benzyl-4*H*-pyran **1** isomerizes on irradiation to give 2-benzyl-2*H*-pyran **2** which, on addition of HCl, is converted quantitatively into 1,2,3,5-tetraphenylbenzene **4**. The transformation into the benzene system probably occurs by an electrocyclic O/C-2 opening of compound **2** to give the dienone **3**, followed by intramolecular aldol condensation:

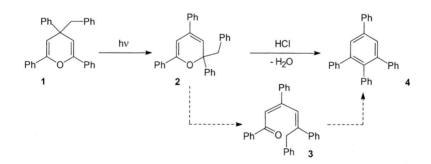

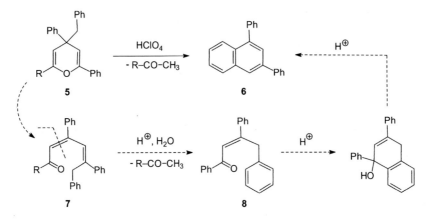

4-Benzyl-2-alkyl-4*H*-pyrans **5** are converted into 1,3-diphenylnaphthalene **6** on treatment with HClO₄. This reaction requires a rearrangement of the 4*H*-pyran **5** as shown above (**1** → **2**) to the corresponding 2*H*-pyran, followed by O/C-2 ring-opening. The resulting dienone **7** undergoes a retro-aldol reaction forming the methyl ketone R–CO–CH₃ and enone **8** which cyclizes to produce **6** by way of an intramolecular S_EAr reaction and dehydration of the intermediate **9**.

C The synthesis of 4*H*-pyrans is carried out by

(1) Cyclocondensation of β-disbustituted enone systems with β-keto esters via 1,5-dicarbonyl compounds **10**:

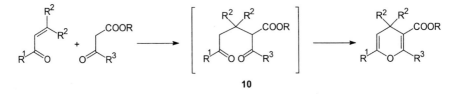

(2) Elimination reactions involving 2-acyloxy substituted dihydropyrans, e.g. **11**. The latter are accessible by a DIELS-ALDER reaction with inverse electron demand between enones and vinyl esters. One of the earliest syntheses of the parent compound is based on this reaction:

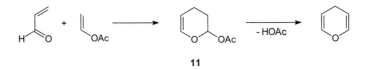

The most important derivative of 4*H*-pyran is the corresponding carbonyl system, namely the 4*H*-pyran-4-one. It is discussed in chapter 6.10.

6.10   *4H*-Pyran-4-one

A The structure of 4*H*-pyran-4-ones was elucidated by microwave spectroscopy and shows (see Fig. 6.7) bond parameters of a cross-conjugated, localized cycloenone system (cf. the data for 2*H*-pyran-2-one, p 233):

Fig. 6.7 Bond parameters of 4*H*-pyran-4-one
(bond lengths in pm, bond angles in degrees)

The ¹H- and ¹³C-NMR data of 4*H*-pyran-4-one are more in keeping with an α,β-unsaturated carbonyl system than with a delocalized heteroarene:

¹H-NMR (CDCl₃)   ¹³C-NMR (CDCl₃)
δ (ppm)   δ (ppm)

H-2: 7.88   C-2: 155.6   C-4: 179.9
H-3: 6.38   C-3: 118.3

The dipole moment of 4*H*-pyran-4-ones (ca. 4 D) and their greater basicity ($pK_a = 0.1$) than enones indicate that a betaine canonical form **1** makes only a minor contribution to the structure:

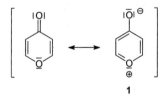

1

B The reactions of 4*H*-pyran-4-ones show some simililarities to the 2-isomer. On irradiation, 4*H*-pyran-4-one isomerizes to 2*H*-pyran-2-one. The following mechanism is proposed: Electrocyclization leads to the zwitterionic intermediate **2**, which rearranges to the epoxycyclopentadienone **3**; eventually migration of oxygen in **3** leads via **4** to 2*H*-pyran-2-one.

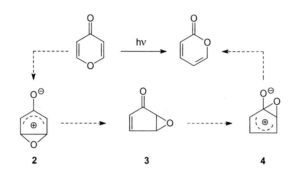

2   **3**   **4**

4*H*-Pyran-4-ones form salts with strong acids and undergo *O*-alkylation (see p 260). Electrophilic substitution occurs at various ring positions, especially in the presence of activating groups.

Nucleophiles attack 4*H*-pyran-4-ones on C-2 or C-4. With aqueous alkali, 1,3,5-triketones are formed by addition of hydroxide ions to C-2. For instance, 2,6-dimethyl-4*H*-pyran-4-one **5** gives heptane-2,4,6-trione **6**:

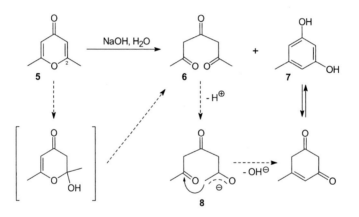

The formation of orcin **7** in addition to **6** is the result of a subsequent deprotonation of **6** to the enolate **8** which recyclizes by an intramolecular aldol condensation. This reaction gave rise to the hypothesis of the biosynthesis of aromatic compounds from polyketides (COLLIE 1907) [34].

4-Substituted pyrylium salts are obtained from GRIGNARD reagents and 4*H*-pyran-4-ones in a 1:1 proportion, followed by treatment with strong acid. Excess of RMgX, however, yields 4,4-disubstituted 4*H*-pyrans:

C According to retrosynthetic analysis, the cyclocondensation of 1,3,5-triketones is a flexible method for the synthesis of 4*H*-pyran-4-ones.

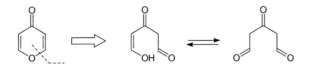

They are made by γ-acylation of 1,3-diketones with carboxylic esters by means of KNH_2 via dianions **9**. Their cyclization to give 4H-pyran-4-ones is brought about by strong acids:

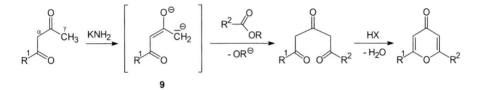

The lithium enolates of 4-methoxybut-3-en-2-ones **10**, when treated with acid chlorides, yield the cross-conjugated enol system **11** by C-acylation. As a masked triketone it cyclizes to form the 2,6-disubstituted 4H-pyran-4-ones on addition of catalytic amounts of CF_3COOH:

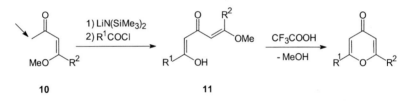

D   **4H-Pyran-4-one**, mp 32°C, is a colourless, crystalline compound. It forms a hydrochloride with HCl, mp 139°C. 4H-Pyran-4-one is prepared by decarboxylation of chelidonic acid **12**. This is made by α,α'-diacylation of acetone with oxalic ester and proceeds via the acid-catalyzed cyclization of the triketo compound **13**:

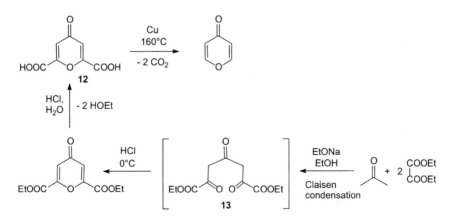

2,6-Dimethyl-4*H*-pyran-4-one, mp 132°C, is water soluble. It is obtained by various methods, the simplest involving the use of acetic anhydride and polyphosphoric acid at elevated temperature:

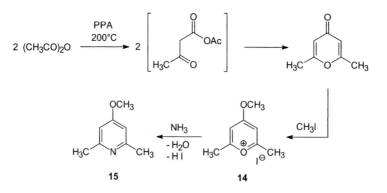

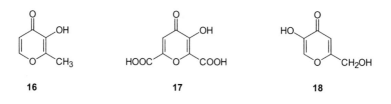

15   **14**

2,6-Dimethyl-4*H*-pyran-4-one is *O*-alkylated by CH₃I, this was established by converting the resulting pyrylium salt **14** into 4-methoxy-2,6-dimethylpyridine **15** with NH₃ (BAEYER 1910).

Apart from chelidonic acid **12**, which occurs in the roots of celendine (*Chelidonium majus*), a few other 4*H*-pyran-4-one derivates are of importance. Maltol **16** occurs in the bark of larch trees and is formed by the dry distillation of starch and cellulose. Meconic acid **17** occurs in opium. Kojic acid **18** is produced by many microorganisms and was first isolated from *Aspergillus oryzae*, a microorganism used in Japan in the production of sake.

16   **17**   **18**

6.11 4*H*-Chromene

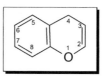

A-D 4*H*-Chromene is characterized by the following ¹H-NMR data: δ = 6.44 (2-H), 4.63 (3-H) and 3.36 (4-H) (CDCl₃). 4*H*-Chromenes absorb at ≈ 280 nm in the UV spectrum. They can be distinguished from 2*H*-chromenes, whose phenyl-conjugated double bond absorbs at longer wavelength (≈ 320 nm).

4*H*-Chromenes are prepared by reacting *o*-acyloxybenzyl bromides **1** with two equivalents of a phosphonium ylide:

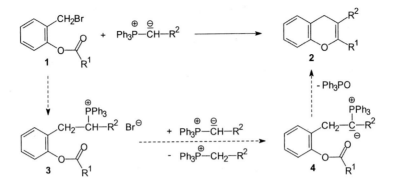

The first step results in alkylation of the ylide to give the phosphonium salt **3** which deprotonates to form another ylide **4**; intramolecular WITTIG reaction leads to the 2,3-disubstituted 4H-chromene **2**.

Important derivatives of 4H-chromene are chromone **5** (4H-1-benzopyran-4-one), flavon **6** (2-phenyl-4H-1-benzopyran-4-one) and xanthene **7** (note the exception to systematic numbering). They will be discussed in chapter 6.12.

5

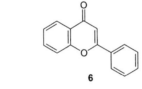

6

7

6.12 4*H*-Chromen-4-one

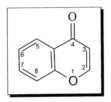

| **A** | 4H-Chromen-4-one possesses the trivial name chromone. It is characterized by the following NMR data (cf. 4H-pyran-4-one, p 257): |

¹H-NMR (CDCl₃)
δ (ppm)

^{1}H-NMR (CDCl$_3$)
δ (ppm)

^{13}C-NMR (CDCl$_3$)
δ (ppm)

H-2: 7.88	H-6: 7.42	C-2: 154.9	C-5a: 124.0
H-3: 6.34	H-7: 7.68	C-3: 112.4	C-8a: 156.0
H-5: 8.21	H-8: 7.47	C-4: 176.9	

Chromones differ from coumarins ($v_{C=O} \approx 1710$ cm^{-1}) by the position of the C=O absorption band in the IR ($v_{C=O} \approx 1660$ cm^{-1}). Chromones, especially flavones, are characterized by two UV absorption bands in the region of 240-285 and 300-400 nm.

B Chromones and flavones show analogies in their reactions to 4*H*-pyran-4-one, i.e. they behave as masked 1,3-dicarbonyl systems. Protonation and alkylation occur on oxygen. Electrophilic attack takes place at the deactivated pyran-4-one ring in the 3-position, e.g. aminomethylation can be brought about under MANNICH conditions.

The chromone system behaves as a MICHAEL acceptor towards nucleophiles. Normally, attack occurs at C-2, but is less likely on C-4, and after addition leads frequently to ring transformations. For instance, the 4*H*-pyran-4-one ring is opened by aqueous alkali, owing to H$_2$O addition to **1**, to form *o*-hydroxyphenyl-1,3-diketones **2**. Subsequent acid cleavage produces either *o*-hydroxyphenyl ketones and carboxylic acids or salicyclic acid and ketones:

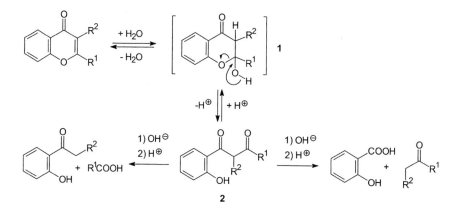

By ring cleavage, primary and secondary amines also lead to enaminones **3**, which on acid treatment revert to the chromone system:

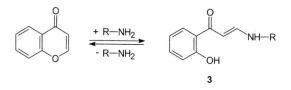

Chromones behave similarly towards both hydroxylamine and hydrazines. Ring opening via C-2 followed by recyclization giving azoles is of preparative interest. This is demonstrated by the reaction of chromone with phenylhydrazine, which leads either to the phenylhydrazone **5** or (via the enehydrazine **4**) to the pyrazole **6**:

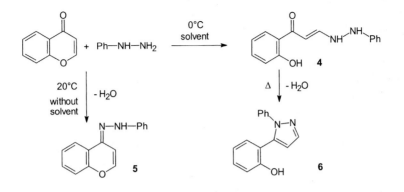

The reversible ring-opening of the chromone system giving *o*-hydroxyphenyl-1,3-diketones **2** (see above) can also occur by acid catalysis. If a further OH function is present in position 5, it also takes part in the recyclization. Chromones and especially flavones with an unsymmetrically substituted benzene ring undergo isomerization by this route when treated with strong acids (*Wesseley-Moser rearrangement*), e.g. **7** → **8**:

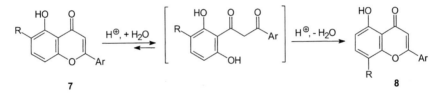

C | The syntheses of chromones and flavones usually start from *o*-hydroxyacetophenones.

(1) The most frequently used method is the acid-catalysed cyclization of *o*-hydroxyaryl-1,3-diketones **9**, which are obtained from *o*-hydroxyacetophenones, especially in their *O*-silyl protected form [35], by a CLAISEN condensation:

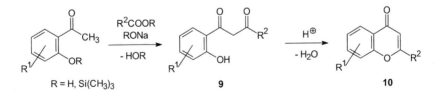

An alternative route to β-diketones **9** is the base-catalysed isomerization of *o*-acyloxyacetophenones **11** (*Baker-Venkataraman rearrangement*), which are readily obtained by *O*-acylation of *o*-hydroxy-acetophenones:

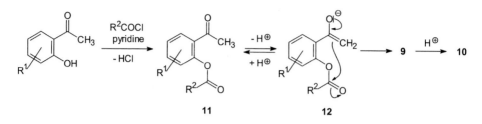

11   **12**

The BAKER-VENKATARAMAN rearrangement can be regarded as a 1,5-acyl migration in the enolate **12** and is of great value in the synthesis of flavones [36].

(2)   A versatile flavone synthesis consists of the oxidative cyclization of chalcones **13** by selenium dioxide in higher alcohols:

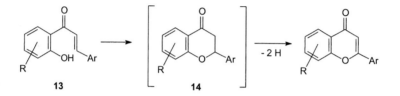

13   **14**

It involves an intramolecular, possibly acid-catalysed, MICHAEL addition of the phenolic OH group resulting in a cyclization to the flavanone **14** which undergoes dehydrogenation to give a flavone.

(3)   Chromones and flavones **16** are formed by Pd(0)-catalyzed carbonylative cyclization of *o*-hydroxy- or *o*-acetoxy iodo arenes with terminal alkynes in the presence of Et_2NH. Intermediates are the corresponding *ortho*-functionalyzed aryl alkynyl ketones **15**, which can be synthesized separately and cyclized by Et_2NH [37].

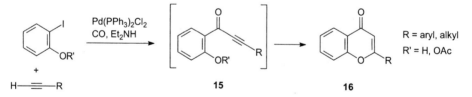

15   **16**

D **Chromone** and **flavone** crystallize in colourless needles of mp 59 and 97°C, respectively. Chromone forms a hydrochloride with HCl. It has a higher pK_a (2.0) than $4H$-pyran-4-one.
Hydroxy-substituted flavones occur widely as yellow pigments in plants, where they are present as O- and C-glycosides [38]. Examples are apigenin **17** and luteolin **18**. 3-Hydroxyflavones, e.g. quercetin **19** and kaempferol **20** are known as flavonols and those with a hydrogenated heterocyclic ring as flavanols. Catechin **21** is the most important representative of flavanols forming the monomeric building-block contained in the so-called condensed tannins. In former times luteolin and quercetin were used as dyestuffs of textile fibres. Flavones were found to possess a wide spectrum of pharmacological actions; for instance, polyhydroxyflavones possess anti-inflammatory and flavone-8-acetic acid **22** antitumour properties [39].

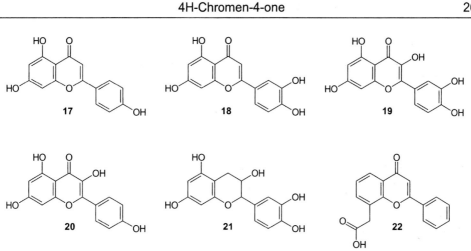

17   **18**   **19**

20   **21**   **22**

Xanthene 24 forms colourless crystals of mp 100°C and is prepared by thermal dehydration of 2,2'-dihydroxydiphenylmethane **23**, its oxidation with HNO_3 leads to xanthone **25**:

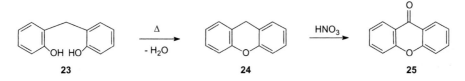

23 $\xrightarrow[-H_2O]{\Delta}$ **24** $\xrightarrow{HNO_3}$ **25**

A number of dyes are derived from the xanthene structure (xanthene dyes) such as fluorescein **26**, eosin **27** and pyronine G **28**:

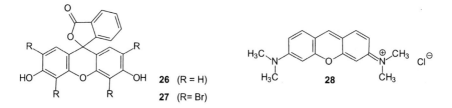

26 (R = H)
27 (R= Br)

28

Pyronine G **28** is obtained by the oxidation, in the presence of acids, of the leuco compound **29**, which in turn is made by an acid-catalysed cyclocondensation of 3-dimethylaminophenol with formaldehyde:

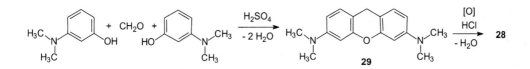

$+ CH_2O +$ $\xrightarrow[-2 H_2O]{H_2SO_4}$ **29** $\xrightarrow[-H_2O]{[O] \atop HCl}$ **28**

Rotenone **30**, the toxic principle from the roots of *derris elliptica*, is a natural insecticide. Its structure shows an unique combination of benzannelated dihydropyran, dihydro-4-pyrone and dihydrofuran ring systems.

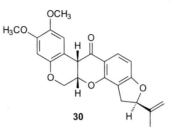

30

6.13 Chroman

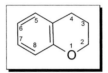

A Chroman (3,4-dihydro-2*H*-1-benzopyran) is derived from 3,4-dihydro-2*H*-pyran by annulation of a benzene ring. Its structural isomer is the isochroman **1**. 2-Phenylchroman **2** is known as flavan.

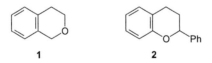

1   **2**

Chroman is characterized by the following NMR data:

¹H-NMR (CDCl₃) δ (ppm)	¹³C-NMR (CDCl₃) δ (ppm)	
H-2: 3.82	C-2: 74.0	C-5a: 121.7
H-3: 1.70	C-3: 32.8	C-8a: 147.5
H-4: 2.28	C-4: 22.6	

NMR examination shows that the chroman system exists in a half-chair conformation.

| B | Reactions with electrophiles, e.g. S_EAr processes, occur on the benzene moiety on C-6 as expected. 2,2-Disubstituted chroman-4-ones undergo ring opening by attack of alkali hydroxide on C-2 and formation of α,β-unsaturated o-hydroxyphenyl ketones **3**. Acids bring about recyclization. Flavanones are ring opened by traces of alkali. |

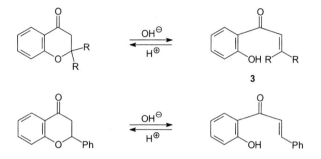

Isochroman **1** undergoes substitution and ring-opening, provided suitable leaving groups are present on the reactive C-1 benzyl position. The transformations of the 1-bromo compound **4**, which involve the carboxonium ion **5**, serves as a characteristic example [40]:

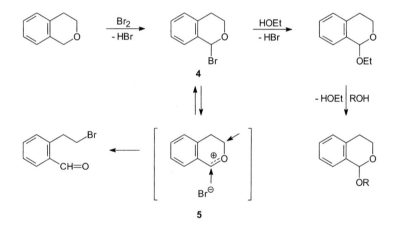

| C | The synthesis of chromans is possible according to several routes. |

(1) 1-Halogeno- or 1-hydroxy-substituted 3-(o-hydroxyphenyl)propanes **6** are cyclized to give chroma-nes, the alcohols by strong acids (H_2SO_4/HOAc, polyphosphoric acid), the halides by bases. These processes occur in an S_N fashion with retention or inversion of configuration [41].

(2) 1-Chloro-3-phenoxypropanes **7** are cyclized by means of *Lewis* acids. The regioselectivity of this principle can be improved by a PARHAM cycloalkylation. In this procedure, 1-bromo-3-(2-bromophenoxy)propanes **8** are cyclized by *n*-butyllithium in a halogen–metal exchange followed by intramolecular alkylation [42].

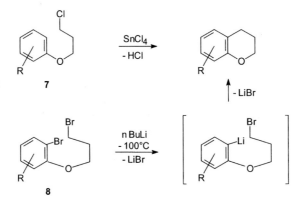

(3) In a Pd-catalyzed intramolecular cross-coupling reaction aryl halides containing an *ortho*-(γ-hydroxy)alkyl side chain are cyclized to give 2,2-disubstituted chromanes [43].

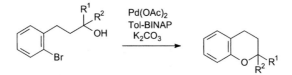

Chroman-4-ones are obtained by acid-catalyzed cyclization of α,β-unsaturated *o*-hydroxyphenyl ketones **9**:

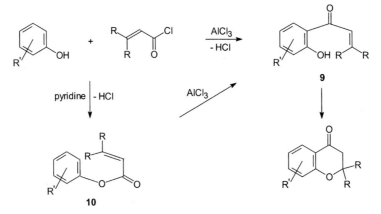

The enones **9** are made either by a FRIEDEL-CRAFTS reaction of phenols and α,β-unsaturated acid chlorides or by a FRIES rearrangement of the phenol esters **10**.

Isochroman is formed from 2-phenylethan-1-ol and formaldehyde in the presence of hydrochloric acid, with 2-phenyl-1-ethyl chloromethyl ether **11** as intermediate:

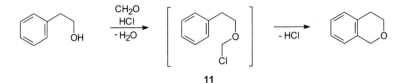

11

D **Chroman** is a colourless, steam-volatile oil, bp 214°C, of a peppermint-like odour.
The chroman moiety is a constituent of some important natural products, e.g. the tocopherols and canabinoids. α-Tocopherol **12** (vitamine E) occurs in wheatgerm oil. It contains three asymmetric centres (C-2, C-4' and C-8') which each have an (*R*)-configuration. α-Tocopherol was prepared by a stereoselective synthesis [44].

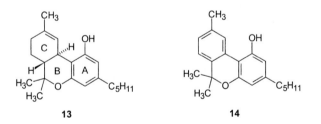

12

Among the constituents of hemp (*Cannabis sativa*), the tricyclic Δ⁹-*trans*-tetrahydrocannabinol **13** possesses the strongest hallucinogenic activity (hashish, marijuana). Cannabinol **14**, which possesses an aromatic C-ring, is psychomimetically inactive.

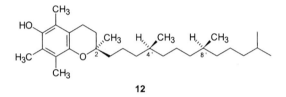

13   **14**

6.14 Pyridine

A Pyridine is the simplest heterocycle of the azine type. It is derived from benzene by replacement of a CH group by an N-atom. The pyridinium ion is isoelectronic with benzene. The positions are indicated as α-, ß-, γ- or numbered 2-, 3-, 4-. The univalent radical is known as pyridyl. Methylpyridines are known as picolines, dimethylpyridines as lutidines and 2,4,6-trimethylpyridine as collidine. The carbonyl systems based on pyridine, namely 2-pyridone and 4-pyridone, are covered in section 6.15.

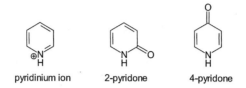

pyridinium ion   2-pyridone   4-pyridone

Pyridine has certain analogies to benzene in its molecular geometry and spectroscopic properties. According to microwave spectroscopy, the pyridine ring is a slightly distorted hexagon. Its C–C bond lengths, as well as its C–N (147 pm) and C=N (128 pm) bond distances are to a large extent averaged (see Fig. 6.8).

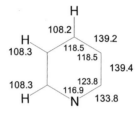

Fig. 6.8   Bond parameters of pyridine   cf. benzene: C-C 139.7
            (bond lengths in pm, bond angles in degrees)   C-H 108.4
                                                                              C-C-C 120°

Pyridine has the following UV and NMR data (cf. also pyridinium ion, p 223):

UV (ethanol) λ (nm) (ε)	^{1}H-NMR (CDCl$_3$) δ (ppm)	^{13}C-NMR (CDCl$_3$) δ (ppm)
251 (3.30) π → π*	H-2/H-6: 8.59	C-2/C-6: 149.8
270 (sh) n → π*	H-3/H-5: 7.38	C-3/C-5: 123.6
	H-4: 7.75	C-4: 135.7

Found for benzene:

208 (3.90) 262 (2.41) π → π* (hexane)	7.26	128.5

These data confirm pyridine to be a delocalized 6π-heteroarene with a diamagnetic ring current. Due to the anisotropic effect of the nitrogen, the individual ring positions have differing π-electron densities. The chemical shifts of the pyridine protons as well as of the ring C-atoms show the α-position to be the most deshielded. The γ-position suffers less deshielding relative to the β-position, which has values closest to benzene (see above). Pyridine can be described mesomerically by canonical structures in which the π-electron density is lowest on the 2, 4 and 6 C-atoms, and highest on the N-atom:

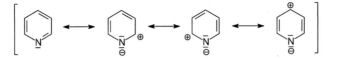

The pyridine system is characterized by an empirical resonance energy of $\Delta E_\pi = 134$ kJ mol^{-1} and a DEWAR resonance energy calculated by the SCF/MO method of 87.5 kJ mol^{-1} (cf. benzene $\Delta E_\pi = 150$ kJ mol^{-1}, Dewar resonance energy = 94.6 kJ mol^{-1}).

The electronic structure of the pyridine system can also be described by means of MO theory [45]. All the ring atoms are sp^2-hybridized. The linear combination of the six $2p_z$ atomic orbitals lead to six delocalized π-MOs, three of which are bonding and three antibonding (see Fig. 6.9).

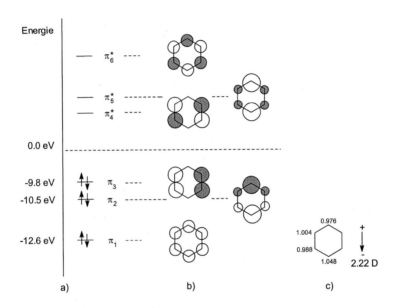

Fig. 6.9   Electronic structure of pyridine
(a) energy level scheme of the π-MOs and electronic occupation
(b) π-MO (the N-atom occupies the lowest corner of the hexagon)
(c) π-electron densities calculated by *ab initio* MO methods [46] and dipole moment

In contrast to benzene, π_2 and π_3, as well as π_4^* and π_5^*, are not degenerate because a nodal plane on the one hand bisects the σ-framework between the C-atoms 2,3 and 5,6 and on the other hand passes between the N-atom and C-4. Each ring atom contributes one electron to the cyclic conjugated system. The six electrons occupy the three bonding π-MOs in pairs. The ionization potentials, and hence the orbital energies of π_1, π_2 and π_3 were ascertained from the photoelectronic spectra (see Fig. 6.9). When compared with the values of benzene ($\pi_1 = -12.25$ eV, $\pi_2 = \pi_3 = -9.24$ eV), it is evident that the N-atom in pyridine lowers the energy of the delocalized π-MOs, which results in a stabilization of the π-system.

Differing values are obtained for the π-electron densities q according to the MO method applied. However, they all show the same rank order. On the N-atom, q is largest, followed by the C-atoms 3 and 5. On the C-atoms 2, 4 and 6, $q < 1$ (see Fig. 6.9). In agreement with these values, and in contrast to benzene ($q = 1.000$ on each C-atom), pyridine possesses a dipole moment, found to be 2.22 D and with its negative end pointing towards the N-atom. This is a consequence of the greater electronegativity of nitrogen compared with carbon. Compared to pyrrole (see p 87), pyridine belongs to the group of *π-deficient nitrogen heterocycles*.

B For pyridine the following reactions can be predicted on the basis of its electronic structure:

- Electrophilic reagents attack preferably at the N-atom and at the β-C-atoms, while nucleophilic reagents prefer the α- and γ-C-atoms;
- Pyridine undergoes electrophilic substitution reactions (S_EAr) more reluctantly but nucleophilic substitution (S_NAr) more readily than benzene;
- Pyridine undergoes thermal as well as photochemical valence isomerizations analogous to benzene.

Electrophilic reactions on nitrogen

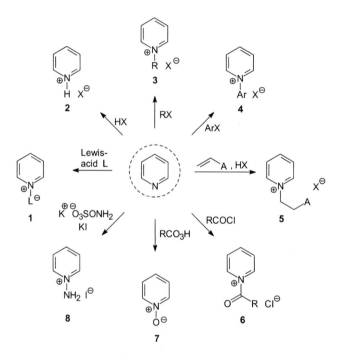

LEWIS acids such as $AlCl_3$, $SbCl_5$, SO_3, etc., form stable N-adducts of type **1**. The SO_3 adduct serves as a mild sulfonating reagent (see p 55). BRÖNSTED acids form salts **2** with pyridine. Some pyridinium salts are used as synthetic reagents; for instance, pyridinium chlorochromate **9** and pyridinium dichromate **10** are oxidizing agents (prim. alcohol → aldehyde, sec. alcohol → ketone), and pyridinium perbromide **11** acts as a brominating agent.

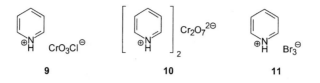

9   **10**   **11**

Alkyl halides, alkyl tosylates or dialkyl sulfates bring about *N*-quaternization of pyridine to give *N*-alkylpyridinium salts **3**, while activated haloarenes, e.g. 1-chloro-2,4-dinitrobenzene, form *N*-arylpyridinium salts **4**. *N*-Alkylation also involves the acid-induced MICHAEL addition of pyridine to acrylic acid derivatives (e.g. to give **5**, A = CN, CO_2R) in the presence of HX, as well as the synthetically useful KING-ORTOLEVA reaction (see p 309) between reactive methyl or methylene compounds and pyridine in the presence of I_2, e.g.:

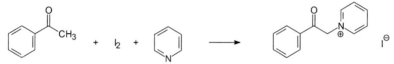

With pyridine, acid chlorides and acid anhydrides yield *N*-acyl pyridinium salts **6** which are very reactive and sensitive to hydrolysis, unlike the quaternary salts **2** - **5**; they are involved in the acylation of alcohols and amines in pyridine as solvent (EINHORN variant of the SCHOTTEN-BAUMANN reaction). However, 4-dimethylaminopyridine **12** (STEGLICH reagent) and 4-(pyrrolidin-1-yl)pyridine **13** are better acylation catalysts by a factor of 10^4 [46a]. Reactions with sulfonyl chlorides in pyridine proceed via *N*-sulfonylpyridinium salts **14**.

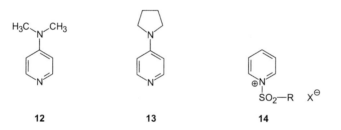

12   **13**   **14**

Peroxy acids react with pyridine to give pyridine *N*-oxide **7** by an electrophilic oxygen transfer (for reactions, see p 285). Other functions can also be electrophilically transferred to the pyridine nitrogen, e.g. CN by cyanogen bromide or NH_2 by potassium hydroxylamine-*O*-sulfonate/KI to **8**.

Electrophilic substitution reactions

S_EAr reactions proceed much more slowly with pyridine than with benzene. They usually demand drastic conditions and occur exclusively at the 3-position [47]. The reactivity of pyridine is comparable to that of nitrobenzene ($\approx 10^{-7}$ relative to benzene). In S_EAr reactions occuring in strongly acidic media (nitration, sulfonation), this reactivity is similar to that 1,3-dinitrobenzene ($< 10^{-15}$). The basicity of the pyridine nitrogen is crucial in deciding whether the S_EAr reactions in an acid medium involve the free pyridine base or the further deactivated pyridinium ion; for instance, pyridines with a $pK_a > 1$ are nitrated via the protonated species, while in the case of pyridines with a $pK_a > 2.5$, the free base is involved. As expected, donor substituents increase the S_EAr reactivity.

The experimentally observed orientations can be rationalized by considering the stability of the σ-complexes (WHELAND intermediate) involved. For instance, when comparing the σ-complexes resulting from the addition of electrophiles to the 2-, 3- and 4-positions in pyridine, it is found that only electrophilic attack on the 3-position avoids the energy-rich nitrenium canonical form. Dications are postulated as intermediates for reactions involving pyridinium ions. Among these, the product resulting from attack on the 3-position has the most favourable electronic stability.

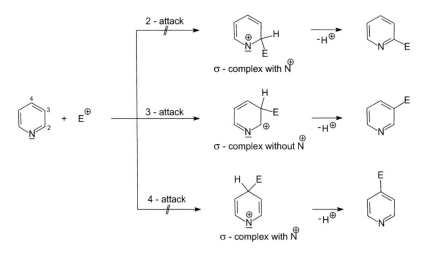

Some S_EAr reactions on pyridine show interesting features with respect to product formation, reactivity and orientation.

Nitration of pyridine with HNO_3/H_2SO_4 occurs under drastic conditions (ca. 300°C), but yields only ca. 15% of 3-nitropyridine **16**. N_2O_5, however, gives rise to good yields of **16** (ca. 70%) when reacted in CH_3NO_2 or SO_2. A mechanism including primary *N*-attack (via intermediate **15**) and an addition/elimination sequence of SO_2 or HSO_3^- is suggested [48].

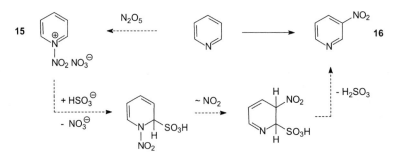

For the activation of pyridine, more than one alkyl group is necessary. Thus, on nitration picolines suffer extensive oxidation of the side chain, but 2,6-lutidine and 2,4,6-collidine are converted smoothly into the 3-nitro products.

　　Hydroxy- and aminopyridines show special features. 3-Hydroxypyridine undergoes nitration exclusively in the 2-position and, when the 2-position is blocked, in the 4-position:

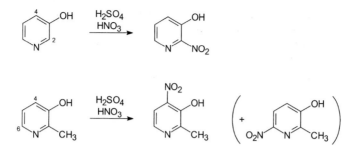

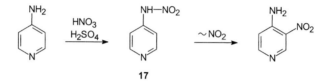

Amino groups activate and direct in an analogous way to the OH function. However, nuclear substitution is usually preceded by *N*-substitution leading to an (acid-catalysed) BAMBERGER-HUGHES-INGOLD rearrangement. The nitration of 4-aminopyridine giving 4-amino-3-nitropyridine via the *N*-nitroamine **17** illustrates this point:

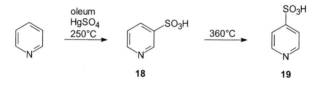

Sulfonation of pyridine with oleum at 250°C with Hg(II) catalysis yields the pyridine-3-sulfonic acid **18** in 70% yield:

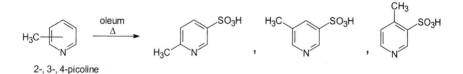

The effect of the Hg(II) in this reaction is attributed to *N*-coordination and suppression of the strongly deactivated *N*-protonation. Sulfonation of pyridine at 360°C or heating of **18** at this temperature yields the pyridine-4-sulfonic acid **19**, which is indicative of a thermodynamic control of the 4-substitution.

Alkyl-substituted pyridines show some peculiarities. Thus, sulfonation of 2-, 3- and 4-picoline always produces the 5-sulfonic acid:

2,6-Lutidine does not undergo nuclear substitution with SO_3, but yields **20** by *N*-addition. However, 2,6-di-*tert*-butylpyridine is converted into the 3-sulfonic acid **21** under mild conditions (SO_3, liq. SO_2, -10°C); with oleum at higher temperature the sulfone **22** is obtained as well as **21**.

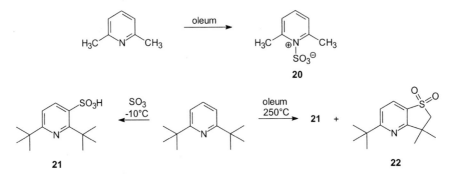

The ready formation of **21** indicates steric hindrance towards attack of the N-atom by the bulky *tert*-butyl groups; accordingly, only nuclear substitution to the alkyl-activated base occurs.

Halogenation of pyridine occurs with elemental chlorine or bromine at high temperature. 3-Halo- and 3,5-dihalopyridines are formed at ca. 300°C as a result of an ionic S_EAr process:

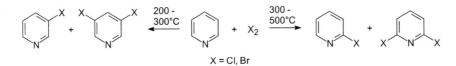

At higher than 300°C, 2-halo- and 2,6-dihalopyridines are produced, probably by a radical mechanism.

Comparatively little is known about other electrophilic substitution reactions of pyridine. Exceptions are activated systems, e.g. 3-hydroxypyridine **23** which undergoes azo-coupling, carboxylation and hydroxymethylation. Its *O*-ethyl ether **24** can be ring-alkylated by a FRIEDEL-CRAFTS method [49]:

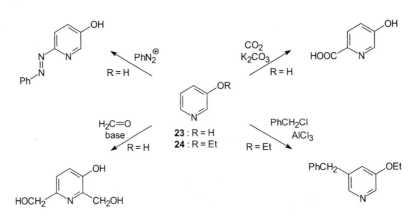

Nucleophilic substitution reactions and metalation

As expected, N-, O-, S- and C-nucleophiles attack the ring C-atoms of pyridine. Addition of the nucleophile and elimination of a pyridine substituent as leaving group occur in a two-step process, i.e. in an S_NAr reaction with regeneration of the heterarene system. S_NAr reactions in pyridine occur preferably in the 2- and 4-positions and less readily in the 3-position, as indicated by studies of relative reactivity of halopyridines (e.g. chloropyridine + NaOEt in EtOH at 20°C: relative reaction rates 2-Cl ≈ 0.2, 4-Cl = 1, 3-Cl ≈ 10^{-5}).

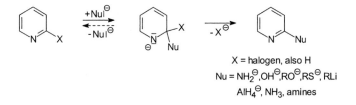

X = halogen, also H
Nu = $NH_2^{\ominus}$, $OH^{\ominus}$, $RO^{\ominus}$, $RS^{\ominus}$, RLi
$AlH_4^{\ominus}$, NH_3, amines

Strong nucleophiles, e.g. amides, organolithium compounds and hydroxides, react at higher temperature according to this S_NAr scheme even with unsubstituted pyridines, although the hydride ion is a poor leaving group.

With 3-halopyridines, the nucleophilic substitution takes place by an aryne mechanism. For instance, reaction of 3-chloropyridine with KNH_2 in liquid NH_3 yields a mixture of 3- and 4-aminopyridine, which is indicative of a 3,4-dehydropyridine ('hetaryne') as an intermediate (for comparison, cf. p 279).

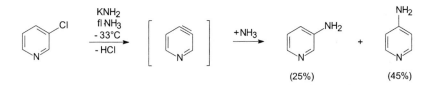

(25%)   (45%)

S_NAr reactions in *N*-alkylpyridinium ions possessing leaving groups on the ring C-atoms proceed much faster than in pyridine. This applies especially to the 2-position (e.g. chloro-*N*-methylpyridinium salts + NaOEt in HOEt at 20°C: relative reaction rates 2-Cl ≈ 10^{11}, 4-Cl ≈ 10^6, 3-Cl ≈ 10^5; rate = 1 for 4-chloropyridine).

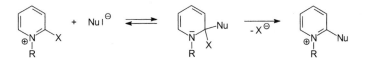

The *Chichibabin reaction* is historically the first S_NAr reaction known for pyridine. It involves reaction with sodium amide (in toluene or dimethylaniline) and produces 2-aminopyridine **25** regioselectively:

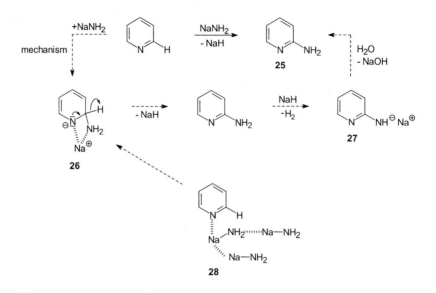

The (simplified) mechanism of the CHICHIBABIN reaction as shown takes into account the following features: loss of the 2-hydrogen as a hydride ion, control of the regioselectivity by Na coordination in the addition complex **26** and formation of the intermediate amide **27**. The reaction sequence is probably more complex and starts with coordination of the pyridine to the NaNH$_2$ surface (structure **28**). Formation of **26**, as well as a single electron transfer (SET) to the heterocycle, is a possibility. This sequence is suggested by the formation of products arising from a radical dimerization, e.g. in the CHICHIBABIN reaction on acridine (see p 354) [50].

Reaction of alkyl- or aryllithiums with pyridine, i.e. the *Ziegler reaction*, also occurs with 2-substitution giving the products **29**:

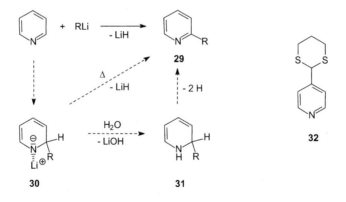

In the reaction of phenyllithium with pyridine, the primary adduct **30** and its protonated product, namely 1,2-dihydropyridine **31**, can be isolated. The complex **30** (R = Ph) is transformed into 2-phenylpyridine (**29**, R = Ph) when heated to 100°C, while 1,2-dihydropyridine **31** (R = Ph) has to be dehydrogenated with O$_2$ (**29**, R = Ph).

In contrast, 2-lithio-1,3-dithiane gives with pyridine the product of 4-substitution **32**.

Organolithium compounds also are capable to react with pyridines by metalation, i.e. by H-metal exchange. The direct lithiation is assisted by substituents (a) which favour deprotonation by inductive effects, like halogens, (b) which stabilize the pyridyl lithium compound by chelating effects, like alkoxy or amide residues, both operating for metalation in *ortho*-position to the given substituent. The regioselectivities observed often are in accord with the finding, that H-D exchange in pyridine itself (MeOD/MeONa, 160°C) occurs in all pyridine positions, but in relative rate of $\alpha : \beta : \gamma \approx 1 : 9 : 12$. This is illustrated by the following examples for metalations of substituted pyridines followed by electrophilic transformations of the pyridyl lithium compounds formed (**33-35** → **36-38**) [50a]:

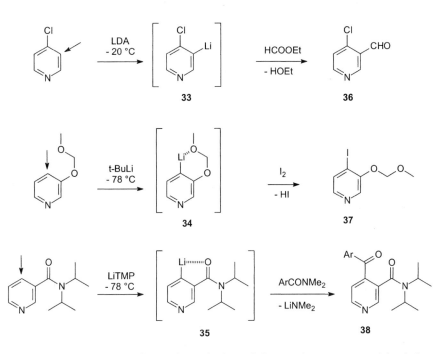

It should be noted that lithiopyridines alternatively and frequently are prepared by halogen-metal exchange reactions of halogenopyridines with organolithiums, e.g. :

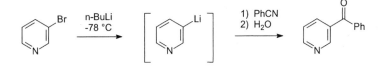

Pyridine shows complex behaviour towards GRIGNARD reagents. By analogy to the ZIEGLER reaction, mainly 2-substitution products are obtained with ether-free RMgX. For instance, with BuMgI a mixture of 2- and 4-*N*-butylpyridines **39** and **40** is formed (**39/40** > 100: 1), and in the presence of excess of Mg, the proportion of **40** rises to 3:1. However, if butyl chloride and Mg are made to react in boiling pyridine, then the 4-substitution product is obtained almost exclusively (**39/40**, > 1: 100). Other metals (e.g. Li, Na) catalyse this reaction.

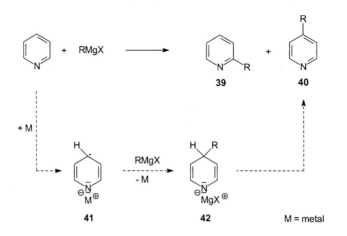

A change in the reaction mechanism to a radical process is thought to be the reason for the reversal in the regioselectivity of the alkylation. A single electron transfer from the metal to pyridine produces the radical anion **41**, which reacts with RMgX to give the 1,4-dihydropyridine **42** and finally the product **40** [51].

Additions of nucleophiles to pyridinium ions

N-Alkylpyridinium ions add hydroxide ions reversibly and exclusively in the 2-position, forming 2-hydroxy-1,2-dihydro-*N*-alkylpyridines ('pseudobases', e.g. **43**) which can be oxidized by mild oxidizing agents to *N*-alkyl-2-pyridones, e.g. **44**:

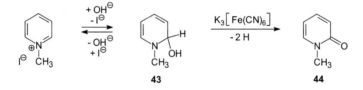

Pyridinium ions **45** with *N*-acceptor substituents also add O- and N-nucleophiles via C-2 to give **46**. This is followed by ring opening, probably in an electrocyclic process, at N/C-2 resulting in the formation of 1-azatrienes **47** (*Zincke reaction*, cf. the corresponding transformations of the pyrylium ion, see p 225):

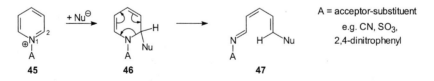

The aldehyde **49** is obtained from the *N*-2,4-dinitrophenylpyridinium salt **48** with aqueous alkali, and by its hydrolysis 5-hydroxypenta-2,4-dienal (glutacondialdehyde) **50** is produced. Ring opening with aniline and an additional amine exchange produces the bisanil **51**:

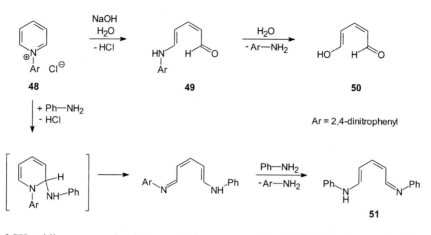

48   **49**   **50**

Ar = 2,4-dinitrophenyl

+ Ph—NH₂
- HCl

51

Anions of CH-acidic compounds add to pyridinium ions mainly at C-4. Substituents already present on pyridine can bring about interesting reaction sequences. This is exemplified by the formation of the 2,7-naphthyridine derivative **53** from the quaternary salt of the nicotinic acid amide **52** and malonic ester:

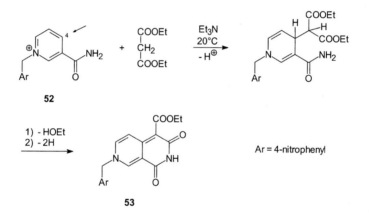

52

Ar = 4-nitrophenyl

53

Side-chain reactivity of pyridine

Alkylpyridines undergo reactions analogous to benzene such as side-chain halogenation and oxidative functionalization (cf. p 291). In addition, C–H bonds directly attached to the heterocycle display a kinetic acidity which is greater by a factor of > 10^5 compared to the corresponding benzene derivatives. This is more pronounced in the 2- and 4-positions than in the 3-position. H/D-Exchange experiments of 2-, 3- and 4-picoline with a relative exchange rate of 130 : 1 : 1810 (MeOD/MeONa at 20°C, cf. toluene ≈ 10^{-5}) demonstrate this point.

Deprotonation of the 2- and 4-pyridyl C–H bonds is easier than at the 3-position. This is due to the resonance stabilization of the corresponding carbanions **54** and **55**, with participation of the ring nitrogen, which is not available for the 3-pyridyl anion **56**:

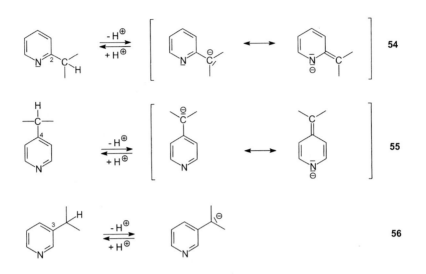

2- and 4-Pyridylcarbanions **54** and **55** are generated in equilibrium either by strong bases (e.g. alkali metal amides, organolithium compounds) in an aprotic medium in situ, or by weaker bases (e.g. hydroxides, alcoholates, amines) in a protic medium.

Under proton or Lewis acid catalysis, 2- and 4-alkylpyridines of the above type are in equilibrium with the tautomeric methylene bases **57** and **58**, which can function as enamines:

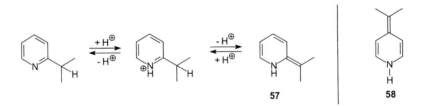

Therefore, alkylpyridines and *N*-alkylpyridinium ions undergo base- or acid-catalysed reactions with electrophilic reagents preferably at the 2- and 4-'heterobenzylic' positions.

For instance, the CH$_3$ group of 2- or 4-picoline can be alkylated (**59**), carboxylated (**60**), and acylated (**62**) by a CLAISEN-like condensation; aldol addition (**61**), multiple aldol addition (**65**), and aldol condensation (**63**, **64**) are also possible. When several alkyl groups are present, deprotonation takes place at the most acidic pyridyl C–H bond. For instance, the synthesis of 3,4-diethylpyridine **67** is effected by selective deprotonation of the 4-methyl group in 3-ethyl-4-methylpyridine **66** followed by alkylation.

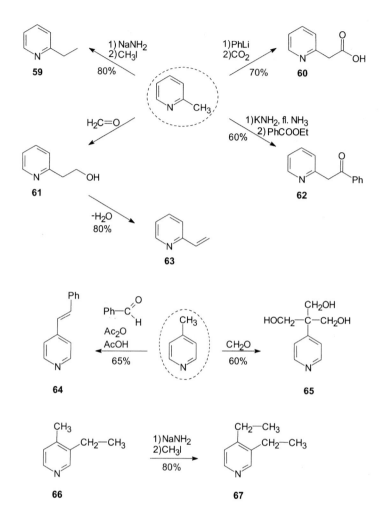

The α-C–H bonds of alkyl groups on a quaternary pyridine nitrogen possess increased CH-acidity. As a consequence, base-catalysed reactions with electrophiles via pyridinium betaines **68** are possible, especially if they are stabilized by acceptor substituents:

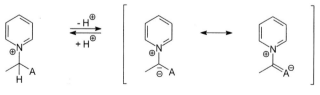

68

This is illustrated by the behaviour of the 1-phenacylpyridinium ion **69**, which is deprotonated by weak bases, e.g. Na$_2$CO$_3$/H$_2$O, to give the stable pyridinium phenacylide **70**. The ylide C-atom can be smoothly alkylated and acylated, and the resulting pyridinium compounds are transformed into ketones or β-diketones by reductive removal of the pyridine moiety (KRÖHNKE 1941). In its chemical behaviour, the ylide **70** formally shows analogy to acetoacetic ester [52] because **69** is hydrolysed in a strongly basic medium (NaOH/H$_2$O) to give the 1-methylpyridinium ion and benzoate:

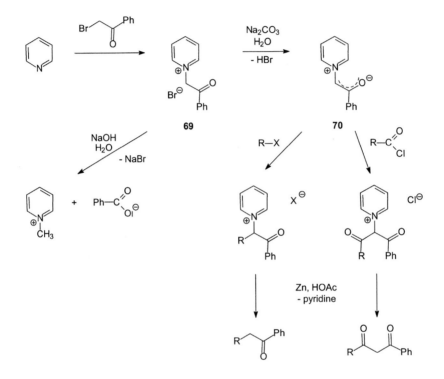

The *N*-phenacylpyridinium ion **71**, derived from 2-picoline, cyclizes on treatment with base giving phenylindolizine **73**. Evidently, the 2-methyl group is first deprotonated (→ **72**) and the product **73** is formed by a subsequent intramolecular aldol condensation with the phenacyl carbonyl group [53].

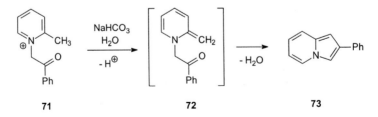

In a reversal of this intramolecularized side-chain reactivity (effecting annulation of a five-membered ring to the N/C-2-position of pyridine) the *Baylis-Hillman reaction* of pyridine-2-aldehydes with acceptor-substituted alkenes, e.g. acrylates, in the presence of DABCO gives rise to formation of products **74**, which can be cyclized to 2-substituted indolizines, e.g. indolizine-2-carboxylates **75** [54].

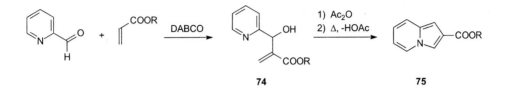

74   **75**

Reactions of pyridine N-oxides

A series of preparatively interesting reactions on pyridine can be carried out by means of pyridine *N*-oxides (OCHIAI 1943, DEN HERTOG 1950), such as the introduction of certain functions into the ring and side-chain which cannot be achieved in the parent system by direct methods [55].

N-Oxidation of pyridine is carried out with peroxy acids (cf. p 272) and deoxygenation back to pyridine is brought about by a redox reaction with phosphorus(III) compounds such as PCl_3, $P(C_6H_5)_3$ or $P(OC_2H_5)_3$:

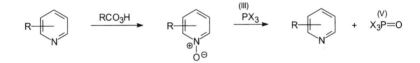

Pyridine-1-oxide undergoes electrophilic and nucleophilic substitution reactions at the 2- and 4-positions, as predicted from its resonance description **76** :

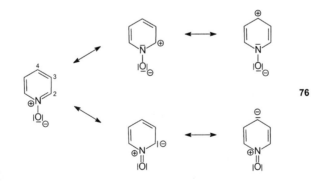

76

The greater reactivity of pyridine-1-oxide towards nitration as compared to pyridine is noteworthy. It proceeds smoothly in nitrating acid, via the free base and the σ-complex **77**, yielding 4-nitropyridine-1-oxide **78** (see p 274). Other S_E reactions (sulfonation, halogenation) require drastic conditions.

In spite of the low nucleofugal tendency of the NO$_2$ group, nucleophilic substitution can be carried out on **78**, e.g. alcoholates give rise to 4-alkoxypyridine-1-oxide **80**. Deoxygenation of **78** with PCl$_3$ furnishes 4-nitropyridine **81**, and its catalytic reduction with H$_2$/Pd-C yields first 4-aminopyridine-1-oxide and finally 4-aminopyridine **79**. The pyridines **79** and **81** are not accessible by direct substitution reactions of pyridine.

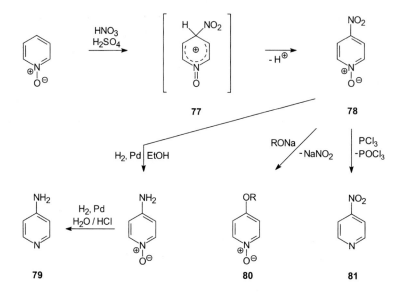

Pyridine-*N*-oxides form stable 1:1-complexes with LEWIS acids. The SbCl$_5$ complexes **82**, on thermolysis followed by hydrolysis, yield 2-pyridones by a regioselective transfer of oxygen to the α-position via **83**.

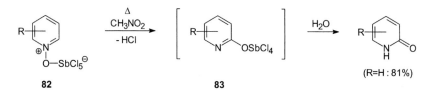

N-Oxides of quinolines and isoquinolines also undergo this type of reaction [56].

Pyridine-*N*-oxides undergo alkylation and acylation at oxygen. *O*-Alkylation occurs most readily with benzyl halides under mild conditions and leads to 1-benzyloxypyridinium ions, e.g. **84**,

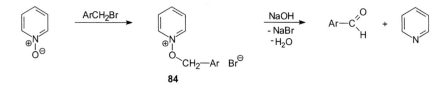

which in a basic medium, e.g. aqueous alkali, undergo disproportionation into arylaldehydes and pyridine (BOEKELHEIDE 1957). The pyridine-*N*-oxide brings about the conversion of a primary haloalkane into an aldehyde (R-CH$_2$-X → R-CH=O), analogously to the KORNBLUM oxidation [57].

O-Acylation is effected by acid anhydrides and inorganic chlorides (SOCl$_2$, POCl$_3$). Subsequently, the pyridine ring is attacked by nucleophilic addition of the remaining part of the acylating agent at the 2-position (**85**). This is followed by loss of the acylated oxygen from the pyridine *N*-oxide because of the formation of a good leaving group. Accordingly, treatment of pyridine-1-oxide with acetic anhydride produces 2-acetoxypyridine **86**, which can be hydrolysed to give 2-pyridone:

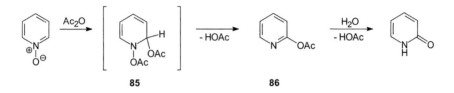

85   **86**

N-Oxides of 2- and 4-alkylpyridines are functionalized in the side-chain by acylating agents (acid anhydrides, sulfonyl chloride, POCl$_3$). For instance, 2- or 4-picoline-1-oxide with acetic anhydride yields a mixture of acetoxy compounds in which the methyl-substituted products, namely 2- and 4-acetoxymethylpyridine, dominate (**87** and **90**). Some nuclear-substituted products (**88**, **89**, **91**) are also formed:

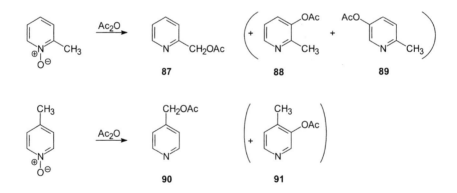

As shown for 2-picoline-1-oxide, these transformations occur by an ion-pair mechanism via the central intermediate **94**; this is formed from the *N*-oxide by *O*-acylation and acetic acid elimination (via **92** and **93**). In some cases a radical mechanism involving a radical pair analogous to **94** is assumed to operate.

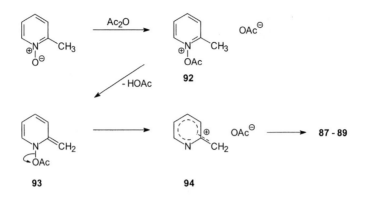

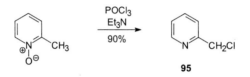

93 **94**

Introduction of a function into the side chain with POCl₃ or with *p*-toluenesulfonyl chloride usually occurs with high product selectivity, such as the formation of 2-chloromethylpyridine **95** from 2-picoline-1-oxide [58]:

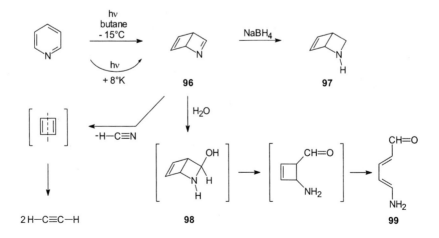

Thermal and photochemical reactions of pyridine

Pyridines show a number of thermal and photochemical transformations which are analogous to the valence tautomerization of benzene, i.e. formation of DEWAR benzene, prismane and benzvalene.

On irradiation in butane at -15°C, pyridine is converted into 'DEWAR pyridine' **96** (half-life 2.5 min at 25°C). Its structure has been spectroscopically confirmed and it was chemically characterized by reduction with NaBH₄ to give the bicyclic azetidine **97**, as well as by hydrolysis to the 5-amino-2,4-pentadienal **99** (via the hemiaminal **98**). Matrix photolysis of pyridine at 8K leads to acetylene and HCN as the result of a [2+2] cycloreversion of the DEWAR pyridine **96**, resulting in cyclobutadiene formation.

Alkylpyridines isomerize on gas-phase photolysis. For instance, 2-picoline is transformed into a photostationary equilibrium with 3- and 4-picoline, which is indicative of an intermediate 'azaprismane' **100**:

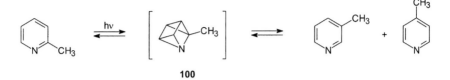

100

Photolysis of highly substituted pyridines gives rise to isolable and stable valence isomers. For instance, pentakis(pentafluoroethyl)pyridine **101** is converted almost quantitatively into the symmetrical 1-aza-DEWAR pyridine **102**. Further irradiation brings about its rearrangement to the corresponding azaprismane derivative **104** [59]:

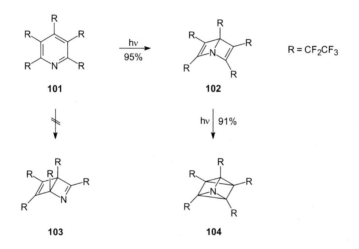

The fact that **102**, and not the isomeric DEWAR pyridine **103**, is formed can be ascribed to substituent interactions in the cyclobutene ring of the [2.2.0] bicycle. This destabilizes the 2-aza system **103** (four large substituents) compared to the 1-aza system **102** (three large substituents). The remarkable stability of **102/104** (only extended heating reconstitutes the pyridine **101**) is ascribed to the reduced steric hindrance of the bulky substituents when compared to a planar structure.

In contrast to **101**, photolysis of the highly substituted pyridine **105** leads to the formation of both possible DEWAR pyridines **106** and **107** with excess of the 2-aza isomer:

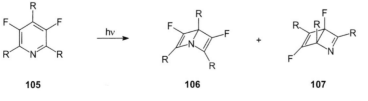

R = CF(CF$_3$)$_2$

Photolysis of pyridinium salts in water or methanol leads to derivatives of 6-azabicyclo[3.1.0]hexene (e.g. **108** from 1-methylpyridinium chloride). The formation of these products may be due to azoniabenzvalenes (e.g. **109**) as precursors:

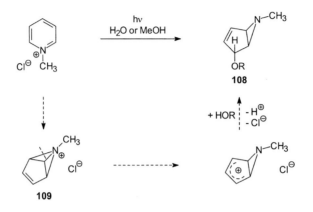

Pyridine-*N*-oxides display varied photochemical behaviour. As with all heterocyclic *N*-oxides, photolysis in the gas-phase leads to deoxygenation to the heteroarene via a triplet state. In solution photolysis, the oxygen can be transferred to a solvent molecule undergoing C–H insertion or addition to a double bond, e.g.:

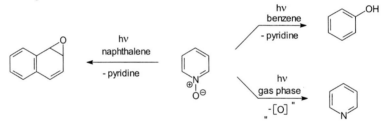

Photolysis of pyridine-*N*-oxides carried out via the singlet excited state in a polar medium leads to isomerization in high yield to 2-pyridones, with oxaziridines as postulated intermediates, e.g.:

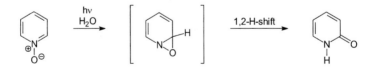

The isoelectronic pyridinium *N*-ylides, e.g. **110**, behave in an analogous manner to the *N*-oxides. Photolysis in an inert solvent leads to the diazepine **112** as the sole product, probably with intermediate formation of a diaziridine **111**, followed by electrocyclic N/C-1 cleavage. Photolysis in benzene yields, in addition to the diazepine **112**, the azepine **113** because of a competitive N–N cleavage and interaction of the solvent with the resulting nitrene (by a [2+1] cycloaddition and electrocyclic opening of the diazanorcaradiene):

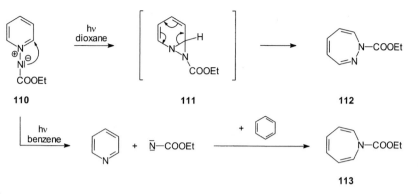

110   **111**   **112**

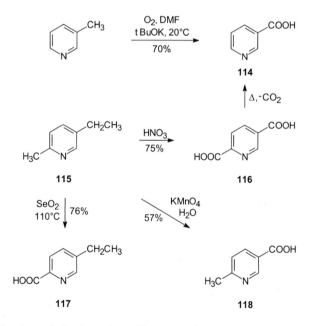

Oxidation

The pyridine ring is remarkably stable towards oxidation. Therefore, pyridine serves as a solvent for oxidation reactions, e.g. in the COLLINS oxidation with CrO_3. Aqueous $KMnO_4$, preferably in a basic medium, oxidizes pyridine to CO_2; peroxy acids bring about *N*-oxidation to pyridine-1-oxide (see p 285).

Alkylpyridines can be oxidized to give pyridine carboxylic acids by a number of methods. For instance, nicotinic acid **114** is produced commercially by oxidation of 5-ethyl-2-methylpyridine **115** with HNO_3, followed by selective thermal decarboxylation of the dicarboxylic acid **116** [60]. Selective side-chain oxidation is also possible, as shown in the examples **117** and **118**:

Oxidative functionalization of the benzyl position can also lead to carbonyl compounds, e.g. the dehydrogenation of picolines in the gaseous phase to the corresponding aldehydes, or the oxidation of 2-benzylpyridine to 2-benzoylpyridine.

Reduction

Pyridines are more readily reduced than benzene derivatives. Catalytic hydrogenation, which requires pressure and high temperatures for benzene, occurs in pyridine at normal pressure and room temperature to produce piperidine quantitatively:

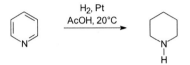

Reactions with complex metal hydrides do not proceed uniformly. LiAlH$_4$ with pyridine yields only a complex **119**, which contains two 1,2- and two 1,4-dihydropyridine units:

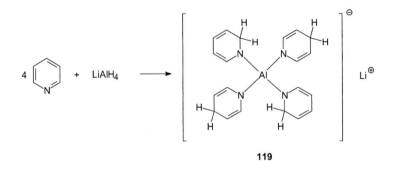

119

NaBH$_4$ does not react with pyridine. However, if the pyridine ring contains electron-attracting substituents, reduction with NaBH$_4$ leads to di- and tetrahydropyridines. 3-Cyanopyridine, for instance, yields **120** and **121**:

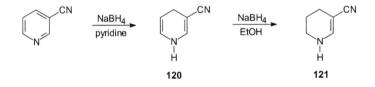

120   **121**

N-Alkylpyridinium ions are smoothly reduced by NaBH$_4$. Product formation can be controlled by the pH of the reaction medium and the nature of the substituents. Thus, 1-methylpyridinium chloride **122** is converted into 1-methyl-1,2-dihydropyridine **123** by NaBH$_4$ in H$_2$O at pH > 7, but at pH 2-5 it is transformed into 1-methyl-1,2,3,6-tetrahydropyridine **124**:

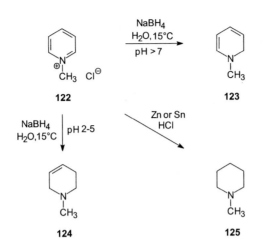

122   **123**

124   **125**

Electron-attracting substituents favour formation of dihydropyridines, as is shown by the reduction of 3-nitropyridinium ions **126** and **127**:

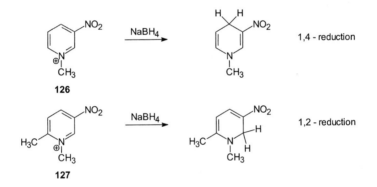

126

127

Many biologically important redox reactions proceed with reversible hydrogen transfer (formally: $+H^+$, $+2\,e^-$) via the 4-position of the *N*-quaternary systems **128** and **129** in the coenzyme $NAD^\oplus$ (see p 306):

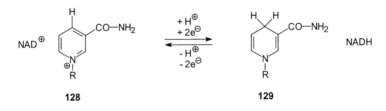

128   **129**

N-Alkyl-substituted nicotinic acid amides used as NAD models are selectively reduced in the 1,4-positions by $Na_2S_2O_4$, e.g. **130** to **131**. Initially, reaction with dithionite to give the base-stable sulfinate occurs:

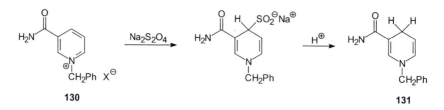

130   **131**

Finally, pyridine derivatives can be reduced by metals. Piperidine is obtained from pyridine with sodium in alcohols (*Ladenburg reaction*); piperidines, as well as tetrahydropyridines, are obtained from 4-alkylpyridines. *N*-Alkylpyridinium ions are converted into *N*-alkylpiperidines by metals such as Zn or Sn in an acidic medium, or by electrochemical reduction (**122** → **125**, see p 293).

Reduction of pyridine with Na in a protic medium is interpreted as being analogous to the BIRCH reaction of arenes, i.e. a two-step, single-electron transfer involving the radical anion **132** followed by a 1,2- or 1,4-addition of hydrogen:

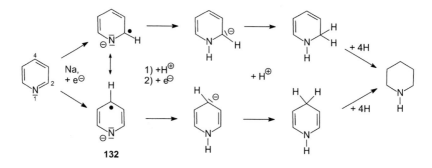

132

Reduction with Na in aprotic medium leads to 'oxidative dimerization' giving rise to 4,4'-bipyridyl **133**, since the pyridyl radical anion **132** undergoes dimerization via 4-position followed by dehydrogenation:

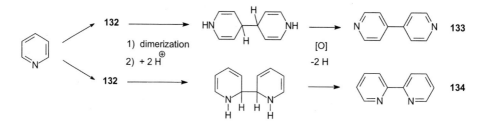

Interestingly, reduction with RANEY nickel in aprotic medium leads to 2,2'-bipyridyl **134** [61]. This 2,2'-mode of dimerization may be rationalized by *N*-chelation of the dimerized species to the Ni surface.

Pyridinium ions can be reductively dimerized. Thus 1-methylpyridinium chloride yields the dimer **136** either with Na/Hg or by cathodic reduction via the radical **135**. The dimer **136** can be oxidized to the bipyridinium dication **137** known as the herbicide paraquat [62].

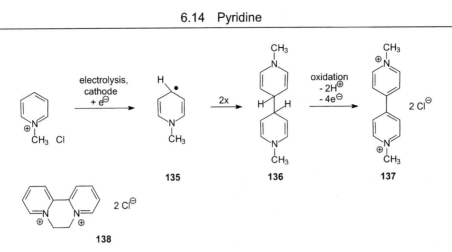

135   **136**   **137**

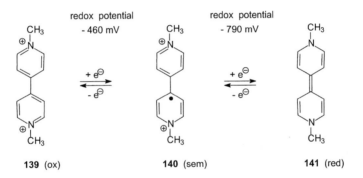

Herbicidal activity is found in other bipyridinium salts, e.g. in diquat **138**. Dications of this type undergo reversible and pH-independent single-electron transfer to form resonance-stabilized radical cations (e.g. **140**). These radical cations are transformed by a more negative potential, in a further pH-dependent single-electron step, into a quinonoid species (e.g. **141**). The latter can be reoxidized to the dications (e.g. **139**):

139 (ox)   **140** (sem)   **141** (red)

Ox = oxidized form, Red = reduced form of a reversible two-electron redox system,
Sem = single-electron intermediate (derived from semiquinone)

Reversible redox systems of this type are known for many heterocyclic structures and have been extensively investigated (HÜNIG 1978) [63].

C The *retrosynthetic analysis* of pyridine (see Fig. 10) can be carried out in several ways.

- If the azine structure is considered by itself, then the retrosynthetic analysis can start at the imine structural element (H$_2$O addition O → C-2, retrosynthetic path **a**). Suggestions for the cyclocondensation of various intermediates arise based on the 5-aminopentadienal or -one system **145**, and further (path **g**, NH$_3$ loss) on pent-2-endial (glutaconic dialdehyde) or its corresponding diketone **146**. Consideration of a 'retro-cycloaddition' (operation **c**) leads to the conclusion that a synthesis of pyridines by a cocyclooligomerization of alkynes with nitriles is possible.
- As a general principle, dihydro- and tetrahydropyridines can be converted into pyridines by dehydrogenation and elimination reactions. As is evident from a study of the retrosynthetic

operations **b**, **d–f**, it should be possible to obtain the hydropyridines **142–144** from [4+2] cycloadditions of azadienes with activated alkynes or alkenes, and also from 1,3-dienes with imines.

- The 1,4-dihydropyridine **148** (as well as the 3,4-dihydropyridine **147** which is retrosynthetically equivalent) can be linked with the retrosynthetic process **a**. As a double enamine, it could be derived from the enaminone **149** (via **h**), which in turn should be accessible (via **i**) from the 1,5-dicarbonyl system **150** and NH$_3$ by cyclocondensation. The systems **149** and **150** are retroanalytically interlinked, not only by dehydrogenation of **145** and **146** respectively, but also by a retro-MICHAEL addition involving enamines or enolates with α,β-unsaturated carbonyl systems as starting materials for the synthesis of 1,4-dihydropyridines.

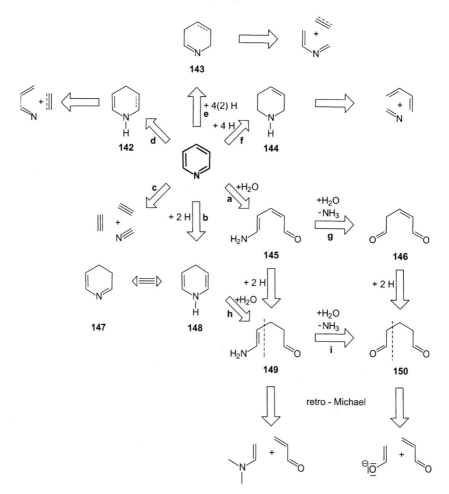

Fig. 6.10   Retrosynthesis of pyridine

Accordingly, the syntheses of pyridine make use of cyclocondensations, cycloadditions or ring transformations of other heterocyclic systems for the construction of the azine system.

Pyridine synthesis by cyclocondensation

(1) 5-Aminopentadienones **151**, formed by addition of esters and nitriles of 3-aminobut-2-enoic acid (3-aminocrotonic acid) to acetylene-aldehydes or -ketones, cyclize thermally with loss of H_2O to the nicotinic acid derivatives **152**:

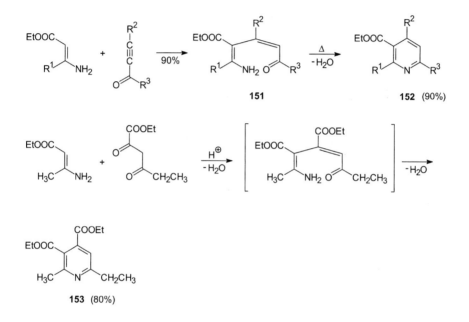

151

152 (90%)

153 (80%)

Cyclocondensation of 3-aminocrotonic ester with β-dicarbonyl compounds also occurs via 5-amino-pentadienone intermediates to give e.g. **153**.

(2) Pent-2-endial (glutaconic dialdehyde) undergoes cyclization with ammonia to give pyridine. It yields pyridine-1-oxide with hydroxylamine and *N*-substituted pyridinium ions with primary amines. With acids, the pyrylium ion is formed reversibly. Hence pyridine derivatives (see p 230) can also be obtained by reacting pyrylium ions with NH_3 and primary amines.

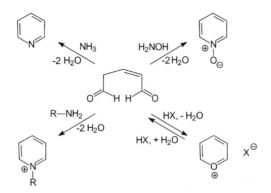

Pent-2-ene-1,5-diones (e.g. **154**) are intermediates in the formation of 3-acylpyridines (e.g. **155**) from 1,3-diketones and ammonia. They result from a KNOEVENAGEL condensation of two molecules of β-diketone:

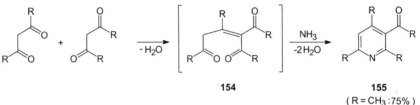

154   **155**
                                  (R = CH$_3$: 75%)

(3)   1,5-Dicarbonyl compounds, on treatment with ammonia, undergo cyclocondensation to give 1,4-dihydropyridines **156**, which on dehydrogenation yield 2,6-disubstituted pyridines **157** :

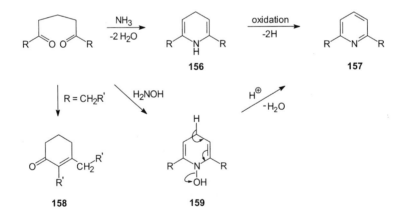

156   **157**

158   **159**

If the substituent R contains an α-CH$_2$ group, an intramolecular aldol condensation to give cyclohexanone derivatives **158** competes with 1,4-dihydropyridine formation. This can be avoided by using hydroxylamine for the cyclocondensation; moreover, the dehydrogenation becomes superfluous, because the *N*-hydroxy intermediate **159** allows H$_2$O elimination, yielding the pyridine derivative **157** directly. The synthesis of the dihydrocyclopenta[b]pyridine **160** provides an example:

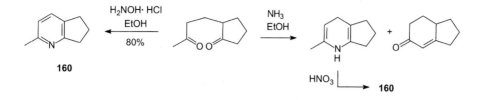

160

1,5-Diketones are obtained by MICHAEL addition of enolates or enamines (see above) to α,β-unsaturated carbonyl compounds.

The 1,5-dicarbonyl functionality is also represented by 2-alkoxy-3,4-dihydro-2*H*-pyrans (e.g. **161**, a masked 5-ketoaldehyde), which are obtained by hetero-DIELS-ALDER reaction of alkenones and vinyl ethers (see p 241). On treatment with hydroxylamine, they afford pyridines (e.g. **162**):

(4) Numerous alkylpyridines are formed in the gas phase by the interaction of carbonyl compounds with ammonia. The mechanism of these reactions is frequently complex and not completely elucidated. However, some of the reactions are of preparative interest in view of their product selectivity and yield. For instance, 2,6-diethyl-3-methylpyridine **163** is formed in 90% yield from diethyl ketone and NH_3 [64].

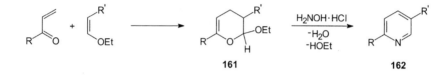

5-Ethyl-2-methylpyridine **164** (the educt in the industrial synthesis of nicotinic acid, see p 291) is formed from acetaldehyde or but-2-enal (crotonaldehyde) and aqueous NH_3 in a similar process.

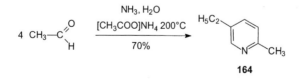

(5) The *Hantzsch synthesis* of pyridine is a method of considerable scope and flexibility. In a condensation of four components, two molecules of a β-dicarbonyl compound react with an aldehyde and ammonia giving 1,4-dihydropyridines **165** which can be dehydrogenated to pyridines **166**:

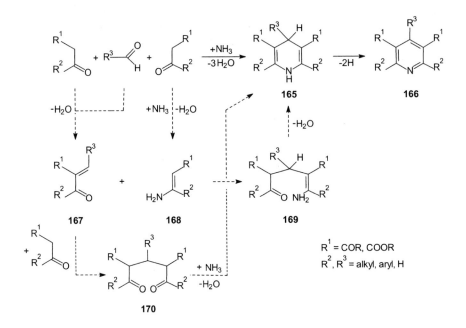

$R^1 = COR, COOR$
$R^2, R^3 = alkyl, aryl, H$

Formation of the 1,4-dihydropyridines **165** occurs by two routes. In the first, NH$_3$ and the β-dicarbonyl compound combine to give a β-enaminone **168**, whereas the aldehyde and dicarbonyl compound interact to produce an α,β-unsaturated ketone **167** as the result of a KNOEVENAGEL condensation; **167** and **168** now undergo a MICHAEL addition to give 5-aminopent-4-enones **169**, followed by cyclocondensation. Alternatively, the two molecules of the β-dicarbonyl compound may interact with the aldehyde by KNOEVENAGEL condensation followed by MICHAEL addition to the 1,5-dicarbonyl system **170**, which undergoes cyclocondensation with NH$_3$.

In modifications of the HANTZSCH synthesis [65], β-enaminones replace one molecule of the β-dicarbonyl compound, and enones undergo cyclocondensation with β-enaminones or 1,5-dicarbonyl compounds with NH$_3$. Thus, for instance, as an alternative to the classical synthesis (2 mol acetoacetate, benzaldehyde and NH$_3$), the 1,4-dihydropyridine-3,5-dicarboxylate **171** is obtained from acetoacetate, benzaldehyde and β-aminocrotonate (1:1:1), as well as from benzylidene acetoacetate and aminocrotonate (1:1).

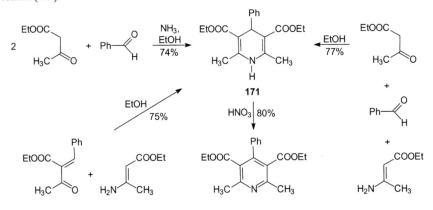

In a recent approach [66], 2,4,6-trisubstituted or 2-aminosubstituted pyridines **173/174** can be prepared in a one-pot cyclocondensation of methyl phosphonate anion with nitriles, aldehydes and enolates of methyl ketones or nitrile α-carbanions (followed by dehydrogenation) via in-situ generated α,β-unsaturated imines **172**.

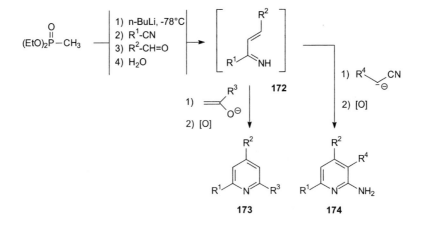

Pyridine synthesis by cycloaddition

(1) The Co(I)-catalysed co-oligomerization of nitriles with alkynes (BÖNNEMANN 1978 [67]) is important in preparative chemistry. Reaction of 2 mol acetylene with nitriles results in a virtually quantitative yield of 2-substituted pyridines. The competing cyclotrimerization of acetylene to benzene can be suppressed by using an excess of nitrile:

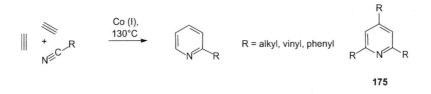

With terminal alkynes, a mixture of 2,3,6- and 2,4,6-trisubstituted pyridines is obtained, with the symmetrical products **175** predominating.

Likewise, α,ω-diynes can be used in this cycloaddition; for instance, octa-1,7-diyne **176** affords the 5,6,7,8-tetrahydroisoquinoline **177** with nitriles:

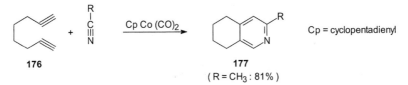

The mechanism for this nitrile-alkyne co-oligomerization is assumed to be a catalytic cycle involving the metallacyclopentadienes **178** and **179**:

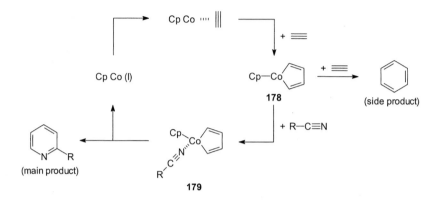

Accordingly, stable azazirconiacyclopentadienes **180** (formed from nitriles, alkynes and Cp$_2$ZrEt$_2$) react with alkynes in a Ni-catalyzed process to give pyridines **181** which formally result from cyclotrimerization of nitriles with two different alkynes [68].

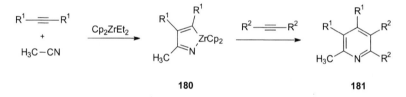

(2)   The [4+2] cycloaddition is of increasing utility for pyridine synthesis [69]. The pyridine nitrogen is introduced either via the diene component [1- or 2-azadienes, oxazoles (cf. p 131)] or via the dienophile (activated imine or nitrile).

For instance, α,β-unsaturated *N*-phenylaldimines react as 1-azadienes with maleic acid affording tetrahydropyridines **182**.

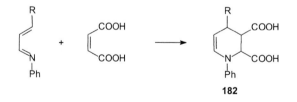

Thermal isomerization of amino-substitued 2*H*-azirines (e.g. **183** → **184**) results in the formation of 2-azadienes. They react with activated alkynes, e.g. with acetylenedicarboxylic ester (ADE), in a hetero-DIELS-ALDER reaction, giving dihydropyridines (e.g. **185**), which aromatize with elimination of amine yielding pyridine-3,4-dicarboxylic esters (e.g. **186**).

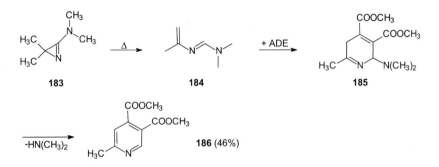

183   **184**   + ADE   **185**

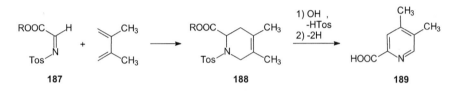

186 (46%)

Imines with acceptor substitutents, for instance the *N*-tosylimine of the glyoxylic ester **187**, add to 1,3-dienes (e.g. 2,3-dimethylbuta-1,3-diene) to produce tetrahydropyridines (e.g. **188**). Elimination of sulfinic acid, saponification and dehydrogenation lead to pyridine-2-carboxylic acids (e.g. **189**):

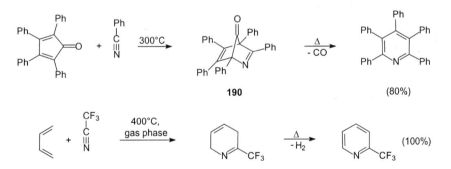

187   **188**   **189**

The weakly dienophilic C≡N bond of nitriles undergoes a [4+2] cycloaddition with conventional 1,3-dienes only under harsh conditions and when activated by strong acceptor substituents. For instance, benzonitrile and tetraphenylcyclopentadienone give rise to pentaphenylpyridine by way of the bicyclic intermediate **190** followed by decarbonylation. Trifluoroacetonitrile and buta-1,3-diene afford 2-trifluoromethylpyridine after thermal dehydrogenation of the initially formed dihydropyridine :

190   (80%)

(100%)

Pyridine synthesis via ring transformations

Furans **191** with an acyl- or carboxylic acid functionality in the 2-position are transformed into 2-substituted 3-hydroxypyridines **193** by the action of NH_3 in the presence of ammonium salts. 5-Aminodienones **192** are presumed to be the intermediates:

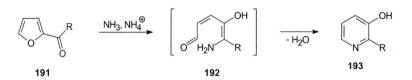

| | | |
| 191 | 192 | 193 |

An elegant sequence starting from 5-alkylfuran-2-carboxylic esters **194** leads to 6-substituted 3-hydroxypyridines **200**. It proceeds by an electrolytic 1,4-methoxylation of the furan system giving **195**, followed by reduction of the corresponding carboxylic amide with LiAlH₄ to the amine **196** and its acid-catalysed ring enlargement (probably by way of the 5-aminodienone **199**). On treatment with aqueous acid, the tetrahydrofuran **198**, obtained by hydrogenation of the intermediate **195**, furnishes the 3-hydroxypyridone **197** [70]:

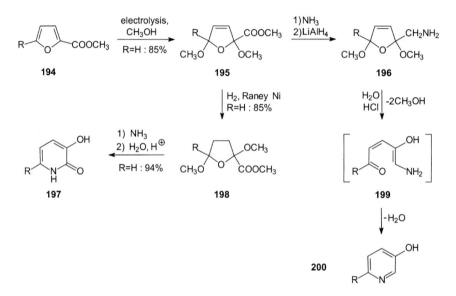

The ring enlargement of pyrroles providing 3-chloropyridines has already been described (see p 93). Oxazoles react as masked 2-azadienes with alkenes yielding pyridine derivatives of various types (see p 131). With enamines and ynamines, diazines and triazines undergo DIELS-ALDER reactions with inverse electron demand (see p 441). This leads to pyridines (e.g. **201**) by the enamine cycloaddition of the 1,2,4-triazine, as shown below:

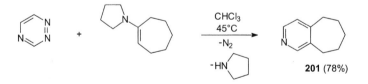

Various diazepines form pyridines by ring contraction. For instance, thermolysis of 1-ethoxycarbonyl-4-methyl-1*H*-1,2-diazepine **202** leads to 4-methylpyridyl-2-carbamate **204** by undergoing valence isomerization to diazanorcaradiene **203** followed by N–N cleavage with aromatization. In contrast, sodium ethoxide opens the seven-membered ring of **202** leading to the (*Z,Z*)-cyanodiene **205**, which recyclizes to give 2-amino-3-methylpyridine **206** with fission of the urethane functionality:

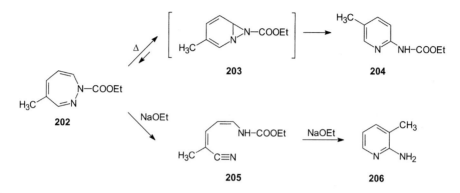

D **Pyridine**, mp -42°C, bp 115°C, is a colourless, water-miscible liquid of an amine-like odour. Pyridine is poisonous, and inhalation of its vapour causes damage to the nervous system. It is a weak base (pK_a = 5.20, cf. aliphatic amine $pK_a \approx 10$, aniline pK_a = 4.58). Pyridine, as well as picolines and lutidines, are constituents of coal-tar and bone-oil.

Nicotinic acid (pyridine-3-carboxylic acid), mp 236°C, was first obtained by oxidation of the alkaloid nicotine by $KMnO_4$. It is produced commercially from 5-ethyl-2-methylpyridine (see p 291). Nicotinic acid and its amide belong to the B group of vitamins (vitamin B_5). The daily requirement of an adult is ca. 20 mg. Deficiency of nicotinic acid causes pellagra, a skin disease.

Pyridine-3-carboxamide (nicotinamide, niacin), mp 130°C, is commercially produced by ammoxidation of 3-picoline, followed by partial hydrolysis of the intermediate 3-cyanopyridine:

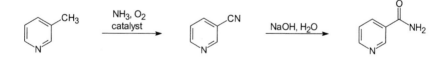

The pyridine alkaloids nicotine (**207**, R = CH_3), nornicotine (**207**, R = H), nicotyrine **208** and anabasine **209** are some of the natural products derived from pyridine [71]:

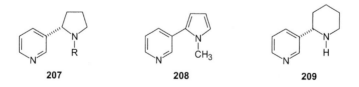

Pyridoxol **210** (pyridoxine, 3-hydroxy-4,5-bis(hydroxymethyl)-2-methylpyridine, vitamin B_6) was formerly known as adermine (KUHN 1938) because vitamin B_6 deficiency causes skin diseases in animals. Pyridoxal (**211**, R = CHO) and pyridoxamine (**211**, R = CH_2NH_2) also belong to the vitamin B group. Pyridoxal phosphate **212** is a coenzyme for many of the enzymes involved in the metabolism of amino acids. **Nicotinamide adenine dinucleotide 213** ($NAD^\oplus$, reduced form NADH) is a component of oxidoreductases (for its action see p 293, synthesis see p 131).

Betalaines are cyanine-related dyes with terminal tetrahydropyridine groups. They occur only rarely as glycosides in higher plants (*Caryophyllales* family). For instance, the dye present in beetroot contains the aglycon betanidin **214**.

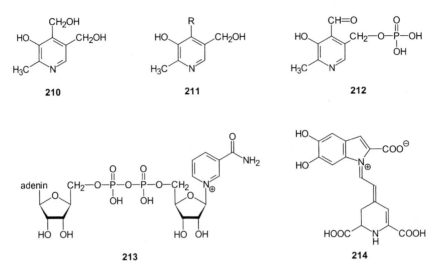

Pyridine derivatives are important as pharmaceuticals. Nicotinic acid derivatives are used as vasodilators, anticoagulants and hypolipidaemic agents. Derivatives of isonicotinic acid (pyridine-4-carboxylic acid) such as isoniazide **215** and ethionamide **216** are used as tuberculostatics, and of 2-benzylpyridine in the form of pheniramine **217** as antihistamine.

Nifedipine **218** and other related 1,4-dihydropyridines are important as antihypertensive agents (Ca antagonists). They are prepared by a classical HANTZSCH synthesis from acetoacetic ester, arylaldehyde and NH$_3$ [72]. Sulfapyridine **219** was one of the first sulfonamide antibacterial agents to be used. Cerivastatin (Lipobay) **220**, a powerful HMG-CoA reductase inhibitor, is applied for treatment of primary hypercholesteremia types IIa and b.

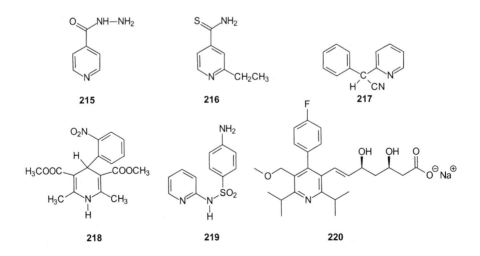

Other pyridine derivatives with biological activity are the herbicides paraquat and diquat (see p 295).

E Pyridines are widely used as building blocks and as intermediates for carrying out synthetically useful transformations.

(1) Ring fission of pyridinium ions with acceptor substituents, e.g. N-2,4-dinitrophenylpyridinium salts **221** (see p 280) leads, on treatment with secondary amines, to pentamethine cyanines **222** (KÖNIG's salts). They condense with sodium cyclopentadienide to yield vinylogous aminofulvenes **223**, which on cyclization produce azulene (azulene synthesis according to ZIEGLER and HAFNER, see p 229 [73]):

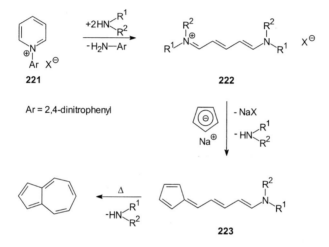

(2) The cyanide-catalysed condensation of two N-methylpyridinium ions **224** affords the herbicide paraquat **227** [74] after a sequence of additions to C-4 (see p 281), HCN elimination and oxidation via the intermediates **225** and **226**.

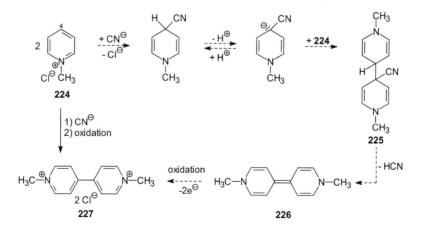

The transformation **224** → **227** can be regarded as a heterocyclic analogue of a benzoin condensation.

(3) 2-Chloro-1-methylpyridinium iodide **228** promotes esterification of carboxylic acids with alcohols as well as lactone formation of hydroxy acids in a basic medium (MUKAIYAMA 1977 [75]) :

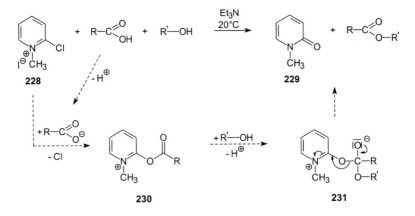

2-Acyloxypyridinium ions **230** are formed as intermediates from **228** by an S_NAr reaction with carboxylate. In these ions, the acyloxy group is activated and transferable to suitable nucleophiles (in the example shown above to alcohols with ester formation, or to primary and secondary amines with amide formation). The potentiality of the 1-methyl-2-pyridone **229** as a leaving group is essential in this addition **231**.

(4) The bis-2-pyridyl disulfide **232** is obtained by oxidation of the corresponding thione **234** with I_2 in a basic medium. In combination with triphenylphosphane, it activates carboxylic acids in the formation of amides and esters. In this reaction, carboxylic acids are converted into pyridine-2-thiol esters **233** as intermediates:

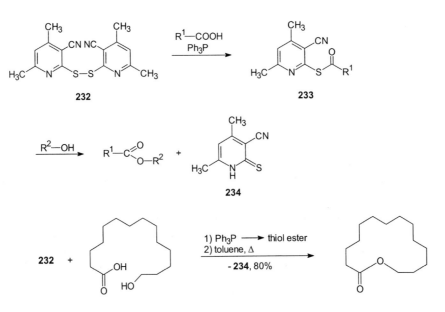

With the aid of this reagent, a method has been developed for the synthesis of peptides [76]. It proceeds under mild conditions and largely without racemization. The combination of reagents provides a method for the synthesis of macrolides (macrocyclic lactones) from long-chain ω-hydroxycarboxylic acids, as demonstrated by the formation of pentadecanolide from 15-hydroxypentadecanoic acid [77].

(5) *N*-Benzylpyridinium ions **235** condense with 4-nitroso-*N*,*N*-dimethylaniline to form nitrones **236** which can be hydrolysed to yield aldehydes (*Kröhnke reaction*):

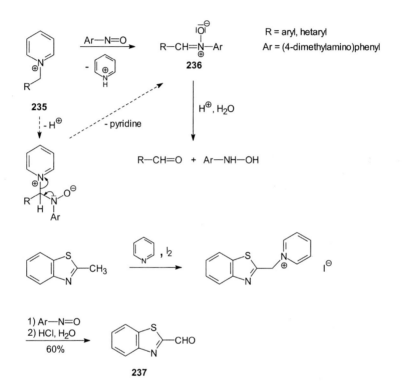

The required pyridinium salts can be obtained by the KING-ORTOLEVA reaction (see p 273), as is shown by the synthesis of benzothiazole-2-carbaldehyde **237**. In the KRÖHNKE reaction, the heterocycle pyridine ensures the targeted conversion CH$_3$ → CH=O of methyl groups in arenes or heteroarenes [78].

(6) 1-Alkyl-2,4,6-triphenylpyridinium salts **238** transfer the *N*-substituent R-CH$_2$ onto a range of halogen-, O-, S-, N- and C-nucleophiles in a thermal reaction [79]. This dealkylation reaction occurs e.g. with iodides and bromides at 200-300°C, with chlorides and fluorides at 80-120°C, and furnishes the corresponding haloalkanes RCH$_2$X. With **238**, carboxylates yield esters RCH$_2$OCOR'.

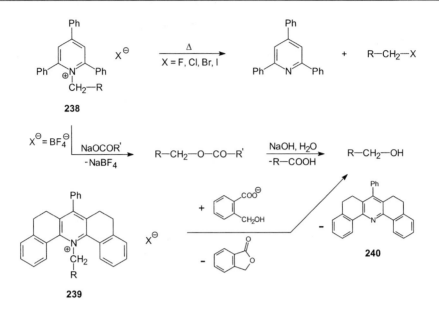

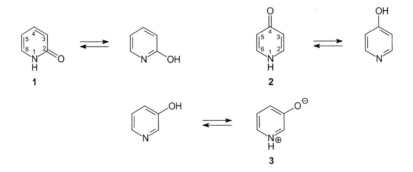

The pyridinium salt **239** yields alcohols RCH$_2$OH and the pyridine **240** with *o*-hydroxymethylbenzoates, because the initially formed *o*-hydroxymethylbenzoates undergo transesterification with separation of phthalide [80]. Since the pyridinium salts **238** and **239** are derived by interaction of primary amines with pyrylium salts (see p 230), they make possible the transformation of primary amines into the corresponding primary alkyl halides or alcohols (RCH$_2$NH$_2 \rightarrow$ RCH$_2$X or RCH$_2$OH).

6.15 Pyridones

A   2-Pyridone **1** [pyridin-2(1*H*)-one] and 4-pyridone **2** [pyridin-4(1*H*)-one] are tautomeric with the corresponding 2- and 4-hydroxypyridines, respectively.

Structural investigation of the pyridones indicates that the keto form predominates in the solid phase, and the hydroxy form in the gaseous state. The keto form is also favoured in most solvents by solvation, except in petroleum ether at high dilution.

3-Hydroxypyridine is in equilibrium with a 3-oxidopyridinium betaine structure **3** depending on the solvent (see [81] regarding the pyridone–hydroxypyridine equilibrium).

2- and 4-Pyridone show the following ^1H and ^{13}C-NMR data. They confirm that 2- and 4-pyridone must be regarded as π-delocalized systems with aromatic character.

	^{1}H-NMR (CDCl$_3$) δ (ppm)		^{13}C-NMR (DMSO-d$_6$) δ (ppm)		
2-Pyridone:	H-3: 6.60	H-5: 6.60	C-2: 162.3	C-3: 119.8	C-5: 104.8
	H-4: 7.30	H-6: 7.23		C-4: 140.8	C-6: 135.2
4-Pyridone:	H-2/H-6: 7.98		C-2/C-6: 139.8	C-4: 175.7	
	H-3/H-5: 6.63		C-3/C-5: 115.9		

B | 2- and 4-Pyridone are weak acids (p$K_a \approx 11$). Deprotonation in a basic medium produces ambident anions which can be attacked by electrophiles on O, N and C. For instance, the anion **5**, derived from 2-pyridone, undergoes acylation by acid chlorides usually on oxygen with ester formation **4**, while carboxylation by CO_2 occurs on C-5 with formation of the carboxylic acid **6**.

The position of alkylation of **5** depends on the solvent, on the counterion and on the spatial demand of the alkylating agent: a nonpolar medium, Ag ions and a large spatial requirement favour *O*-alkylation, whereas Na or K ions and a small spatial requirement promote *N*-alkylation, as is shown by the following examples [82]:

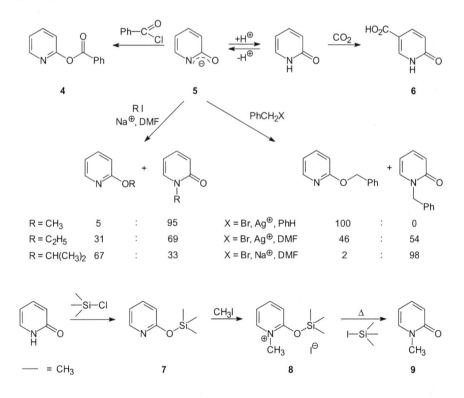

R = CH$_3$	5	:	95	X = Br, Ag$^\oplus$, PhH	100	:	0
R = C$_2$H$_5$	31	:	69	X = Br, Ag$^\oplus$, DMF	46	:	54
R = CH(CH$_3$)$_2$	67	:	33	X = Br, Na$^\oplus$, DMF	2	:	98

—— = CH$_3$ **7**   **8**   **9**

Selective *N*-methylation of 2-pyridones can be effected indirectly, namely via the exclusively *O*-directed silylation: *N*-methylation of the silyl ether **7**, followed by desilylation of **8**, leads to 1-methyl-2-pyridone **9**. Specific *N*-alkylation is also observed with dialkyl phosphates [83].

S$_E$Ar reactions (halogenation, nitration, sulfonation) with 2- and 4-pyridones occur in the 3- and/or 5-positions more readily than in pyridine.

Acid chlorides and anhydrides, as well as POCl$_3$ or PCl$_5$, attack the 2-pyridone oxygen electrophilically. The resulting presence of a leaving group in the 2-position of a pyridinium ion **10** allows a nucleophilic substitution by halogen leading to 2-chloropyridines:

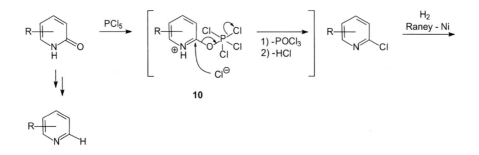

Since 2-chloropyridines are readily dehalogenated by reduction, e.g. by H$_2$/RANEY-Ni, the above reaction sequence offers a preparatively useful conversion of pyridones into pyridines.

2-Pyridones can act as diene components in [4+2] cycloadditions. The DIELS-ALDER reaction with activated alkenes or alkynes lead to derivatives of 2-azabicyclo[2.2.2]octa-5,7-dien-3-one **12** by the addition of acetylenedicarboxylate to 1,4,6-trimethyl-2-pyridone **11**, as shown below [84]:

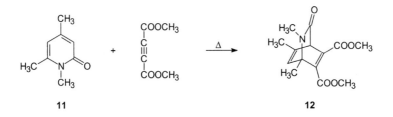

The mesoionic 3-oxido-*N*-alkylpyridinium betaine **13** undergoes 1,3-dipolar cycloadditions as the dipoles **a** and **b**. Thus, acetylenedicarboxylate addition via O and C-2 is followed by electrocyclic cleavage of the pyridine N/C-2 bond in the intermediate **16**, resulting in the formation of furan derivatives **17**. In contrast, phenylacetylene and acrylic ester add to the betaine **13** via C-2 and C-6 producing 8-azabicyclo[3.2.1]octane derivatives **14** and **15** [85] :

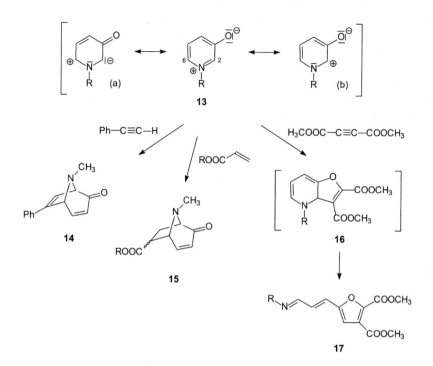

13

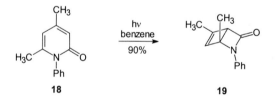

On irradiation in dilute solution, *N*-alkyl- und *N*-aryl-2-pyridones are converted into DEWAR pyridones (2-azabicyclo[2.2.0]hex-5-en-3-ones). For instance, 4,6-dimethyl-1-phenyl-2-pyridone **18** affords product **19**:

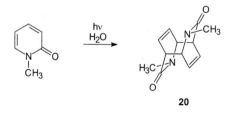

Apart from this intramolecular photoelectrocyclization, an intermolecular photodimerization can also occur, e.g. 1-methyl-2-pyridone is converted into **20**:

20

> **C** *Retroanalysis* of 2-pyridones via routes **a** and **b** suggests β-dicarbonyl compounds **21** or β-substituted acrylic acids **22** as building blocks. These 1,3-biselectrophiles have to be linked by cyclization to activated methyl or methylene moieties or to enamines.

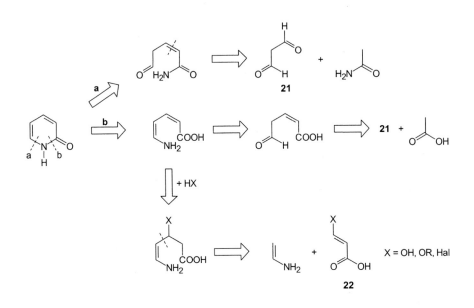

The cyclocondensation of 1,3-diketones with cyanoacetamide catalysed by base yielding 3-cyano-pyridones **23** (*Guareschi synthesis*) finds wide application:

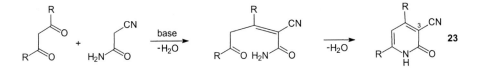

With unsymmetrical 1,3-diketones, the reactivity of the carbonyl groups determines the outcome of the reaction, which proceeds via an initial KNOEVENAGEL condensation. The two cyanopyridines **24** and **26** are formed from 1-ethoxypentane-2,4-dione **25** in a proportion of 15:75 because of the competition between the two different electrophilic carbonyl groups. In contrast, the diketo ester **27** yields exclusively the cyanopyridone **28**, because it reacts via the more strongly activated carbonyl function adjacent to the ester group:

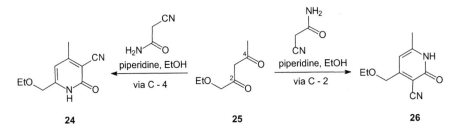

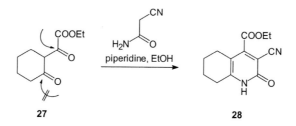

27   **28**

Likewise, β-keto aldehydes, malonaldehyde, acetylene ketones and α-acyl ketone-*S,S*-acetals can be employed for the GUARESCHI synthesis [86].

Cycloaddition of aroyl isocyanates and trimethylsilyl ketene generates 4-trimethylsilyloxy-1,3-oxazin-6-ones **29** in situ, which smoothly react with cycloalkanone enamines to give 3,4-cycloalkeno-2-pyridones **30** [87]:

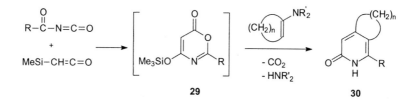

29   **30**

Another type of bicyclic 2-pyridones **31** results from the SnCl$_4$-promoted cyclocondensation of β-enamino nitriles with malonates [88]:

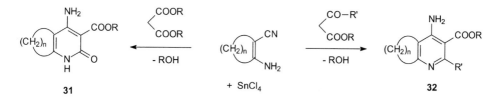

31   + SnCl$_4$ **32**

With β-ketoesters, 5,6-cycloalkeno-4-aminopyridine-3-carboxylates **32** are obtained.

The synthesis of 4-pyridones proceeds (according to the retroanalysis) by cyclocondensation of 1,3,5-tricarbonyl compounds **33** or their 1,5-bisenol ethers **34** with ammonia or primary amines:

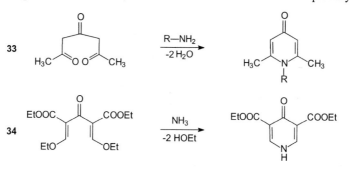

Cyclocondensation of dianions of β-dicarbonyl compounds **35** with nitriles offers a possibility for the formation of asymmetrically 2,6-disubstituted 4-pyridones **36**, e.g. :

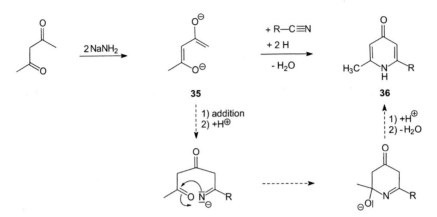

Finally, 4-pyridones are also obtained from 2-azadienes of type **37** and 1,1-carbonyldiimidazole in the presence of BF$_3$ etherate in a [5+1] heterocyclization [89]:

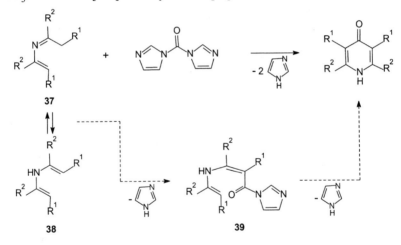

Cyclization probably occurs in the course of double acylation of the tautomeric enamine **38** via the intermediate **39**. The azadienes **37** are easily accessible [90].

6.16 Quinoline

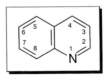

The topology of pyridine allows three benzene-annulated products, namely quinolizinium ion **1** (benzo[a]pyridinium ion), quinoline **2** (benzo[b]pyridine) and isoquinoline **3** (benzo[c]-pyridine):

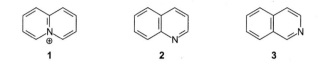

1   **2**   **3**

The neutral systems **2** and **3** will be discussed first.

A   Quinoline is derived from naphthalene by replacement of one of its α-CH groups by nitrogen. 2-
and 3-Methylquinoline (quinaldine and lepidine), 2-quinolone (carbostyril), 4-quinolone and the
quinolinium ion are important derivatives of quinoline.

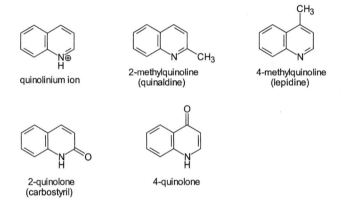

quinolinium ion

2-methylquinoline
(quinaldine)

4-methylquinoline
(lepidine)

2-quinolone
(carbostyril)

4-quinolone

Quinoline has many analogies with naphthalene and pyridine in its molecular geometry, its bond
parameters and its spectral and energy data. The bond parameters cited below are the values found in
the quinoline complex Ni[S$_2$P(C$_2$H$_5$)$_2$](C$_9$H$_7$N). Their bond variation is comparable to that of
naphthalene
(C-1/C-2 136.1 pm, C-2/C-3 142.1 pm) (see Fig. 6.11).

Comparison with
naphthalene :

Fig. 6.11   Bond parameters of quinoline
(bond lengths in pm, bond angles in degrees)

Quinoline shows the following UV and NMR spectral data, which correspond closely to those of
naphthalene:

UV (H$_2$O) λ (nm) (ε)	^{1}H-NMR (CDCl$_3$) δ (ppm)		^{13}C-NMR (CDCl$_3$) δ (ppm)		
226 (4.36)	H-2: 8.81	H-6: 7.43	C-2: 150.3	C-6: 126.3	C-4a: 128.0
275 (3.51)	H-3: 7.26	H-7: 7.61	C-3: 120.8	C-7: 129.2	C-8a: 148.1
299 (3.46)	H-4: 8.00	H-8: 8.05	C-4: 135.7	C-8: 129.3	
312 (3.52)	H-5: 7.68		C-5: 127.6		

Comparison with naphthalene:

220 (5.01)	302 (3.50)	H$_\alpha$: 7.66	C$_\alpha$: 128.0	C-4a/C-8a: 133.7
275 (3.93)	310 (2.71)	H$_\beta$: 7.30	C$_\beta$: 126.0	
(heptane)				

Quinoline is characterized by an empirical resonance energy ΔE_π of 222 kJ mol^{-1} and by a DEWAR resonance energy of 137.7 kJ mol^{-1}, calculated by the SCF/MO method. The corresponding stabilization energies in naphthalene are ΔE_π = 292 and 127.9 kJ mol^{-1} for the DEWAR resonance energy.

B For the reactions of quinoline, addition and substitution processes are to be expected in view of its similarity to naphthalene and pyridine. It is of interest to note the extent to which the fused benzene ring influences the positional and relative reactivities.

Electrophilic substitution reactions

As in pyridine, it is the nitrogen in quinoline which undergoes protonation, alkylation, acylation and, with peroxyacids, oxidation to the *N*-oxide. S$_E$Ar reactions occur on the ring C-atoms, preferentially on those of the more activated benzene moiety. The relative reactivities of the individual heteroaromatic positions were determined by an acid-catalysed H–D exchange with D$_2$SO$_4$. This demonstrated that the S$_E$Ar process occurs via the conjugated acid, i.e. the quinolinium ion, and the positional selectivity C-8 > C-5/C-6 > C-7 > C-3 was observed.

Nitration (in contrast to pyridine, see p 274) takes place with nitrating acid under mild conditions. Since the strongly acidic medium causes complete *N*-protonation, monosubstitution occurs exclusively in positions C-5 and C-8 with formation of the products **4** and **5**:

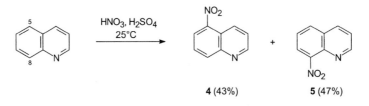

4 (43%)   5 (47%)

Nitration of quinoline with HNO$_3$ in acetic anhydride furnishes only small amounts of identifiable products such as **4** and **5** (< 1 %) and 3-nitroquinoline **6** (6%), probably resulting from an initial 1,2-addition (cf. the bromination, p 319). Transition metals can exert special directional effects. For instance, Zr(NO$_3$)$_4$ nitrates quinoline to give the 7-nitro derivative **7**:

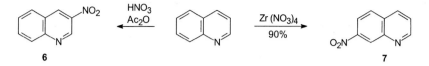

6   **7**

The following relative reactivity of quinoline compared to pyridine and naphthalene is observed on nitration: benzene $= 1$, naphthalene $\approx 10^5$, pyridinium ion $\approx 10^{-12}$, quinolinium ion $\approx 10^{-6}$, and for comparison, N,N,N-trimethylanilinium ion $\approx 10^{-8}$. Thus quinoline possesses a higher $S_E Ar$ reactivity than pyridine.

Halogenation proceeds by various methods and mechanisms. For instance, quinoline, when brominated with Br_2 in H_2SO_4 in the presence of Ag_2SO_4, affords 5- and 8-monosubstituted products **8** and **9** in the proportion $\approx 1:1$:

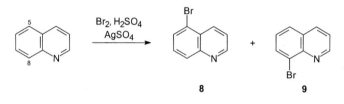

8   **9**

Interaction of bromine with the $AlCl_3$ complex of quinoline leads predominantly to the 5-substitution product **8**, owing to steric hindrance of the 8-position; the postulated mechanism is an $S_E Ar$ process involving a quinolinium cation. However, if bromine is made to react with quinoline in the presence of pyridine, 3-bromoquinoline **10** is obtained as the sole product. An addition/elimination mechanism with an initial N/C-2 addition of bromine is presumably responsible for this anomalous product formation:

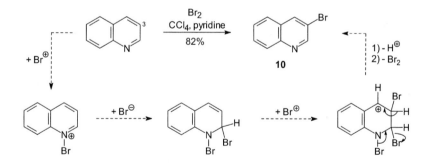

10

Sulfonation of quinoline produces different products depending on the reaction temperature. At 90°C, the 8-sulfonic acid **11** is formed predominantly; raising the temperature increases the proportion of 5-sulfonic acid **13**, the sole product at 170°C in the presence of $HgSO_4$ (steric hindrance at the 8-position by N/Hg^{2+} coordination). At 300°C, 6-sulfonic acid **12** is the sole product. On heating to 300°C, **11** and **13** are converted into the acid **12** which represents the thermodynamically favoured sulfonation product. Thus in the sulfonation process, quinoline resembles naphthalene in its kinetic and thermodynamic product control.

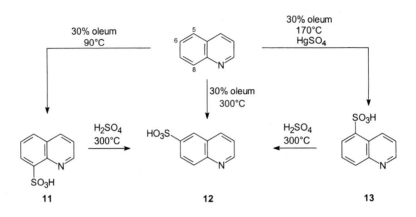

11   12   13

Nucleophilic substitution reactions

As expected, nucleophilic substitution of quinoline occurs in the hetero ring, as a rule in the 2- or 4-position. S_NAr processes proceed faster in quinoline than in pyridine, because the fused benzene ring stabilizes the addition products by conjugation.

The CHICHIBABIN amination (see p 277) proceeds with alkali amides in liquid ammonia. In this reaction, quinoline provides a mixture of the 2- and 4-amino compound **14** and **15**, whereas 2-phenylquinoline yields the 4-amino compound :

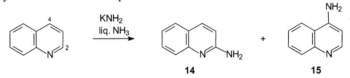

14   15

The ZIEGLER reaction (see p 278) with organolithium compounds leads exclusively to 2-alkyl- or 2-arylquinolines. For instance, **20** is produced with *n*-butyllithium. After hydrolysis, the primary adducts (e.g. **16**) yield stable 1,2-dihydroquinolines (e.g. **17**) which can be dehydrogenated by nitro compounds. It is obvious that control of the RLi addition occurs through coordination, because even 2-substituted quinolines (e.g. **18**) yield mainly 2-addition products (e.g. **19**).

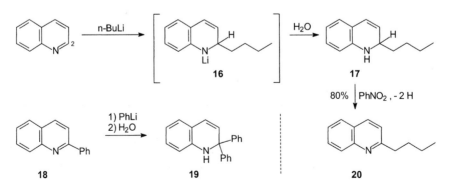

18   19   20

The formation of 2-quinolone from quinoline, which is brought about by the action of alkali hydroxides at high temperature, can be formally regarded as an S_NAr process with a hydride ion as leaving group:

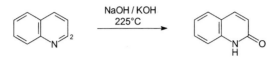

S_NAr reactions of quinoline, as for pyridine (see p 277), take place very readily in the presence of leaving groups, e.g. halogen, on those ring C-atoms which are α- or γ- to nitrogen (ring positions 2 or 4).

Nucleophilic additions to quinoline occur readily after quaternization of the nitrogen and are often followed by preparatively useful reactions. For instance, cyanide adds to the 4-position in 1-methyl-quinolinium iodide and the adduct **21** can be oxidized with *N*-demethylation to give 4-cyanoquinoline **22**:

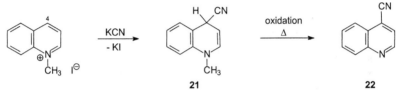

Quinoline adds cyanide [as KCN or $(CH_3)_3SiCN$] in the 2-position in the presence of acylating agents, usually C_6H_5COCl or ClCOOR, producing *N*-acyl-2-cyano-1,2-dihydroquinolines **23** (*Reissert reaction*):

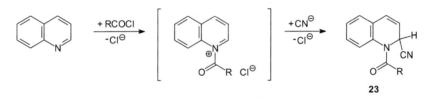

The REISSERT reaction of quinoline of $Me_3SiCN/RCOCl$ can be conducted in an enantioselective fashion in the presence of a chiral BINAP-based catalyst to give enantiomeric nitriles of type **23** [91]. The REISSERT compounds **23** are hydrolysed in an acid medium giving aldehydes and quinoline-2-carboxylic acid **24**:

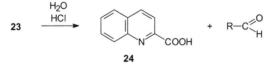

The synthetic potential of the REISSERT reaction will be discussed in detail together with the corresponding isoquinoline derivatives (see p 338).

2-Cyanoquinoline is obtained from quinoline 1-oxide by treatment with cyanide/benzoyl chloride with deoxygenation and α-substitution. *O*-Acylation of the *N*-oxide is the first step; this is followed by addition of cyanide to the 1-(benzoyloxy)quinolinium ion **25** to give the 1,2-dihydroquinoline **26** which eliminates benzoic acid (see p 287):

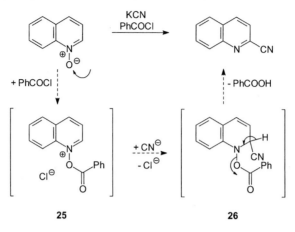

25 **26**

Side-chain reactivity

CH$_3$ groups attached to the hetero moiety of quinoline possess CH-acidity according to the sequence 4-CH$_3$ > 2-CH$_3$ >> 3-CH$_3$. Thus 2- and 4-methylquinoline, analogously to pyridine (see p 281), undergo base or acid catalysed C–C forming condensation reactions such as aldol condensation (e.g. **28**), CLAISEN condensation (e.g. **27** and **30**) or MANNICH reactions (e.g. **29**):

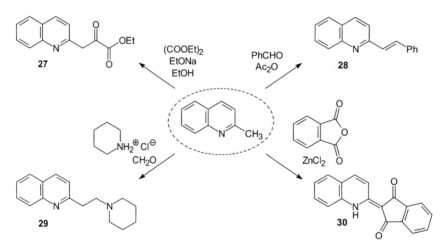

27   **28**

29   **30**

2,4-Dimethylquinoline **32** undergoes regioselective C–C condensations which are controlled by the nature of the base. Lithium alkyls coordinate to quinoline nitrogen and preferentially deprotonate the 2-CH$_3$ group, whereas lithium dialkylamides complex the lithium preferentially with the amide nitrogen resulting in deprotonation of the more acidic 4-CH$_3$ group [92]. The capture of the carbanions formed by addition to benzophenone yielding the products **31/33** demonstrates these conclusions.

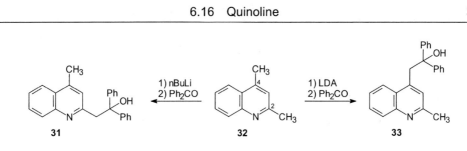

31   **32**   **33**

N-Quaternization increases the CH-acidity of the 2- and 4-CH$_3$ groups in quinoline even further. The 'anhydrobases' formed after deprotonation (e.g. **34**) are usually stable and, as enamines, are subject to electrophilic attack on the CH$_2$ group, e.g. alkylation (**34** → **35**):

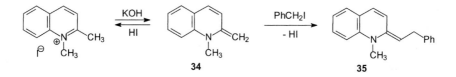

34   **35**

The same reaction principles apply to the formation of cyanine dyes with heterocyclic end groups which play an important part as sensitizers in colour photography. Pinacyanol **37**, made by a base-catalysed condensation of orthoformic ester with two moles 1-ethyl-2-methylquinolinium iodide **36** via the anhydrobase **38**, is an example of quinoline as a terminal heterocycle:

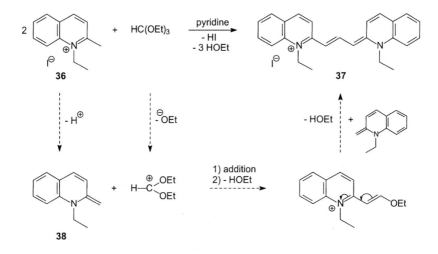

Oxidation

Oxidation affects both rings of the quinoline system. Alkaline permanganate causes oxidative cleavage of the benzene ring in quinoline and 2-substituted quinolines to give pyridine-2,3-dicarboxylic acid **39** (R = H: quinolinic acid). In contrast, acidic permanganate oxidizes the pyridine ring with formation of *N*-acylanthranilic acid **40**:

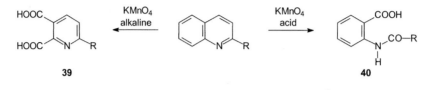

Alkylquinolines are oxidized with dichromate/H_2SO_4 providing the corresponding quinolinecarboxylic acids. SeO_2 oxidizes 2- and 4-methyl groups in quinoline to formyl groups. This side-chain conversion can also be effected regioselectively, e.g. 2,3,8-trimethylquinoline can be oxidized to the aldehyde **41** [93]:

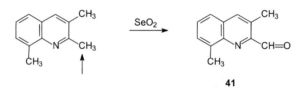

Reduction

Quinoline is reduced at high temperature and pressure by RANEY nickel as catalyst to give the decahydro system **44** (*cis/trans* mixture). At atmospheric pressure, only the pyridine moiety is affected yielding 1,2,3,4-tetrahydroquinoline **43**. Selective hydrogenation of the benzene ring giving 5,6,7,8-tetrahydroquinoline **42** occurs in CF_3CO_2H with PtO_2 as catalyst.

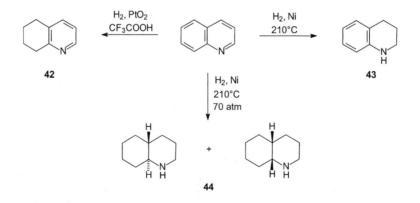

Quinoline is reduced by LiAlH$_4$ or diethylaluminium hydride, with N-coordination and hydride transfer to the 2-position yielding 1,2-dihydroquinoline, and by Li or Na in liquid NH$_3$ yielding 1,4-dihydroquinoline:

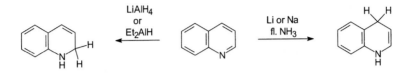

On treatment with zinc in acetic anhydride, quinoline is readily transformed to the product **46** showing a pentacyclic quinobenzazepine skeleton. This formal reductive dimerization process is likely to be initiated by SET (single electron transfer) of the N-acetylchinolinium intermediate **45** followed by radical dimerization via 4-position [94].

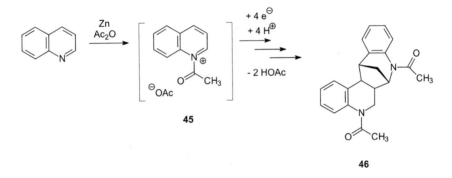

Quinaldine was found to undergoe a similar reductive dimerization with Zn/acetic acid [95].

C The *retrosynthesis* of quinoline (see Fig. 6.12) can be considered in an analogous way to pyridine (see p 296).

- Bond fission of the imino structure leads to the *o*-aminocinnamic aldehyde **47**, or after reduction to the *o*-aminophenylpropanal **48**. These products are suggested as starting materials for a cyclocondensation. The starting material **48** can be obtained from the 1,4-dihydroquinoline **49** (retrosynthetic operation **a-d**).

- The retrosynthetic operations **e/f** lead to the dihydroquinolines **51** and **52**. For **52**, a synthesis involving [4+2] cycloaddition of alkynes to N-phenylimines is retroanalytically relevant. Retroanalysis via **51** and **52** (operations **h** and **i**) suggests intermediates which, after a bond fission at the C-4/benzene ring (**j** and **k**) facilitate an S$_E$Ar cyclization. The aniline derivatives **53** and **54** emerge as starting materials that possess an electrophilic centre (C–X, C=O) in the γ-position to the nitrogen.

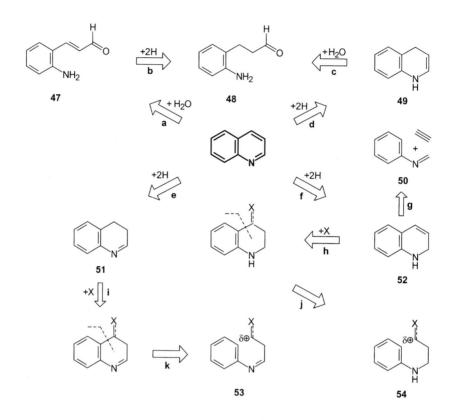

Fig. 6.12 Retrosynthesis of quinoline

The following syntheses of quinoline follow from the above retrosynthetic considerations.

(1) *o*-Aminocinnamoyl compounds obtained by reduction from the corresponding *o*-nitro compounds often cyclize in situ to give quinoline derivatives. For instance, *o*-nitrocinnamic aldehydes or -ketones yield quinolines **55** and *o*-nitrocinnamic acid 2-quinolones **56**. Occasionally, the reducing agent influences product formation; for instance, *o*-nitrocinnamic ester **58** and Zn/acetic acid provide the quinolone **57**, whereas triethylphosphite yields the alkoxyquinoline **59**:

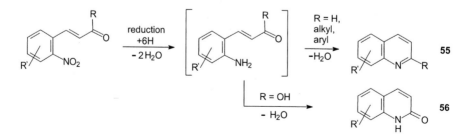

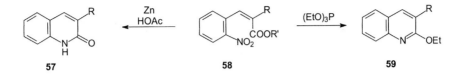

57   58   59

Recently, an interesting principle of formation for 2-substituted quinoline derivatives was realized in the Pd-catalysed transfer hydrogenation/heterocyclization of (*o*-aminophenyl)ynones **60** in the presence of tertiary amin/formic acid to give **61** [96] :

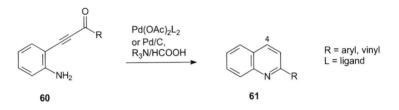

60   61

Ynones **60** also can be cyclized by addition of nucleophiles HNu (e.g. HOR, HSR etc.) to give quinolines analogous to **61** with an additional Nu-substituent in the 4-position [97].

o-Aminocinnamic acid derivatives **63** are obtained by a different method, namely from the sulfonium ylide **62**, easily accessible via *N*-methylaniline, by a [2,3] sigmatropic rearrangement and elimination of benzenethiol. They cyclize giving methyl-2-quinolones **64** [98]:

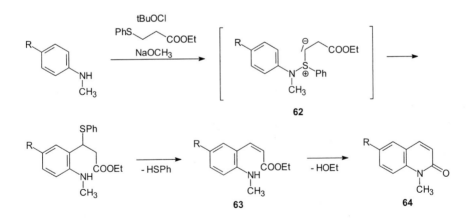

62

63   64

(2) Cyclocondensation of *o*-aminoaryl ketones, or aldehydes with ketones possessing an α-CH$_2$ group (*Friedländer synthesis*) involves formation of *o*-aminocinnamoyl systems, e.g. **65** [99]:

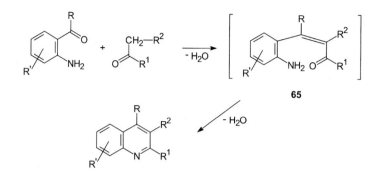

The required aldol condensation, whether base- or acid-catalysed, exerts regioselectivity in the case of unsymmetrical ketones. For instance, the quinoline formation from methyl ethyl ketone and 2-benzoylaniline **66** occurs by base catalysis via the enolate giving the 2-ethyl-4-phenylquinoline **67**. However, under acid catalysis via the enol, it affords 2,3-dimethyl-4-phenylquinoline **68**:

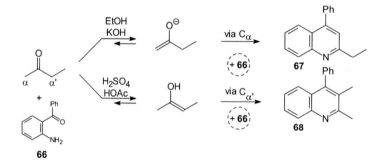

N-Phenylimines can be used in place of the often unstable *o*-aminoaryl aldehydes or ketones.

The reaction of methylene ketones with 2,3-dihydroindol-2,3-dione (isatin) in an alkaline medium (*Pfitzinger synthesis*) is the most important variant of the FRIEDLÄNDER synthesis:

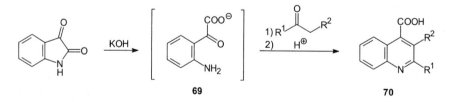

The alkali first converts isatin into the salt **69** of isatinic acid which, being a carbonyl compound, cyclocondenses via the α-keto function with methylene ketones to give derivatives of quinoline-4-carboxylic acid [100].

N-Acetylisatin, after ring-opening by base to the α-keto carboxylate **71**, undergoes an intramolecular aldol condensation producing 2-quinolone-4-carboxylic acid **72**. In the same way, *o*-*N*-acylaminoaryl ketones **73** can be made to cyclize to give either 2- or 4-quinolones (*Camps synthesis*):

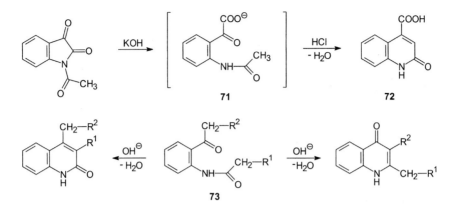

(3) Some of the quinoline preparations based on intramolecular S_EAr processes belong to the classics of heterocycle synthesis. For instance, primary arylamines with a free *ortho* position can be made to undergo cyclocondensation with β-diketones or with β-keto aldehydes in a strongly acidic medium (*Combes synthesis* [101]):

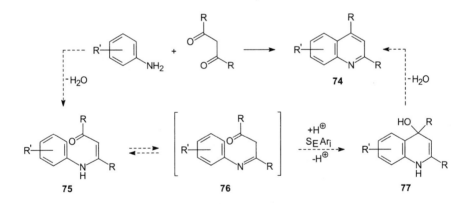

In the first step, β-enaminones **75** or *N*-arylimines **76** are formed which, after C=O protonation (possibly by way of an O–N diprotonated species), undergo intramolecular hydroxyalkylation and dehydration (via **77**) to afford quinolines **74**. Regioselectivity and product formation in unsymmetrical β-dicarbonyl compounds can be influenced by the acidity of the reaction medium and by the reaction temperature, e.g.:

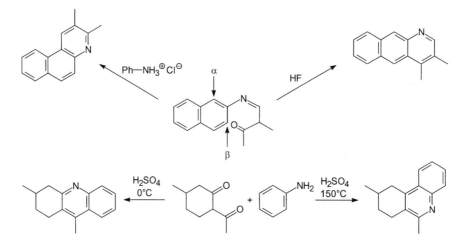

Primary arylamines and β-keto esters condense in the presence of strong acids to form 2-quinolones (*Knorr synthesis*) via β-keto anilides (e.g. **78**). A thermal reaction involving β-anilinoacrylic esters (e.g. **79**) yields 4-quinolones (*Konrad-Limpach synthesis*):

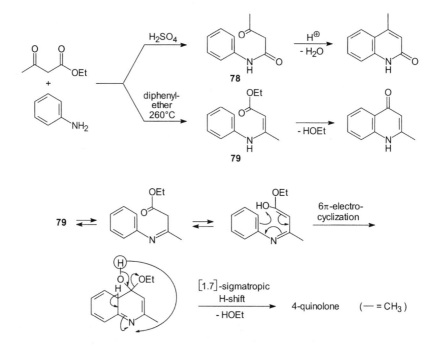

While the 2-quinolone formation is due to an S_EAr process analogous to a COMBES synthesis, the 4-quinolone formation is thought to be the result of an electrocyclization. The KONRAD-LIMPACH synthesis is very flexible with regard to the arylamine and β-keto ester substituents, as well as to the use of alkoxymethylene malonic esters and dianilides of malonic aicid

(4) In the *Skraup* and *Doebner-Miller synthesis* of quinoline, primary arylamines with an unsubstituted *ortho* position react with α,β-unsaturated carbonyl compounds in an acid medium in the presence of an oxidizing agent (nitroarene, As_2O_5)*:

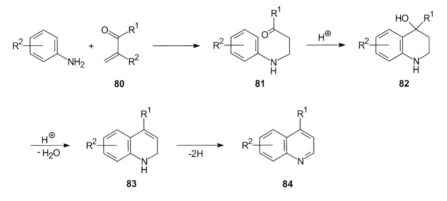

The sequence of this reaction was established by the isolation of intermediates, the distribution of substituents in the products and ^{13}C-labeling. The first step is the MICHAEL addition of the amine to the enone system **80** with formation of β-amino ketones **81**. They cyclize to give **82** with intramolecular hydroxyalkylation via the protonated C=O group. Dehydration leads to 1,2-dihydroquinoline **83**, which on dehydrogenation affords quinoline **84**.

2-Methylquinoline is formed from aniline and but-2-enal (crotonaldehyde), whereas 4-methylquinoline is obtained from methyl vinyl ketone; the 1,2-dihydroquinoline **85** is the end product when 4-methylpent-3-en-2-one (mesityl oxide), a terminally disubstituted enone, is used.

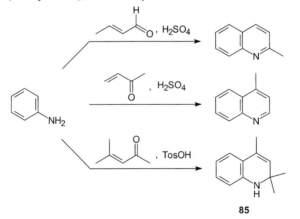

As the amino component can be varied in many ways, the SKRAUP synthesis is widely applicable to the preparation of quinolines unsubstituted in the heterocyclic moiety and especially to the preparation of polyheterocycles. This is illustrated by the synthesis of 1,10-phenanthroline **86**, a much-used chelating ligand, starting from 8-aminoquinoline :

---

* The *Skraup synthesis* consists of the reaction with acrolein which is formed in situ from glycerine/H_2SO_4. The *Doebner–Miller synthesis* was originally limited to the reaction with crotonaldehyde to give quinaldines, but is increasingly used as a generic term for this reaction.

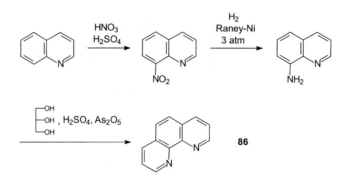

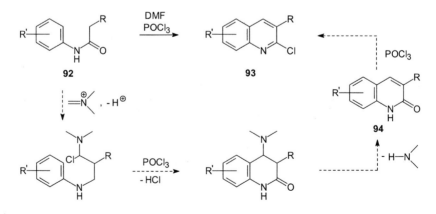

(5) Some lesser known quinoline syntheses also use cyclization of aniline derivatives with nitrogen side-chains rendered electrophilic by protic or LEWIS acids. For instance, either 3,4-dihydro-2-quinolones **87** or 2-quinolones **89** can be obtained from cinnamic acid anilides **88**, whereas 3-(aryl-amino)propionic acids **90** produce 2,3-dihydro-4-quinolones **91**:

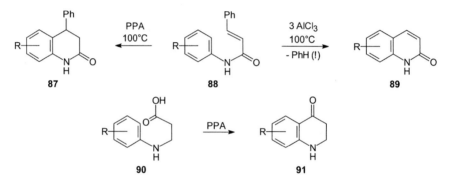

Acetanilides **92** are converted into 3-substituted 2-chloroquinolines **93** by DMF/POCl$_3$ (*Meth-Cohn synthesis* [102]). The α-methylene group in the anilide **92** first undergoes a VILSMEIER reaction. This is followed by an S$_E$Ar cyclization with amine elimination to give the 2-quinolone **94** which, under the conditions of the reaction, is converted into **93** ([103], cf p 312):

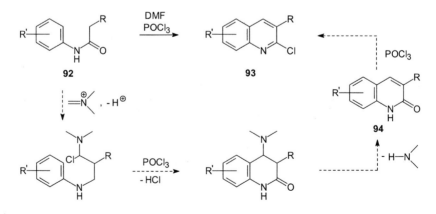

(6)   The preparation of the quinoline system by pericyclic reactions is less important than the previous syntheses. Some cyclization/elimination reactions of N-phenylimines **95** and **96** with alkynes, enol ethers or enamines are of preparative interest. Although they correspond to [4+2] cycloadditions, they are likely to proceed by S_EAr mechanisms and are usually catalysed by LEWIS acids. In the enolether cycloaddition to imines **95** ytterbium(III)triflate proved to be particularly effective [104].

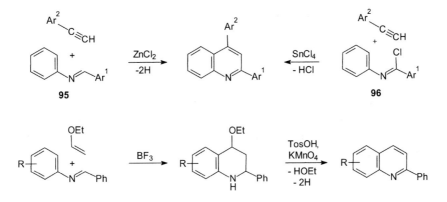

The principle underlying the thermal cyclization of N-(2-vinylphenyl)imines **97**, which by dehydrogenation yields quinoline **98** [105], is of interest.

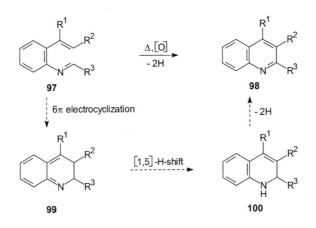

The mechanism postulated for this reaction involves 6π-electrocyclization of the 2-azahexatriene system in **97**. This is followed by a [1,5] sigmatropic hydrogen shift in the quinonoid intermediate **99**. Finally aromatization and oxidation of the 1,2-dihydroquinoline **100** [106] occurs.

(7)   Quinolines are accessible by ring transformation of other heterocyclic systems. Thus, N-aryl-azetidinones **101** isomerize to 1,2,3,4-tetrahydroquinoline-4-ones **102** under acid catalysis:

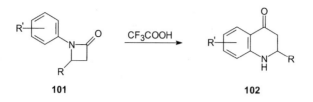

101   102

Indoles yield 3-chloroquinolines by addition of dichlorocarbene and HCl elimination (see p 101). A number of ring enlargement reactions of isatins into quinolines are known (as example, see the PFITZINGER synthesis, p 328, and [107/108]).

Oxazoles and 4,5-dihydrooxazoles can be transformed into quinolines. For instance, 5-(o-acylamino-aryl)oxazole-4-carboxylic esters **104**, made from benzoxazinones **103** and isocyanoacetic ester, are converted into 3-amino-4-hydroxy-2-quinolones **105** in an acid medium [109].

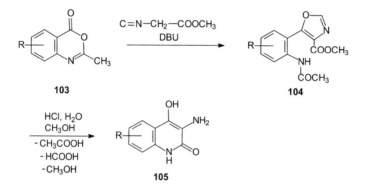

103   104

105

4,5-Dihydrooxazoles (e.g. **106**) yield 3-substituted 2-quinolones (e.g. **107**) in an acid-catalysed condensation with o-chlorobenzaldehydes by the following mechanism [110]:

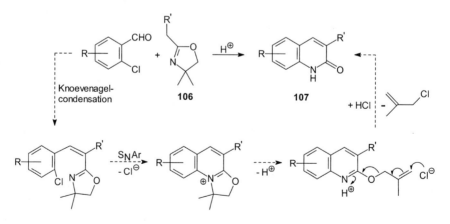

106   107

Knoevenagel-
condensation

S_NAr

D Quinoline, bp 237°C, is a colourless steam-volatile liquid of characteristic odour. Quinoline was first obtained by alkaline degradation of the alkaloid cinchonine (see below, GERHARD 1842). It possesses a lower dipole moment ($\mu = 2.16$ D) than pyridine ($\mu = 2.22$ D) and isoquinoline ($\mu = 2.60$ D). Quinoline, because of the nitrogen link with the fused benzene ring, is a weaker base ($pK_a = 4.87$) than isoquinoline ($pK_a = 5.14$).

8-Hydroxyquinoline ('oxine'), mp 75°C, contains an intramolecular hydrogen bond and is used as a complexing and precipitating reagent for many metallic ions in analytical chemistry. It is obtained by a SKRAUP synthesis starting from *o*-aminophenol.

4-Hydroxyquinoline-2-carboxylic acid (kynurenic acid) was isolated from the urine of dogs (LIEBIG 1853). It is a metabolic product of tryptophan.

The alkaloids of the cinchona bark are natural products containing a quinoline structure [111]. Examples are the diastereoisomeric pairs quinine/quinidine and cinchonidine/cinchonine **108** and **109** in which a 4-methylquinoline unit is bonded to a vinyl-substituted quinuclidine system (1-azabicyclo[2.2.2]octane). Camptothecin **110**, a highly toxic polycyclic quinoline alkaloid, was isolated from the stem wood of the Chinese tree *Camphoteca acuminata* (Nyssaceae).

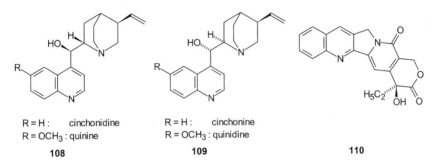

R = H :   cinchonidine   R = H :   cinchonine
R = OCH₃ : quinine   R = OCH₃ : quinidine

108   **109**   **110**

Many quinoline derivatives are important biologically active agents. 8-Hydroxyquinoline and some of its halogenated derivatives are used as antiseptics. Chloroquine **111** is one of the older but still important antimalarials. *N*-Alkyl-4-quinolone-3-carboxylic acid and systems derived therefrom are constituents of antibacterials (gyrase inhibitors [112]) such as nalidixic acid **112**, ciprofloxazin **113** and moxifloxazin **114**. The quinoline-8-carboxylic acid derivative **115** (quinmerac) is employed as a herbicide for *Galium aparine* and other broad-leaved weeds. Methoxatin **116**, known as coenzyme PQQ is a heterotricyclic mammalian cofactor for lysyl oxidase and dopamine β-hydroxylase [113].

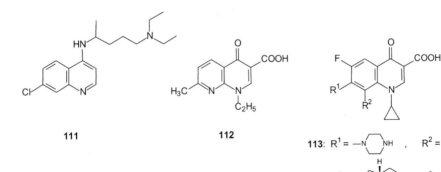

111   **112**

113: R¹ = —N⟨ ⟩NH   ,   R² = H

114: R¹ = —N⟨ ⟩   ,   R² = OCH

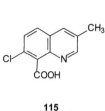

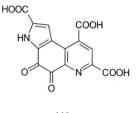

115   **116**

6.17 Isoquinoline

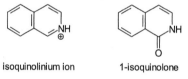

A   Isoquinoline is derived from naphthalene by replacement of a β-CH group by nitrogen. The isoquinolinium ion and 1-isoquinolone [isoquinolin-1(2*H*)-one, isocarbostyril] are important derivatives of isoquinolines.

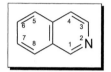

isoquinolinium ion   1-isoquinolone

Isoquinoline has structural (see Fig. 6.13) and spectroscopic analogies to naphthalene and pyridine.

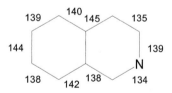

Fig. 6.13   Bond parameters of the isoquinolinium ion (bond lengths in pm)

Isoquinoline shows the following UV and NMR data:

UV (hexane) λ (nm) (ε)	^{1}H-NMR (CDCl$_3$) δ (ppm)		^{13}C-NMR (CDCl$_3$) δ (ppm)		
216 (4.91)	H-1: 9.15	H-6: 7.57	C-1: 152.5	C-6: 130.6	C-4a: 135.7
266 (3.61)	H-3: 8.45	H-7: 7.50	C-3: 143.1	C-7: 127.2	C-8a: 128.8
306 (3.35	H-4: 7.50	H-8: 7.87	C-4: 120.4	C-8: 127.5	
318 (3.56)	H-5: 7.71		C-5: 126.5		

B The reactions of isoquinoline closely parallel those of quinoline and pyridine. Protonation, alkylation, acylation and oxidation with peroxy acids occur on nitrogen. S_EAr and S_NAr reactions take place on the ring C-atoms. As in quinoline, the fused benzene ring influences the reaction site and reactivity.

Electrophilic substitution reactions

S_EAr reactions occur preferentially in the 5- or 8-position of isoquinoline. Nitration with HNO_3/H_2SO_4 at 25°C affords the 5- and 8-nitro compounds **1** and **2**, bromination in the presence of strong protic acids or of $AlCl_3$ leads to the 5-bromo compound **3**, and sulfonation with oleum at temperatures up to 180°C yields the 5-sulfonic acid **4**:

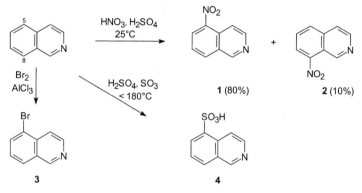

The relative reactivity of isoquinoline compared with pyridine and quinoline was determined from nitration experiments: benzene = 1, pyridinium ion $\approx 10^{-12}$, quinolinium ion $\approx 10^{-6}$, isoquinolinium ion $\approx 10^{-5}$. Accordingly, isoquinoline has a greater S_EAr reactivity than pyridine.

Nucleophilic substitution reactions

Nucleophilic reactions take place on the hetero ring of isoquinoline, preferably in the 1-position. For instance, the CHICHIBABIN amination with $NaNH_2$ in liquid ammonia yields 1-aminoisoquinoline **5**. The ZIEGLER reaction with *n*-butyllithium furnishes the 1-substituted product **7**; as with quinoline, benzene ring annulation stabilizes the primary addition product **6** (1,2-dihydroisoquinoline), which can be isolated and dehydrogenated to **7** by nitro compounds:

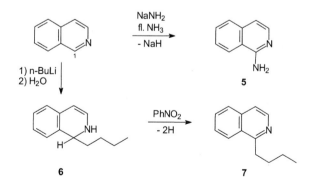

Isoquinolines halogenated in the 1-position are particularly reactive towards S$_N$Ar replacement. For instance, 1,3-dichloroisoquinoline is selectively substituted in the 1-position by methoxide to form the 1-methoxy compound **8**. Reductive dehalogenation also removes the 1-halogen substituent and produces 3-chloroisoquinoline **9** which, by a S$_N$Ar reaction with methoxide, yields **10**:

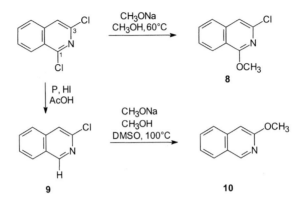

Isoquinoline, like quinoline, undergoes the REISSERT reaction with KCN or (CH$_3$)$_3$SiCN in the presence of acylating agents such as C$_6$H$_5$COCl or ClCO$_2$R affording 2-acyl-1-cyano-1,2-dihydroisoquinoline **11**:

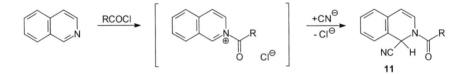

The 1,2-dihydroisoquinoline **11** undergoes hydrolysis in acid medium to give aldehydes and isoquinoline-1-carboxylic acid. This disproportionation reaction, which is used for an aldehyde synthesis, probably involves formation of an aminooxazolium ion **13**, followed by ring-opening to the amide **14**:

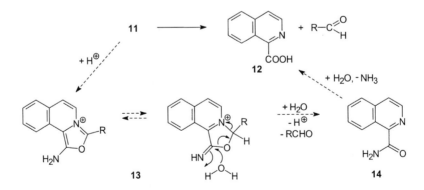

A number of synthetically useful transformations have been carried out with the REISSERT compound **15** (POPP 1967 [114]). Strong bases bring about deprotonation to give the anion **16**, which rearranges to 1-benzoylisoquinoline **17** after loss of cyanide and a 1,2-acyl shift. Electrophilic reagents, e.g. alkyl halides, carbonyl compounds or MICHAEL acceptors, convert the anion **16** into 1-substituted isoquinolines (examples **18-20**), again with cyanide elimination:

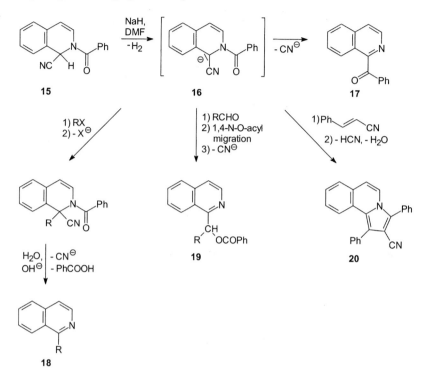

Side-chain reactivity

As with quinoline, CH_3 groups attached to the hetero moiety of isoquinoline possess CH-acidity (note: 1-CH_3 >> 3-CH_3) and can, therefore, undergo base- or acid-catalysed C–C bonding reactions (**21, 22**):

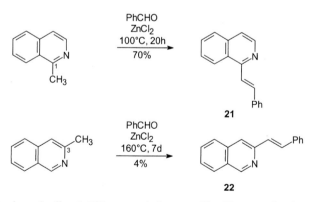

The CH-acidity of the isoquinoline 1-CH_3 group is increased by *N*-quaternization. Anhydro bases (e.g. **24**), obtained by deprotonation of *N*-alkylisoquinolinium ions (e.g. **23**), are usually stable and as enamines can undergo electrophilic reactions on the β-C-atom (see p 323):

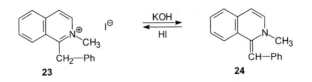

23   **24**

Oxidation

Oxidation of isoquinoline with alkaline permanganate yields a mixture of phthalic acid and pyridine-3,4-dicarboxylic acid, while KMnO$_4$ in a neutral medium does not affect the benzene ring affording phthalimide as oxidation product:

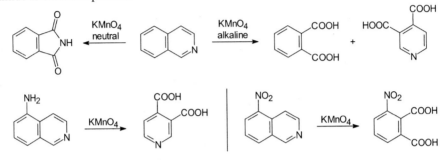

Substituents on the benzene ring can influence the outcome of the oxidation. For instance, oxidation of 5-aminoisoquinoline with KMnO$_4$ affects only the benzene ring, whereas with 5-nitroisoquinoline, only the pyridine ring is oxidized.

Reduction

Reduction of isoquinoline is carried out by catalytic hydrogenation, by hydride reagents or by metals. Catalytic hydrogenation is controlled by the acidity of the reaction medium: selective reduction of the pyridine ring to the 1,2,3,4-tetrahydro compound **26** occurs in CH$_3$CO$_2$H, whereas in concd HCl, the benzene moiety is selectively reduced to afford the 5,6,7,8-tetrahydro compound **25**. Further hydrogenation leads to the decahydroisoquinoline **27** (*cis/trans* mixture):

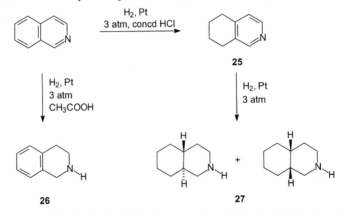

When treated with diethylaluminium hydride, isoquinoline yields its 1,2-dihydro derivative, as does quinoline. Metal reductions, e.g. with sodium in liquid NH$_3$ or with tin/HCl, afford the tetrahydroisoquinoline **26**. Isoquinolinium ions are reduced to 1,2,3,4-tetrahydroisoquinolines by NaBH$_4$. This reaction is important for establishing the structure of alkaloids. Since the reduction of the heterocycle with sodium borohydride proceeds rapidly, other reducible substituents such as the carbonyl group remain unaffected, as example **28** shows.

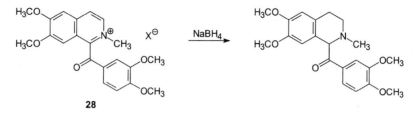

28

For further synthetic applications of the reduction of isoquinolinium ions, see [115].

C   The *retrosynthesis* of isoquinoline (see Fig. 6.14) is approached in a way similar to that of quinoline (see p 326).

- Bond fission of the imine moiety (retrosynthetic steps **a** and **c**) suggest compounds **29** and **30** as starting materials for a cyclocondensation. The aminocarbonyl compounds **31** and **35** are relevant to the synthesis of dihydroisoquinolines **32** and **34**, if reductions (**b,d,f,g**) are taken into consideration.
- Bond fission at positions C-4 or C-1 (**i/j**) in the 3,4-dihydroisoquinoline **33** lead to formation of the synthons **36** or **37**. They are suitable starting compounds for a S$_E$Ar cyclization as β-electrophilic imines or α-electrophilic amines.

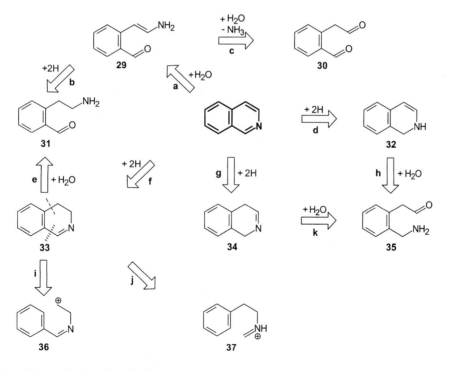

Fig 6.14   Retrosynthesis of isoquinoline

The synthesis of isoquinoline is considered under the same headings as the retrosynthesis.

(1) (2-Formylphenyl)ethanal **38** and analogous dicarbonyl compounds undergo cyclization with ammonia to give isoquinolines. Primary amines produce *N*-substituted isoquinolinium ions, hydroxylamine yields isoquinoline *N*-oxides, and hydrazines give isoquinolinium *N*-betaines:

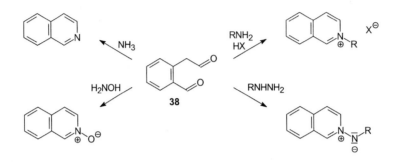

As shown in examples **40** and **42**, this synthetic principle can be applied to biselectrophilic systems with functions similar to **38** but in a higher oxidation state, such as *o*-formylphenyl or *o*-cyanophenyl derivatives of acetic acid **39** and **41**:

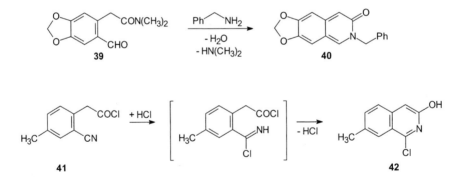

The dioxolane-protected ketones **44**, obtained from *o*-cyanobenzyl ketones **43**, are accessible by CLAISEN condensation of *o*-cyanotoluene with arylcarboxylic esters. They are converted into 3-arylisoquinolines **45** by treatment with acids followed by dehydrogenation of the intermediate 1,4-dihydroisoquinoline.

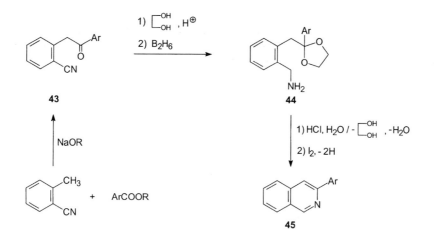

(2) The isoquinoline syntheses, which utilize intramolecular S_EAr reactions for building the heterocycle, are of greater preparative importance [116]. For instance, on treatment with strong protic acids (H_2SO_4, polyphosphoric acid), LEWIS acid or $POCl_3$, N-(2-arylethyl)amides **46** cyclize giving 1-substituted 3,4-dihydroisoquinolines **47** (*Bischler-Napieralski synthesis*):

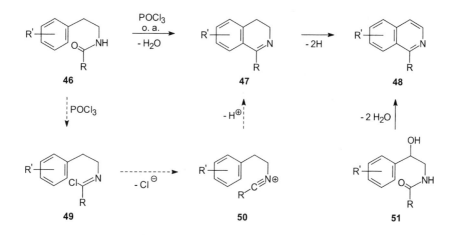

The first step of this cyclodehydration brought about by $POCl_3$ is akin to the VILSMEIER reaction. It proceeds via chloroimines **49** (or their conjugate acids) and the electrophilic nitrilium ions **50**, which are isolable as hexachloroantimonates. The 3,4-dihydroisoquinolines can be transformed to isoquinolines **48** by conventional methods, e.g. catalytic dehydrogenation.

Electron-releasing substituents in the *meta*-position of the N-(2-arylethyl)amide facilitate ring closure and lead to 6-substituted 3,4-dihydroisoquinolines, whereas *para*-substituents can prevent the cyclization. The scope of the BISCHLER-NAPIERALSKI synthesis is illustrated by the cyclization of the amide **52** to produce hexahydroisoquinolines **53**, styrylisocyanates **54** to yield 1-isoquinolones **55**, and 2-acylaminobiphenyls which provides phenanthridines (see p 357):

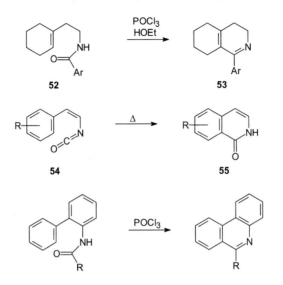

52 → **53**

54 → **55**

2-Hydroxy-substituted *N*-(2-arylethy)amides **51** lead to isoquinolines **48** under BISCHLER-NAPIERALSKI conditions. In this case, dehydrogenation is not required (*Pictet-Gams synthesis*), because H_2O elimination occurs at the dihydroisoquinoline step. The starting materials **51** are available e.g. from aryl-4,5-dihydrooxazoles (e.g. **57**).

The synthesis of the alkaloid papaverine **56** serves as an example of the two synthetic methods:

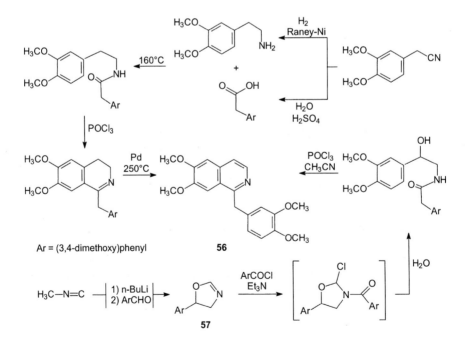

(3)   Arylaldehydes react with aminoacetaldehyde acetals to give isoquinolines (*Pomeranz-Fritsch synthesis*):

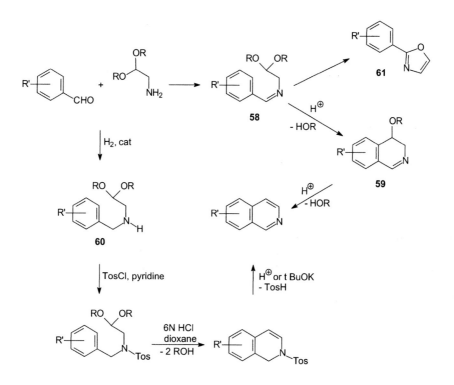

Firstly, (benzylidenamino)acetaldehyde acetals **58** are formed; they are cyclized to give 4-alkoxy-3,4-dihydroisoquinolines **59**, which are converted into isoquinolines by loss of ROH. The electrophilic cyclization is favoured by *meta*- and *ortho*-electron-releasing substituents in the arylaldehyde. Electron-withdrawing substituents (especially NO_2 groups) favour an alternative ring closure leading to oxazoles **61** with dehydrogenation.

A preparative improvement is achieved by combining the arylaldehyde and aminoacetal to give the secondary amine **60** in a reductive amination. Its tosylate **62** cyclizes in an acid medium to the 1-tosyl-1,2-dihydroisoquinoline **63** with elimination of ROH. Subsequent loss of toluenesulfinic acid leads to isoquinoline. The synthesis of compound **64** shows that 1,2,3,4-tetrahydroisoquinolines can also be obtained by a POMERANZ-FRITSCH methodology:

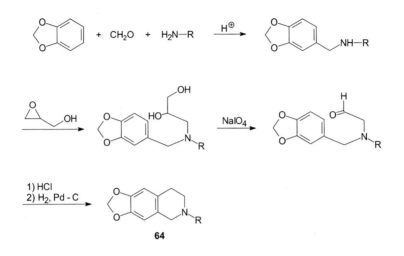

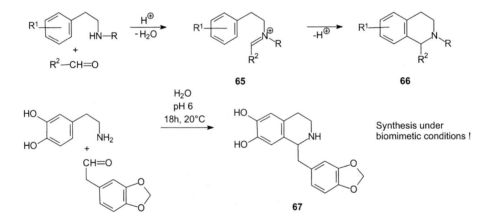

(4)   The acid-catalysed cyclocondensation of 2-arylethylamines with aldehydes is the most important method for the preparation of 1,2,3,4-tetrahydroisoquinolines **66** (*Pictet-Spengler synthesis*):

In the first step, analogous to a MANNICH reaction, an iminium ion **65** is formed which is responsible for the electrophilic cyclization. This process occurs under mild conditions, if donor substituents are present in the amine component (e.g. **67**), and is of fundamental importance for the biogenesis of isoquinoline alkaloids (see p 348).

In a variant of the PICTET-SPENGLER synthesis imines of type **68** are cyclized to give N-acyltetra-hydroisoquinolines **70** by acylation with acid chlorides in the presence of AlCl$_3$ via the chloroamine **69**:

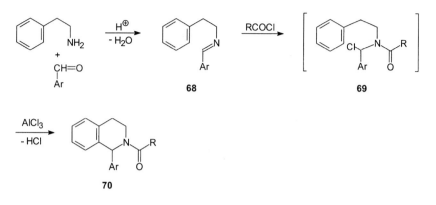

68   **69**

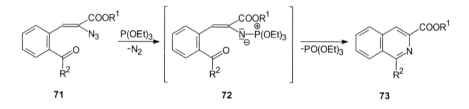

70

(5)   An interesting isoquinoline synthesis uses an intramolecular aza-WITTIG reaction to establish the C=N bond in *o*-acylstyryl ylides **72**, producing 1-substituted isoquinoline-3-carboxylic esters **73** [117]:

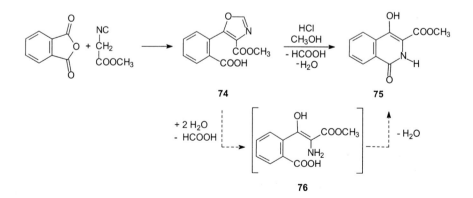

71   **72**   **73**

The aza ylides **72** are intermediates in the reaction of azidocinnamic esters **71** with P(OEt)$_3$. Ylide formation and the aza-WITTIG reaction occur under mild and neutral conditions (20-35°C).

(6)   As for quinolines, isoquinoline derivatives can be prepared by ring transformation of other heterocycles. For instance, oxazole-4-carboxylic ester **74**, obtained from phthalic anhydride and isocyanoacetic ester, is converted into the 1-isoquinolone-3-carboxylic ester **75** in an acidic medium, probably by hydrolysis to the enaminocarboxylic acid **76** followed by cyclization.

74   **75**

76

D **Isoquinoline**, mp 26°C, bp 243°C, is a colourless substance with a pleasant smell. Isoquinoline occurs in coal-tar and in bone-oil. For its basicity, see p 335.

Isoquinoline derivatives occur widely in nature [118]. There are over 600 known isoquinoline alkaloids, one of the largest groups of alkaloids known. With the exception of a few simple systems of the anhalonium type, e.g. anhalamine **77** and anhalonidine **78**, constituents of the peyote cactus,

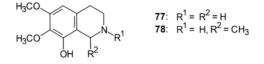

77: $R^1 = R^2 = H$
78: $R^1 = H, R^2 = CH_3$

isoquinoline alkaloids are derived from 1-benzylisoquinoline. The MANNICH cyclization, used in the PICTET-SPENGLER synthesis (see p 346), is the key step of biogenesis. It combines as building blocks [2-(3,4-dihydroxyphenyl)ethyl]amine **79** and (3,4-dihydroxyphenyl)acetaldehyde **80**, derived from the proteinogenic amino acid tyrosine, yielding the benzyltetrahydroisoquinoline derivative norlaudanosoline **81**:

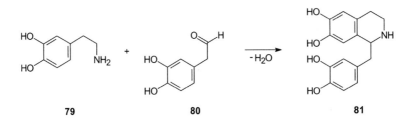

| 79 | 80 | 81 |

Other isoquinoline alkaloids are derived from the biogenetic key compound **81** by further transformations, mainly by oxidative phenol coupling of the isoquinoline ring and on the benzyl residue. A number of structural types of isoquinoline alkaloids are known, e.g. the systems **82-87** (for individual examples and further details, see textbooks of natural products).

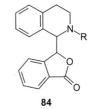

| 82 | 83 | 84 |
| benzylisoquinoline - type | protoberberine - type | phthalide isoquinoline - type |

85

pavine - type

86

aporphine - type

87

morphinane - type

A number of pharmaceuticals are derived from isoquinoline. The long-known isoquinoline alkaloid papaverine **56** (synthesis on p 344) is still important as a spasmolytic. The antidepressant nomifensin **88** and the antibilharzia drug praziquantel **89** are derived from 1,2,3,4-tetrahydroisoquinoline.

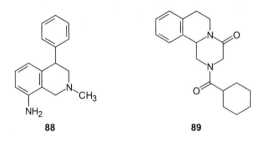

88

89

6.18 Quinolizinium Ion

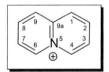

A The quinolizinium ion is structurally derived from naphthalene by substituting one C-atom for an azonia-nitrogen in the fused position (4a). In a departure from the established nomenclature (see p 9), the reduced systems **1, 2** and **3** are known as quinolizines and the perhydrosystem **4** (4a-azadecahydronaphthalene) as quinolizidines:

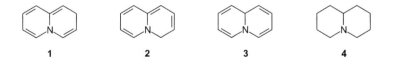

1   **2**   **3**   **4**

The quinolizinium ion is considered to be aromatic from its spectroscopic data (cf. quinoline/isoquinoline p 317/336). The strong shielding of the protons in the 4- and 6-positions next to the azonia-nitrogen are characteristics in the ^{1}H and ^{13}C-NMR spectra of the quinolizinium ion:

UV (H$_2$O) λ (nm) (ε)	^{1}H-NMR (D$_2$O) δ (ppm)	^{13}C-NMR (D$_2$O) δ (ppm)
226 (4.25)	H-1/H-9: 8.69	C-1/C-9: 127.9
272 (3.42)	H-2/H-8: 8.43	C-2/C-8: 138.0
310 (4.03)	H-3/H-7: 8.14	C-3/C-7: 125.0
323 (4.23)	H-4/H-6: 9.58	C-4/C-6: 137.0
(perchlorate)	(bromide)	(bromide)

B Reactions of the quinolizinium ion have analogies with the pyridinium ion (see p 277/280). The quinolizinium ion as a deactivated heteroarene is resistant to electrophilic reactions. Bromination is one of the few exceptions. Firstly, it leads to the perbromide **5**, then only under drastic S$_E$Ar conditions to the substitution product **6**:

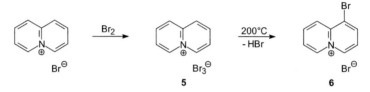

Some nucleophilic substitution reactions occur with nucleophiles, if a halogen is present at the 4-position of the quinolizinium ion; for instance, the thione **7** is obtained with sodium sulfide and the methylene compound **8** with sodium malonic ester:

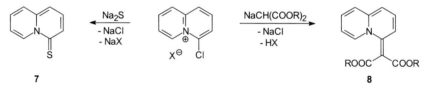

Nucleophilic ring-opening reactions are also observed. They occur, for instance, with secondary amines to give aminodienes **9** or with aryl GRIGNARD compounds yielding aryl-1,3-dienes **10**:

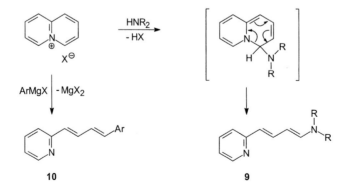

Finally, 2- or 4-methyl groups present in the quinolizinium ion have side-chain reactivity as in the pyridinium ion (see p 282) and facilitate C–C bond formation.

| C | The *retrosynthesis* of the quinolizinium ion is based on operations a/b indicating retro-aldol reactions which suggest **11/12** or **13/14** as educts for synthesis

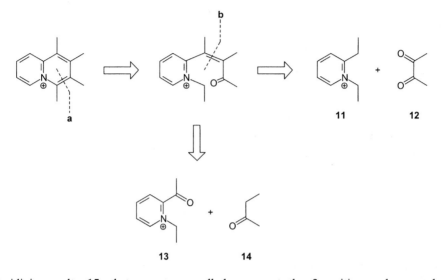

(1) Pyridinium salts **15**, that possess an alkyl group at the 2-position and an acylmethyl-, alkoxycarbonylmethyl- or cyanomethyl group at the 1-position, cyclocondense with 1,2-dicarbonyl compounds under base catalysis giving quinolizinium ions **16** (*Westphal synthesis*):

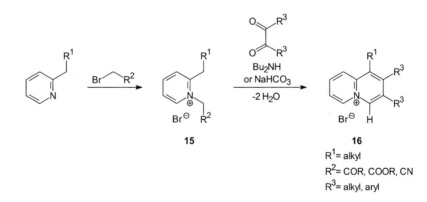

In this reaction, a double aldol condensation with acid fission of the R² residue takes place.

(2)   A modification of the WESTPHAL synthesis is based on 2-substituted pyridines **18**. They are obtained by *O*-acylation of the aldol adducts **17** obtained from pyridine-2-carbaldehyde and methylene ketones. The pyridines **18** are quaternized with bromoacetic ester or bromoacetonitrile to **19**, followed by base-catalysed cyclization giving 2,3,4-trisubstituted quinolizinium ions **20**:

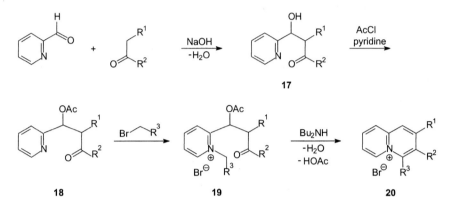

(3)   Cyclocondensation of 2-(acylmethyl)pyridines with alkoxymethylenemalonic esters or with the corresponding nitro compounds yields derivatives of the 1-quinazolone **21**.

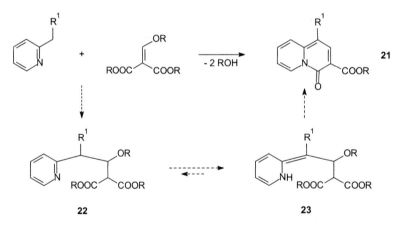

The reaction sequence is thought to consist of a MICHAEL addition leading to **22**, followed by tautomerization to **23** and lactam formation.

D   Quinolizidine alkaloids occur in the species *Lupinus*, *Cytisus* and *Genista* of the *papilionaceae*. Lupinine **24**, cytisine **25** and sparteine **26**, the main alkaloid in common broom (*Cytisus scoparius*), are typical examples. Sparteine is used in the treatment of cardiac arrhythmias.

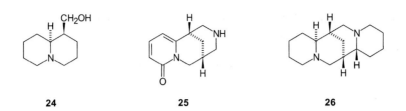

| 24 | 25 | 26 |

The dibenzo[*a,g*]quinolizinium derivative coralyn **27** (SCHNEIDER 1920) was known before the synthesis of the parent compound (BÖKELHEIDE 1954). Coralyn possesses antileukaemia properties [119].

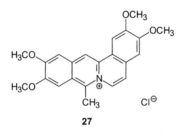

27

6.19 Dibenzopyridines

A Two neutral dibenzannulated products are derived from pyridine: the linear annulated acridine **1** (dibenzo[b,e]pyridine, an exception to systematic numbering) and the angular annulated phenanthridine **2** (dibenzo[*b,d*]pyridine):

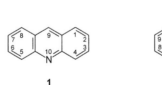

| 1 | 2 |

Acridine and phenanthridine have the following UV and ^{13}C-NMR spectroscopic data:

	UV (ethanol) λ (nm) (ε)	^{13}C-NMR (CDCl$_3$) δ (ppm)			
acridine	249 (5.22)	C-1: 129.5	C-9: 135.9		
	339 (3.81)	C-2: 128.3	C-1a: 126.6		
	351 (4.0)	C-3: 125.5	C-4a: 149.1		
	379 (3.44)	C-4: 130.3			
phenanthridine	245 (4.65)	C-6: 153.1	C-1/C-10:	121.0/121.3	
	289 (4.01)	C-4a: 144.1	C-2/C-9:	127.0/126.6	
	330 (3.27)	C-7a: 142.0	C-3/C-8:	128.3/128.2	
	346 (3.27)		C-4/C-7:	130.4/129.8	
			C-1a/C-10a:	126.6/123.7	

The UV data of the two dibenzopyridines correlate with those of anthracene [(hexane, nm, lg ε): 252 (5.34), 339 (3.74), 356 (3.93), 374 (3.93)] and those of phenanthrene [(hexane, nm, lg ε): 251 (4.83), 292 (4.17), 330 (2.40), 346 (2.34)].

B Reactions of dibenzopyridines show analogies with pyridine, quinoline and isoquinoline. Acridine and phenanthridine are *N*-protonated by strong protic acids, *N*-alkylated by alkyl halides and *N*-oxidized by peroxy acids. Electrophilic substitutions of acridine often result in disubstitution at the 2- and 7-positions (e.g. nitration giving **3**), whereas those of phenanthridine occur at different positions (e.g. nitration mainly at the 1- and 10-position yielding **4** and **5**):

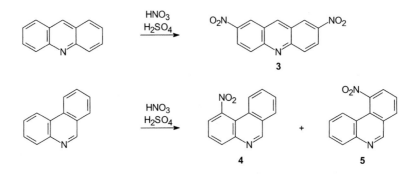

Nucleophilic substitution of phenanthridine always takes place at the 6-position with organolithium compounds, as in the CHICHIBABIN amination and ZIEGLER reaction.

However, acridine shows variable regiochemistry towards nucleophiles. CHICHIBABIN amination with NaNH$_2$ in liquid ammonia leads exclusively to 9-aminoacridine **6**, whereas in *N,N*-dimethylaniline, the main product is 9,9'-bi(9,10-dihydroacridinyl) **7**.

An SET mechanism (see p 278) is responsible for the formation of **7** [50].

Organolithium compounds and CH-acidic compounds add to acridine at the 9-position forming 9,10-dihydroacridines. For instance, with nitromethane in a basic medium, the adduct **8** is formed, whereas phenylmethylsulfone yields 9-methylacridine **10** via the adduct **9**, which eliminates phenylsulfinic acid:

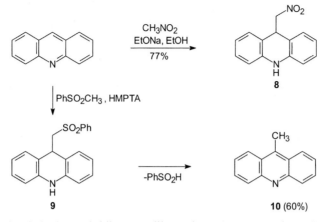

9-Haloacridines and 6-halophenanthridines readily undergo S$_N$Ar reactions (by analogy with the corresponding pyridine, quinoline and isoquinoline compounds, see e.g. p 338).

Phenanthridine with KCN/C$_6$H$_5$COCl forms REISSERT compounds at the C-6/N bond (see p 321/338) and is reduced to 5,6-dihydrophenanthridine by the action of Sn/HCl or by catalytic hydrogenation with RANEY-Ni. Reduction of acridine occurs in the pyridine ring (with Zn/HCl) yielding 9,10-dihydroacridine **11** and in the benzene rings (by catalytic hydrogenation with Pt in hydrochloric acid) providing the pyridine **12**:

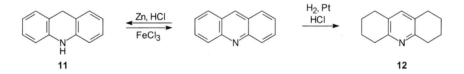

Acridine is oxidized by dichromate in acetic acid giving 9-acridone **13** and degraded by permanganate in an alkaline medium forming quinoline-2,3-dicarboxylic acid **14**:

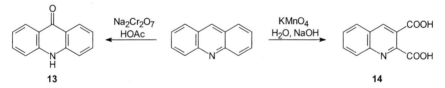

Phenanthridine is oxidized to 6-phenanthridone **15** and quinoline-3,4-dicarboxylic acid **16** by the action of ozone, followed by treatment with alkaline H$_2$O$_2$ solution:

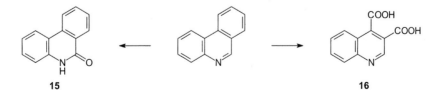

C The following methods are of preparative value for the synthesis of acridines and phenanthridines.

(1) Primary arylamines yield acridines **17** (*Ullmann synthesis*) by condensation with aldehydes in the presence of strong mineral acids (H_2SO_4, HCl) followed by dehydrogenation:

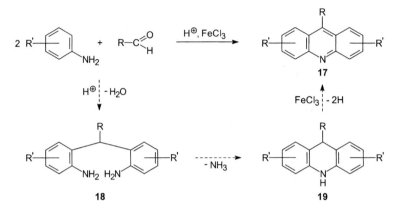

Bis(*o*-aminophenyl)methanes **18** are formed as intermediates. They cyclize by an unknown mechanism to 9,10-dihydroacridines **19**; final dehydrogenation, e.g. by $FeCl_3$, yields acridines **17**.

(2) Diphenylamine reacts with carboxylic acids in the presence of LEWIS acids (e.g. $AlCl_3$, $ZnCl_2$) forming 9-substituted acridines **20** (*Bernthsen synthesis*):

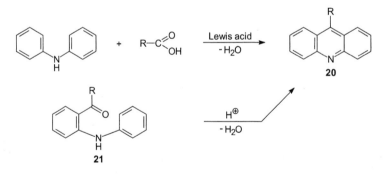

Alternatively, acridines **20** are obtained from *o*-arylaminophenyl ketones **21** by cyclodehydration with strong acids (H_2SO_4, polyphosphoric acid). Both cyclizations are intramolecular S_EAr reactions.

(3) (*o*-Arylamino)benzoic acids **22** cyclize to acridones **23** under the influence of strong acids, and to 9-chloroacridines **25** when treated with $POCl_3$:

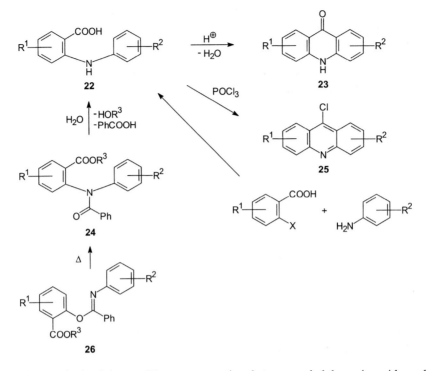

The acids **22** are obtained by an ULLMANN reaction between *o*-halobenzoic acids and primary arylamines in the presence of Cu powder in a basic medium. They are also prepared from *N*-aryl-*N*-benzoyl-*o*-aminobenzoates **24** which are accessible by a CHAPMAN rearrangement of imidates **26**.

(4) Phenanthridines **28** are synthesized by cyclodehydration of 2-(acylamino)biphenylenes **27** with POCl$_3$:

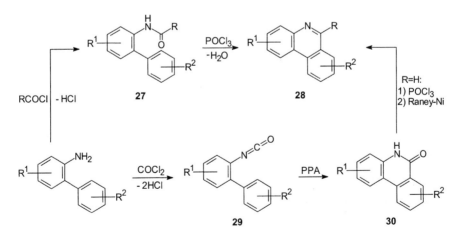

This reaction corresponds to the BISCHLER-NAPIERALSKI synthesis of isoquinoline (see p 343) and proceeds as an intramolecular S_EAr process by way of nitrilium ions as intermediates. By an anlogous reaction, biphenyl-2-isocyanates **29**, derived from 2-aminobiphenyl, on treatment with polyphosphoric acid, cyclize to yield phenanthridones **30**.

(5) On acid catalysis, 2-aminobiphenyl-2'-carboxylic acids **31** cyclize affording phenanthridones **32**, which are reduced by zinc or LiAlH$_4$ to phenanthridines **34**. The aminobiphenylcarboxylic acids **31** are accessible from the corresponding nitro compounds **33** by reduction.

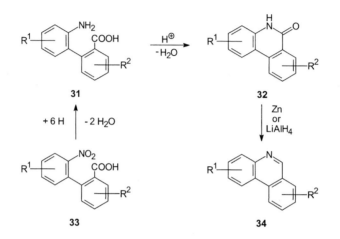

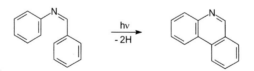

Acridine, mp 110°C on sublimation, bp 346°C, like phenanthridine, is isolated from the anthracene fraction of coal-tar. It forms colourless needles and shows an intensively blue fluorescence in solution. Acridine is a weak base (pK_a = 5.60).

Phenanthridine, mp 108°C, is a colourless, basic compound (pK_a = 4.52). It is prepared by cyclization of 2-(formylamino)biphenyl with POCl$_3$/SnCl$_4$ in nitrobenzene, or by photolysis of benzylidenaniline:

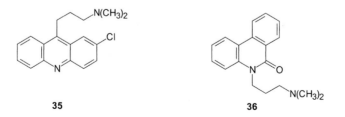

Some pharmaceuticals are derived from acridine and phenanthridine, for instance the antidepressants and tranquilizers clomacran **35** and fantridon **36**:

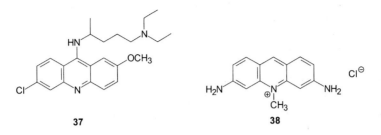

37   **38**

Atebrin **37** was used as an antimalarial during World War II, but is no longer important because of its side-effects. Acriflavine hydrochloride A **38** is an antiseptic.

Acridine dyes are amino derivatives of acridine. Acridine yellow G **40** is prepared by a similar method to pyronine G (see p 266) from 2,4-diaminotoluene via the leuco compound **39** by a BERNTHSEN synthesis.

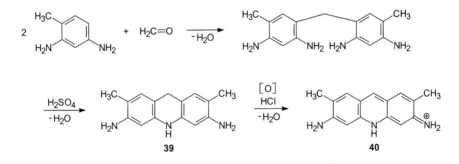

39   **40**

The bisacridinium salt **41** (lucigenin), on oxidation with an alkaline H_2O_2 solution, displays an intensely green chemiluminescence. The light emission is due to the excited state of *N*-methylacridone **43**, formed by an electrocyclic ring opening of the initially formed dioxetane **42** (see p 47):

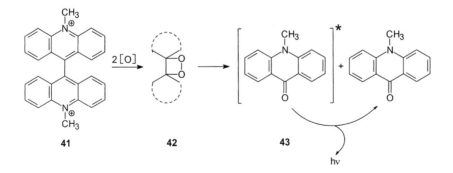

41   **42**   **43**

6.20 Piperidine

A Structural examination of some of its crystalline derivatives shows that the piperidine molecule exists in a chair form with torsion angles of 53-56°. Therefore, it is slightly more folded than cyclohexane. The N–H bond in piperidine prefers the equatorial position. The enthalpy difference compared with the axial N–H bond form is ≈ 2 kJ mol⁻¹; the enthalpy of activation for ring inversion (equatorial/axial) is ≈ 25 kJ mol⁻¹.

In 1-methylpiperidine, the chair conformation with an equatorial N-CH$_3$ group is more stable by 11.3 kJ mol⁻¹ than with an axial group.

Piperidine has the following ¹H and ¹³C-NMR spectroscopic data:

¹H-NMR (CDCl₃)
δ (ppm)

¹³C-NMR (CDCl₃)
δ (ppm)

H-2/H-6: 2.77 H-3/H-4/H-5: 1.52   C-2/C-6: 47.5   C-3/C-5: 27.2 C-4: 25.5

C The synthesis of piperidines [120] can be carried out by

(1) cyclization of 1-amino-5-haloalkanes:

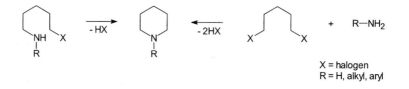

X = halogen
R = H, alkyl, aryl

This involves an intramolecular S$_N$ reaction. Thus, piperidines are also formed by cyclizing 1,5-dihalo-alkanes with primary amines.

(2) reductive cyclization of pentanediamides or -dinitriles with H$_2$ over Cu chromite:

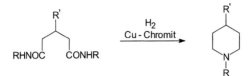

(3) catalytic hydrogenation of pyridine derivatives:

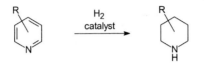

4-Piperidones are important piperidine derivatives. Other methods are available for their synthesis, e.g. the DIECKMANN cyclization of the diesters **1**, the THORPE-ZIEGLER cyclization of the dinitriles **2** and cyclization of dialkyl ketones or acetone dicarboxylic esters with aldehydes and primary amines in a MANNICH reaction:

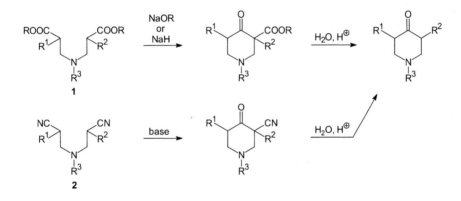

|D| **Piperidine** is a colourless liquid, miscible with water, and of an unpleasant smell with a bp of 106°C. It has properties typical of a secondary amine and is a stronger base (pK_a = 11.2) than pyridine (pK_a = 5.2). It is produced commerically by catalytic hydrogenation of pyridine.

Chemical degradation with fission of the piperidine ring was once used for establishing the constitution of piperidine derivatives (see below). Three methods were applied.

(1) Exhaustive methylation at the piperidine nitrogen followed by a HOFMANN degradation of the resulting quaternary ammonium hydroxide:

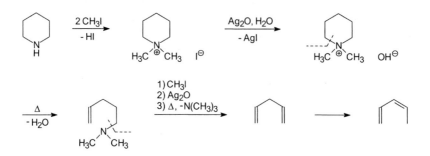

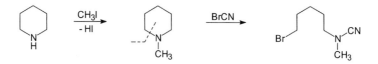

If this reaction sequence is carried out on piperidine, then penta-1,4-diene is formed, which rearranges into penta-1,3-diene under the reaction conditions.

(2)   Fission of 1-methylpiperidine with cyanogen bromide (V. BRAUN 1900):

(3)   Fission of 1-benzoylpiperidine with phosphorus tribromide (V. BRAUN 1910):

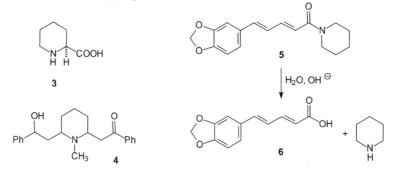

Piperidine alkaloids occur in only a few species of higher plants. However, (S)-pipecolic acid **3** [121] is widely distributed in higher plants, microorganisms and animals. Lobeline **4** is the main constituent of lobelia alkaloids (*Lobelia inflata, Campanulaceae*). It is a respiratory stimulant in mammals. Piperine **5** is the active ingredient in black pepper (*Piper nigrum*). On hydrolysis it furnishes piperic acid **6** and piperidine (hence its name) [122]:

Other piperidine alkaloids are isopelletierine **7** (occurring in the pomegranate tree, *Punica granatum*), coniine **8** (the toxic constituent of hemlock, *Conium maculatum, Apiaceae*), arecoline **9** (found in the betel nut, the seed of the palm tree, *Areca catechu*) and anabasine (see p 305). Piperidine alkaloids occurring in animals are pumiliotoxin *B* **10** and histrionicotoxin **11**. The toxin **10** influences the cationic permeability of cell membranes and facilitates Ca uptake. Toxins of type **11** are interesting because of their interaction with the acetylcholine receptor. A number of sterol alkaloids occurring in solanaceae contain piperidine rings as constituents of a tetrahydrofuran spirostructure (spirosolane type **12**) or of a perhydroindolizine structure (solanidane type **13**). Nojirimycin **14** is an aminosugar, in which nitrogen occupies the position of the pyranose oxygen. Both **14** and its reduction product deoxynojirimycin are potent β-glucosidase inhibitors.

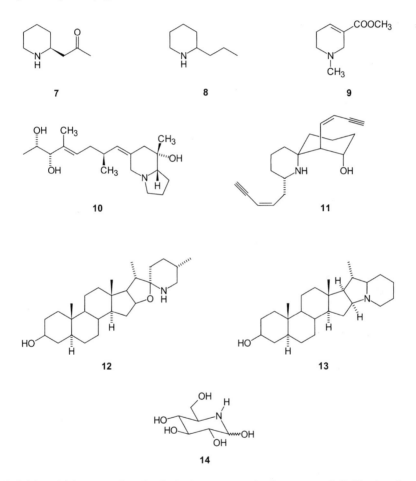

Tropane alkaloids, which occur abundantly in nature, contain the saturated C_5/C_6 ring framework of tropane (*N*-methyl-8-azabicyclo[3.2.1]octane) **15**. Atropine and hyoscyamine (ester of tropine **16** with racemic tropic acid and *L*-tropic acid **17**) are important examples. They occur in deadly nightshade (belladonna), henbane and thornapple. They are parasympatholytic drugs (anticholinergics), blocking the action of acetylcholine at postganglionic cholinergic nerve endings. Cocaine **18**, the benzoate of ecgonine methyl ester **19** (configuration 2*R*,3*S*), is the main alkaloid constituent of the South American coca plant. It is a local anaesthetic and narcotic and inhibits the uptake of noradrenaline in membranes.

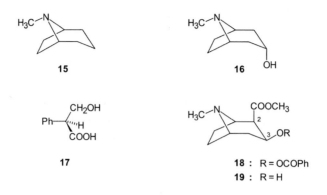

15

16

17

18 : R = OCOPh
19 : R = H

The piperidine ring is probably the most common heterocycle occurring in pharmaceuticals. Piperidine is often used as a secondary amine in the synthesis of drugs. The local anaesthetic bupivacaine **20** (used as the racemate or the (*S*)-enantiomer levobupivacaine) is a pipecolinic acid derivative. The antihistamine bamipine **21** and the analgesic fentanyl **22** are derived from 4-aminopiperidines. The antidiarrhoeal diphenoxylate **23** and the antiparkinsonian agent budipine **24** are 4-arylpiperidine derivatives.

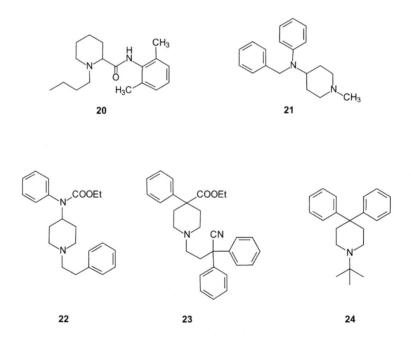

20

21

22

23

24

6.21 Phosphabenzene

Formal replacement of a CH group by trivalent phosphorus in benzene results in the formation of λ^3-phosphinine **1**, also known as phosphabenzene. Replacement by pentavalent phosphorus yields λ^5-phosphinine **2**. Saturated six-membered rings **3** with trivalent phosphorus are known as phosphinanes.

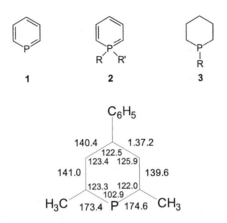

Fig. 6.15 Bond parameters of the phosphabenzene system in 2,6-dimethyl-4-phenyl-λ^3-phosphinine (bond lengths in pm, bond angles in degrees)

A Phosphabenzene has the geometry of an expanded planar hexagon, as shown by the X-ray structure of the 2,6-dimethyl-4-phenyl derivative (see Fig. 6.15). By its NMR data, phosphabenzene satisfies the criteria of a delocalized 6π-heteroarene with a diamagnetic ring current. However, the electron distribution between β- and γ-positions is the reverse of that of pyridine (see p 270):

^{1}H-NMR (CDCl$_3$)
δ (ppm)

^{13}C-NMR (CDCl$_3$)
δ (ppm)

H-2/H-6: 8.61   H-4: 7.38
H-3/H-5: 2.72

C-2/C-6: 151.1   C-4: 128.8
C-3/C-5: 133.6

A comparison of the electronic spectra of their 2,4,6-tri-*tert*-butyl derivatives shows further differences between phosphabenzene and pyridine. The longest wave UV absorption of both systems is almost identical (262 nm), but phosphabenzene can be ionized (8.0 and 8.6 eV) more easily than pyridine (8.6 and 9.3 eV). Supported by MO calculations, it was concluded that the HOMO in phosphabenzene is a π-orbital, whereas that in pyridine is the n-orbital of nitrogen [123]. Phosphabenzene can be regarded as a π-donor system and pyridine as a (π-deficient) σ-donor system.

B The reactions of the phosphabenzene system [124] confirm these conclusions. Phosphabenzenes have low basicity towards 'hard' acids. They are not protonated by CF_3CO_2H nor alkylated by trialkyloxonium salts. However, 'soft' acids attack at phosphorus. For instance, 2,4,6-triphenyl-phosphabenzene forms compounds **4** with the hexacarbonyl derivatives of Cr, W and Mo in which the phosphorus coordinates to the metal, possibly with metal–P back-donation. The complexes **4** rearrange photochemically or thermally affording the 6π-heteroarene complexes **5**. Although 2,4,6-triphenyl-pyridine is protonated on nitrogen, it undergoes complex formation with chromium hexacarbonyl exclusively on the phenyl moieties yielding the η^6-arene complexes **6** [125].

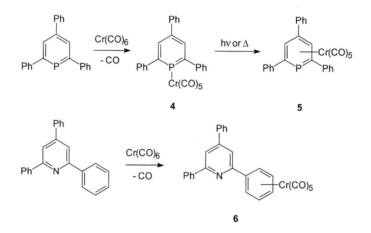

Almost all other reactions of phosphabenzene occur with involvement of phosphorus. For instance, hexafluorobut-2-yne undergoes a [4+2] cycloaddition furnishing the 1-phosphabicyclo[2.2.2]octa-2,5,7-triene **7**:

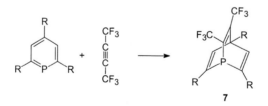

Nucleophilic and radical reaction partners transform λ^3-phosphinines into λ^5-phosphinines. For instance, with alcohols or amines in the presence of Hg(II) acetate, 1,1-dialkoxy- or 1,1,-diamino-λ^5-phosphinines **8** are formed:

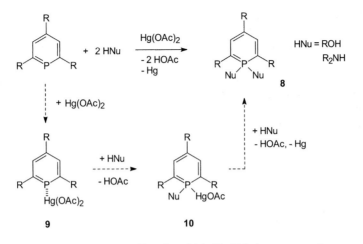

The adduct **9** is postulated as an intermediate in which Hg(OAc)$_2$ acts as a LEWIS acid towards the heterocycle; **9** reacts in two steps with the nucleophile, namely with reduction of Hg(II) to Hg(0) and conversion of **10** into the product **8**.

Organolithium compounds add to phosphabenzenes producing deep-blue phosphabenzene anions **11** which undergo alkylation on P with alkyl halides yielding λ^5-phosphinines **12**:

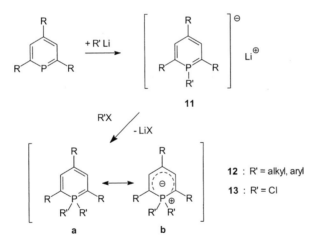

Photochlorination of phosphabenzenes brings about a radical halogen addition to phosphorus forming 1,1-dichloro-λ^5-phosphinines **13** which undergo further substitution.

λ^5-Phosphinines react as cyclic phosphonium ylides, e.g **12a** ↔ **12b**. In contrast to 4-methylpyridine or 4-methylpyridinium salts, the system **14** does not undergo deprotonation. However, it suffers hydride abstraction forming discrete carbenium–phosphonium salts **15**. Subsequent reaction with nucleophiles affect the side-chain:

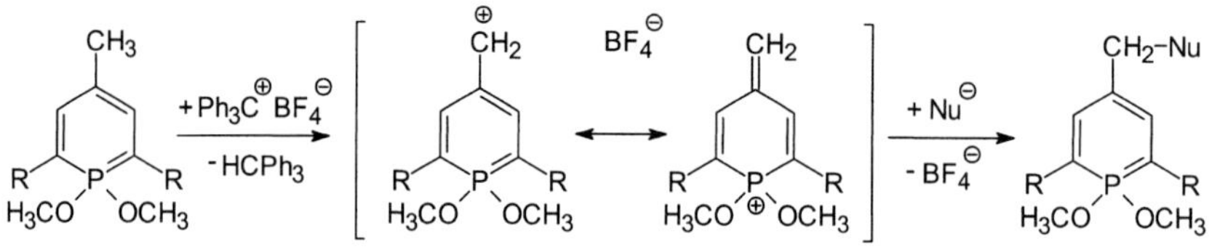

14 **15**

C There are two convenient methods for the synthesis of phosphabenzenes.

(1) 2,4,6-Trisubstituted pyrylium salts yield 2,4,6-trisubstituted phosphabenzenes **16** (MÄRKL 1966) when treated with phosphanes such as tris(hydroxymethyl)phosphane, tris(trimethylsilyl)phosphane or phosphonium iodide (according to the reaction principle on p 225):

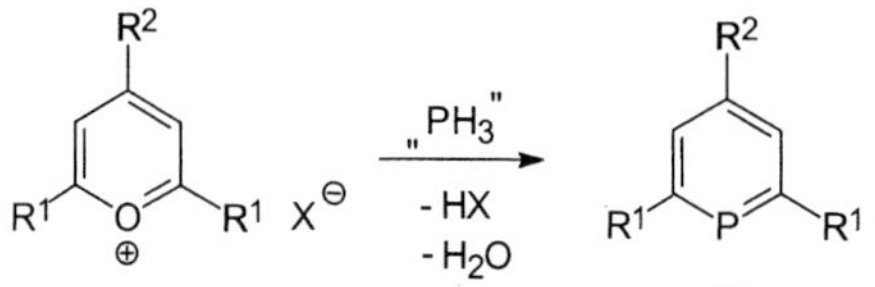

16

(2) The synthesis of the parent compound is not possible by method 1. It is achieved by reaction of penta-1,4-diyne **17** with dibutyltin hydride forming stannacyclohexa-1,4-diene **18** by a cyclization/ addition sequence. Tin–phosphorus exchange by phosphorus tribromide converts **18** into dihydro-λ^3-phosphinine **19** which, on dehydration with DBU (ASHE 1971), yields the phosphabenzene:

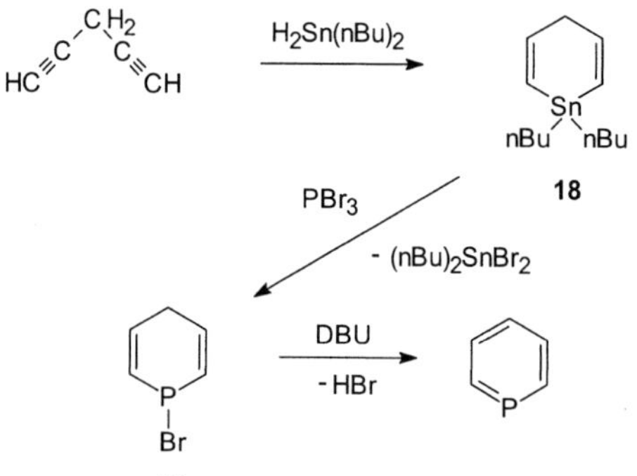

19

The corresponding six-membered heterocycles with As, Sb and Bi (arsabenzene, stibabenzene and bismabenzene) have been prepared by this method.

4-Substituted phosphabenzenes **23** are made from 4-methoxy-1,1-dibutylstannacyclohexa-1,4-dienes **20** by an extension of this method: fission of the C–Sn bond by butyllithium gives 1,5-dilithiopenta-1,4-dienes **21** which, on heterocyclization with butoxydichlorophosphane and $LiAlH_4$ reduction of the resulting dihydrophosphinines **22**, yields **23**:

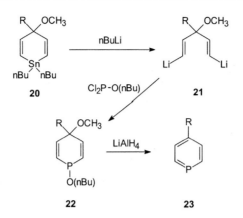

20   **21**   **22**   **23**

λ^3-Phosphinines with a fused benzene ring, such as phosphanaphthalene **24**, phosphaanthracene **25** and phosphaphenanthrene **26**, have also been prepared.

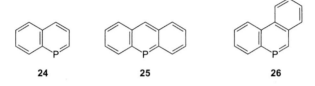

24   **25**   **26**

6.22 1,4-Dioxin, 1,4-Dithiin, 1,4-Oxathiin

The parent compounds 1,4-dioxin **1** and 1,4-dithiin **2** are known; 1,4-oxathiin **3** is not known.

1 : X = Y = O
2 : X = Y = S
3 : X = O, Y = S

A   X-Ray analysis (e.g. **2**: bond lengths C=C 129 pm, C–S 178 pm; bond angles C–S–C 100.2°, S–C–C 122.6° and 124.4°) shows that the systems **1** and **2** have the planar geometry of a slightly distorted hexagon. The NMR data (e.g. ^{1}H-NMR, **1**: δ = 5.50, **2**: δ = 6.13) prove **1** and **2** to be a cyclic vinyl ether and a vinyl thioether, respectively, with no ring current effects.

X-Ray crystal analyses of the dibenzo derivatives dibenzo[1,4]dioxin **4**, thianthrene **5** and phenoxathiin **6**, show that **4** is planar but that **5** and **6** have a folded structure.

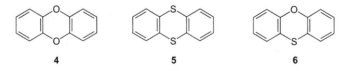

4   **5**   **6**

B Among the reactions of 1,4-dioxins and 1,4-dithiins, oxidation of their tetraphenyl and dibenzo derivatives is important. This reaction leads to the formation of radical cations **7** and dications **8** by one-electron transfers which are polarographically defined:

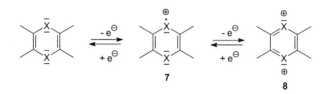

Tetraphenyl-1,4-dioxin, with $SbCl_5$, forms a blue-violet radical cation. On further oxidation, a green dication is obtained. However, tetraphenyl-1,4-dithiin with $SbCl_5$ yields a violet dication as a hexachloroantimonate. The dications **8** can be regarded as HÜCKEL aromatic 6π-systems, in contrast to the antiaromatic parent compounds **1 - 3**.

C The syntheses of the tetraaryl and dibenzo systems are of interest.

(1) Tetraaryl-1,4-dioxins **9** are formed by cyclocondensation of α-hydroxy ketones with HCl in CH_3OH followed by reductive elimination of the intermediate acetals **10** and **11** with Zn/acetic anhydride:

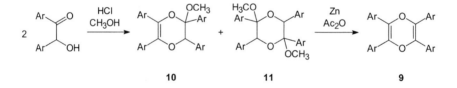

Tetraaryl-1,4-dithiins **13** are obtained by photolysis of 4,5-diaryl-1,2,3-thiadiazoles. This involves N_2 elimination and ring contraction to afford thiirenes **12** (see p 196), which dimerize forming the dithiin system.

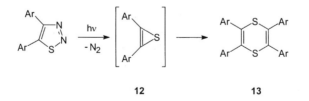

(2) Dibenzo-1,4-dioxins are prepared from *o*-halophenols on treatment with Cu powder in the presence of K_2CO_3 or from 2-(2-hydroxyphenoxy)diazonium salts [126]:

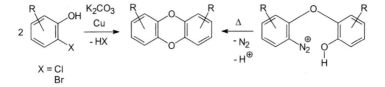

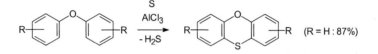

Phenoxathiins are obtained from the interaction of diphenyl ether with sulfur in the presence of $AlCl_3$:

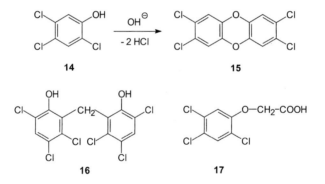

D Polychlorinated dibenzo-1,4-dioxins are extremely toxic compounds [127] which are difficult to degrade biologically. This especially applies to 2,3,7,8-tetrachlorodibenzo[1,4]dioxin **15** (Seveso dioxin, TCDD) which is teratogenic. TCDD (LD_{50} = 45 µg kg^{-1} in rats) is formed as a byproduct in the commercial synthesis of 2,4,5-trichlorophenol **14**. It is an intermediate in the production of the germicide hexachlorophene **16** and the herbicide (2,4,5-trichlorophenoxy)acetic acid **17**:

6.23 1,4-Dioxane

A According to electron diffraction studies, 1,4-dioxane exists in a chair conformation with a dihedral angle of 57.9°, i.e. with more folding than cyclohexane. The bond distances are C–C 152 pm and C–O 142 pm; the angles are C–C–C 105° and C–O–C 112°.

The enthalpy of activation for the ring inversion of 1,4-dioxane is 40.6 kJ mol^{-1}. The H-atoms resonate at $\delta = 3.70$ in the ^{1}H-NMR spectrum (CDCl$_3$) and the C-atoms at $\delta = 67.8$ in the ^{13}C-NMR spectrum (CDCl$_3$). The difference in the chemical shift between the equatorial and axial hydrogens is 1.74 Hz.

Halogen atoms in 1,4-dioxane prefer to occupy an axial position, favoured by an anomeric effect. For this reason, *trans*-2,3-dichloro-1,4-dioxane prefers the diaxial geometry **1**:

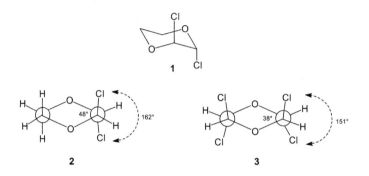

Moreover, the dihedral angle between the two C–halogen bonds becomes smaller than 180° and the dioxane ring is flattened (**2**) because of repulsive synclinal interaction between the halogen and oxygen atoms. This effect is more pronounced with *trans,cis,trans*-tetrachloro-1,4-dioxane (**3**) (dihedral angle 151°; cf. 162° in **2**).

B 1,4-Dioxanes behave like cyclic ethers. For instance, 1,4-dioxane forms peroxides and undergoes chlorination yielding the 2,5-dichloro and 2,3-dichloro compounds. The action of HBr causes fission of the ether furnishing bis(2-bromoethyl)ether **6**. Ring-opening also occurs by the action of acid chlorides in the presence of AlCl$_3$ forming 1-acyloxy-2-chloroethane **7**; acetic anhydride in the presence of FeCl$_3$ affords ethylene glycol diacetate **8** and bis(2-acetoxyethyl)ether **9**.

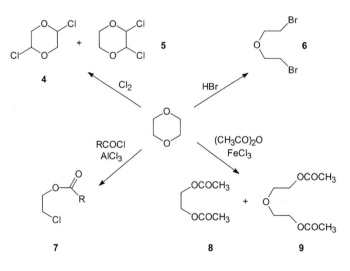

C The syntheses of 1,4-dioxanes are carried out by the classical methods used for dialkyl ethers as exemplified for the parent system:
- by acid-catalysed cyclodehydration of 3-oxapentane-1,5-diol **10**
- by the action of bases on 5-halo-3-oxapentan-1-ols **11** (i.e. by an S_N reaction of a WILLIAMSON synthesis)
- by the reaction of 3-oxa-1,5-dihalopentane **12** with alkali hydroxides
- by acid-catalysed cyclocondensation of ethylene glycol with oxirane:

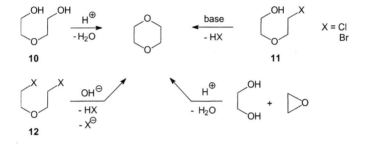

D **1,4-Dioxane**, mp -12°C, bp 101°C, is a colourless liquid of pleasant odour, hygroscopic, and miscible with water and most organic solvents. It is widely used as a solvent. It is prepared by a cyclizing dehydration of ethylene glycol or diglycol, and by dimerization of oxirane, under acid catalysis in each case:

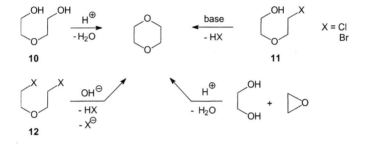

E As a donor of an n-electron pair, dioxane forms stable adducts with LEWIS acids and other acceptor molecules. These adducts are important as reagents in organic synthesis. For instance, the dioxane–SO$_3$ adduct **13** is used for the sulfonation of alcohols and alkenes, the dioxane–BH$_3$ adduct **14** is a hydroboration agent, and the dioxane–Br$_2$ adduct is employed for controlled bromination e.g. of furan (see p 55).

13 : L = SO$_3$ (L = Lewis acid)

14 : L = BH$_3$

15 : L = Br$_2$

6.24 Oxazines

A-C Oxazines are derived from 2*H*- and 4*H*-pyrans by replacing either a CH$_2$ unit by NH or a CH unit by N. Hence eight oxazines **1 - 8** and nine benzoxazines exist. The corresponding sulfur compounds are known as thiazines.

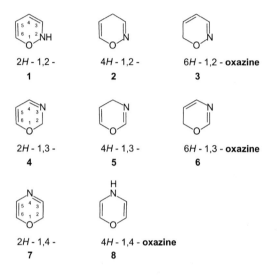

2*H* - 1,2 -
1

4*H* - 1,2 -
2

6*H* - 1,2 - **oxazine**
3

2*H* - 1,3 -
4

4*H* - 1,3 -
5

6*H* - 1,3 - **oxazine**
6

2*H* - 1,4 -
7

4*H* - 1,4 - **oxazine**
8

The following compounds are of general importance: 5,6-dihydro-4*H*-1,3-oxazines **9**, some systems derived from 4*H*- and 6*H*-oxazine **5** and **6**, respectively, and from 4*H*-3,1-benzoxazine **10**, dibenzo[1,4]oxazine **11** (phenoxazine) and the corresponding sulfur compound **12** (phenothiazine).

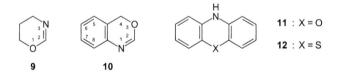

11 : X = O

12 : X = S

9

10

5,6-Dihydro-4*H*-1,3-oxazines are obtained by cyclization of *N*-acyl-(3-hydroxyxpropyl)amines with acid, or of *N*-acyl-(3-bromopropyl)amines with base:

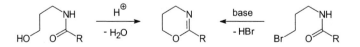

The most important synthesis is the acid-catalysed cyclocondensation of alkyl cyanides with propanediols, e.g. **13**:

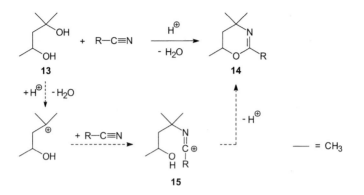

This process is similar to the RITTER reaction for the preparation of amines from nitriles and olefins. The first step is the regioselective formation of a carbenium ion from the diol **13** followed by addition to the nitrile. The resulting nitrilium ion **15**, following attack on the remaining OH group, cyclizes and gives dihydro-1,3-oxazine **14**.

As imido esters, dihydro-1,3-oxazines of type **14** display reactivity similar to 2-oxazolines (see p 135). The α-C–H bonds of the 2-alkyl group are CH-acidic. They are metalated by reaction with butyllithium and then undergo C–C forming reactions with electrophiles such as haloalkanes, oxiranes or carbonyl compounds. The dihydro-1,3-oxazines **16**, which are obtained from **15** by α-metalation and alkylation, undergo reduction with $NaBH_4$ forming **17**. Acid hydrolysis cleaves the cyclic aminal function of **17** yielding aldehydes **18**:

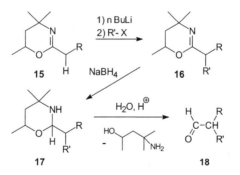

Therefore, the incorporation of a nitrile into a dihydro-1,3-oxazine leads to its α-functionalization and conversion into an aldehyde ($RCH_2C{\equiv}N \rightarrow RR'CHCH{=}O$) [128].

1,3-Oxazinones are accessible in several structural variants. For instance, 2-substituted 1,3-oxazin-4-ones **19** result from the cycloaddition of ketenes to isocyanates. A second molecule of ketene acylates the OH function of the product **19** yielding the corresponding *O*-acetyl derivative :

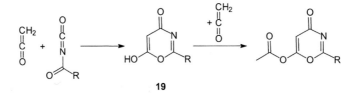

19

The 1,3-benzoxazin-4-ones **20** related to **19** are prepared by cyclodehydration of *O*-acylsalicylamides with acidic reagents:

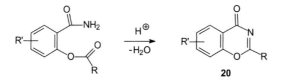

20

1,3-Oxazin-6-ones **23** are prepared by thermal cyclization of 3-(acylamino)acrylic esters **21** or by the interaction of diarylcyclopropenones **24** with nitrile oxides:

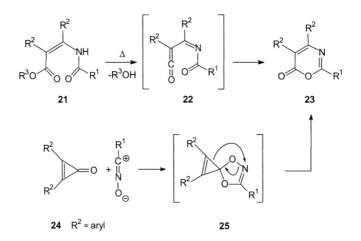

21   **22**   **23**

24 R^2 = aryl   **25**

The reaction probably involves loss of alcohol from **21** followed by electrocyclization of the acylimino ketene intermediate **22** furnishing **23**. From **24**, the heterocycle **23** is produced by a 1,3-dipolar cycloaddition of the nitrile oxide to the carbonyl group via **25**.

3,1-Benzoxazin-4-ones 26 are formed by cyclodehydration of *N*-acylanthranilic acids with acetic anhydride, POCl$_3$ or SOCl$_2$ [128a] :

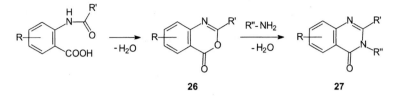

26   27

Benzoxazinones **26** react with primary amines by exchanging the ring oxygen for nitrogen forming quinazolinones **27**.

3,1-Benzoxazin-2,4-diones 28 (isatoic anhydrides) are prepared by cyclocondensation of anthranilic acid with phosgene or by a BAEYER-VILLIGER oxidation of indole-2,3-diones (isatins) with peroxy acetic acid:

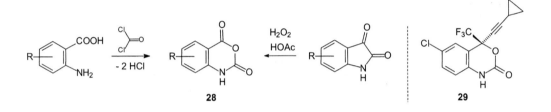

28   29

The parent system (**28**, R = H) serves as a building block, e.g. for polyheterocyclic systems (see p 474). It is commercially produced by a modified HOFMANN degradation of phthalimide.

Efavirenz **29** is an anti-HIV drug of the 3,1-benzoxazin-2-one type acting as a non-nucleocidic reverse transcriptase inhibitor (cf. p 416).

1,3-Oxazinium salts 30 are the 3-aza analogues of pyrylium salts. They are obtained by cyclocondensation of β-chlorovinyl ketones with nitriles or by cyclocondensation of alkynes with N-acylmethanimidoyl chlorides catalysed by $SnCl_4$:

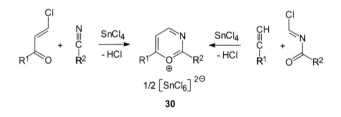

30

The reactivity of 1,3-oxazinium salts is comparable to that of pyrylium salts. Nucleophiles are added at C-6 and, after ring fission and recyclization, incorporated into the ring as shown in the scheme on p 225. Thus 1,3-thiazinium salts **31** are obtained with H_2S and subsequent treatment with $HClO_4$:

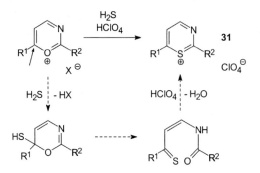

31

Phenoxazine 11 is obtained by thermolysis of *o*-aminophenol in an acid medium, such as a mixture of *o*-aminophenol and its hydrochloride. **Phenothiazine 12** is prepared by heating diphenylamine with sulfur in high boiling solvents (*Bernthsen synthesis*):

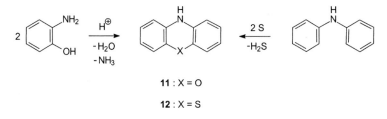

11 : X = O

12 : X = S

Phenoxazines and phenothiazines **32** and **33** are also made by reductive cyclization of *o*-nitrodiphenyl ethers or *o*-nitrodiphenyl sulfides, respectively, with triethyl phosphite.

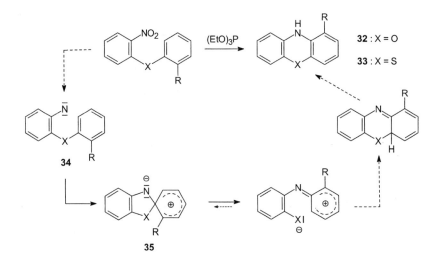

32 : X = O

33 : X = S

The nitro group is deoxygenated to a nitrene **34** which forms a spirobetaine intermediate **35** by an ipso attack on the other benzene ring. The conversion of **35** into the products **32** or **33** occurs by expansion of the spiro five-membered ring into a six-membered ring. The reaction can be regarded as analogous to a SMILES rearrangement.

D The phenoxazine moiety occurs in pigments of fungi, lichen and butterflies, and also in the ommochromes which are responsible for the colour of the pigments in some insects. Xanthommatin **36**, which contains a quinonoid phenoxazone chromophore isolated from the secretion of the fox-moth, serves as an example:

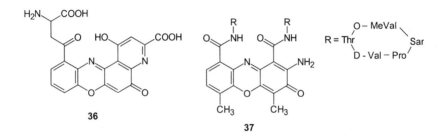

The orange-red actinomycins **37** (WAKSMAN 1940) found in various types of streptomyces are 2-aminophenoxazone-1,9-dicarboxylic acids which are linked as amides to cyclic pentapeptide units. They are able to intercalate into DNA and are employed as cytostatic agents in tumour therapy.

The phenothiazine structure is found in a number of pharmaceuticals used as antihistamines, antipsychotics, sedatives and antiemetics. Examples are the neuroleptic chlorpromazin **38** and the sedative promethazine **39**:

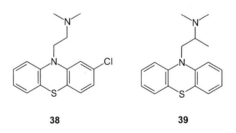

Phenoxazine and phenothiazine dyes are basic dyes which possess a delocalized quinone iminium chromophore analogous to phenoxazone. They are derived from the classical dyes Meldola Blue **40** and Lauth's Violet **41**:

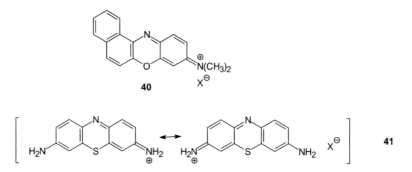

40

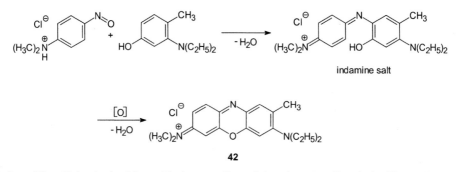

41

Capri Blue 42 is prepared by condensation of *N,N*-dimethyl-4-nitrosoanilinium chloride with 3-diethylamino-4-methylphenol. The intermediate is the salt of a substituted phenylquinone diimine (indamine), which is oxidized by excess of the nitroso compound:

indamine salt

42

Methylene Blue 43 is obtained by oxidative coupling of 4-amino-*N,N*-dimethylaniline and *N,N*-dimethylaniline with $Na_2Cr_2O_7$ in the presence of $Na_2S_2O_3$. Again, the intermediate is an indamine salt which undergoes cyclization giving the salt of 3,7-bis(dimethylamino)phenothiazine (leuco form of Methylene Blue). Its dehydration produces the dye (CARO 1876, BERNTHSEN 1885):

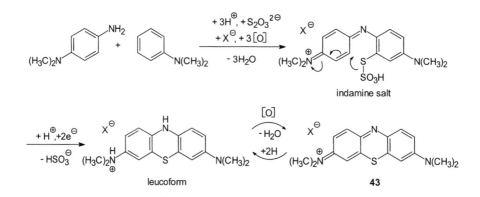

indamine salt

leucoform

43

Reducing agents transform Methylene Blue into the yellow leuco form which is reconverted into the dye by aerial oxygen. For this reason it is used as a redox indicator.

6.25 Morpholine

A Morpholine (tetrahydro-1,4-oxazine) adopts a chair conformation. The enthalpy difference between the equatorial and axial position of the NH group (2.63 kJ mol^{-1}) is similar to that for piperidine (see p 360) and favours the equatorial conformation.

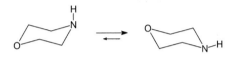

The barrier to ring inversion of 4-methylmorpholine (48.1 kJ mol^{-1}) is comparable to the value (49.8 kJ mol^{-1}) found for 1-methylpiperidine.

The morpholine protons resonate at δ = 3.65 (H-2) and 2.80 (H-3) in the ^{1}H-NMR spectrum, and the C-atoms at δ = 68.1 (C-2) and 46.7 (C-3) (CDCl$_3$) in the ^{13}C-NMR spectrum.

C Morpholine and *N*-substituted morpholines **1** are obtained by cyclization of bis(2-aminoethyl) ethers or by cyclocondensation of bis(2-chloroethyl) ethers with ammonia or primary amines:

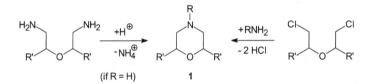

2,3-Disubstituted morpholines **2** are prepared from *N*-benzyl protected 2-aminoethanols and oxirane in the presence of 70% H$_2$SO$_4$ followed by catalytic debenzylation:

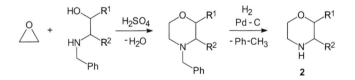

2-Oxomorpholine derivatives **3** are made by cyclocondensation of α-aminocarboxylic esters with oxirane or from *N*-substituted 2-aminoethanols with α-bromocarboxylic esters:

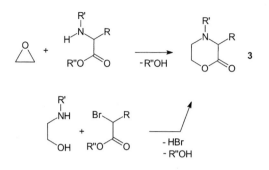

Morpholine, bp 128°C, is a colourless, hygroscopic liquid, miscible with water, and as a result of the inductive influence of the oxygen atom, it possesses lower basicity (pK_a = 8.4) than piperidine (11.2) and piperazine (9.8). Morpholine is synthesized either by acid-catalysed cyclodehydration of bis(2-hydroxyethyl)amine or from bis(2-chloroethyl) ether and NH_3. Morpholine, like piperidine, is used as a basic condensing agent and as a solvent. In industry, it is employed as an additive to the feed-water inhibiting corrosion in steam boilers.

Like piperidine, morpholine is used in pharmaceuticals as a secondary amine component. One of the drugs derived from morpholine as a heterocycle is viloxazine **5**, which can be obtained from **4** by reaction with ethanolamine hydrogensulfate. Structurally related to **5** is reboxetine **6** (in the form of the (*S,S*)-(+)-stereoisomer). Both **5** and **6** are effective as antidepressants.

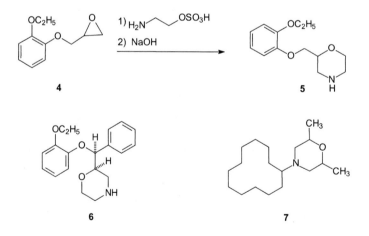

Several *N*-alkyl-2,6-dimethylmorpholines, e.g. dodemorph **7** (acetate or benzoate), are used as fungicides and bactericides.

6.26 1,3-Dioxane

| **A** |
A 1,3-Dioxanes are cyclic acetals or ketals. Their five-membered ring homologues are the 1,3-dioxolanes (see p 118).

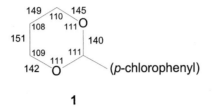

1

Fig. 6.16   Bond parameters of the 1,3-dioxane system in 2-(*p*-chlorophenyl)-1,3-dioxane
             (bond lengths in pm, bond angles in degrees)

According to X-ray studies, 1,3-dioxanes adopt a chair conformation, especially in 2-substituted derivatives, as for instance 2-(*p*-chlorophenyl)-1,3-dioxane **1** (see Fig. 6.16). The grouping O/C-2/O, which has a dihedral angle of 60-63°, is puckered more strongly than the alicyclic part C-4/C-5/C-6, which has a dihedral angle of 53-55°. With 149-151 pm, the C–C bonds are significantly shorter in 1,3-dioxanes than in cyclohexane (153.3 pm). Hence the dioxane ring is more compact than the cyclohexane ring. The energy of activation for ring inversion was calculated to be 41 kJ mol^{-1} and is almost the same as the value for 1,4-dioxane (see p 371).

1,3-Dioxane shows the following ^{1}H and ^{13}C-NMR spectroscopic data:

^{1}H-NMR (CDCl$_3$) ^{13}C-NMR (CDCl$_3$)
δ (ppm)   δ (ppm)

H-2: 4.70, H-4/H-6: 3.80, H-5: 1.68   C-2: 95.4, C-4/C-6: 67.6, C-5: 26.8

Alkyl and aryl groups in the 2-position of the 1,3-dioxane ring prefer the equatorial position. The axial position is more destabilized (ΔG ≈ 13-17 kJ mol^{-1}) than in the alicyclic part of the molecule (ΔG ≈ 3-5 kJ mol^{-1}). This is due to stronger folding of the O–C–O apex of the 1,3-dioxane ring and the ensuing greater diaxial 1,3-interaction (**2**). The axial position is favoured by electronegative substituents at C-2 by ca. 2 kJ mol^{-1} due to an anomeric effect (see p 244), e.g. with OCH$_3$. Electron-attracting groups in the 5-position, such as F (not, however, Cl and Br), SOR, SO$_2$R, sulfonium and ammonium, prefer the axial arrangement. This effect may be due to electrostatic interaction with the oxygen atoms (**3**).

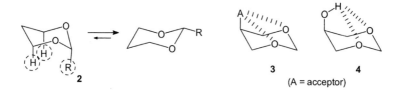

3   4

(A = acceptor)

5-Hydroxy groups also occupy an axial position in 1,3-dioxane because they form hydrogen bonds to the oxygen atoms of the ring (**4**).

B Aqueous acids hydrolyse 1,3-dioxanes as cyclic acetals or ketals forming 1,3-diols and carbonyl derivatives:

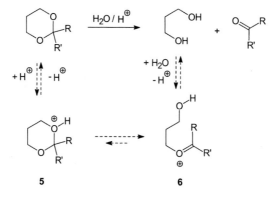

5   6

An oxonium ion is formed (**5 → 6**) in the rate-determining step after a reversible *O*-protonation and O/C-2 fission. Its formation is also responsible for the epimerization of 5-substituted 1,3-dioxanes, and is brought about by the action of nonaqueous protic or LEWIS acids. It is important for the conformational analysis of these systems (by equilibration of the conformers **7** and **8** which occupy an equatorial and axial position, respectively).

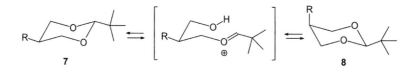

7   8

1,3-Dioxanes are liable to undergo thermal ring fission. For instance, at 350°C 1,3-dioxanes **10** react with fission of the O-3/C-4/6 bond and formation of the ester **9**. Over SiO_2 or pumice, fissions occur mainly at the O/C-2 bond producing β-alkoxyaldehydes **11**:

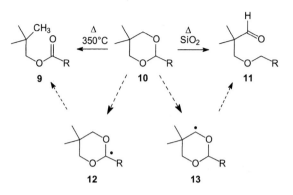

It is assumed that first H-abstraction from C-2 and C-4 or C-6 takes place, producing the oxygen-stabilized radicals **12** and **13**. O–C Fission and H-transfer yield the products **9** and **11**.

C | The acid-catalysed cyclocondensation of aldehydes or ketones with 1,3-diols is the most important synthesis of 1,3-dioxanes. The use of *p*-toluenesulfonic acid yields the best results.

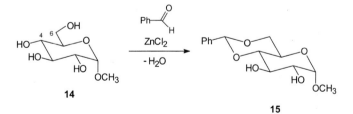

This reaction, like all acetalizations, is reversible. It is the basis for the wide application of dioxanes as protective groups for carbonyl compounds, as well as for systems with a 1,3-diol group, for instance the conversion of hexoses into so-called benzylidene derivatives [129] by the action of benzaldehyde in the presence of LEWIS acids. The 6- and 4-OH groups are blocked simultaneously, forming a 1,3-dioxane ring, e.g. the α-methylglycoside **14** is converted into the dioxane derivative **15**:

The formation of 1,3-dioxanes often competes with that of 1,3-dioxolanes in their formation from open-chain hydroxy compounds (see p 118), as for instance in the reaction of glycerol with formaldehyde:

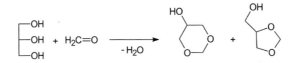

1,3-Dioxanes can also be produced by the acid-catalysed condensation of alkenes with aldehydes, preferably with formaldehydes (*Prins reaction*):

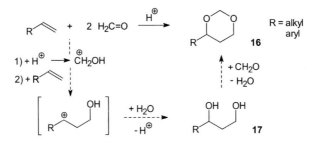

Electrophilic addition of the protonated aldehyde to the alkene yields a carbenium ion which reacts with water yielding a 1,3-diol **17**. Acetal formation with another molecule of aldehyde affords the 1,3-dioxane **16**.

(5-Methylene)-1,3-dioxane-4-ones **19** are formed from aryl acrylates and aliphatic aldehydes in the presence of DABCO via the products **18** of a BAYLIS-HILLMAN reaction (cf. p 284) [130]:

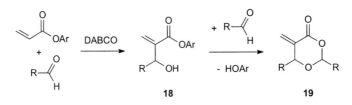

1,3-Dioxane is a colourless liquid, bp 105°C, readily soluble in all common solvents. It is prepared from formaldehyde and propane-1,3-diol in the presence of an acid catalyst.

2,2-Dimethyl-1,3-dioxane-4,6-dione **20** (MELDRUM's acid) is obtained by cyclocondensation of acetone with malonic acid. MELDRUM's acid represents a useful reagent in organic synthesis due to its reactivity as a malonate equivalent [131].

Several systems derived from 1,3-dioxane, for instance (2R,6R)-2-isopropyl-6-methyl-1,3-dioxan-4-one **21** and (R)-2-*tert*-butyl-6-methyl-1,3-dioxin-4-one **23**, serve as chiral building-blocks for stereoselective transformations (SEEBACH 1986).

For instance, the compound **21**, derived from (*R*)-3-hydroxybutyric acid, on treatment with alkyl halides in the presence of lithium diisopropylamide, is α-alkylated affording the products **22**. These show a high *anti*-selectivity (de greater than 98%) with respect to the position of the newly introduced 5-substituent in the 1,3-dioxane:

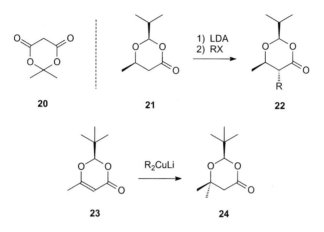

The system **23**, derived from ethyl acetoacetate, undergoes MICHAEL additions with dialkyl cuprate stereoselectively (de greater than 95%) yielding products **24**.

A further application of chiral 1,3-dioxanes of type **21** is the stereoselective fission of the 1,3-dioxane ring, e.g. of **25**, by silyl nucleophiles in the presence of TiCl$_4$ or other titanium compounds:

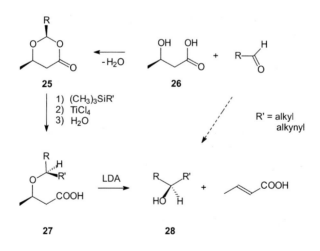

Thus, the chiral 1,3-dioxane system mediates the enantioselective addition of nucleophiles to an aldehyde [132], since the resulting β-alkoxycarboxylic acids **27** can be cleaved by lithium diisopropylamide to give the chiral alcohols **28** with elimination of crotonic acid.

The chiral 1,3-dioxanes **21** and **25** are obtained from the biopolymer poly[(*R*)-3-hydroxybutyric acid] (PHB) by hydrolysis providing the chiral β-hydroxy acid **26** which is condensed with aldehydes.

6.27 1,3-Dithiane

A-C | 1,3-Dithiane is a ring homologue of 1,3-dithiolane (see p 122) and a dithioacetal of formaldehyde.

1,3-Dithianes **1** substituted at position 2 are obtained by cyclocondensation of aldehydes or ketones with propane-1,3-dithiol:

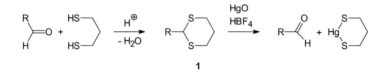

1

1,3-Dithianes are not hydrolysed by aqueous alkali or by dilute acids. Their fission (dethioacetalization) is preferably carried out in the presence of heavy metal compounds, e.g. mercuric oxide in the presence of tetrafluoroboric acid.

Although 1,3-dithianes have been known for some time, their importance for organic synthesis has only been recognized since 1965. COREY and SEEBACH found that 1,3-dithiane and 2-monosubstituted 1,3-dithianes, on treatment with *n*-butyllithium in THF at -40°C, yield relatively stable and reactive 2-lithio-1,3-dithianes **2**:

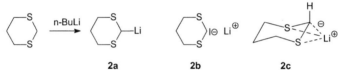

Formula **2a** represents a covalent C–Li bond, whereas **2b** shows the ionic limiting case. Stabilization of the carbanion is effected by interaction between the occupied $2p_z$ orbital of the C-atom and unoccupied 3d orbitals of the S-atoms (p_π–d_π-bonding). ELIEL has shown that the equatorial H-atom is preferably substituted by lithium, as shown in the formula **2c**.

1-Lithio-1,3-dithianes react with electrophiles, especially by nucleophilic substitution of alkyl halides:

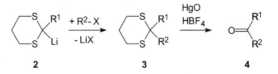

Dethioacetalization of product **3** yields the ketone **4**.

E | The concept of 'umpolung' has been developed on the basis of such reactions. It is defined as the change in the polarity of an atom in a functional group through derivatization. The 2-lithio-

1,3-dithiane behaves like an aldehyde that has undergone umpolung: In this case, although aldehydes as electrophiles do not directly react with alkyl halides, the reaction proceeds according to route $1 \rightarrow 2 \rightarrow 3 \rightarrow 4$.

The ketone **4** is formally produced by alkylation of an acyl anion with an alkyl halide. For this reason, 2-lithio-1,3-dithianes are also regarded as synthetic acyl anion equivalents:

$$R^1\!-\!\overset{\overset{\textstyle O}{\|}}{C}{}^{\ominus} \quad + \quad R^2\!-\!X \quad \longrightarrow \quad R^1\!-\!\overset{\overset{\textstyle O}{\|}}{C}\!-\!R^2 \quad + \quad X^{\ominus}$$

Numerous synthetic applications have been described using 1,3-dithianyl anions [133]. For instance, the lithium compounds **2** undergo nucleophilic additions with oxiranes and with carbonyl compounds:

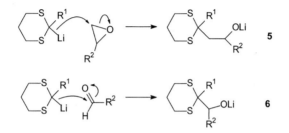

β-Hydroxy ketones are obtained from the addition product **5** and α-hydroxy ketones from **6**.

6.28 Cepham

A-C Cepham is a condensed ring system consisting of a 1,3-thiazan and an azetidine ring. The numbering deviates from IUPAC rules, as is the case for penam (see p 159). The molecule is chiral.

Cephalosporins are derived from cepham. Cephalosporin C, a metabolite of *Cephalosporium acremonium*, was isolated by NEWTON and ABRAHAM in 1955. Its structure was established in 1961 by chemical methods and X-ray diffraction. It is an acid amide derived from (*R*)-α-aminoadipic acid and 7-aminocephalosporic acid. The latter is (6*R*,7*R*)-3-acetoxymethyl-7-amino-8-oxoceph-3-em-4-carboxylic acid.

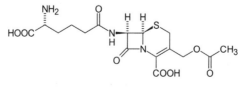

cephalosporin C

Like the penicillins, cephalosporins possess an azetidinone ring, both being β-lactam antibiotics (see p 159). The fused ring is a 1,3-thiazolidine in the penicillins and a dihydro-1,3-thiazine system in the cephalosporins. The asymmetric C-atoms of the 2-azetidinone ring have an (R)-configuration in both penicillins and cephalosporins, and this β-lactam ring is responsible for the biological activity. It causes irreversible acylation of amino groups in enzymes, mainly in the transpeptidases that are responsible for the synthesis of the peptidoglycan in the bacterial cell wall.

The cephalosporins have a broader spectrum of activity than penicillins and resistance is less common (see p 159). The low toxicity of the β-lactam antibiotics is due to the fact that the mammalian cell walls lack peptidoglycans and for this reason also lack the relevant enzymes.

The total synthesis of cephalosporin C by WOODWARD, which was the subject of his NOBEL address in 1965 [134], is a milestone in the synthesis of complex natural products. The sixteen-step synthesis starts from cysteine and the principal features are outlined below:

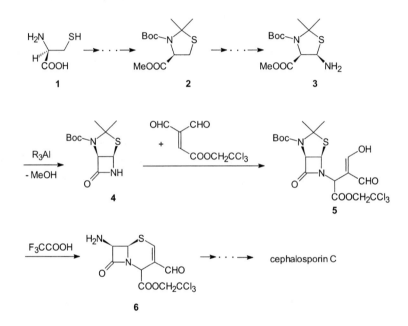

The choice of (+)-(R)-cysteine **1** as starting material determines the R-configuration at the C-7 atom in the final molecule. The NH$_2$ and SH groups are blocked by cyclocondensation with acetone and reaction with *tert*-butoxycarbonyl chloride yielding **2**. Introduction of an amino group **3** gives rise to the formation of the azetidin-2-one ring **4**. This is followed by two steps (**4** → **5** → **6**) leading to the formation of the dihydro-1,3-thiazine ring. The final steps produce the required substituents.

Other total syntheses of cephalosporins have since been achieved. However, they cannot compete with the commercial extraction from moulds or with semisynthetic methods. There are two possible semisynthetic ways:

(1)   Conversion of cephalosporin C into 7-aminocephalosporanic acid **8**:

Attempts at removing the δ-(α-aminoadipoyl) substituent by hydrolysis invariably result in fission of the lactam ring. However, the following method has proved successful:

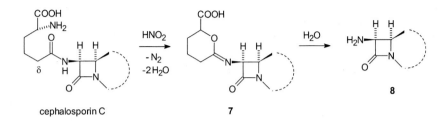

cephalosporin C   7

On treatment with nitrous acid, cephalosporin C yields the imino lactone 7. Its C=N bond is more reactive than the β-lactam ring and favours the hydrolytic formation of 8 in good yield.

(2)   Conversion of semisynthetic penicillins, e.g. 9, into 7-acylamino-3-methyl-8-oxoceph-3-em-4-carboxylic ester 12 (penicillin rearrangement, MORIN reaction 1969) [135].

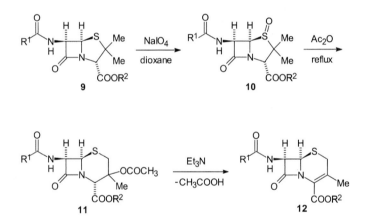

This interesting reaction sequence, which is a type of PUMMERER reaction, is important because the production of penicillin from mould cultures can be carried out on a large scale, and more economically than that of cephalosporins.

Starting from compounds with a cepham structure obtained by a partial synthesis, numerous semisynthetic cephalosporins have been prepared, mainly by

- variation of the acylamino substituent at position 7
- variation of the side chain at position 3
- introduction of a second substituent at position 7

They are much more effective than cephalosporin C. Three examples serve as an illustration (see chapter 5, ref. [106]):

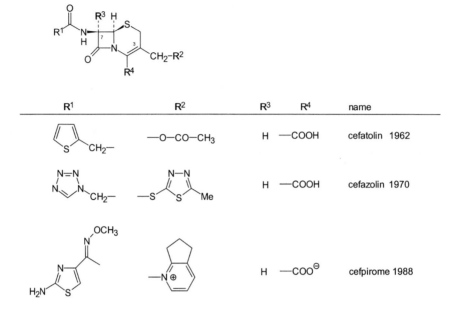

R¹	R²	R³	R⁴	name
(2-thienyl)CH₂—	—O—CO—CH₃	H	—COOH	cefatolin 1962
(tetrazolyl)CH₂—	—S—(thiadiazolyl)—Me	H	—COOH	cefazolin 1970
(aminothiazolyl) with =N—OCH₃	—N⊕ (cyclopenta-pyridinium)	H	—COO⊖	cefpirome 1988

The development began in 1962 with the synthesis of cefatolin by acylation of **8** with (2-thienyl)ethanoyl chloride. The product is active against gram-positive bacteria and against penicillin-resistant staphylococci. In cefazolin, the side-chain at the 3-position is changed. The compound is also active against gram-negative bacteria. Cefpirome belongs to the new generation of cephalosporins. They are distinguished by a high efficacy coupled with a wide spectrum of activity even against bacteria of the pseudomonas group. Their stability to β-lactamases and metabolic stability are high.

Cephalosporins are usually administered parenterally. These drugs are the most frequently used for bacterial infections because of their broad activity and low toxicity. The β-lactam antibiotics account for about 60% of the annual worldwide sales of antibiotics amounting to 11 thousand million US dollars. Of this amount, cephalosporins claim 40% and penicillins 20%.

The biosynthesis of cephalosporins is analogous to that of the penicillins. It is derived from the peptide δ-(α-aminoadipoyl)cysteinylvaline (see p 160). First, the β-lactam ring is formed. This is followed by a C–H bond fission, not at the β-position but at the γ-position of the valine building-block. However, enzymatic conversion of the penam system into the ceph-3-em system cannot be excluded.

6.29 Pyridazine

The structurally isomeric diazines pyridazine (1,2-diazine), pyrimidine (1,3-diazine) and pyrazine (1,4-diazine) are derived from pyridine by appropriate substitution of a CH group by nitrogen.

 Pyridazine has the geometry of a planar, slightly distorted hexagon (see Fig. 6.17), as deduced from electron diffraction and microwave spectroscopy.

Fig. 6.17   Bond parameters of pyridazine
(bond lengths in pm, bond angles in degrees)

The N–N bond has single bond character. Therefore, pyridazine can be described as a resonance hybrid with limiting structures **a** and **b**, the canonical form **a** making the major contribution:

The ^{1}H- and ^{13}C-NMR spectra show similarities with pyridine. The additional nitrogen is responsible for a greater downfield shift of the ring protons and C-atoms at positions 3 and 6:

UV (hexane) λ (nm) (ε)	^{1}H-NMR (CDCl$_3$) δ (ppm)	^{13}C-NMR (CDCl$_3$) δ (ppm)
241 (3.02) 251 (3.15) 340 (2.56)	H-3/H-6: 9.17 H-4/H-5: 7.52	C-3/C-6: 153.0 C-4/C-5: 130.3

B   Reactions of pyridazine also show analogies to pyridine [136]. Electrophiles attack the ring N-atoms, for instance in protonation, alkylation or N-oxidation. S$_E$Ar reactions at the ring C-atoms are difficult to carry out, even in the presence of activating substituents due to the deactivation by the additional N-atom. However, N-oxidation facilitates the substitution in some cases.

Reactions with nucleophiles occur at ring position C-4 (e.g. with GRIGNARD reagents [137]), or at C-3 (with organolithium compounds). They are preparatively not as important as the CHICHIBABIN or ZIEGLER reaction in pyridine:

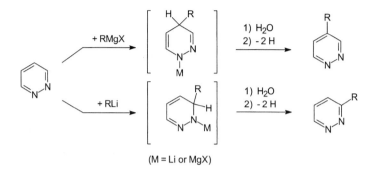

(M = Li or MgX)

3-Substituted pyridazine-1-oxides react with cyanide by analogy to a REISSERT reaction (see p 321), functionalizing the C-6 position:

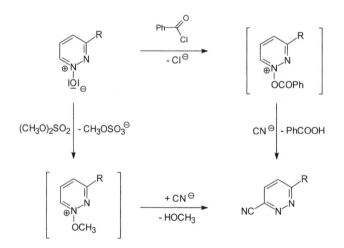

Some S$_N$Ar reactions (e.g. with amides, amines and alcoholates) also proceed smoothly with pyridazines with a leaving group at position 3 or 6. Side-chain reactivity, which is characteristic for pyridines and benzopyridines (see p 281), is also observed in 3- and 4-methylpyridazines.

A number of thermal and photochemical transformations of the 1,2-diazine system are noteworthy. Thermolysis isomerizes pyridazines into pyrimidines and/or pyrazines, as has been demonstrated for a series of perfluoro and perfluoroalkyl derivatives (see p 288). Pyridazine is converted into pyrimidine at 300°C. A process of valence isomerism is thought to be responsible for the thermal isomerization of pyridazines via diazabenzvalenes (simplified as **3** and **4**).

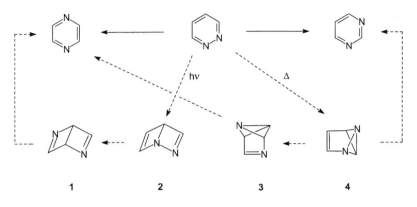

1   2   3   4

In contrast to this reaction, photolysis of pyridazines yields mainly pyrazines. The photochemical reorganization of the 1,2-diazine framework presumably involves intermediates of the DEWAR benzene type **1** and **2**.

C The synthesis of pyridazines follows from a simple *retrosynthetic* consideration, leading to the 1,4-dicarbonyl systems **5** and **6** and hydrazine as starting materials:

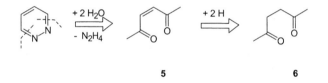

5   6

(1) Saturated and α,β-unsaturated 1,4-dicarbonyl compounds undergo cyclocondensation with hydrazine via hydrazones yielding 1,4-dihydropyridazines **8** or pyridazines **7**:

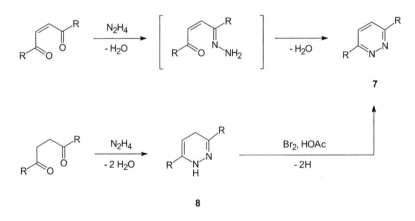

Dehydrogenation of dihydropyridazines **8** to **7** is brought about by Br_2 in glacial acetic acid. Condensation is carried out in the presence of mineral acids so as to prevent the competing formation of *N*-aminopyrroles.

γ-Ketocarboxylic acids or their esters furnish 6-substituted pyridazine-3(2*H*)-ones **10** on cyclocondensation with hydrazine and dehydrogenation of the dihydro compound **9**:

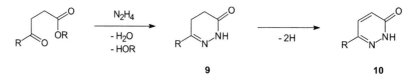

(2) The cyclocondensation of 1,2-diketones, reactive α-methylene esters and hydrazines leads, in its simplest form as a one-pot reaction, to pyridazin-3(2*H*)-ones **11** (*Schmidt–Druey synthesis*):

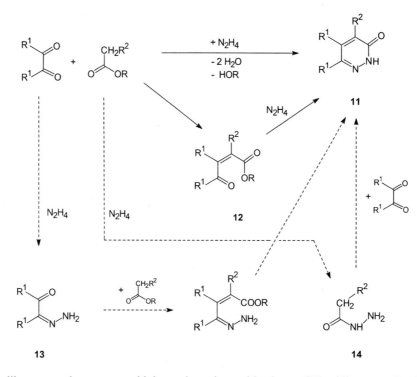

The 1,2-diketone undergoes an aldol condensation with the α-CH-acidic ester furnishing the intermediate **12** which is cyclized by N_2H_4, similar to the formation of **9**. In another combination of this three-component system, the monohydrazones of the 1,2-dicarbonyl compounds **13** or hydrazides with a reactive α-CH_2 group **14** can be employed for the pyridazin-3(2*H*)-one synthesis.

(3) Pyridazine derivatives are also obtained by [4+2] cycloadditions. For instance, the tetrahydropyridazines **15** are produced by a DIELS-ALDER reaction from 1,3-dienes with azodicarboxylic ester. The pyridazines **17** are produced by addition of alkynes to 1,2,4,5-tetrazines followed by a retro-DIELS-ALDER reaction of the adducts **16** with N_2 elimination:

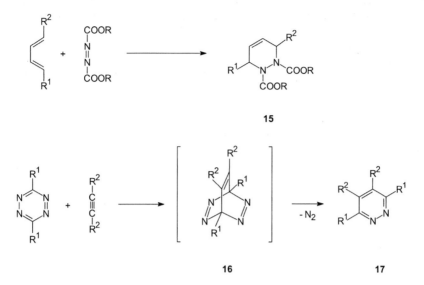

15

16   **17**

D **Pyridazine**, mp -8°C, bp 208°C, is a colourless liquid, soluble in water and alcohols but insoluble in hydrocarbons (H-bridges, due to acceptor function of the N-atoms). Of the diazines, pyridazine has the highest basicity (pK_a = 2.3, pyrimidine = 1.3, pyrazine = 0.4), but in common with all diazines, it is much less basic than pyridine (pK_a = 5.2). Its dipole moment (μ = 3.95 D) is higher than that of pyrimidine (μ = 2.10 D). Pyrazine has no dipole moment.

The enthalpies of formation for diazines have been calculated as 4397.8 kJ mol^{-1} (1,2-diazine), 4480.2 kJ mol^{-1} (1,3-diazine) and 4480.6 kJ mol^{-1} (1,4-diazine). Thus, the thermodynamic stability of pyridazine is lower than that of pyrimidine and pyrazine by 83 kJ mol^{-1}.

Applying the synthetic principle (1) (see p 395), pyridazine itself can be prepared from maleic anhydride. Its reaction with hydrazine yields maleic hydrazide **18** which is converted with POCl$_3$/PCl$_5$ into 3,6-dichloropyridazine **19** due to an azinone–hydroxyazine tautomerism (see p 310); **19** gives rise to pyridazine by reductive dehalogenation with H$_2$/Pd-C:

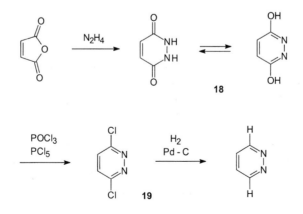

18

19

Relatively few pyridazine derivatives occur in natur, e.g. the quaternary salt pyridazinomycin **20**. Some pyridazine derivatives show biological activity and are applied as herbicides and anthelmintics, e.g. maleic hydrazide **18** and the chlorinated pyridazinones **21** (pyrazon)/**22** (pyridaben). The tetrahydro-pyridazinone derivative levosimendan **23** is an innovative myofilament calcium sensitizer applied as cardiotonic in treatment of heart failure.

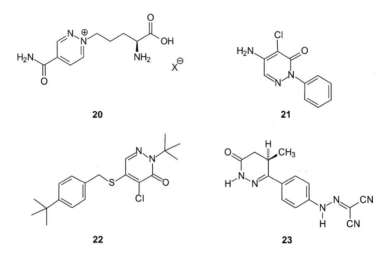

6.30 Pyrimidine

A According to X-ray diffraction studies (see Fig. 6.18), pyrimidine exists as a distorted hexagon:

Fig. 6.18 Bond parameters of pyrimidine
(bond lengths in pm, bond angles in degrees)

As in pyridazine, the additional ring nitrogen causes stronger deshielding of the ring protons and C-atoms than in pyridine (see p 270). This is shown by the NMR spectroscopic data of pyrimidine:

UV (H$_2$O) λ (nm) (ε)	^{1}H-NNR (CDCl$_3$) δ (ppm)	^{13}C-NMR (CDCl$_3$) δ (ppm)
238 (3.48) 243 (3.50) 272 (2.62)	H-2: 9.26 H-4/H-6: 8.78 H-5: 7.36	C-2: 158.4 C-4/C-6: 156.9 C-5: 121.9

B In its reactions, pyrimidine behaves as a deactivated heteroarene. Its reactivity is comparable to that of 1,3-dinitrobenzene or 3-nitropyridine. Electrophiles attack on nitrogen in protonating and alkylating reactions. Electrophilic substitutions at carbon are not observed in the parent compound. Electron-donating substituents (OH, NH$_2$) increase the S$_E$Ar reactivity in the pyridmidine system (two substituents give a reactivity corresponding to benzene, three to that of phenol) and enable nitration, nitrosation, aminomethylation and azo-coupling to take place at the 5-position, e.g.:

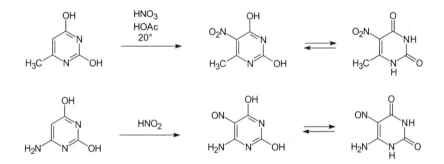

Nucleophilic attack occurs at the 2-, 4- and 6-positions. Only a few examples are known for pyrimidine itself, e.g. addition of some organometallic compounds to 3,4-dihydropyrimidines that can then be dehydrogenated to give pyrimidines:

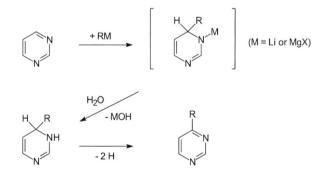

Nucleophilic substitutions (e.g. with amides, amines, alcoholates, sulfides) in 2-, 4- or 6-halo-substituted pyrimidines are widely applied for the preparation of functionalized pyrimidines, e.g.:

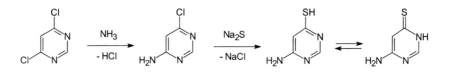

These S_NAr processes occur faster at C-4/C-6 than at C-2. Dehalogenation is possible at all positions of pyrimidines by means of H_2/Pd in the presence of weak bases such as $CaCO_3$ or MgO. This can be carried out selectively at C-4 and C-6 with Zn dust in H_2O or in a weakly alkaline medium.

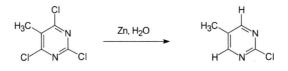

2- and 4(6)-Alkylsulfanyl groups are readily introduced by S-alkylation of the corresponding thiones, and are substituted by nucleophiles, e.g. by amines or H_2O:

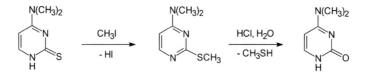

The pyrimidine system shows side-chain reactivity typical of azines. CH_3 groups at the 2-, 4- or 6-position undergo aldol condensations (with aldehydes in the presence of LEWIS acids) or CLAISEN condensations (with esters in the presence of strong bases) with marked preference at C-4, e.g.:

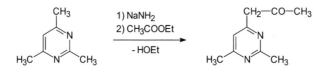

The regioselectivity of the side-chain reactions on pyrimidine can be controlled by reaction conditions. This was found in the bromination of 4,5-dimethylpyrimidine [138]. Bromination under ionic conditions (Br_2 in HOAc) favours substitution of the 4-methyl group, but under radical conditions (NBS in CCl_4), the 5-methyl group is favoured:

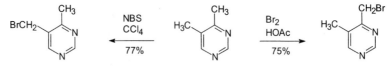

C For the *retrosynthesis* of pyrimidine (see Fig. 6.19) it is essential that C-2 is at the oxidation level of a carboxylic acid, i.e. that an amidine moiety is present in the diazine system.

Hence the retroanalytical operation **a** (addition of H_2O to C-4 and C-6, and bond fission at N-1/C-6 and N-3/C-4) suggests 1,3-dicarbonyl compounds and N–C–N systems of an amidine type as building blocks for a pyrimidine synthesis. The retroanalytical operation **b** (addition of H_2O at C-2 followed by bond disconnection at N-1/C-2 or N-3/C-2 giving **1** or **2**) points to the alternative starting materials, diaminoalkenes **3** and carboxylic acids.

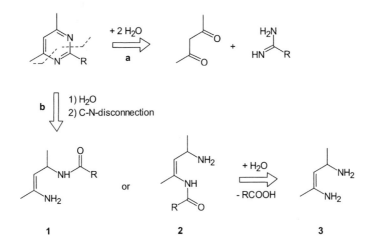

Fig. 6.19   Retrosynthesis of pyrimidine

The following methods are used for the synthesis of pyrimidine [139]:

(1)   In a standard method, 1,3-diketones are cyclocondensed with amidines, ureas, thioureas and guanidines giving 4,6-di- and 2,4,6-trisubstituted pyrimidines **4**, 2-pyrimidones **5**, 2-thiopyrimidones **6** and 2-aminopyrimidines **7** respectively (*Pinner synthesis*) [140] :

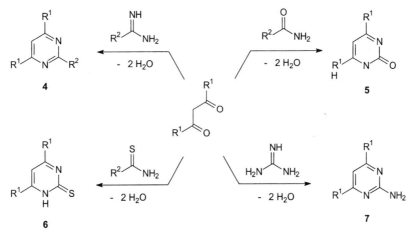

β-Keto esters react in an analogous way, yielding pyrimidin-4(3*H*)-ones **8**:

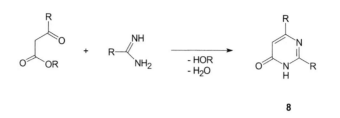

8

Important variations are the base-catalysed condensations of N–C–N building blocks (e.g. urea) with malonic esters producing derivatives of barbituric acid (e.g. **9**) or with cyanoacetic esters via cyanoacetyl compounds (e.g. **10**) furnishing 6-amino-2-hydroxypyrimidin-4(3*H*)-ones (e.g. **11**):

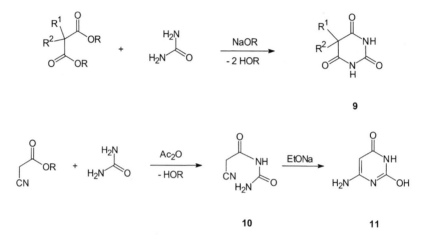

9

10 **11**

Masked β-dicarbonyl systems can also be used as 1,3-bis-electrophiles. The PINNER cyclocondensation of ethyl 2-cyano-3-ethoxypropenoate with acetamidine demonstrates that the direction of cyclization at the intermediate **12** can be controlled by the reaction medium: an acid medium favours pyrimidine formation **13** via the ester function, whereas a basic medium leads to **14** via the nitrile function:

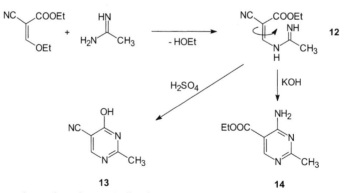

13 **14**

(2) Pyrimidine syntheses based on 1,3-diaminopropenes or propanes are less important. The base-catalysed cyclocondensation of malonamides with carboxylic esters leads to 6-hydroxypyrimidin-4(3*H*)-ones **15** (*Remfry-Hull synthesis*):

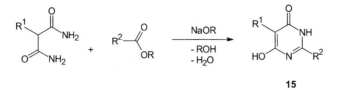

15

Uracil and thymine derivatives are obtained from cyanoacetic acid and *N*-alkylurethanes. An amide **16** is formed first, which condenses with orthoformic ester to give the enol ether **17**; aminolysis of **17** leads to **18** which on cyclization yields 5-cyanouracil **19** (*Shaw synthesis*):

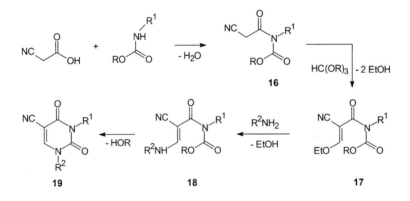

19 **18** **17**

(3) In a novel principle of 1,3-diazine formation phosphazenes like **20** containing an amidine moiety undergo an aza-WITTIG reaction with α,β-unsaturated aldehydes followed by oxidative electrocyclic ringclosure to pyrimidines **21** [141]:

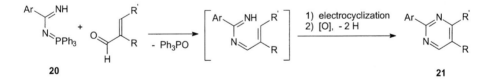

20 **21**

D **Pyrimidine**, mp 22,5°C, bp 124°C, is a water-soluble, weak base that forms a sparingly soluble complex with $HgCl_2$. Pyrimidine is prepared by condensation of 1,1,3,3-tetraethoxypropane with formamide over montmorillonite (*Bredereck synthesis*) at 210°C in the gas phase [142]:

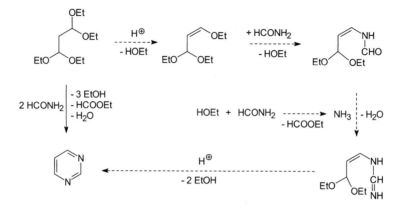

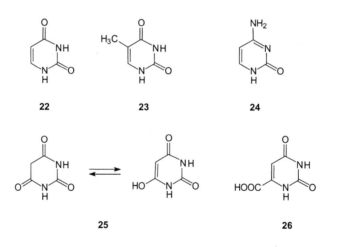

Uracil **22**, thymine **23**, cytosine **24**, barbituric acid **25** and orotic acid **26** are important pyrimidine derivatives.

These compounds exist as tautomers. Hydroxypyrimidines adopt the lactam form whereas aminopyrimidines prefer the enamine form (see p 413).

Barbituric acid, mp 245°C, $pK_a = 4$, a cyclic ureido derivative, is prepared from urea by condensation with malonic acid/$POCl_3$ or with malonic ester and sodium ethoxide [143].

Orotic acid is the key compound in the biosynthesis of almost all naturally occurring pyrimidine derivatives. Its synthesis is achieved by cyclocondensation of diethyl 2-oxobutanedioate with *S*-methyl-isothiourea, followed by direct or oxidative hydrolysis of the 2-methylsulfanyl intermediate **27**:

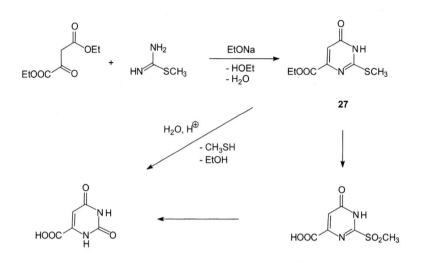

Many natural products are derived from pyrimidine. The 'pyrimidine bases' thymine, cytosine and uracil (22-24) are important as building-blocks for the nucleic acids. 5-Methylcytosine (obtained from hydrolytic extracts of tubercle bacilli) and 5-(hydroxymethyl)cytosine (obtained from the bacteriophage of *Escherichia coli*) are rarer.

Aneurin 28 (thiamine, vitamin B_1) occurs in yeast, in rice polishings and in various cereals (cf. p 154). Aneurin is prepared commercially by various synthetic methods [144].

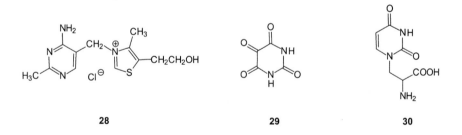

28	29	30

Alloxan 29 is formed in organisms by the oxidative degradation of uric acid (see p 414). Willardiin 30 is a nonproteinogenic α-amino acid with a uracil structure which is present in the seeds of some types of acacia. A few pyrimidine antibiotics, especially those isolated from streptomyces, possess potent antitumour properties, e.g. the structurally complex bleomycins 31. The commercial product is a mixture of various bleomycins, main components being bleomycins A_2 (55-70%) and B_2 (25-32%).

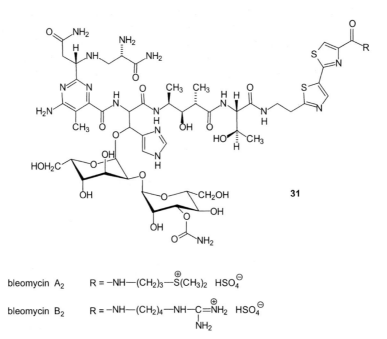

31

bleomycin A₂ $R = -NH-(CH_2)_3-\overset{\oplus}{S}(CH_3)_2$ $HSO_4^{\ominus}$

bleomycin B₂ $R = -NH-(CH_2)_4-NH-\overset{\oplus}{C}=NH_2$ $HSO_4^{\ominus}$
                  $\underset{NH_2}{\big|}$

The pyrimidine ring is a constituent of many pharmaceuticals, such as the chemotherapeutics trimethoprim **32** and sulfadiazine **33**, the dihydrofolate reductase inhibitor pyrimethamine **34** and hexetidine **35** derived from hexahydropyrimidine and used as an oral antiseptic.

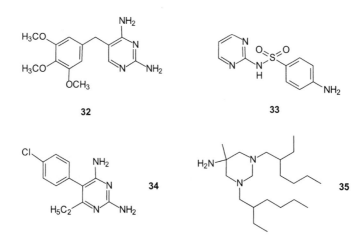

32

33

34

35

5,5-Disubstituted derivatives of barbituric acid **25** (barbiturates) are used therapeutically. 5,5-Diethylbarbituric acid **36** was the first barbiturate to be used (E. FISCHER 1903, veronal, barbital). Barbital and its Na salts are sedatives, as are phenobarbital **37** and hexobarbital **38**, which are more potent [145]. Methylphenobarbital **39** is an antiepileptic drug and the thiobarbiturate thiopental **40** is a short-acting anaesthetic. Sedative barbiturates have now been replaced by other drugs because of their toxicity and dependence problems.

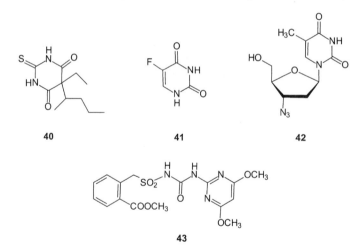

36 : $R^1 = R^2 = C_2H_5$, $R^3 = H$

37 : $R^1 = C_2H_5$, $R^2 = C_6H_5$, $R^3 = H$

38 : $R^1 = CH_3$, $R^2 =$ ⬡ , $R^3 = CH_3$

39 : $R^1 = C_2H_5$, $R^2 = C_6H_5$, $R^3 = CH_3$

5-Fluorouracil **41** is an antineoplastic agent, zidovudin **42** (3'-azido-2',3'-dideoxythymidine, AZT) is an important drug used in the treatment of HIV and orotic acid **26** is used in the treatment of metabolic disorders.

Sulfonylureas derived from pyrimidine (and 1,2,4-triazine) are of considerable importance among the more recently introduced growth regulators for herbicides, e.g. bensulfuronmethyl **43**.

40   **41**   **42**

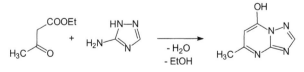

43

Most photographic materials contain 7-hydroxy-5-methyl-1,2,4-triazolo[1,5-*a*]pyrimidine **44** as an emulsion stabilizer. It is prepared by cyclocondensation of ethyl acetoacetate with 3-amino-1,2,4-triazole [146].

44

6.31 Purine

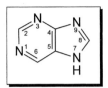

A For historical reasons, the numbering of purine (imidazo[4,5-d]pyrimidine) does not comply with IUPAC rules. The X-ray structure of purine shows that the imidazole part is planar and the fused pyrimidine ring deviates from a coplanar arrangement (see Fig. 6.20). Purine exists in two tautomeric forms, namely 7*H*-purine **1** and 9*H*-purine **2**, which are in equal concentration in solution (annular tautomerism). In the solid state, the 7*H*-form is dominant:

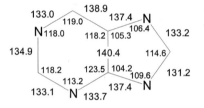

Fig. 6.20   Bond parameters of purine
(bond lengths in pm, bond angles in degrees)

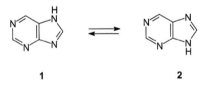

         1   **2**

The NMR spectroscopic data of the pyrimidine ring (see p 398) are comparable with those of the parent system. However, the protons are strongly shifted downfield in the imidazole ring when compared with the parent system.

UV (methanol) λ (nm) (ε)	^{1}H-NMR (DMSO-d$_6$) δ (ppm)	^{13}C-NMR (DMSO-d$_6$) δ (ppm)	
363 (3.88)	H-2: 8.99	C-2: 152.1	C-6: 145.5
	H-6: 9.19	C-4: 154.8	C-8: 146.1
	H-8: 8.68	C-5: 130.5	

B Purine undergoes reactions with electrophiles and nucleophiles [147]. Protonation of purine occurs at N-1, but alkylation at N-7 and/or N-9. This is shown by its reactions with dimethyl sulfate, iodomethane and vinyl acetate:

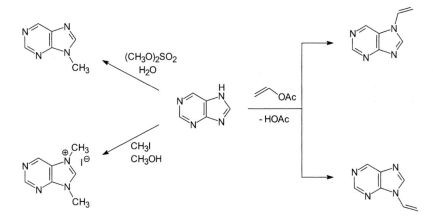

Nucleophilic substitution (S_NAr) can be carried out on purine itself as well as on halo-, alkoxy- and alkylsulfanylpurines. The CHICHIBABIN reaction (see p 278) of purine with KNH_2 in liquid ammonia leads to 6-aminopurine (adenine) **3**:

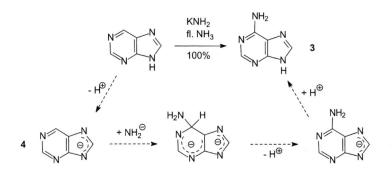

Formation of the 7,9-anion **4**, the result of electrostatic effects, directs addition of the amide into the 6-position of the pyrimidine ring.

Halogen substituents in the purine system display a leaving group ability which decreases in the series C-6 > C-2 > C-8. This is demonstrated by the synthetic steps leading from 2,6,8-trichloropurine **5** to guanidine **8**:

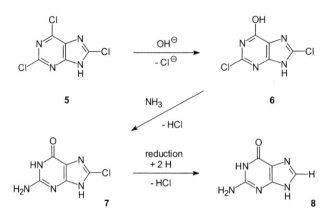

5

OH$^\ominus$ / - Cl$^\ominus$

6

NH$_3$ / - HCl

reduction / + 2 H / - HCl

7

8

Action of NaOH on **5** leads to substitution of 6-Cl by OH and formation of 2,8-dichloro-6-hydroxypurine **6**. Reaction of **6** with ammonia introduces the NH$_2$ group at C-2 giving 8-chloroguanine **7**, which is reductively dehydrogenated to guanine **8**.

Selective attack by nucleophiles on the pyrimidine ring of **5** does not occur, as expected, from the coefficients of the frontier orbitals in purine calculated on the basis of the FUKUI FMO theory. It is considered (see above) to be due to primary formation of the anion **9**:

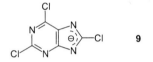

9

If anion formation is not possible, e.g. in the 9-alkyl-2,6,8-trichloropurines, the initial nucleophilic attack occurs predictably on C-8 in the imidazole ring.

In the purine system, ring transformations are also observed. For instance, 6-imino-1-methyl-1,6-dihydropurine **10** isomerizes to give 6-(methylamino)purine **12** by the action of aqueous alkali:

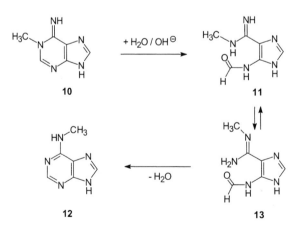

10

+ H$_2$O / OH$^\ominus$

11

12

- H$_2$O

13

The first step is H$_2$O addition at C-2 with fission of the pyrimidine ring forming the imidazole derivative **11**. Its tautomer **13** recyclizes to the purine system **12** in a DIMROTH rearrangement [148] (see also p 202).

C It is important to note that in the *retrosynthesis* of purine, C-2 and C-8 are at the oxidation level of formic acid. Therefore, two types of bond fission, namely **a** or **b** can be considered:

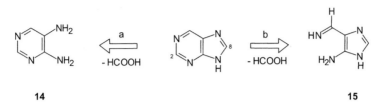

Thus, the synthesis of purine may start from 4,5-diaminopyrimidine **14** to which the imidazole moiety is added, or from 4(5)-amino-5(4)(iminomethyl)imidazole **15** onto which the pyrimidine part is cyclized.

(1) In the standard method for purine synthesis, 4,5-diaminopyrimidines are subjected to cyclocondensation with formic acid or with formic acid derivatives (*Traube synthesis*, 1910):

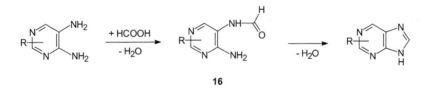

The formamides **16** are intermediates. Formamide (BREDERECK variant of the TRAUBE synthesis [149]), formamidine, orthoformic ester, diethoxymethyl acetate, VILSMEIER reagent (from DMF and POCl$_3$) and dithioformic acid are used as formic acid derivatives. 4,5-Diaminopyrimidines can be obtained from 4-aminopyrimidines **17** by nitrosation with HNO$_2$ followed by reduction of the nitroso compounds **18**:

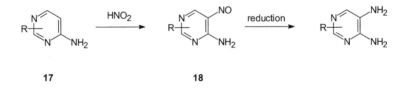

(2) In some cases, purines are prepared from 4,5-disubstituted imidazoles by cyclocondensation with formic acid or its equivalents, e.g. the 9-alkylpurin-6(1*H*)-one **20** from 1-alkyl-5-aminoimidazole-4-carboxamide **19** and formic acid :

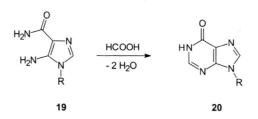

19   **20**

(3)   Purines can also be derived from simple acyclic building blocks. For instance, purine is formed by the thermolysis of formamide, or by interaction of *N*-(cyanomethyl)phthalimide with tris(formamino)methane, or with formamidine acetate in formamide or butanol:

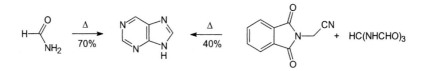

Hypoxanthine (see below) is obtained in a similar one-pot reaction, namely by heating the acetamidocyanoacetate **21** with ethanolic NH_3, ammonium acetate and orthoformic ester:

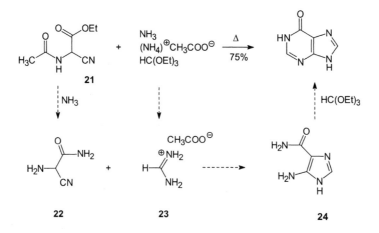

22   **23**   **24**

Formation of the purine system probably occurs by way of the aminoimidazolecarboxamide **24** according to method 2; **24** results from the interaction of aminocyanoacetamide **22** and formamidine acetate (methanimidamidium acetate) **23** derived from NH_3, ammonium acetate and $HC(OR)_3$.

This type of purine formation is a so-called abiotic synthesis [150]. This is defined as the synthesis of compounds essential for living organisms from nonbiological material in the course of chemical evolution.

D	**Purine**, mp 216°C, was first obtained by E. FISCHER (1899) by reduction of 2,6,8-trichloropurine. Purine is water soluble and acts as a weak base ($pK_a = 8.9$).

Hypoxanthine 25 (6-hydroxypurine) exists predominantly in the lactam form in the solid state and in solution:

25

lactim form   lactam form

It is prepared in a one-pot reaction from **21** (see p 412) or according to TRAUBE via 4,5-diamino-6-hydroxypyrimidine (existing in the lactam form **28**) :

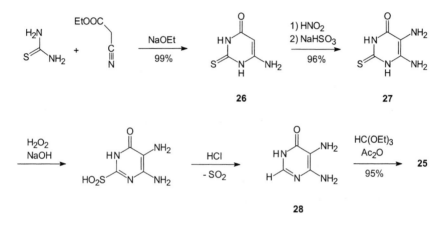

Ethyl cyanoacetate is cyclocondensed with thiourea under base catalysis giving 4-amino-2-thiouracil **26**, onto which the second amino group is introduced (**27**). Oxidative desulfurization yields a sulfinic acid and SO_2 elimination giving **28** is followed by condensation with orthoformic ester yielding **25**.

Adenine 29 (6-aminopurine) and **guanine 8** (2-amino-6-hydroxypurine) are obtained from the hydrolytic cleavage of nucleic acids (see below). Deamination with HNO_2 converts adenine into hypoxanthine, guanine into xanthine. Guanine is synthesized from 2,6,8-trichloropurine (see p 410) or from cyanoacetic ester and guanidine via 2,4-diamino-6-hydroxypyrimidine **30**:

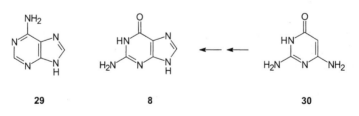

29   **8**   **30**

Xanthine 32 (2,6-dihydroxypurine), mp 262°C, is chemically similar to uric acid (see below). It is soluble in bases and acids and forms a sparingly soluble perchlorate. Xanthine is prepared from cyanoacetic ester and urea by way of 4-aminouracil **31** (according to the TRAUBE principle), or from uric acid **33** by heating with formamide (BREDERECK 1950):

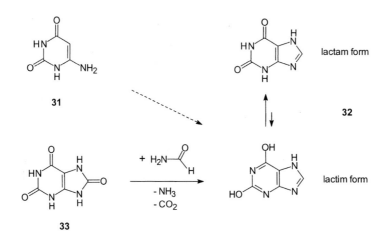

31   **32**

lactam form

lactim form

33

+ H_2N—CHO

- NH_3
- CO_2

The natural products theophylline **34** (1,3-dimethylxanthine), theobromine **35** (3,7-dimethylxanthine) and caffeine **36** (1,3,7-trimethylxanthine) are derived from the lactam form of xanthine. Theophylline occurs in tea leaves and is a diuretic and a coronary vasodilator. Cocoa beans contain ca. 5% theobromine, which is a stronger diuretic than theophylline or caffeine.

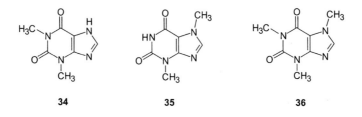

34   **35**   **36**

Caffeine, mp 263°C, sublimes on heating and is more soluble than dimethylxanthines. Caffeine has a stimulating effect on the central nervous system. It is extracted from green coffee beans with liquid CO_2 [151]. Caffeine is synthesized by methylation of xanthine, theophylline or theobromine with methyl iodide or dimethyl sulphate [149].

Uric acid 33 (2,6,8-trihydroxypurine) was isolated from urinary calculi by SCHEELE (1774). After ascertaining its molecular formula (LIEBIG and MITSCHERLICH 1834), its structure was established by oxidative degradation and structural correlation with barbituric acid, alloxan, allantoin and hydantoin (LIEBIG, WÖHLER, BAEYER):

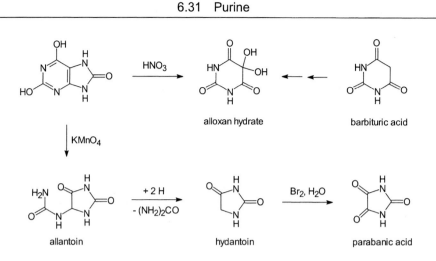

alloxan hydrate   barbituric acid

allantoin   hydantoin   parabanic acid

Its structure was established by an independent synthesis (E. FISCHER and ACH, 1895) from malonic acid and urea giving barbituric acid (see p 404) and its 5-amino derivative **37**. On treatment with potassium cyanate, **37** yielded pseudouric acid **38** which on cyclodehydration gave uric acid **33**:

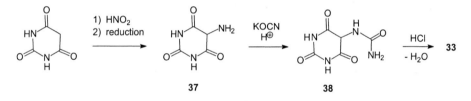

The synthesis of uric acid by cyclization of 4,5-diaminouracil with ethyl chloroformate according to TRAUBE is more convenient.

Uric acid is only sparingly soluble in water and a weak dibasic acid (pK_{a1}= 5.4, pK_{a2}= 10.6). A violet colour appears when uric acid is heated with HNO_3, cooled and treated with NH_3 (Murexide reation). The action of $POCl_3$ on uric acid produces 2,6-dichloro-8-hydroxypurine, and under drastic conditions yields 2,6,8-trichloropurine.

Purine derivatives are of great biological importance. Adenine and guanine occur in the free state, and also as *N*-glycosides (nucleosides, e.g. adenosine **39**) and phosphorylated *N*-glycosides [nucleotides, e.g. adenosine-5-phosphate **40** (AMP), -5-diphosphate **41** (ADP) and -5-triphosphate **42** (ATP)]. They are also constituents of ribo- and deoxyribonucleic acids.

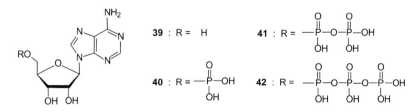

The plant growth regulators (plant hormones) known as cytokinins are adenine derivatives substituted at the 6-amino group [152]. Zeatin **43** is a natural cytokinin isolated from maize. The highly active 6-(benzylamino)purine (cf. **43/44**, R = CH$_2$Ph) has been used commercially (verdan) for keeping vegetables in fresh condition. Kinetin **44** is another synthetic cytokinin used as dermatologic for treatment of age-related photodamage of skin.

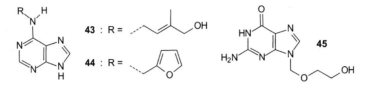

Another medicinally relevant purine is the guanine derivative acyclovir **45**, which is used as an antiviral agent in the treatment of Herpes infections.

Isosteres of purine (e.g. pyrazolo- and pyrolopyrimidines) are of importance as building units of pharmaceuticals. Examples are allopurinol **46** (gout therapeuticum [153]) and sildenafil **47** (inhibitor of type V cGMP phosphodiesterase, Viagra).

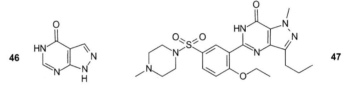

Numerous nucleoside antibiotics are known. For instance, 9-(β-D-arabinofuranosyl)adenine **48** possesses antiviral and antitumor activity, puromycin **49** is an inhibitor of protein biosynthesis in bacteria and mammalian cells, tubercidin **50** (as example of a purine isoster, isolated from Streptomyces) possesses anticancer activity.

Carbocyclic analogues of nucleosides are also biologically active. Aristeromycin **51** possesses antimicrobial and neplanocin **52** antineoplastic activity [154]. Abacavir **53**, one of the most potent synthetic anti-HIV agents, is a selective inhibitor of HIV-1 and HIV-2 replication [155].

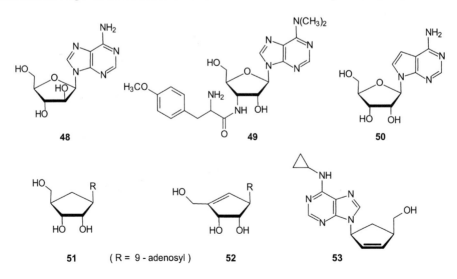

6.32 Pyrazine

A The structure of pyrazine has been determined by X-ray analysis and by electron diffraction. Pyrazine is a planar hexagon with D_{2h} symmetry, the C–C bond lengths being very similar to those in benzene (139.7 pm; see Fig. 6.21).

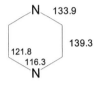

Fig. 6.21 Bond parameters of pyrazine
(bond lengths in pm, bond angles in degrees)

The symmetrical 1,4-diazine structure is reflected in the ^{1}H- and ^{13}C-NMR spectra (CDCl$_3$) with one signal for the ring protons ($\delta = 8.60$) and one for the ring C-atoms ($\delta = 145.9$). Pyrazine shows UV maxima at 261 (3.81), 267 (3.72) and 301 nm (2.88) (H$_2$O).

B As in the other diazines, the reactions of pyrazine are determined by the presence of N-atoms of the ring. They are attacked by electrophiles, e.g. protonation and *N*-oxidation, but strongly deactivate the ring C-atoms. Hence, very few S$_E$Ar processes, e.g. halogenation, take place, and if so, only in moderate yields. Donor substituents, e.g. amino groups, activate the heteroarene and are *ortho*- and *para*-directing. However, this depends on the presence of other substituents. For instance, 2-aminopyrazine **1** readily undergoes nuclear halogenation to give **2**; 3-aminopyrazine-2-carboxylic acid **3** undergoes only *N*-halogenation to **4** and no nuclear substitution:

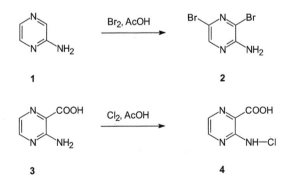

Pyrazine is more reactive than pyridine towards nucleophiles. Although the CHICHIBABIN amination is unsatisfactory, substitution of the halogen in 2-halopyrazine occurs readily by ammonia, amines, amide, cyanide, alkoxide and thiolate.

It is noteworthy that these substitutions often do not occur by a simple addition–elimination pattern.

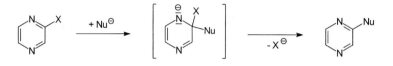

Thus, the reaction of 2-chloropyrazine **5** with sodamide in liquid NH_3 yields 2-aminopyrazine **6** in addition to 2-cyanoimidazole **7**. If N-1 in the starting material **5** is labelled, then the label is found to be in the exocyclic amino group of the product **6**, and not in the ring nitrogen. This shows that the nucleophile does not attack at the ring C-atom which carries the leaving group, but that the substitution proceeds according to an ANRORC mechanism (**A**ddition of **N**ucleophile, **R**ing **O**pening and **R**ing **C**losure) [156]:

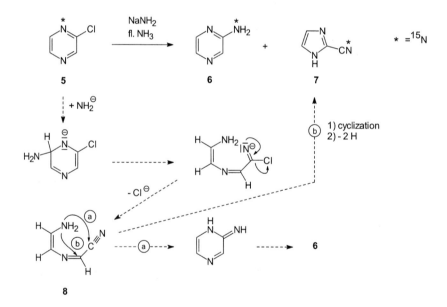

After addition of the amide ion at C-6, fission of the pyrazine ring occurs with elimination of chloride. This produces the cyanoaza-1,3-diene **8** which recyclizes to the pyrazine **6** or to the imidazole **7**. In the case of 2-chloro-3,6-diphenylpyrazine **9**, the intermediate existence of a MEISENHEIMER complex **10** has been demonstrated, which then converts into **11** prior to product formation (**12**):

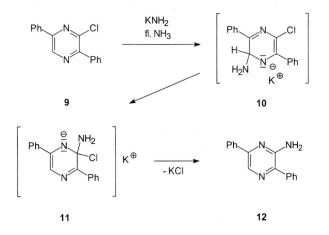

9   **10**

11   **12**

Like the other diazines and pyridine, alkylpyrazines can undergo base catalysed C–C bond-forming reactions of the CH groups adjacent to the heteroatom. 2-Methylpyrazine, after deprotonation with NaNH$_2$ in liquid NH$_3$ via anion **13**, can be alkylated, acylated and nitrosated:

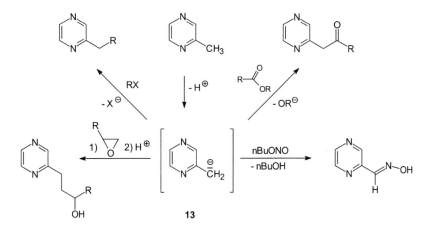

13

Anionization with organolithium compounds is difficult because of competing addition reactions on the ring.

C The *retrosynthesis* (see Fig. 6.22) of the pyrazine system is based on principles proven valid for the other azines. Bond fission at the imine function (retrosynthetic step **a**) leads to 1,2-dicarbonyl compounds **14** and 1,2-diaminoethenes **15** as starting materials for the direct pyrazine synthesis by cyclocondensation. The dihydropyrazines **16** and **18** (from the retroanalysis operation **b** and **c**) are alternative starting materials which are accessible from the 1,2-diaminoethanes **17** and **14** or from the α-amino ketones **19**.

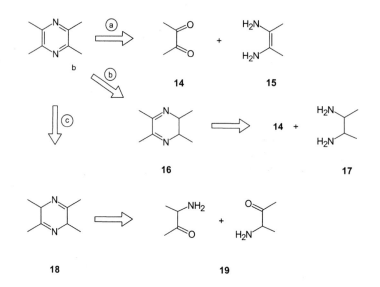

Fig. 6.22   Retrosynthesis of pyrazine

The following methods are convenient for the synthesis of pyrazine.

(1)   1,2-Dicarbonyl compounds and 1,2-diaminoethanes cyclocondense (with double imine formation) to afford 2,3-dihydropyrazines **20** which are conveniently oxidized to pyrazines **21** by CuO or MnO$_2$ in KOH/ethanol:

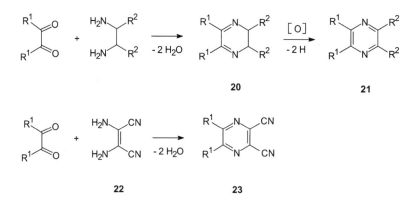

This method yields the best results with symmetrical starting compounds. Diaminomaleonitrile **22** as a diaminoethene condenses directly with 1,2-diketones or reacts with β-keto sulfoxides followed by dehydrogenation [157] yielding 2,3-dicyanopyrazines **23**.

(2)   The classical pyrazine synthesis involves the self-condensation of two molecules of an α-amino-carbonyl compound furnishing 3,6-dihydropyrazine **24** followed by oxidation, usually under mild conditions, to pyrazine **25**:

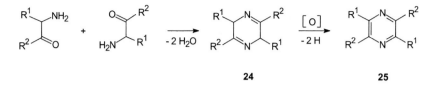

24   25

The required α-amino aldehydes or ketones are usually prepared in situ because of their instability. They are obtained from α-hydroxycarbonyl compounds and ammonium acetate or by catalytic reduction of α-hydroxyimino- or α-azido-carbonyl compounds.

An alternative synthesis of the dihydropyrazine **24** utilizes a cyclizing aza-WITTIG reaction of two molecules of α-phosphazinyl ketones **26** which are accessible from α-azido ketones and triphenylphosphane:

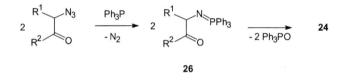

26

(3) Pyridazine derivatives can also be obtained from dioxopiperazines (see p 423) by oxidation with trialkyloxonium salts forming bislactim ethers **27**. On oxidation with DDQ, they are dehydrogenated providing 2,5-dialkoxypyridazines **28**.

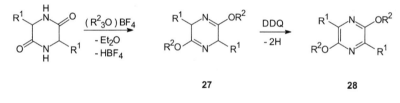

27   28

(4) A regioselective synthesis of alkylpyrazines starts from α-hydroxyimino ketones **29** (BÜCHI 1991 [158]):

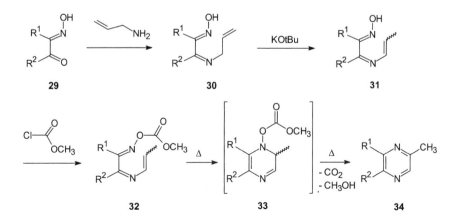

29   30   31

32   33   34

They are condensed with allylamines to the imines **30** that isomerize to the corresponding 1-hydroxy-1,4-diazahexatrienes **31** followed by *O*-acylation (**32**) with chloroformate. Thermal 6π-electro-cyclization of **32** yields the dihydropyrazines **33**, which aromatize to alkylpyrazines **34** by loss of CO_2 and CH_3OH.

D | **Pyrazine**, mp 57°C, bp 116°C, and the simple alkylpyrazines are colourless, water-soluble compounds. Pyrazine has the lowest basicity ($pK_a = 0.6$) of the diazines.

Alkylpyrazines occur frequently as flavour constituents in foodstuffs that undergo heating, e.g. coffee and meat. They are probably formed by a MAILLARD reaction between amino acids and carbohydrates. Alkylpyrazines also act as alarm pheromones in ants. Coelenterazine **35**, a bioluminiscent natural product isolated from a jelly fish, is used in bioassays [159].
A number of pyrazin-2(1*H*)-ones and 1-hydroxypyrazin-2(1*H*)-ones, e.g. **36**, have antibiotic properties.

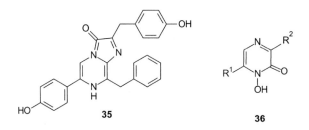

35   **36**

6.33 Piperazine

The structure of piperazine is similar to that of piperidine. Electron diffraction studies show that it prefers a chair conformation with bond lengths C–C 154.0 and C–N 146.7 pm, and bond angles C–C–N 110 and C–N–C 109°. The N–H bonds favour the equatorial position and the same applies to *N*-substituents in *N*-alkyl and *N,N*-dialkylpiperazines.

   The free activation enthalpy of ring inversion is 43.1 kJ mol⁻¹ (piperidine: 43.5; cf. 1-methyl-piperazine 48.1, 1-methylpiperidine 49.8 kJ mol⁻¹). The protons appear at $\delta = 2.84$ in the ¹H- and the carbon atoms at $\delta = 47.9$ in the ¹³C-NMR spectrum ($CDCl_3$). The difference in the chemical shifts between the equatorial and axial protons is 0.16 ppm.

D,E | **Piperazine**, mp 106°C, bp 146°C, behaves as a secondary amine and is a weaker base ($pK_a = 9.8$) than piperidine ($pK_a = 11.2$; cf. morpholine 8.4) due to the inductive effect of the second heteroatom.

   The commercial synthesis of piperazine is carried out from 2-aminoethan-1-ol in the presence of NH_3 at 150–220°C and 100–200 bar, or alternatively from ethylenediamine and oxirane:

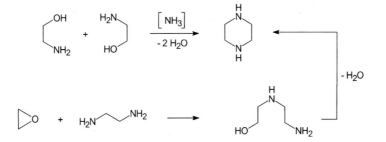

2,5-Dioxopiperazines 1 (diketopiperazines) are formed by a dimerizing cyclocondensation of α-amino acids or their esters

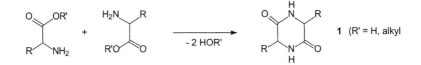

or by cyclization of dipeptide esters resulting in unsymmetrical substitution (**2**):

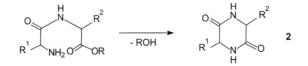

N-Acylated dioxopiperazines undergo base-catalysed C-3 alkylation, e.g.:

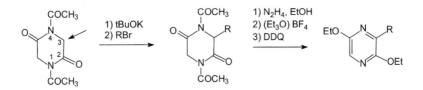

This reaction, when applied to dioxopiperazines, can lead to the synthesis of 2,5-dialkoxypiperazines (see p 421).

O-Alkylation of dioxopiperazines with oxonium salts yields bislactim ethers, e.g. **4**, which are used as reagents for the asymmetric synthesis of amino acids (bislactimether method, SCHÖLLKOPF 1979). The chiral bislactim ether **4** is converted into the 6π-anion **5** (under kinetic control) by *n*-butyllithium. Alkylation proceeds with high stereoselectivity (greater than 95%). Acid hydrolysis of the alkylation product **6** leads to (unnatural) (*R*)-amino acids **7** and recovery of the chiral auxiliary (*S*)-valine, from which the starting material dioxopiperazine **3** was derived [160].

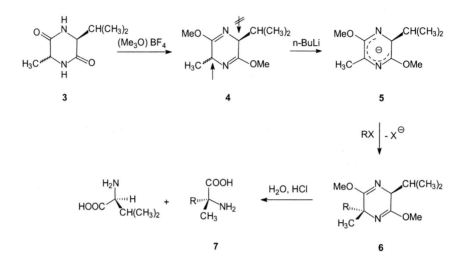

3   **4**   **5**

RX $\Big|$ -X$^\ominus$

7   **6**

Salts of piperazine (citrate, phosphate, adipate) or its hexahydrate are used as anthelminitics. The piperazine ring is frequently used as a building block for pharmaceuticals. Benzhydrylpiperazines, such as cinnarizine **8** and its difluoro analogue flunarizine **9**, possess considerable activity as peripheral and cerebral vasodilators. Hydroxyzine **10** is used as a tranquilizer:

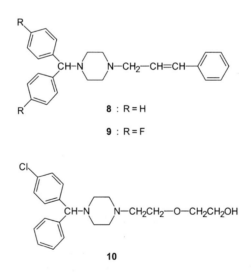

8 : R = H

9 : R = F

10

6.34 Pteridine

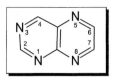

A X-ray structural analysis shows that the two fused six-membered rings in pteridine (pyrazino[2,3-*d*]pyrimidine) are virtually coplanar. Its C–C and C–N bond lengths and C–N–C bond angles (with values less than 120°) are comparable to those of pyrimidine and pyrazine (see Fig. 6.23):

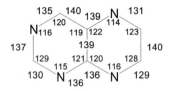

Fig. 6.23   Bond parameters of pteridine
(bond lengths in pm, bond angles in degrees)

The spectroscopic data, especially the strong downfield shifts of the H- and C-atoms in the NMR spectra, are characteristic of pteridine as a strong π-deficient heteroaromatic system. Pteridines are coloured (yellow to red), the parent compound being yellow.

UV (Cyclohexan) λ (nm) (ε)	^{1}H-NMR (CDCl$_3$) δ (ppm)	^{13}C-NMR (CDCl$_3$) δ (ppm)	
263 (3.10)	H-2: 9.65	C-2: 159.2	C-7: 153.0
296 (3.85)	H-4: 9.80	C-4: 164.1	C-9: 154.4
302 (3.87)	H-6: 9.15	C-6: 148.4	C-10: 135.3
390 (1.88)	H-7: 9.33		

B Pteridine is inert [161] towards electrophiles, but reacts with nucleophiles. Because of the reversible addition of water across the N-3/C-4 bond, pteridine can act as an acid or a base in an aqueous system. The amidine configuration in the pyrimidine ring (diazaallyl system) stabilizes the positive charge after protonation and the negative charge after deprotonation:

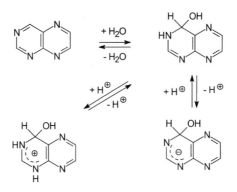

C The synthesis of pteridines usually starts from 4,5-diamino-substituted pyrimidine derivatives (see p 411).

(1) 4,5-Diaminopyrimidines cyclocondense with symmetrical 1,2-dicarbonyl compounds to give pteridines **1** (*Gabriel-Isay synthesis*):

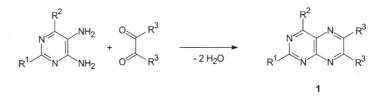

Unsymmetrical 1,2-dicarbonyl compounds lead to isomeric mixtures; α-keto aldehydes react regioselectively and the corresponding α,α-dichloro ketones behave in a complementary manner, e.g.:

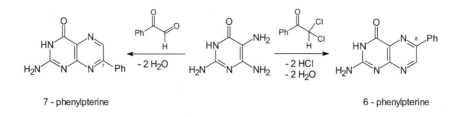

7 - phenylpterine   6 - phenylpterine

(2) 4-Amino-5-nitrosopyrimidines **2** cyclocondense with systems possessing an activated CH_2 group (aldehydes, ketones, nitriles, esters, reactive methylene compounds, phenacylpyridinium salts) providing pteridines **3** on base catalysis (*Timmis synthesis*):

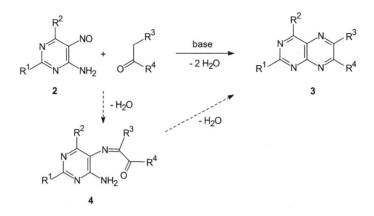

Condensation occurs chemoselectively involving the nitroso group giving the imine **4**, which cyclizes via the remaining NH₂ group with the terminal side chain (C=O, ester, nitrile etc.). The regioselectivity of the Timmis synthesis is demonstrated by condensation of the 4-amino-5-nitrosouracil **6**, leading to 1,3-dimethyl-6-phenyllumazine **5** with phenylacetaldehyde or to the 7-phenyl-isomer **7** with acetophenone:

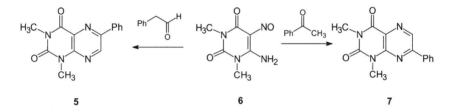

<table>
<tr><td>5</td><td>6</td><td>7</td></tr>
</table>

D   **Pteridine**, mp 139°C, forms yellow crystals and is soluble in all common solvents from petroleum ether to water. Pteridine in an aqueous system (see above) is both weakly acidic (pK_a = 11.2) and weakly basic (pK_a = 4.8).

    Pterin (2-amino-4-hydroxypteridine, which occurs in the lactam form **8**) and **lumazin** (2,4-dihydroxypteridine, which exists as the 1,2,3,4-tetrahydropteridine-2,4-dione structure **9**) are the basic structures of the naturally occurring pteridine derivatives:

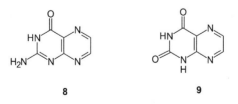

<table>
<tr><td>8</td><td>9</td></tr>
</table>

Pteridines occur as pigments in the wings and eyes of butterflies and other insects (WIELAND, SCHÖPF 1925-1940) and in the skin of fish, amphibians and reptiles. Other examples of butterfly pigments are xanthopterin **10** (brimstone butterfly), leucopterin **11** (cabbage butterfly), and erythropterin **13** (ruby kip butterfly).

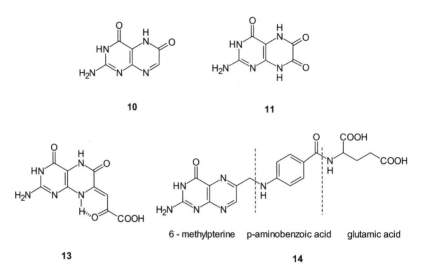

10   **11**

13   **14**

6 - methylpterine   p-aminobenzoic acid   glutamic acid

Folic acid 14 consists of 6-methylpterine, *p*-aminobenzoic acid and (*S*)-glutamic acid residues. It belongs to the vitamin B complexes group and has been isolated from spinach leaves. Folic acid is a growth hormone and effective in the treatment of certain types of anaemia. It is important for the metabolism of amino acids, proteins, purines and pyrimidines. Folic acid forms orange, water-insoluble crystals, mp 250°C (dec). Its synthesis is carried out by condensation of 6-hydroxy-2,4,5-triaminopyrimidine, 1,1,3-trichloroacetone and *N*-(4-aminobenzoyl)-(*S*)-glutamic acid at pH 4-5 and in the presence of NaHSO$_3$:

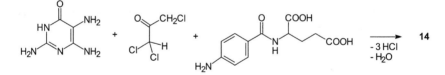

N-(4-Aminobenzoyl)-(*S*)-glutamic acid is obtained by acylation of (*S*)-glutamic acid with 4-nitrobenzoyl chloride followed by reduction of the NO$_2$ group by catalytic hydrogenation.

Pteridines are also used as pharmaceuticals. Triamterene **15** is a diuretic that promotes excretion of Na$^+$ and retains K$^+$. The folic acid antagonist methotrexate **16** (amethopterin) is of considerable importance as an antineoplastic agent in cancer chemotherapy.

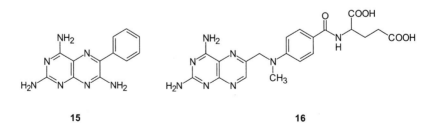

15   **16**

From **benzo[g]pteridine 17** a number of compounds of biological importance is derived.

17

Alloxazine 18 is formed by the cyclocondensation of *o*-phenylenediamine with alloxan (KÜHLING 1891). The tautomeric isalloxazin **19** (flavin) is the parent compound of the 10-substituted derivatives.

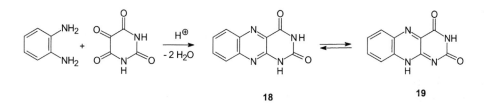

18   19

Alloxazine occurs as yellow needles decomposing above 200°C. It is a weak acid. Lumiflavin (7,8,10-trimethylisoalloxazine) is an irradiation product of lactoflavin (see below) and of the yellow respiratory enzyme.

   Lactoflavin 20 (riboflavin, vitamin B$_2$) was established by KUHN and KARRER (1935). It is synthesized by condensation of 3,4-dimethylaniline with D-ribose giving the imine **21**, followed by reduction to **22** and coupling with benzenediazonium chloride producing the azo compound **23**; reaction of **23** with barbituric acid, after loss of aniline and cyclocondensation of the intermediate **24**, yields lactoflavin **20**.

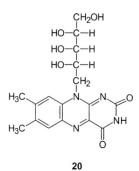

20

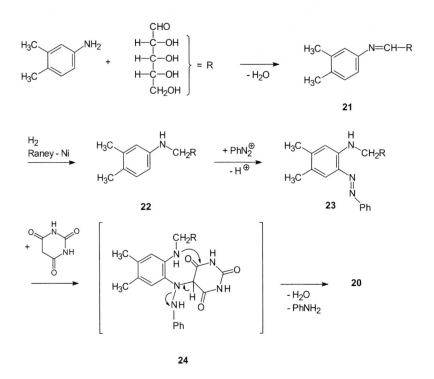

6.35 Benzodiazines

| A |

Cinnoline 1 (benzo[*c*]pyridazine), **phthalazine 2** (benzo[*d*]pyridazin), **quinazoline 3** (benzo[*d*]-pyrimidine) and **quinoxaline 4** (benzopyrazine) are benzene-fused diazines. Of the dibenzo-annulated systems, only **phenazine 5** (dibenzopyrazine) will be discussed.

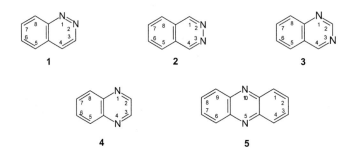

The spectroscopic features of the benzodiazines resemble those of the parent diazines, quinoline and isoquinoline. This is evident from the UV and NMR data:

	UV λ (nm) (ε)	^{1}H-NMR (CDCl$_3$) δ (ppm)	^{13}C-NMR (CDCl$_3$) δ (ppm)	
Cinnoline	(cyclohexane) 276 (3.45) 308 (3.30) 322 (3.34) 390 (2.40)	H-3: 9.15 H-4: 7.75	C-3: 146.1 C-4: 124.7 C-5: 127.9 C-6: 132.3 C-7: 132.1	C-8: 129.5.3 C-4a: 126.8 C-8a: 151.0
Phthalazine	(cyclohexane) 259 (3.67) 290 (3.00) 303 (2.95)	H-1/H-4: 9.44	C-1/C-4: 152.0 C-5/C-8: 127.1 C-6/C-7: 133.2 C-4a/C-8a: 126.3	
Quinazoline	(heptane) 267 (3.45) 299 (3.29) 311 (3.32)	H-2: 9.23 H-4: 9.29	C-2: 160.5 C-4: 155.7 C-5: 127.4 C-6: 127.9	C-7: 134.1 C-8: 128.5 C-4a: 125.2 C-8a: 150.1
Quinoxaline	(heptane) 304 (3.71) 316 (3.78) 339 (2.83) 375 (2.00)	H-2/H-3: 9.74	C-2/C-3: 145.5 C-5/C-8: 129.8 C-6/C-7: 129.9 C-4a/C-8a: 143.2	
Phenazine	(methanol) 248 (5.09) 362 (4.12)	AA'BB' system	C-1/C-4/C-6/C-9: 130.9 C-2/C-3/C-7/C-8: 130.3 C-4a/C-5a/C-9a/C-10a: 144.0	

B Reactions of benzodiazines show no exceptional features compared with the simple diazines. Reactivity towards electrophiles is less than in quinoline and isoquinoline. If S$_E$Ar reactions take place, they lead to substitution of the benzene ring. As a rule, nucleophilic substitution of benzodiazines occur in the diazine ring, particularly if substituted by halogen. The quinazoline system displays C-4 regioselectivity, e.g. in the reactions of 2,4-dichloroquinazoline with amines or alcohols:

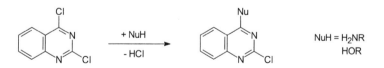

The C-4 reactivity of quinazoline, which is responsible for the anomalous basicity (see p 434), applies to other nucleophiles, e.g. the addition of GRIGNARD compounds, enolates, cyanide or hydrogensulfite:

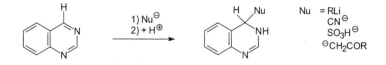

Side-chain reactivity is also observed in benzodiazines. For instance, 4-substituents in quinazolines react selectively, as is shown by the MANNICH reaction of 2,4-dimethylquinazoline:

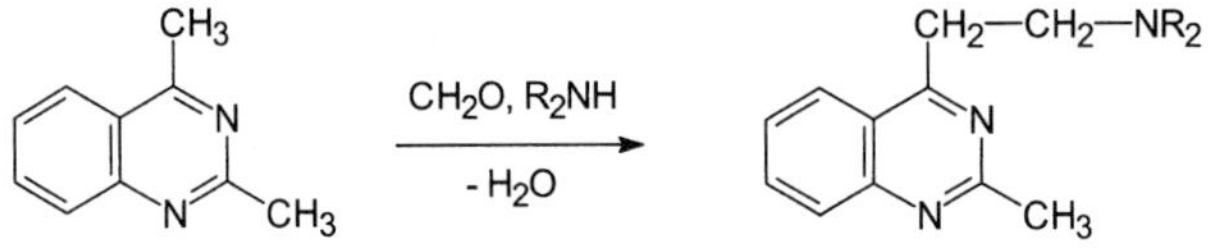

C The following methods are convenient for the synthesis of benzodiazines.

(1) Cinnolines are prepared by an intramolecular cyclization of *o*-alkenyl or *o*-alkynyl aryldiazonium salts. For instance, (*o*-aminophenyl)alkynes **6** or alkyl(*o*-aminophenyl)ketones **8** (via the enol form) yield 4-hydroxycinnolines **7** (the *v. Richter* and *Borsche syntheses*, respectively). *o*-Aminostyrenes **10** afford 3,4-disubstituted cinnolines **11** (*Widman–Stoermer synthesis*):

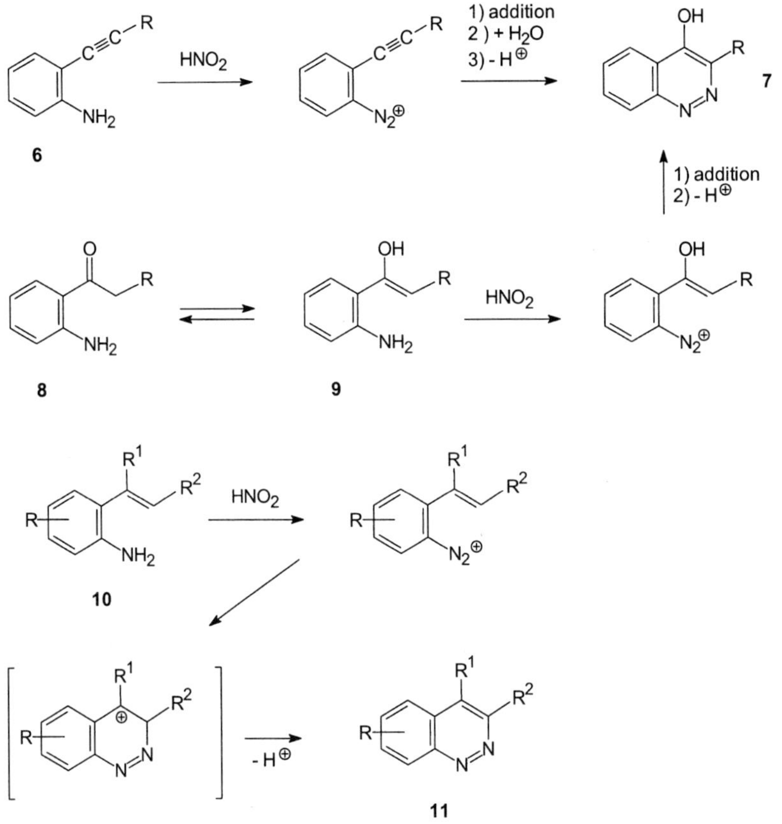

Among these methods, only the WIDMAN-STOERMER synthesis is of preparative importance.

(2) Phthalazine and its 1,4-disubstituted derivatives **13** are obtained by cyclocondensation of *o*-diacylbenzenes **12** with hydrazine; e.g. the parent compound is produced from benzene-1,2-dicarbaldehyde (phthaldialdehyde) and hydrazine hydrate:

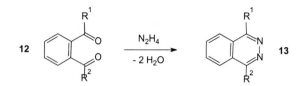

(3) Quinazolines can be prepared by several methods. *N*-Acylanthranilic acids undergo cyclization with ammonia or primary amines via amides **14** forming quinazolin-4(3*H*)-ones **15** (*Niementowski synthesis*). Alternatively *o*-(acylamino)benzaldehydes or -acetophenones react with ammonia producing quinazolines **16** (*Bischler synthesis*):

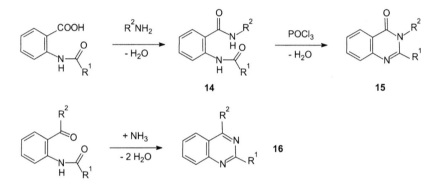

In a recent approach 5-acceptor-substituted 2-fluorobenzaldehydes **17** combine with amidines to give quinazolines **18** in an intramolecular $S_N Ar$ ringclosure process [162].

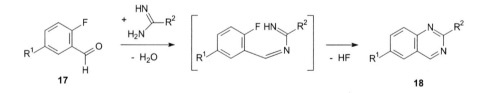

(4) Quinoxalines are obtained by cyclocondensation of 1,2-diaminoarenes with 1,2-dicarbonyl compounds or with α-halocarbonyl compounds followed by dehydrogenation of the dihydroquinoxalines **19**:

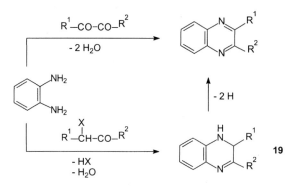

The regioselectivity of the condensation between 1,2-diamines **20** and 2-oxo-2-phenylethanal depends on the basicity of the amino groups and on the conditions of the reaction:

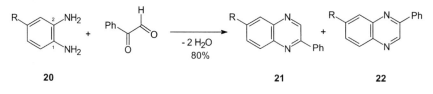

For instance, in a neutral or acid medium, **20** (R = NO$_2$) produces the regioisomeric phenylquinoxalines **21** and **22** (R = NO$_2$) in a ratio of 5:1, because condensation of the aldehyde function with the more basic 2-NH$_2$ group is favoured. However, the diamine **20** (R = OCH$_3$) yields the products **21** and **22** (R = OCH$_3$) in a ratio of 1:8 in a neutral medium, because protonation deactivates the more basic 1-NH$_2$ group [163].

| D,E |

Cinnoline (mp 40°C), **phthalazine** (mp 90°C), **quinazoline** (mp 48°C) and **quinoxaline** (mp 31°C) are colourless crystalline solids. Benzodiazines are weak bases with basicities comparable with those of the corresponding parent heterocycles, with the exception of quinazoline.

cinnoline	pK$_s$ 2.6	for comparison:	pyridazine pK$_s$ 2.1
phthalazine	3.5		
quinazoline	3.3		pyrimidine 1.1
quinoxaline	0.6.		pyrazine 0.4

Quinazoline is much more basic than pyrimidine. This is due to H$_2$O addition at C-4 of the quinazolinium ion (see p 431). In a nonaqueous medium, a pK$_a$ value of 1.95 is found for the equilibrium between the nonhydrated cation and the neutral quinazoline.

On oxidation with alkaline H$_2$O$_2$ in the presence of haemin, luminol **23** (5-aminophthalhydrazide, 5-amino-1,2,3,4-tetrahydrophthalazine-1,4-dione) displays an intensely blue chemiluminescence (see p 47).

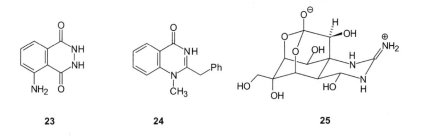

23   **24**   **25**

Several natural products contain the quinazoline structure, e.g. the quinazoline alkaloids obtained from rutaceae (e.g. arborine **24**). Tetrodotoxin **25** derived from decahydroquinazoline and found in the Japanese puffer fish is one of the strongest nonproteinogenic neurotoxins.

Several pharmaceuticals are derived from quinazoline, such as the hypnotic methaqualone **26**, the oral diuretic quinethazone **27**, the analgesic and antirheumatic proquazone **28** and the antihypertensive prazosin **29**:

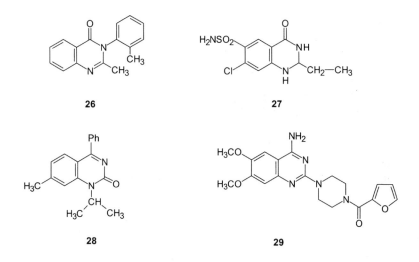

26   **27**

28   **29**

The S_NAr reactivity of the quinazoline system offers synthetic applications for this heterocycle. For instance, 4-chloro-2-phenylquinazoline **30**, by reaction with phenolates, is converted into the 4-(aryloxy)quinazoline **31**. This undergoes a CHAPMAN rearrangement at 300°C with a 1,3-migration of the aryl residue to N-3 forming the 3-arylquinazolin-4(3H)-one **32**. The latter hydrolyses with aqueous acid to form benzoxazinone **33** and a primary arylamine:

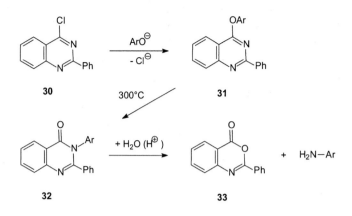

Hence the heterocycle can be used for the conversion of phenols into the corresponding primary arylamines [164], i.e. Ar-OH → Ar-NH₂. The chloroquinazoline **30** is obtained from anthranilic acid and benzamide:

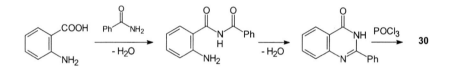

Phenazine crystallizes as bright yellow needles, mp 117°C. It sublimes easily and is a weak base ($pK_a = 1.23$).

Previous syntheses of phenazines proved unsatisfactory. They used the direct condensation of *o*-phenylenediamine with *o*-quinones or catechol, followed by dehydration of the intermediate 5,10-dihydrophenazines **34**:

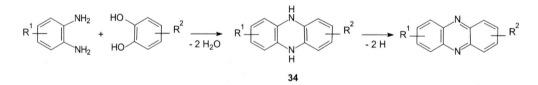

Phenazine derivatives result from 1,2,3-benzoxadiazol-1-oxide and phenols (*Beirut reaction*, cf. p 196 and [165]):

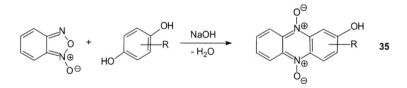

In basic medium and under mild conditions, phenazine-5,10-dioxides **35** are formed. However, from 1,2,3-benzoxadiazol-1-oxide and enolates or enamines quinoxaline-1,4-dioxides arise. The Beirut reaction possesses considerable preparative scope, but its mechanism has not yet been fully elucidated.

Phenazine dyes are of considerable practical importance. Safranine T **37** is synthesized by oxidation of 2,5-diaminotoluene, *o*-toluidine and aniline with sodium dichromate in the presence of hydrochloric acid. The indamine salt **36** is an intermediate:

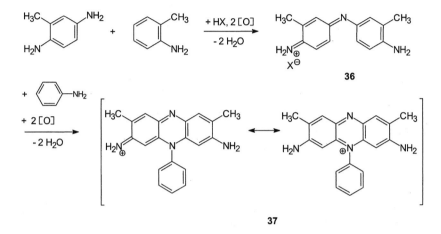

Phenazines and their derivatives have been shown to be of biological significance, especially in the field of photodynamic therapy [166].

6.36 1,2,3-Triazine

A   The parent compound was not synthesized until 1981. X-ray studies of 4,5,6-tris(4-methoxy-phenyl)-1,2,3-triazine show that the 1,2,3-triazine ring is planar (see Fig. 6.24).

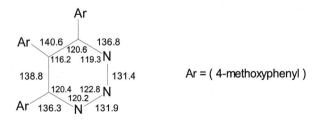

Ar = (4-methoxyphenyl)

Fig. 6.24   Bond parameters of the 1,2,3-triazine system in 4,5,6-tris(4-methoxyphenyl)-1,2,3-triazine (bond lengths in pm, bond angles in degrees)

The NMR spectra of 1,2,3-triazines confirm the symmetrical arrangement of the C–H bonds and ring atoms relative to the three N-atoms.

UV (ethanol) λ (nm) (ε)	^{1}H-NMR (CDCl$_3$) δ (ppm)	^{13}C-NMR (CDCl$_3$) δ (ppm)
233, 288, 325	H-4/H-6: 9.06 H-5: 7.45	C-4/C-6: 149.7 C-5: 117.9

B 1,2,3-Triazines undergo hydrolysis and oxidation. At room temperature, monocyclic 1,2,3-triazines are stable towards aqueous acid; at higher temperature, ring fission occurs to form 1,3-dicarbonyl compounds:

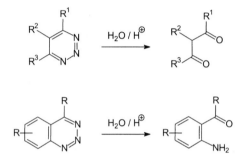

1,2,3-Benzotriazines are ring-opened by aqueous acid at room temperature yielding *o*-aminobenz-aldehydes, whereas 4-substituted derivatives require heating to form *o*-aminoaryl ketones. 1,2,3-Benzotriazin-4(3*H*)-ones **1** suffer acid cleavage of the heterocycle producing the diazonium ion **2** derived from anthranilic acid amide. Subsequent reactions of these ions involving the *N*-amino system **3** lead to the anthranilic acid azide **4** by an internal redox disproportionation:

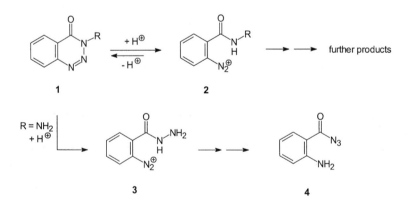

1,2,3-Triazines are oxidized by peroxy acids yielding 1- or 3-oxides. 3-Amino-1,2,3-benzotriazin-4(3*H*)-ones **5** are dehydrogenated at the amino group by lead tetraacetate. The intermediate nitrenes **6** are stabilized by eliminating a molecule of N$_2$ yielding 3*H*-indazolones **7**. Alternatively, by loss of two molecules of N$_2$, they form benzocyclopropenones **8** which are confirmed by trapping [13]:

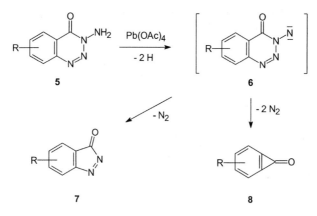

5

6

7

8

C | The following methods are employed for the synthesis of trisubstituted 1,2,3-triazines.

(1)   Oxidation of 1-aminopyrazoles **9**:

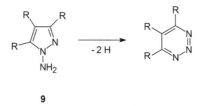

9

This transformation can be effected by lead tetraacetate or nickel peroxide.

(2)   Thermal rearrangement of cyclopropenyl azides **10**:

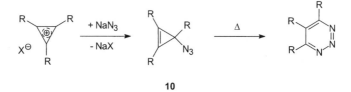

10

This rearrangement takes place even under mild conditions. The cyclopropenyl azides **10** are accessible from cyclopropenylium ions and NaN$_3$.

3-Substituted 1,2,3-benzotriazin-4(3H)-ones, e.g. **12**, in a reversal of the hydrolysis process (cf. p 438), are obtained by cyclization of the diazonium salts derived from the N-substituted anthranilic acid amides **11**:

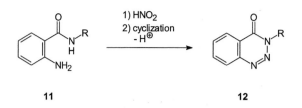

11   12

D 1,2,3-Triazine forms colourless, sublimable crystals, mp 70°C. It is prepared by NiO_2 oxidation of 1-aminopyrazole [167].

The 1,2,3-triazine system has not yet been found in natural products. Some pesticides contain the 1,2,3-benzotriazin-4(3H)-one system, e.g. guthion **13**:

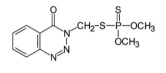

13

6.37 1,2,4-Triazine

A X-ray studies of 5-(p-chlorophenyl)-1,2,4-triazine **1** have shown that the 1,2,4-triazine ring is planar. The structural parameters indicate that the canonical form **a**, with a single bond between N-1 and N-2, is a more important contributor to the resonance hybrid than structure **b** with a double bond between N-1 and N-2.

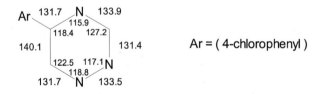

Fig. 6.25 Bond parameters of the 1,2,4-triazine system in 5-(p-chlorophenyl)-1,2,4-triazine (bond lengths in pm, bond angles in degrees)

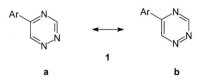

1

a   b

The UV and NMR spectroscopic data of 1,2,4-triazine are characteristic, especially the downfield chemical shift for the 3-position:

UV (methanol) λ (nm) (ε)	^{1}H-NMR (CDCl$_3$) δ (ppm)	^{13}C-NMR (CDCl$_3$) δ (ppm)
248 (3.48)	H-3: 9.73	C-3: 158.1
374 (2.60)	H-5: 8.70	C-5: 149.6
	H-6: 9.34	C-6: 150.8

B Among the reactions of the 1,2,4-triazines, the hetero-DIELS-ALDER reactions with electron-rich alkenes and alkynes are of special importance in preparative chemistry [168]. The heterocyclic ring reacts with enamines, enol ethers and ketene acetals as an electron-deficient 2,3-diazadiene across the ring positions C-3 and C-6:

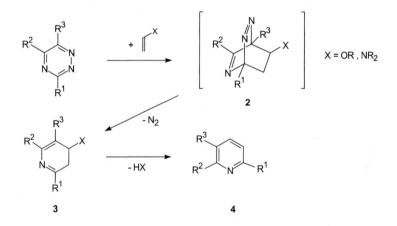

X = OR , NR$_2$

2

3   **4**

The initial product **2** loses N$_2$ in a retro-DIELS-ALDER reaction forming the 3,4-dihydropyridine **3**, which aromatizes giving the pyridine derivative **4** by elimination of amine or alcohol. The geometry of the transition state of this [4+2] cycloaddition with inverse electron demand follows from the reaction of 3- or 6-phenyl-1,2,4-triazine **5** or **8** with enamines of cyclopentanone. It is apparently influenced by the secondary orbital interaction between the amino and phenyl groups. 3-Phenyl-1,2,4-triazine **5** favours the transition state **11**. It leads first to the 3,4-dihydropyridine **6** which, on oxidation followed by a COPE elimination, affords the 2-phenyldihydrocyclopenta[c]pyridine **7**. However, 6-phenyl-1,2,4-triazine **8** favours the transition state **12** leading to 3,4-dihydropyridine **9**. Elimination of amine yields 5-phenyldihydrocyclopenta[c]pyridine **10**:

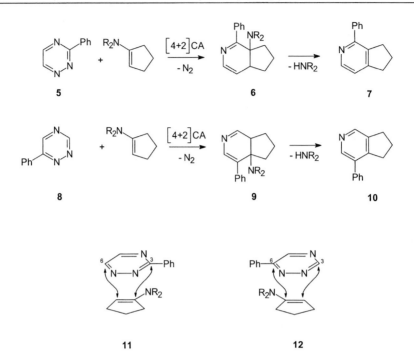

5   6   7

8   9   10

11   12

However, 1,2,4-triazines react as 1,3-diazadienes across the ring positions N-2 and C-5 with alkynylamines, also in a [4+2] cycloaddition:

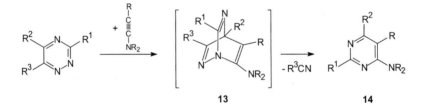

13   14

In a retro-DIELS-ALDER reaction, the initially formed triazabicyclo[2.2.2]octatriene **13** eliminates a unit of nitrile by N–N fission forming the 4-amino-substituted pyrimidine **14**.

> **C** For *retrosynthetic analysis* (see Fig. 6.26) the 1,2,4-triazine system offers two favourable routes.
> Route **a**: After cleavage of the N-4/C-5 bond via the intermediate **15**, route **a** leads to 1,2-dicarbonyls and amidrazones or semicarbazides as possible building-blocks. Alternatively, NH$_3$ elimination from **15** suggests 1,2-dicarbonyl compounds and hydrazides via the intermediate **17** as potential starting materials.
> Route **b**: Fission of the N-1/C-6 bond produces the intermediate **16** as a precursor to hydrazine and α-acylamino ketones, which are suitable starting materials for an oxidative cyclization.

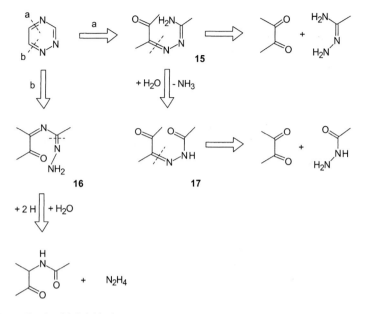

Fig. 6.26   Retrosynthesis of 1,2,4-triazine

The following methods are convenient for the synthesis of 1,2,4-triazines.

(1)   The cyclocondensation of symmetrical 1,2-dicarbonyl compounds with amide hydrazones is universally applicable to the synthesis of 3,5,6-trisubstituted 1,2,4-triazines **18**:

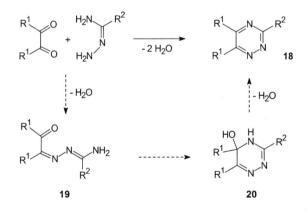

The initial products are the often isolable hydrazones **19**, which undergo cyclization via the hemiaminal **20**. Mixtures of isomeric triazines usually result from unsymmetrical 1,2-dicarbonyl compounds; 3,5-disubstituted triazines **21** are predominantly obtained from 1,2-ketoaldehydes because of their different carbonyl reactivity:

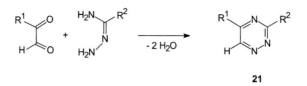

21

Semicarbazide and 1,2-dicarbonyl compounds yield 1,2,4-triazin-3(2*H*)-ones **22**, again via intermediates of the type **19** and **20**:

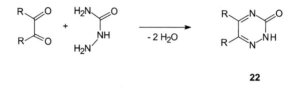

22

Thus α-keto carboxylic acids produce 1,2,4-triazin-5(4*H*)-ones **23** with amide hydrazones, and 2,3,4,5-tetrahydro-1,2,4-triazine-3,5-diones **24** with semicarbazide:

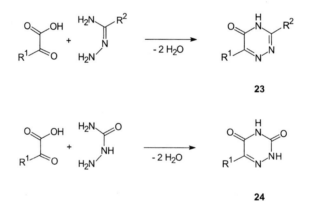

23

24

(2) α-Keto-*N*-acylhydrazones **25**, which are accessible from 1,2-dicarbonyl compounds and acid hydrazides, undergo cyclocondensation with NH$_3$ affording 1,2,4-triazines either directly or after conversion into α-chloroazines **26**.

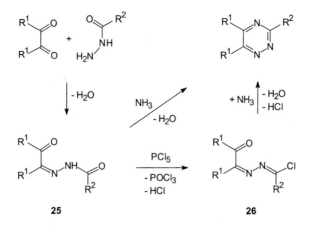

25   **26**

(3) A frequently used method is the cyclocondensation of α-acylamino ketones with hydrazine providing 4,5-dihydro-1,2,4-triazines **27** followed by dehydrogenation to triazines **28**:

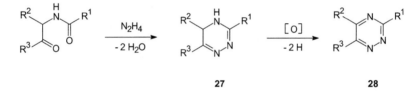

27   **28**

> **D** **1,2,4-Triazine** is a yellow, thermally unstable substance, mp 16°C, bp 158°C. It is prepared from formamidrazone hydrochloride and monomeric glyoxal in the presence of NEt$_3$ [169].

The 1,2,4-triazine system is a constituent of antibiotics of the pyrimido[5,4-*c*][1,2,4]triazine-5,7-dione type, e.g. planomycin **29**. Lamotrigine **30**, a sodium channel blocker, is clinically used as anticonvulsant.

4-Amino-1,2,4-triazin-5(4*H*)-ones are biologically active; some of them are used as herbicides, e.g. metribuzin **31** and metamitron **32**.

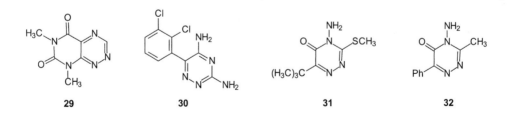

29   **30**   **31**   **32**

6.38 1,3,5-Triazine

A As shown by X-ray studies, the symmetrical 1,3,5-triazine exists as a planar, distorted hexagon with C–N distances of 131.9 pm and bond angles N–C–N of 126.8° and C–N–C of 113.2°.

The H- and C-atoms are strongly deshielded in the NMR spectra (comparable to the 2-position in pyrimidine, see p 398) because of the alternating ring N-atoms [^{1}H-NMR: δ = 9.25; ^{13}C-NMR: δ = 166.1 (CDCl$_3$)]. The UV spectrum (cyclohexane) shows maxima at 218 (2.13) and 272 (2.89) nm. 1,3,5-Triazine has few absorption bands in its IR spectrum due to its D_{3h} symmetry, with four main band at 1555 (E^I), 1410 (E^I), 735 (A$^{II}_2$) and 675 cm^{-1} (E^I).

B 1,3,5-Triazine is inert towards electrophiles. However, nucleophilic attack occurs easily and opens the ring, but in the case of alkyl- and aryl-1,3,5-triazines, only under forced conditions. Typical examples are the reaction of 1,3,5-triazine with primary amines leading to ring fission of the heterocycle into formamidine units **1**,

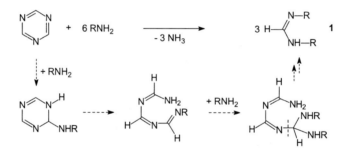

or ring-opening by malonic ester in the presence of secondary amines yielding enamino ester **2**. Thus, the overall reaction corresponds to the transfer of an H–C=N unit onto the reactive methylene component, with an additional amine exchange:

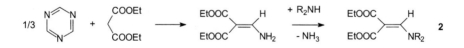

Arenes and heteroarenes are formylated by 1,3,5-triazine in the presence of HCl in an analogous way (HCN-free GATTERMANN synthesis):

For the synthetic potential of 1,3,5-triazine, see [170].

C *Retrosynthetic analysis* reduces the 1,3,5-triazine system by cycloreversion to three HCN or nitrile molecules. Alternatively, bond fission at the C=N units suggests amides and their analogues (imidic esters, amidines) as building-blocks suitable for synthesis.

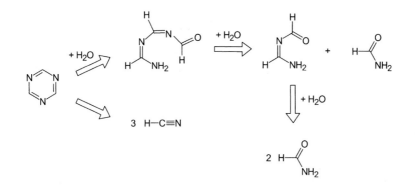

(1)   2,4,6-Trisubstituted 1,3,5-triazines **3** are obtained by acid- (HCl, LEWIS acid) or base-catalysed cyclotrimerization of nitriles:

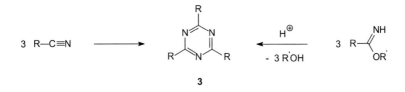

A cyclic transition state is postulated for the acid-catalysed process, whereas the base-catalysed reaction is thought to proceed by a series of nucleophilic additions. Arylcyanides, arylcyanates and cyano compounds with electron-attracting substituents (e.g. ROOC, Cl) are most suitable for the cyclotrimerization. Alkylcyanides trimerize only at high temperature and pressure in moderate yields.

(2)   1,3,5-Triazines of type **3** are also accessible by an acid-catalysed cyclocondensation of imidic esters with elimination of alcohol [see (1)]. The joining of the three imidic ester molecules that function as activated nitriles probably takes place by an electrophilic addition–elimination mechanism involving immonium ions as intermediates. Cyclization of nitriles in the presence of NH_3, catalysed by lanthanum(III) ions, with initial formation of amides, is also effective [171].

(3) 1,3,5-Triazines **5** with three different 2,4,6-substituents are conveniently prepared by cyclocondensation of acylamidines **4** with amidines [172]. Acylamidines result from amides and amide acetals.

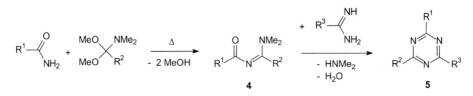

(4) 2,4-Diaryl-1,3,5-triazines **9** are obtained from 5-methoxymethylene-2,2-dimethyl-1,3-dioxane-4,6-dione **6** (easily accessible from MELDRUM's acid, p 386) by sequential reaction with two molecules of arylamidine [173]. Triazine formation is thought to proceed via condensation of the first amidine with **6** to give **7** and addition of the second amidine moiety to give **8**, which unexpectedly cyclizes to **9** with elimination of NH_3 and MELDRUM's acid.

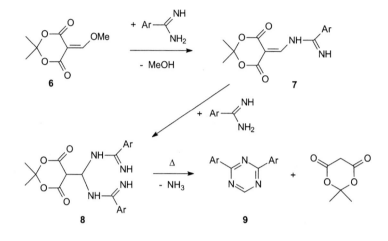

D,E **1,3,5-Triazine**, mp 80°C, forms colourless needles. It is prepared by thermal cyclocondensation of formamidine acetate and triethyl orthoformate [174]:

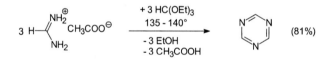

Cyanuric chloride (2,4,6-trichloro-1,3,5-triazine), mp 145°C, bp 190°C, is produced commercially by gaseous trimerization of cyanogen chloride over charcoal catalysts. Cyanuric chloride behaves as a heterocyclic acid chloride analogue. Its reactive chlorine atoms are easily substituted (also stepwise) by S_NAr processes:

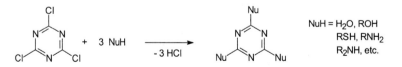

For this reason, cyanuric chloride is used as a chlorinating and dehydrating agent in organic synthesis. It transforms (preferably secondary) alcohols into alkyl halides, carboxylic acids into acid halides, hydroxy carboxylic acids into lactones and aldoximes into nitriles:

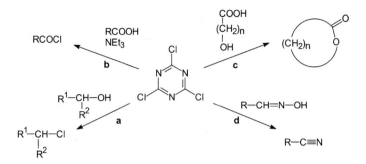

The reactions **a–c** can be interpreted as proceeding by an initial addition of the oxygen nucleophile to the chlorotriazine followed by loss of the alkyl halide with *O*-dealkylation. This sequence leads to the product and to the formation of dichlorotriazin-2(1*H*)-one **10**. Repetition of this sequence involving all available chlorine atoms finally produces cyanuric acid (see below):

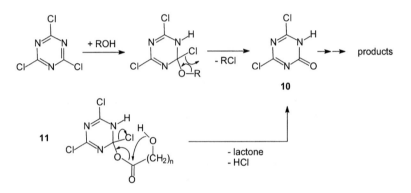

The intermediate **11** is thought to account for the lactone formation (**c**).

Cyanuric chloride is also used for the introduction of a 'link group' into dyes. The S_NAr reactivity of the chlorotriazine system leads to covalent bonding with the OH groups of cellulosic fibres, e.g. cotton (reactive dyes, e.g. cibacron and procion dyes, e.g. **12** [175]):

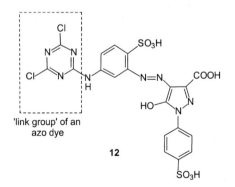

'link group' of an
azo dye

12

Finally, cyanuric chloride is used as educt for the production of the 1,3,5-triazine herbicides which have two (often differently substituted) amino groups in the 2- and 4-positions, and Cl, OCH_3 or SCH_3 substitutents in position 6. Important examples are simazine and atrazine **13** and **14**:

13 : R = C_2H_5

14 : R = $CH(CH_3)$;

1,3,5-Triazine herbicides are photosynthesis inhibitors. They interrupt the light-induced electron transfer of water to NADP$^{\oplus}$.

Melamine (2,4,6,-triamino-1,3,5-triazine) is obtained by trimerization of cyanamide or commercially by thermal cyclocondensation of urea at 400°C with elimination of NH_3 and CO_2. Polycondensation of melamine with formaldehyde produces melamine resins that are used as plastics, glues and adhesives.

Cyanuric acid 15 (2,4,6-trihydroxy-1,3,5-triazine) was obtained by SCHEELE (1776) from pyrolysis of uric acid and was the first 1,3,5-triazine derivative known. Cyanuric acid is synthesized by trimerization of isocyanic acid and is tautomeric with isocyanuric acid **16**:

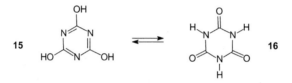

O-Alkyl derivatives of cyanuric acid (cyanurates) are formed by trimerization of cyanates or by the action of alcoholate on cyanuric chloride (see above). *N*-Alkyl derivatives of isocyanuric acid are produced by trimerization of isocyanates. Tris-*O*-allyl cyanurate, which is produced commerically from cyanuric chloride and allyl alcohol, is a co-monomer for the production of polymers used in high temperature insulation.

6.39 1,2,4,5-Tetrazine

| A | According to X-ray studies, 1,2,4,5-tetrazine is planar. Its structural parameters, bond lengths C–N 133.4 and N–N 132.1 pm, bond angles N–C–N 127.2 and C–N–N 115.6°, indicate a delocalized structure, which is best described by the two equal-energy KEKULÉ structures **1a** and **1b**:

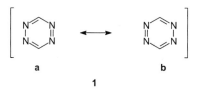

The H- and C-atoms of 1,2,4,5-tetrazine are more strongly deshielded than those of 1,3,5-triazine: ^{1}H-NMR $\delta = 11.05$, ^{13}C-NMR $\delta = 161.2$ (acetone-d$_6$).

1,2,4,5-Tetrazines are coloured (red-violet). They have UV/VIS maxima in the visible region at 520-570 mn (n $\rightarrow \pi^*$ transition), and in the UV region at 250-300 nm ($\pi \rightarrow \pi^*$ transition). They have a characteristic simple fragmentation pattern in the mass spectrum:

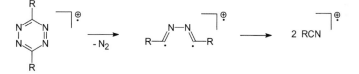

| B | Even more so than 1,2,4-triazines (see p 441), 1,2,4,5-tetrazines display heterodiene activity in their reactions towards electron-rich, multiply-bonded systems. Enol ethers, enamines, ketene acetals, imido esters, alkynylamines and nitriles undergo [4+2] cycloadditions with inverse electron demand across the ring positions C-3 and C-6 [176]. Olefinic dienophiles lead to diverse products depending on their substituents:

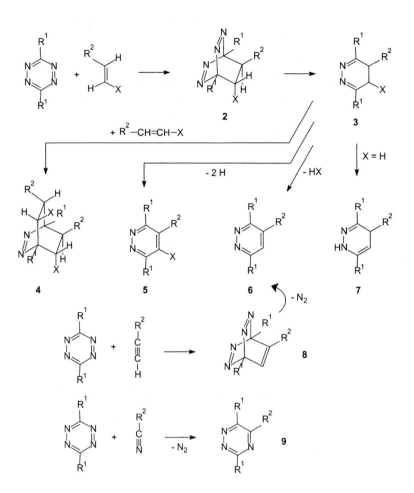

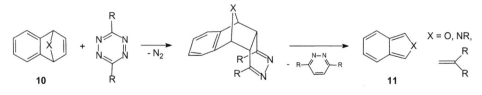

As a rule, the initial hetero-DIELS-ALDER adduct **2** cannot be isolated. It eliminates N_2 in a retro-DIELS-ALDER reaction and is converted into a 4,5-dihydropyridazine **3**. This can be stabilized as a 1,4-dihydropyridazine **7** (especially if X = H) by a 1,5 hydrogen shift or (if X = OR and NR_2) as the pyridazines **5** and **6** by dehydrogenation or HX elimination. As a diazadiene, it can also engage in a further DIELS-ALDER reaction with excess of alkene yielding the stable 2,3-diazabicyclo[2.2.2]oct-2-ene **4**. The initial DIELS-ALDER product tetraazabicyclo[2.2.2]octatriene **8**, which arises from the reaction between alkynes and 1,2,4,5-tetrazines, undergoes a cycloreversion with N_2 elimination affording the pyridazine **6**. With nitriles, 1,2,4-triazines **9** are obtained.

The [4+2] cycloaddition of 1,2,4,5-tetrazines can be used for the synthesis of other heterocycles, e.g. isobenzofuran, isoindole and isobenzofulvene from 1,4-dihydronaphthalenes **10**.

In this reaction, the 1,2,4,5-tetrazine formally effects the extrusion of an acetylene unit and thereby converts the benzenoid starting compound **10** into the quinonoid product **11**.

C   The *retrosynthesis* of 1,2,4,5-tetrazine can be carried out by two routes. In route **a** (hydrolysis as a retrosynthetic operation), hydrazine is eliminated followed by reduction of the remaining N=N bond in **12**. In route **b**, the operations are reversed. Both approaches suggest 1,2-diacylhydrazines and hydrazine as starting materials. Alternatively, the intermediate **13** can be split into hydrazine and carboxylic acid by a C–N bond cleavage.

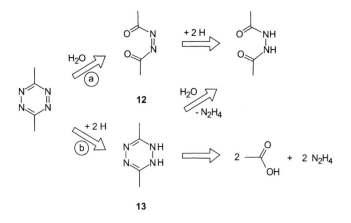

Hence, the synthesis of 1,2,4,5-tetrazines can be undertaken by two methods.

(1)   1,4-Dichloroazines **14** cyclocondense with hydrazine giving dihydro-1,2,4,5-tetrazines **16**, which on dehydrogenation yield symmetrically or unsymmetrically substituted 1,2,4,5-tetrazines **15**:

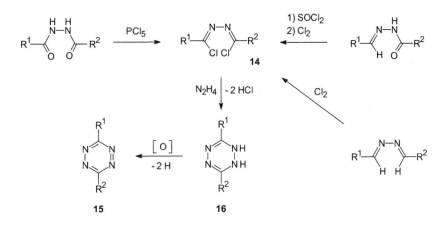

1,4-Dichloroazines are accessible by chlorination of either 1,2-diacylhydrazines with PCl$_5$, of acylhydrazones with SOCl$_2$/Cl$_2$ or of aldazines with Cl$_2$.

(2) Symmetrically substituted 1,2,4,5-tetrazines **20** are obtained by the interaction of nitriles and hydrazine:

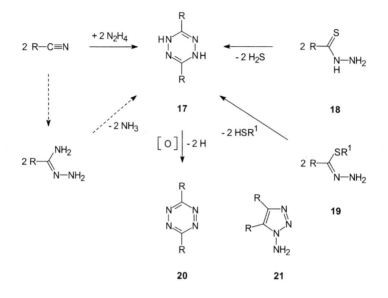

Addition of hydrazine to the nitrile function leads to amidrazones, which cyclocondense by dimerization yielding 1,4-dihydro-1,2,4,5-tetrazines **17**; their dehydrogenation produces **20**. In an alternative preparation, thiohydrazides **18** or *S*-alkyl isothiohydrazides **19** are oxidatively dimerized, also via 1,4-dihydro-1,2,4,5-tetrazines **17**, to afford 1,2,4,5-tetrazines, with 1-aminotriazoles **21** as byproducts.

> **D** **1,2,4,5-Tetrazine**, crimson crystals, mp 99°C, is formed by thermal decarboxylation of 1,2,4,5-tetrazine-3,6-dicarboxylic acid (HANTZSCH 1900).

Verdazyls 24 are intensely green azaallyl radicals of a tetrahydro-1,2,4,5-tetrazine type. They are obtained by the H$^+$-catalysed cyclocondensation of aldehydes, usually formaldehyde, with formazanes **22** and reduction of the initially produced verdazylium ions **23** in a basic medium (KUHN 1963):

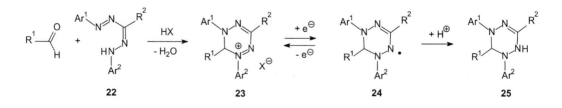

Further reduction of the verdazyls leads to 1,2,3,4-tetrahydro-1,2,4,5-tetrazines **25** [177]. Verdazyls **24** can be reoxidized to the verdazylium ions **23** by *o*-quinones in an SET reaction. The *o*-quinones are thereby converted into the corresponding *o*-semiquinones.

Verdazyls of the type **27** are very stable. They are produced by dehydrogenation of 1,4,5,6-tetrahydro-1,2,4,5-tetrazin-3(2*H*)-ones or -thiones **26** [178]:

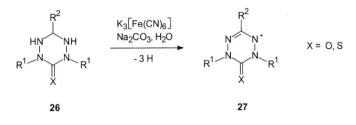

| **26** | **27** |

The imidazotetrazinone temozolomide **28** is medicinally applied as cancer therapeuticum, the cyclic triazene moiety is considered to produce a cascade of ionic and radical species with antitumoral effects [179].

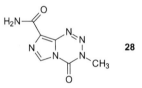

28

Summary of the general chemistry of six-membered heterocycles :

- Pyridine, the diazines, 1,3,5-triazine, 1,2,4,5-tetrazine, quinoline, isoquinoline, quinolizinium ion and the benzodiazines are Hückel-aromatic systems (heteroarenes). 2*H*-Pyrane, 4*H*-pyrane and the corresponding benzo compounds, 1,4-dioxin, 1,4-dithiin and 1,4-oxathiin are not aromatic. However, the cations derived therefrom, like pyrylium and thiinium cations or dioxinium and dithiinium dications are aromatic; the same is true for the corresponding benzo or dibenzo systems (e.g., benzopyrylium). Phosphinin is characterized by a specific heteroaromatic situation.

- An important role play carbonyl-bearing heterocycles, e.g. 2*H*- and 4*H*-pyranones, pyridones, diazinones and the benzo systems derived thereof, as well as hydrogenated heteroarenes like tetrahydropyrane, chromane, piperidine, morpholine and piperazine.

- Pyridine, benzazines, diazines and their benzo derivatives are basic and can be attacked by electrophiles at the N-atoms in alkylation, acylation and *N*-oxide formation.

- Concerning S_EAr reactions, N-atoms deactivate the arene system in pyridine and especially in the diazines. Thus, pyridine substitution requires drastic conditions and occurs in the 3-position; for S_EAr reactions of diazines donor substituents are prerequisite with only few exceptions, and in benzazines only the benzene ring is attacked in electrophilic substitutions.

- Concerning S_NAr reactions, pyridine and the benzazines show enhanced reactivity in comparison to benzene and give regioselective substitution reactions even with hydrogen as leaving group (e.g.

CHICHIBABIN and ZIEGLER reaction). In the presence of good leaving groups, diazines in general are more reactive in S_NAr reactions than pyridine; often substitution does not occur by simple addition/elimination processes, but by ANRORC mechanisms (cf. p 418).

Azines and diazines are capable of preparatively useful metalation reactions with lithium organyls [180].

- Heteroarenes show specific side-chain reactivity at heterobenzylic C-H-bonds in position 2 or 4 to the ring hetero atom. This is responsible for a large number of base-catalyzed electrophilic C-C forming processes like aldol, CLAISEN and MANNICH reactions with appropriately substituted pyrylium and pyridinium ions, pyridines, benzazines and benzodiazines.

- $2H$-Pyranes, $2H$-pyran-4-ones, 1,2,4-triazines and 1,2,4,5-tetrazines may serve as 1,3-diene components in DIELS-ALDER reactions of normal or inverse electron demand, in many cases valuable additional transformations are possible with the primary products.

  Pyridine and the diazines are capable of thermal and photochemical valence tautomerizations in analogy to benzene.

- Among the methods for synthesis of six-membered heteroarenes, cyclocondensations are applied in a nearly universal scope, less frequently used are cycloaddition (e.g. for pyridines) and ring transformation reactions (e.g. for quinolines and isoquinolines). For partially unsaturated and saturated heterocycles special methods (e.g. S_Ni processes for piperidines, 1,4-dioxanes or morpholines) are available.

- Nearly all types of six-membered heterocycles are found as structural units of natural products (e.g. alkaloids) and are used as building blocks for pharmaceuticals and other products of biological activity (biocides). Many other fields for technical application of six-membered heterocycles are known.

- Applications as vehicles for organic syntheses are found, for instance, in the chemistry of pyrylium and pyridinium ions, e.g. syntheses for cyanines, azulenes, specifically substituted benzene derivatives and ring transformations to other six-membered heteroarenes with N, P and S atoms.

References

[1] A. T. Balaban, A. Dinculescu, G. N. Dorofeenko, G. W. Fischer, A. V. Koblik, V. V. Mezheritskii, W. Schroth, *Adv. Heterocycl. Chem.* **1982**, Suppl. 2, 1.

[2] W. Schroth, *Revue Roum. Chim* **1989**, *34*, 271.

[3] A. R. Katriztky, S. A. Henderson, B. Yang, *J. Heterocycl. Chem.* **1998**, *35*, 1123; A. R. Katritzky, X. Lan, J. Z. Yang, O. V. Denisko, *Chem. Rev.* **1998**, *98*, 409.

[4] A. R. Katritzky, P. Czerney, J. L. Levell, *J. Org. Chem.* **1997**, *62*, 8198.

[5] J. Liebscher, H. Hartmann, *Synthesis* **1979**, 241.

[6] L. F. Tietze, Th. Eicher, *Reactions and Synthesis in the Organic Chemistry Laboratory*, University Science Books, Mill Valley, CA 1989, p 348.

[7] F. Klages, H. Träger, *Chem. Ber.* **1953**, *86*, 1327.

[8] *Organic Syntheses*, Coll. Vol. V, **1973**, 1088.

[9] see [6], p 349.

[10] J. R. Wilt, G. A. Reynolds, J. A. van Allan, *Tetrahedron* **1973**, *29*, 795.

[11] P. J. Brogden, C. D. Gabbutt, J. D. Hepworth in: A. R. Katritzky, C. W. Rees, A. J. Boulton, A. McKillop (Eds.), *Comprehensive Heterocyclic Chemistry* Vol.3, p 634ff, Pergamon, Oxford **1984**.

[12] O. L. Chapman, C. L. McIntosh, J. Pacansky, *J. Am. Chem. Soc.* **1973**, *95*, 614.

[13] Th. Eicher, J. L. Weber, *Top. Curr. Chem.* **1975**, *57*, 1.

[14] S. Rousset, M. Abarbri, J. Thibonnet, A. Duchene, J.-L. Parrain, *Chem. Commun.* **2000**, 1987.

[15] Chemistry of dehydroacetic acid: M. Moreno-Manas, R. Pleixats, *Adv. Hetercycl. Chem.* **1992**, *53*, 1

[16] S. F. Martin, H. Rueger, S. A. Williamson, S. Greiszczak, *J. Am. Chem. Soc.* **1987**, *109*, 6124.

[16a] H. M. Sampath Kumar, B. V. Subba Reddy, E. J. Reddy, J. S. Yadav, *Chem. Lett.* **1999**, 857.

[17] J. Sauer, R. Sustmann, *Angew. Chem. Int. Ed. Engl.* **1980**, *19*, 779; Application in natural product synthesis: T. Kametani, S. Hibino, *Adv. Heterocycl. Chem.* **1987**, *42*, 246.

[18] C. Semeyn, R. H. Blaauw, H. Hiemstra, W. N. Speckamp, *J. Org. Chem.* **1997**, *62*, 3426.

[19] L. F. Tietze, T. Brumby, S. Brand, M. Bratz, *Chem. Ber.* **1988**, *121*, 499; see [6], p 442.

[20] L. F. Tietze, *Angew. Chem. Int. Ed. Engl.* **1983**, *22*, 828.

[21] N. Greeves, W.-M. Lee, *Tetrahedron Lett.* **1997**, 6449.

[22] P. T. Ho, *Can. J. Chem.* **1982**, *60*, 90.

[23] A-Pelter, A. Hussain, G. Smith, R. S. Ward, *Tetrahedron* **1997**, *53*, 3879.

[24] J. Staunton in: D. Barton, W. D. Ollis (Eds.), *Comprehensive Organic Chemistry,* Vol. 4., p 629ff, Pergamon, Oxford **1979**.

[24a] T. Sugino, K. Tanaka, *Chem. Lett.* **2001**, 110.

[25] G. A. Kraus, J. O. Pezzanite, *J. Org. Chem.* **1979**, *44*, 2280.

[26] B. M. Trost, F. D. Toste, *J. Am. Chem. Soc.* **1996**, *118*, 6305.

[27] Th. Eicher, H. J. Roth, *Synthese, Gewinnung und Charakterisierung von Arzneistoffen*, p 300, Thieme, Stuttgart **1986**.

[28] R. D. H. Murray, J. Mendez, S. A. Brown, *The Natural Cumarins*, Wiley, New York **1982**; R. D. H. Murray, *Nat. Prod. Reports* **1989**, 591.

[29] see [27], p 160.

[30] E. V. Kuznetsov, I. V. Shcherbakova, A. T. Balaban, *Adv. Heterocycl. Chem.* **1990**, *50*, 158.

[31] see [6], p 351.

[32] J. B. Harborne, H. Baxter (Eds.), *The handbook of natural flavonoids*, Vol 1 and 2, Wiley-VCH, Weinheim **1999**.

[33] E. Bayer, H. Egeter, A. Fink, K. Netker, K. Wegmann, *Angew. Chem. Int. Ed. Engl.* **1966**, *5*, 791.

[34] J. N. Collie, *J. Chem. Soc.* **1907**, 1806.

[35] D. Nagarathan, M. Cushman, *Tetrahedron* **1991**, *47*, 5071.

[36] see [6], p 350; see also H. J. Laas, Th. Eicher, *Journ. Hattori Bot. Lab.* **1989**, *67*, 383.

[37] H. Miao, Z. Yang, *Org. Letters* 2000, *2*, 1765.

[38] H. D. Zinsmeister, H. Becker, Th. Eicher, *Angew. Chem. Int. Ed. Engl.* **1991**, *30*, 130; see also [32].

[39] D. S. Zaharko, C. K. Grieshaber, J. Plowman, J. C. Cradock, *Cancer Treatment Reports* **1986**, *70*, 1415; G. Wurm, *Deutsche Apotheker-Zeitschrift* **1990**, *130*, 2306.

[40] S. Hünig, *Angew. Chem. Int. Ed. Engl.* **1964**, *3*, 348.

[41] N. Cohen, R. J. Lopresti, C. Neukom, *J. Org. Chem.* **1981**, *46*, 2445.

[42] W. E. Parham, L. D. Jones, Y. A. Sayed, *J. Org. Chem.* **1976**, *41*, 1184; K. J. Hodgetts, *Tetrahedron Lett.* **2000**, *41*, 8655.

[43] M. Palucki, J. P. Wolfe, S. L. Buchwald, *J. Am. Chem. Soc.* **1996**, *118*, 10333.

[44] S. Takano, T. Sugikara, K. Ogasawara, *Synlett* **1990**, 451.

[45] K. B. Wiberg, D. Nakaji, C. N. Breneman, *J. Am. Chem. Soc.* **1989**, *111*, 4178; also: electronic structure of diazines, 1,3,5-triazine and 1,2,4,5-tetrazine.

[46] J. E. Del Bene, *J. Am. Chem. Soc.* **1979**, *101*, 6184.

[46a] U. Ragnarsson, L. Grehn, *Acc. Chem. Res.* **1998**, *31*, 494.

[47] A. R. Katritzky, R. Taylor, *Adv. Heterocycl. Chem.* **1990**, *47*, 1;
A. R. Katritzky, W.-Q. Fan, *Heterocycles* **1992**, *34*, 2179.

[48] J. M. Bakke, E. Ranes, J. Riha, H. Svensen, Acta Chem. Scand. **1999**, *53*, 141;
J. M. Bakke, J. Riha, ibid., 356.

[49] F. M. Saidova, E. A. Filatova, *J. Org. Chem. USSR* (Engl.Trans.) **1977**, *13*, 1231.

[50] A. F. Pozharski, A. M. Simonow, V. N. Doron'kin, *Russ. Chem. Rev.* (Engl. Trans.) **1978**, *47*, 1042;
C. K. McGill, A. Rappa, *Adv. Heterocycl. Chem.* **1988**, *44*, 1.

[50a] For further examples see:
B. H. Lipshuts, S. S. Pfeiffer, W. Chrisman, *Tetrahedron Lett.* **1999**, *40* 7889;
P. Gros, Y. Fort, *J. Chem. Soc. Perkin Trans. 1*, **1998**, 3515

[51] D. Bryce-Smith, P. L. Morris, B. J. Wakefield, *J. Chem. Soc., Perkin Trans. 1*, **1976**, 1331.

[52] A. N. Johnson, *Ylid Chemistry*, Academic, New York **1966**, p 260.

[53] see [6], p 335.

[54] E. Ciganek, *Org. React.* **1997**, *51*, 201.

[55] A. R. Katritzky, J. M. Lagowski, *Heterocyclic N-Oxides*, Methuen, London **1971**;
A. R. Katritzky, J. N. Lam, *Heterocycles* **1992**, *33*, 1011.

[56] J. Yamamoto, M. Imagawa, S. Yamauchi, O. Nakazawa, M. Umezu, T. Matsuura, *Tetrahedron* **1981**, *37*, 1871.

[57] W. Feely, W. L. Lehn, V. Boekelheide, *J. Org. Chem.* **1957**, *22*, 1135.

[58] M. L. Ash, R. G. Pews, *J. Heterocycl. Chem.* **1981**, *18*, 939.

[59] M. G. Barlow, R. N. Hazeldine, J. G. Dingwall, *J. Chem. Soc., Perkin Trans. 1*, **1973**, 1542.

[60] K. Weissermel, H.-J. Arpe, *Industrielle Organische Chemie*, 3rd ed., p 202, VCH, Weinheim **1988**.

[61] W. H. F. Sasse, *Org. Synth. Coll. Vol. V*, **1973**, 102.

[62] Bisquaternary salts of 4,4'-bipyridyl type e.g. **137** also serve as electron relays in the photoreduction of water: L. A. Summers et al., *J. Heterocycl. Chem.* **1991**, *28*, 827.

[63] K. Deuchert, S. Hünig, *Angew. Chem. Int. Ed. Engl.* **1978**, *17*, 875.

[64] R. Hamilton, M. A. McCervey, J. J. Rooney, *J. Chem. Soc., Chem. Commun.*, **1976**, 1038.

[65] F. Bossert, H. Meyer, E. Wehinger, *Angew. Chem. Int. Ed. Engl.* **1981**, *20*, 762.

[66] A. S. Kiselyov, *Tetrahedron Lett.* **1995**, *36*, 9297. According to a similar reaction principle 2,3,4,6-substituted pyridines are obtained from phosphonate α-carbanions by sequential addition

of nitriles and α,β-unsaturated carbonyl compounds: F. Palacios, A. M. O. de Retana, J. Oyarzabal, *Tetrahedron Lett.* **1996**, 4577.

[67] H. Bönnemann, *Angew. Chem. Int. Ed. Engl.* **1978**, *17*, 505;
H. Bönnemann, W. Brijoux, *Adv. Heterocycl. Chem.* **1990**, *48*, 177;
A. W. Fatland, B. E. Eaton, *Org. Lett.* **2000**, *122*, 4994.

[68] T. Takahashi, F.-Y. Tsai, M. Kotora, *J. Am. Chem. Soc.* **2000**, *122*, 4994.

[69] M. Ohba, R. Izuta, E. Shimizu, *Tetrahedron Lett.* **2000**, 10251.

[70] N. Clauson-Krus, P. Nedenskov, *Acta Chem. Scand.* **1955**, *9*, 14.

[71] In the synthesis of pyridine alkaloids, recent progress was brought about by palladium chemistry: J. J. Li, *Prog. Heterocycl. Chem.*, Vol. 12, **2000**, p0 37.

[72] S. Goldmann, J. Stoltefuß, *Angew. Chem. Int. Ed. Engl.* **1991**, *30*, 1559; see [27], p 168.

[73] see [6], p 304

[74] R. H. Reuss, L. J. Winters, *J. Org. Chem.* **1973**, *38*, 3993.

[75] K. Saigo, M. Usui, K. Kikuchi, E. Shimada, T. Mukaiyama, *Bull. Chem. Soc. Jpn.* **1977**, *50*, 1863; see [6], p 383.

[76] T. Mukaiyama, R. Matsueda, M. Suzuki, *Tetrahedron Lett.* **1970**, 1901.

[77] U. Schmitt, D. Heermann, *Angew. Chem.* **1979**, *91*, 330.

[78] W. Ried, H. Bender, *Chem. Ber.* **1956**, *89*, 1893;
W. Ried, R. M. Gross, *Chem. Ber.* **1957**, *90*, 2646.

[79] A. R. Katritzky, *Tetrahedron* **1980**, *36*, 679.

[80] A. R. Katritzky, A. Saba, R. C. Pabel, *J. Chem. Soc., Perkin Trans. 1*, **1981**, 1462.

[81] C. D. Johnson in: A. R. Katritzky, C. W. Rees, A. J. Boulton, A. McKillop (Eds.), *Comprehensive Heterocyclic Chemistry*, Vol. 2, p 147ff, Pergamon, Oxford **1984**.

[82] H. Tieckelmann, *Chem. Heterocycl. Compd.* **1974**, *14*, 597.

[83] M. Hayashi, K. Yamuchi, M. Kinoshita, *Bull. Chem. Soc. Jpn.* **1977**, *50*, 1510.

[84] Reactions of this type are also used in natural product synthesis, e.g. C. Herdeis, C. Hartke, *Synthesis* **1988**, 76 (Desethylibogamin).

[85] C. J. Ishag, K. J. Fischer, B. E. Ibrahim, G. M. Ishander, A. R. Katritzky, *J. Chem. Soc., Perkin Trans. 1*, **1988**, 917.

[86] W. Jünemann, H. G. Opgenorth, H. Scheuermann, *Angew. Chem.* **1980**, *92*, 390.

[87] K. Takaoka, T. Aoyama, T. Shioiri, *Tetrahedron Lett.* **1996**, 4977.

[88] A. C. Veronese, R. Callegari, C. F. Morelli, *Tetrahedron* **1995**, *51*, 12277.

[89] J. Barluenga, R. P. Carlon, F. J. Gonzalez, S. Fustero, *Tetrahedron Lett.* **1990**, 3793.

[90] J. Barluenga, J. Joglar, F. J. Gonzalez, S. Fustero, *Synlett* **1990**, 129.

[91] M. Takamura, K. Funabashi, M. Kanai, M. Shibasaki, *J. Am. Chem. Soc.* **2000**, *122*, 6327.

[92] R. Levine, D. A. Dimmig, W. M. Kadunce, *J. Org. Chem.***1974**, *39*, 3834.

[93] A. Godard, P. Duballet, G. Queguiner, B. Pastour, *Bull. Soc. Chim. Fr.* **1976**, 789.

[94] M. G. Dubois, J. C. Gauffre, *Synlett* **1995**, 603.

[95] J. C. Gauffre, M. G. Dubois, *J. Chem. Res.* (S) **1995**, 242.

[96] S. Cacchi, G. Fabrizi, F. Marinelli, *Synlett* **1999**, 401.

[97] A Arcadi, F. Marinelli, E. Rossi, *Tetrahedron* **1999**, 55, 13233.

[98] P. G. Gassman, R. L. Parton, *J. Chem. Soc., Chem. Commun.*, **1977**, 694.

[99] M. Suzuki, K. Tanikawa, R. Sakoda, *Heterocycles* **1999**, *50*, 479;
G. Sabitha, R. S. Babu, B. V. S. Reddy, J. S. Yadav, *Synth. Commun.* **1999**, *29*, 4403.

[100] see [6], p 359

[101] P. Sanna, A. Carta, G. Paglietti, *Heterocycles* **1999**, *51*, 2171.

[102] O. Meth-Cohn, S. Rhouati, B. Tarnowski, *Tetrahedron Lett.* **1979**, 4885.

[103] For a preparative example, see [6], p 361.

[104] Y. Makioka, T. Shindo, Y. Taniguchi, K. Takaki, Y. Fujiwara, *Synthesis* **1995**, 801.

[105] L. G. Quiang, N. M. Baine, *J. Org. Chem.* **1988**, *53*, 4218.

[106] S. Hibino, E. Sugino, *Heterocycles* **1987**, *26*, 1883.

[107] W. Ried, P. Weidemann, *Chem. Ber.* **1971**, *104*, 2484.

[108] L. Capuano, V. Diehl, *Chem. Ber.* **1976**, *109*, 723.

[109] K. Matsumoto, M. Suzuki, N. Yoneda, M. Miyoshi, *Synthesis* **1976**, 805.

[110] H. L. Wehrmeister, *J. Heterocycl. Chem.* **1976**, *13*, 61.

[111] J. P. Michael, *Nat. Prod. Reports* **1991**, 53;
M. F. Grundon, *Nat. Prod. Reports* **1990**, 131;
M. F. Grundon, *Nat. Prod. Reports* **1988**, 293.

[112] Antibacterial 4-quinoline-3-carboxylic acid and its derivatives: D. C. Laysen, M. W. Zhang, A. Hae-mers, W. Bollaert, *Pharmazie* **1991**, *46*, 485 and 557.

[113] E. J. Corey, A. Tramontano, *J. Am. Chem. Soc.* **1981**, *103*, 5599;
J. A. Gainor, S. M. Weinreb, *J. Org. Chem.* **1982**, *47*, 2833;
P. Martin, E. Steiner, K. Auer, T. Winkler, *Helv. Chim. Acta* **1993**, *76*, 1667.

[114] F. D. Popp, J. M. Wefer, *J. Heterocycl. Chem.* **1967**, *4*, 167;
J.-T. Hahn, J. Kant, F. D. Popp, S. R. Chabra, B. C. Uff, *J. Heterocycl. Chem.* **1992**, *29*, 1165;
E. Reimann, H. Benend, *Monatsh.Chem.* **1992**, *123*, 939.

[115] R. V. Stevens, J. W. Canary, *J. Org. Chem.* **1990**, *55*, 2237

[116] T. Kametani, K. Fukumoto, *Chem. Heterocycl. Compd.* **1981**, *38* (1), 139.

[117] D. M. B. Hickey, A. B. McKenzie, C. J. Moody, C. W. Rees, *J. Chem. Soc., Perkin Trans. 1*, **1987**, 921.

[118] M. Hesse, *Alkaloide*, Wiley-VHC, Weinheim **2000**;
K. W. Bentley, *Nat. Prod. Reports* **1988**, 265; **1989**, 805; **1990**, 245; **1991**, 339.

[119] K. J. Zee-Chang, C. C. Cheney, *J. Pharm. Sci.* **1972**, *61*, 969.

[120] A. Mitchinson, A. Nadin, *J. Chem. Soc., Perkin Trans. 1* **1999**, 2553;
for stereoselective synthesis of piperidines see S. Laschat, T. Dickner, *Synthesis* **2000**, 1781.

[121] C. Fernandez-Garcia, M. A. McKervey, *Tetrahedron Asymmetry* **1995**, *6*, 2905.

[122] Synthesis of **6**: see [27], p 303;
enantioselective synthesis of chiral piperidine alkaloids: P. E. Hammann, *Nachr. Chem. Tech. Lab.* **1990**, *38*, 342.

[123] H. Oehling, W. Schäfer, A. Schweig, *Angew. Chem. Int. Ed. Engl.* **1971**, *10*, 656;
C. W. Bird, *Tetrahedron* **1990**, *46*, 5697.

[124] G. Märkl in: *Houben-Weyl, Vol. E 1 (organische Phosphorverbindungen)*, p 72, Thieme, Stuttgart **1982**.

[125] K. Dimroth, H. Kaletsch, *Angew. Chem. Int. Ed. Engl.* **1981**, *20*, 871.

[126] H. H. Lee, W. A. Denny, *J. Chem. Soc., Perkin Trans. 1*, **1990**, 1071.

[127] O. Hutzinger, M. Fink, H. Thoma, *Chemie in unserer Zeit* **1986**, *20*, 165;
J. P. Landers, N. J. Bunce, *Biochem. J.* **1991**, 275;
P. Cicryt, *Nachr. Chem. Tech. Lab.* **1993**, *41*, 550.

[128] A. I. Meyers, *Heterocycles in Organic Synthesis*, p 201, Wiley, New York **1974**.

[128a] G. M Coppola, *J. Heterocycl. Chem.* **1999**, *36*, 563.

[129] T. K. Lindhorst, *Essentials of Carbohydrate Chemistry and Biochemistry*, Wiley-VCH, Weinheim **2000**.

[130] P. Perlmutter, E. Puniani, G. Westman, *Tetrahedron Lett.* **1996**, *37*, 1715;
P. Perlmutter, E. Puniani, *Tetrahedron Lett.* **1996**, *37*, 3755.

[131] B. C. Chen, *Heterocycles* **1991**, *32*, 970.

[132] D. Seebach, R. Imwinkelried, T. Weber in: *Modern Synthetic Methods* (Ed. R. Scheffold), p 191, Springer, Berlin **1986**;
T. Pietzonka, D. Seebach, *Chem. Ber.* **1991**, *124*, 1837;
D. Seebach, U. Miszlitz, P. Uhlmann, *Chem. Ber.* **1991**, *124*, 1845;
further applications of **18**: D. Seebach, J.-M. Lapierre, W. Jaworek, P. Seiler, *Helv.Chim. Acta* **1993**, *76*, 459;

A. S. Kiselyov, L. Strekowski, *Tetrahedron* **1993**, *49*, 2151.

[133] D. Seebach in: *Modern Synthetic Methods* (Ed. R. Scheffold), p 173, Salle und Sauerländer, Frankfurt **1976**;
D. J. Ager in: *Umpoled Synthons* (Ed. T.A.Hase), p 19, Wiley, New York **1987**;
P. C. Bulman Page, M. B. van Niel, J. C. Prodger, *Tetrahedron* **1989**, *45*, 7643.

[134] R. B. Woodward, *Angew. Chem.* **1966**, *78*, 557.

[135] P. G. Sammer, *Chem. Rev.* **1976**, *76*, 113.

[136] M. Tisler, B. Stanovnik, *Adv. Heterocycl .Chem.* **1990**, *49*, 385.

[137] A radical mechanism was postulated as a result of kinetic studies and product distribution: T. Holm, *Acta Scand.Chem., Ser. B.*, **1990**, *44*, 279.

[138] L. Strekowski, R. L. Wydra, L. Janda, D. B. Harden, *J. Org. Chem* **1991**, *56*, 5610.

[139] T. L. Gilchrist, *Contemp. Org. Synth.* **1995**, *2*, 337.

[140] A. M. Manning et al., *J. Med. Chem.* **2000**, *43*, 3995.

[141] E. Rossi, G. Abbiati, E. Pini, *Synlett* **1999**, *6*, 756.

[142] H. Bredereck, R. Gompper, H. Herlinger, *Chem. Ber.* **1958**, *91*, 2832.

[143] H. Wamhoff, J. Dzenis, K. Hirota, *Adv. Heterocycl. Chem.* **1992**, *55*, 129;
J. T. Bogarski, J. L. Mokrosz, H. J. Barton, M. H. Paluchowski, *Adv. Heterocycl. Chem.* **1985**, *38*, 229.

[144] A. Kleemann, H. J. Roth, *Arzneistoffgewinnung*, p 172, Thieme, Stuttgart **1983**.

[145] see [27], p 194.

[146] G. Fischer, *Z. Chem.* **1990**, *30*, 305.

[147] J. A. Joule, K. Mills, *Heterocyclic Chemistry*, 4th edition, p 461, Blackwell Science, Oxford **2000**; for a concise treatment of the basicity/acidity problem of purines see: G. Kampf, L. E. Kapinos, R. Grisser, B. Lippert, H. Sigel, *J. Chem. Soc., Perkin Trans. 2*, **2000**, 1320.

[148] T. Fuji, T. Itaya, *Heterocycles* **1998**, *48*, 359.

[149] see [27], p 213.

[150] S. L. Miller, *Science* **1953**, *117*, 528; *J. Am. Chem. Soc.* **1955**, *77*, 2351.

[151] K. Zosel, *Angew. Chem.* **1978**, *90*, 748.

[152] S. Matoubara, *Phytochemistry* **1980**, *19*, 2239.

[153] For the synthesis of **46**, see [27], p 219.

[154] G. B. Elion, *Angew. Chem. Int. Ed. Engl.* **1989**, *28*, 870.

[155] Annual reports in Medicinal Chemistry, Vol. 35, p 333, Academic Press, New York 2000.

[156] P. J. Lont, H. C. van der Plas, A. J. Verbeek, *Rec. Trav. Chim. Pays-Bas* **1972**, *91*, 949.

[157] S. Kano, Y. Takahagi, S. Shibuya, *Synthesis* **1978**, 372.

[158] G. Büchi, J. Galindo, *J. Org. Chem.* **1991**, *56*, 2605.

[159] K. Jones, M. Keenan, F. Hibbart, *Synlett* **1996**, 509;

J. Chem Soc., Chem. Commun. **1997**, 323.

[160] J. Mulzer, H.-J. Altenbach, M. Braun, K. Krohn, H.-U. Reissig, *Organic Synthesis Highlights*, p 300, VCH, Weinheim **1991**;
U. Groth, T. Huhn, B. Porsch, C. Schmeck, U. Schöllkopf, *Liebigs Ann. Chem.* **1993**, 715.

[161] W. Pfleiderer in: A. R. Katritzky, C. W. Rees, A. J. Boulton, A. McKillop (Eds.), *Comprehensive Heterocyclic Chemistry*, Vol. 3, p 263., Pergamon, Oxford **1984**.

[162] H. Kotsuki, H. Sakai, H. Morimoto, H. Suenaga, *Synlett* **1999**, *6*, 1993.

[163] M. Loriga, A. Nuvole, G. Paglietti, *J. Chem. Res. (S)* **1989**, 202.

[164] R. A. Scherrer, H. R. Bratty, *J. Org. Chem.* **1972**, *37*, 1681.

[165] A. Gasco, A. J. Boulton, *Adv. Heterocycl. Chem.* **1981**, *29*, 251;
T. Takabatake, T. Miyazawa, M. Kojo, H. Hasegawa, *Heterocycles* **2000**, *53*, 2151.

[166] D. F. Gloster, L. Cincotta, J. W. Foley, *J. Heterocycl. Chem.* **1999**, *36*, 25.

[167] A. Oshawa, H. Arai, H. Igeta, *J. Chem. Soc., Chem. Commun.* **1981**, 1174.

[168] H. Neunhöffer in: A. R. Katritzky, C. W. Rees, A. J. Boulton, A. McKillop (Eds.), *Comprehensive Heterocyclic Chemistry*, Vol. 3, p 269., Pergamon, Oxford **1984**.

[169] H. Neunhöffer, H. Hennig, *Chem. Ber.* **1968**, *101*, 3952.

[170] C. Grundmann, *Angew. Chem.* **1963**, *75*, 404;
A. Kreutzberger, A. Tautawy, *Chem.-Ztg.* **1978**, *102*, 106.

[171] H. J. Forsberg, V. T. Spaciano, S. P. Klump, K. M. Sanders, *J. Heterocycl. Chem.* **1988**, *25*, 767.

[172] C. Chen, R. Dagnino, Jr., J. R. McCarthy, *J. Org. Chem.* **1995**, *60*, 8428.

[173] P. Wessig, J. Schwarz, *Monatsh. Chem.* **1995**, *126*, 99.

[174] T. Maier, H. Bredereck, *Synthesis* **1979**, 690.

[175] see [6], p 402.

[176] G. Seitz, H. Wassmuth, *Chem.-Ztg.* **1988**, *112*, 281;
J. Sauer, D. K. Heldmann, J. Hetzenegger, J. Krauthan, H. Sichert, J. Schuster, *Eur. J. Org. Chem.* **1998**, 2885; see also [6], p 364.

[177] For a preparative improvement of the formation of **22-24** see: A. R. Katritzky, S. A. Belyakov, *Synthesis* **1997**, 17.

[178] F. A. Neugebauer, H. Fischer, R. Siegel, *Chem. Ber.* **1988**, *121*, 815.

[179] U. Brunner, B. M. Gensthaler, *Pharm. Ztg.* **1999**, *144*, 1040.

[180] A. Turck, N. Plè, F. Mongin, G. Quéguiner, *Tetrahedron* **2001**, *57*, 4489.

7   Seven-Membered Heterocycles

Oxepin **1**, thiepin **2** and azepine **3** are the parent compounds of the seven-membered heterocycles with one heteroatom. Of the seven-membered ring heterocycles with several heteroatoms, only diazepines e.g. 1,2- and 1,4-diazepines **4** and **5** will be discussed.

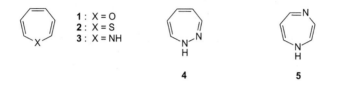

1 : X = O
2 : X = S
3 : X = NH

4

5

7.1   Oxepin

| A,B | Monocyclic oxepins exist in equilibrium with bicyclic isomeric benzene oxides (1,2-epoxybenzenes), e.g. **1** with **4** [1]: |

For this reason, the bond parameters for oxepin and other monocyclic derivatives have not been determined. The spectroscopic data show that oxepins have a polyolefinic structure with localized C=C bonds. Oxepins exist in a nonplanar boat conformation that equilibrates with inversion as in **1a** and **1b** [2]. This process is slowed down by fusion with benzene rings. 13-Methyltribenzoxepin-11-carboxylic acid **6** can be separated into its enantiomers which racemize with an energy barrier $\Delta G^{\neq}_{20.5} = 86.9$ kJ mol^{-1} [3].

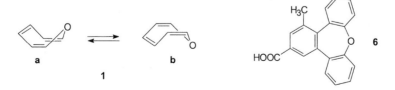

Oxepin **1** and benzene oxide **4** can be distinguished at low temperature by NMR spectroscopy (-130°C) and by their UV absorptions:

^{1}H NMR :	oxepin	$\delta = 5.7$ (H-2/H-3), 6.3 (H-4);
	benzene oxide	$\delta = 4.0$ (H-2), 6.3 (H-3/H-4).
^{13}C NMR :	oxepin	$\delta = 141.8$ (C-2), 117.6 (C-3), 130.8 (C-4);
	benzene oxide	$\delta = 56.6$ (C-2), 128.7 (C-3), 130.8 (C-4).
UV (λ_{max}) :	oxepin 305 nm, benzene oxide 271 nm.	

The spontaneous oxepin–benzene oxide isomerization proceeds as a thermally allowed, disrotatory process according to the WOODWARD-HOFFMANN rules. Because of eclipsing interactions, 2,7-substituents destabilize the benzene oxide structure and favour oxepin formation. If the 2,7-positions are bridged, the size of the bridge influences the oxepin–benzene oxide equilibrium. This is shown by studies of the 2,7-methylene-bridged systems **7** and **8**: If n = 3, only the indane oxide is present; if n = 4, tetrahydronaphthalene oxide predominates in the equilibrium mixture; if n = 5, oxepin and benzene oxide are present in a ratio of 1:1 [4].

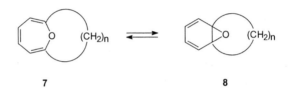

7   8

The possibility of valence isomerization also affects the reactions of the oxepins [5]. For instance, cycloadditions involve the benzene oxide, as shown by the DIELS-ALDER reaction with activated alkynes, and give the epoxybicyclo[2.2.2]octatriene **9** or, in the presence of singlet oxygen, afford the peroxide **10**. The latter isomerizes thermally yielding the *trans*-benzene trioxide **11**:

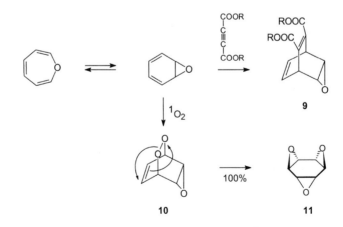

10   11

The acid-catalysed conversion into phenols is achieved in the same way. Deuterium and tritium labeling experiments have established that this rearrangement involves a 1,2-hydride shift from carbon to oxygen (the NIH shift, named after the National Institute of Health, Bethesda,USA, where it was discovered):

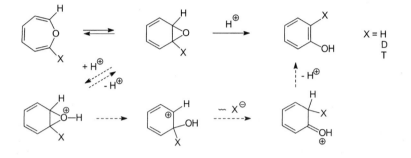

The NIH shift is important in the enzyme-catalysed hydroxylation of arenes in vivo [6].

C   The synthesis of monocyclic oxepins starts from 3,4-dibromo-7-oxabicyclo[4.1.0]heptanes **12**, which are readily accessible from cyclohexa-1,4-dienes by monoepoxidation providing **13**, followed by bromine addition to the remaining double bond. A double dehydrobromination of **12** with methoxide or DBU yields the benzene oxide/oxepin:

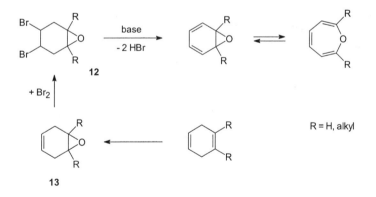

The flexibility of this method is illustrated by the synthesis of oxygen-bridged [10]annulene **16**. This compound is obtained from the tetrabromodecahydronaphthalene epoxide **14** together with the benzo[b]oxepin by way of 9,10-epoxynaphthalene **15** [7]:

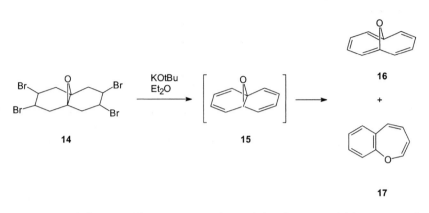

14   15

16

+

17

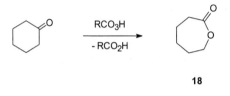

D **Oxepan-2-one 18** (hexano-6-lactone, ε-caprolactone) is of commercial importance for the production of the polyester poly-ε-caprolactone. It is obtained by a BAEYER-VILLIGER oxidation of cyclohexanone with peroxy acids:

18

Amongst natural products, the oxepin structure occurs only in senoxepin **19**, a norsesquiterpene lactone of the groundsel *Senecio platiphylla*; its structure has been established by synthesis [8]. Hydrogenated oxepins and oxepanones are often found in natural products, e.g. in the alkaloids strychnine (**20**, R = H) and brucine (**20**, R = OCH₃), and in the brassinosteroids, e.g. the growth regulator brassinolid **21**.

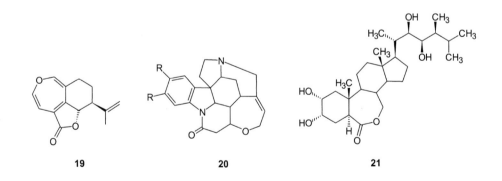

19   20   21

7.2   Thiepin

The parent compound has not yet been synthesized. However, substituted thiepins are stable and accessible by various routes. For instance, 3-amino-substituted thiophenes and activated alkynes undero a [2+2] cycloaddition followed by electrocyclic cyclobutene fission (see p 74) to give the thiepin **1**:

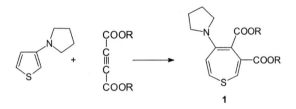

The stable thiepin **5** is obtained from interaction of the thiinium salt **2** and lithiodiazoacetic ester via C-4 addition to **3** and ring enlargement of the carbene intermediate **4** [9]:

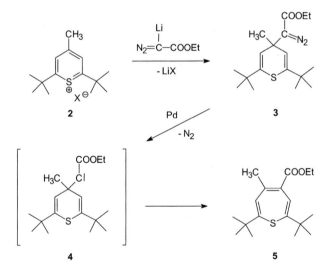

Thiepins are converted by valence isomerism into thiirans **6**, which on desulfurization (see p 25) and ring contraction yield arenes, e.g.:

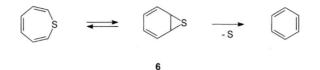

6

Dibenzo[*b,f*]thiepins **8** are obtained by cyclization of [2-(phenylsulfanyl)phenyl]acetic acid **7** with POCl$_3$:

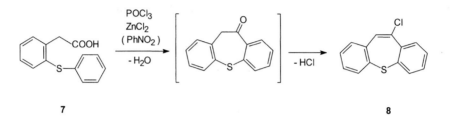

7   **8**

The reaction involves an intramolecular FRIEDEL-CRAFTS acylation followed by halogenation of the intermediate dihydrothiepinone, probably via its enol form.

10,11-Dihydrodibenzo[*b,f*]thiepins of type **9** are used as pharmaceuticals because of their marked CNS-stimulating, antidepressive and anti-inflammatory actions.

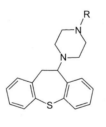

9

7.3   Azepine

A   The azepine system occurs in four tautomeric forms, the 1*H*-, 2*H*-, 3*H*- and 4*H*-azepines **1-4** [10]. The 1*H*- and 3*H*-systems are the most important.

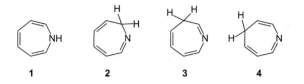

1   **2**   **3**   **4**

Very few 1*H*-azepines are known. The parent compound 1*H*-azepine **1** is an unstable red oil even at -78°C, and rearranges to the more stable 3*H*-azepine **3** in the presence of acid or base. Electron-attracting *N*-substituents increase the stability of the 1*H*-azepines. In 1-(*p*-bromophenylsulfonyl)-1*H*-azepine (Fig. 7.1) the seven-membered ring is boat shaped with alternating C_{sp2}–C_{sp2} single and double bonds.

134

146   127   123   144

124   127

137   121   118   138

116

145   N   143

SO$_2$— (*p*-bromophenyl)

Fig. 7.1   Bond parameters of the 1*H*-azepine system in 1-(*p*-bromophenylsulfonyl)-1*H*-azepine
(bond lengths in pm, bond angles in degrees)

The protons of the 1*H*-azepine system appear in the vinyl region of the ^{1}H NMR spectrum. The parent compound has $\delta = 5.22$ (H-2/H-7), 4.69 (H-3/H-6) and 5.57 (H-4/H-5) (CCl$_4$). Hence, 1*H*-azepines are not 8π-antiaromatic planar molecules, either as solids or in solution, but are atropic nonplanar cyclopolyenes.

3*H*-Azepines with substituents in the 2-position are conformationally mobile and display ring inversion between two boat structures (**5a** and **5b**). The inversion barrier can be ascertained by temperature-dependent NMR spectroscopy and was found to be $\Delta G^{\neq} = 42.7$ kJ mol^{-1} for 2-anilino-3*H*-azepine [11].

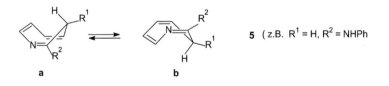

a   **b**   **5** (z.B. R^1 = H, R^2 = NHPh)

B   Azepines display the chemical behaviour of polyenes as shown by pericyclic reactions such as cycloadditions and dimerizations, and also by cheletropic and sigmatropic rearrangements.
The rearrangement of 1*H*-azepines into 3*H*-azepines (see above) corresponds to a 1,5-sigmatropic hydrogen shift. If the 3*H*-azepine is part of strained system, e.g. in **6**, rearrangement giving a 1*H*-azepine is also possible [12]:

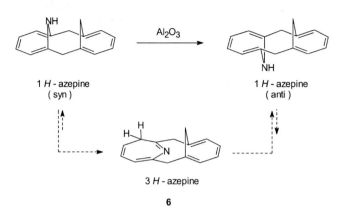

1 *H* - azepine
(syn)

1 *H* - azepine
(anti)

3 *H* - azepine

6

The (4n+2)π-electrocyclization of 1*H*-azepines is analogous to the oxepin–benzene oxide rearrangement. It yields 7-azabicyclo[4.1.0]hepta-2,4-dienes (dihydrobenzazirines) and occurs with electron-accepting N-substituents [13]. An example is the conversion of **7** into **8** at -70°C in a proportion of 9:1. A trimethylene chain bridging the 2,7-positions favours the formation of dihydrobenzazirine (**9**), but a tetramethylene bridge that of azepine (**10**):

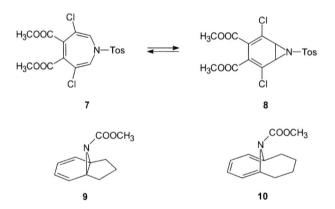

7

8

9

10

1*H*-Azepines with electron-accepting N-substitutents (e.g. **11**) are converted photochemically into 2-azabicyclo[3.2.0]hepta-3,6-dienes (e.g. **12**). The bicyclic compounds undergo thermal ring-opening of the cyclobutene ring, thereby reversing the 4nπ–electrocyclization, reforming the 1*H*-azepines:

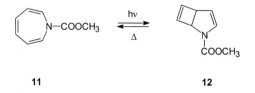

11

12

C   In the synthesis of azepines, the seven-membered heterocyclic ring is produced (a) by ring enlargement of six-membered ring systems, (b) by ring closure of suitable acyclic precursors.

(1)   1*H*-Azepines with electron-accepting N-substituents **14** are obtained by a photoinduced reaction between arenes and azides with electron-withdrawing substituents leading to N_2 elimination (HAFNER, LWOWSKI 1963):

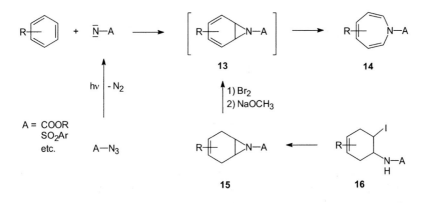

The first step is the photolysis of the azide giving a nitrene which forms a dihydrobenzazirine **13** by a [1+2] cycloaddition with the arene; the latter isomerizes yielding 1*H*-azepine. An alternative synthesis of 1*H*-azepines **14** is based on bicyclic aziridines **15**, which are transformed to **14** by addition of bromine to the double bond, two-fold dehydrobromination and valence tautomerization of the resulting **13**. The bicyclic aziridines **15** are available from cyclohexa-1,4-dienes via 1-alkoxycarbonylamino-2-iodocyclohexenes **16** and their S_{Ni}-cyclization [15].

(2)   2-Amino-3*H*-azepines **17** are formed by thermolysis of arylazides in secondary amines [16]:

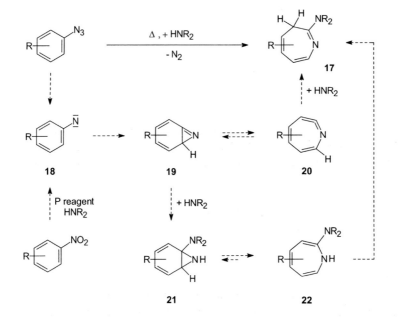

A N_2 elimination from the arylazide leads to an arylnitrene **18** which, following attack on the *ortho*-C-atom, rearranges into 2*H*-benzazirine **19**. The azepine system is formed from the benzazirine by two routes. In the first, valence isomerization of **19** leads to the azacycloheptatetraene **20** which is converted into **17** by amine addition to the heterocumulene system. In the second, the amine adds directly to **19**, producing the dihydrobenzazirines **21** which yields the 1*H*-azepine **22** by electrocyclic ring-opening. 1*H*-/3*H*-Azepine rearrangement affords the product **17**. It has been demonstrated that the highly strained 1-azatetraene **20** is an intermediate in the photolysis of phenylazide at 8 K and that it is in equilibrium with 1*H*-benzazirine in the case of naphthylazide [17].

2-Amino-3*H*-azepines are also obtained by deoxygenation of nitro- or nitrosoarenes with trivalent phosphorus compounds such as $P(OR)_3$, $CH_3P(OR)_2$ or $(C_6H_5)_2POR$ in the presence of secondary amines. Again arylnitrenes **18** are intermediates [18].

(3) Dihydroazepines are obtained by lithium-induced cyclization of azatrienes followed by *N*-acylation [19], e.g. the 4,5-dihydroazepine **25** from **23**, presumably via the corresponding anion **24** :

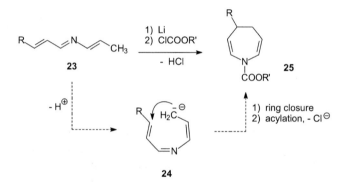

(4) Tetrahydroazepines of several structure types are efficiently prepared by ring-closure metathesis (RCM, [20]) of α,ω-diene or -eneyne systems [21], e.g. the pyrrolidinoazepinone **27** from the pyrrolidine-derived α,ω-diene precursor **26**:

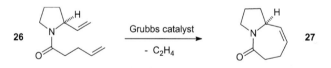

D Azepan-2-one (hexano-6-lactam, ε-caprolactam, **28**) is industrially the most important azepine derivative. It is used in the production of perlon and is synthesized by BECKMANN rearrangement of cyclohexanone oxime. Structurally related to caprolactam is the CNS stimulant pentetrazole **29** (1,5-pentamethylenetetrazole, see p 217).

Dibenzoazepines and their dihydro derivatives are components of a series of psychopharmaceuticals (antidepressants and antiepileptics) such as carbamazepine **31** and desipramine **33**.

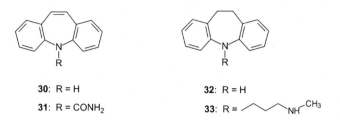

30: R = H **32**: R = H

31: R = CONH₂ **33**: R =

The parent systems **30/32** are accessible from *o*-nitrotoluene via 1,2-bis(2-nitrophenyl)ethane **34**:

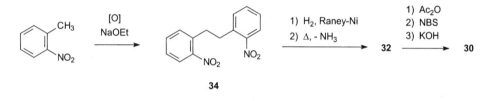

34

A number of natural products contain the azepine framework, e.g. the nonproteinogenic aminoacid muscaflavine **35**, a yellow pigment of the fly agaric, the 2*H*-azepine chalciporon **36**, a pungent substance of the peppery bolete [22], and the ethano-bridged tetrahydroazepine (±)-anatoxin **37**, a toxic principle of blue-green algae.

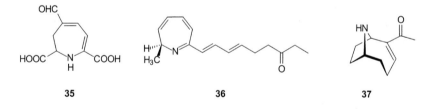

35   **36**   **37**

The synthesis of (±)-anatoxin **37** [23] was elegantly achieved starting from 1,5-cyclooctadiene **38** via a ring-opening / ring-closure reaction sequence of the β-lactam epoxide **39** with methyl lithium to give **40**, which can be further converted to the target molecule **37**.

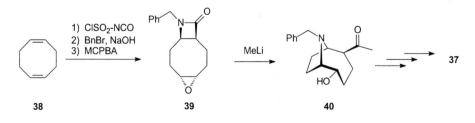

38   **39**   **40**

Galantamin **41**, a natural product from *Galanthus nivalis*, is structurally derived from 2,3-dihydrobenzo[*b*]furan, in which C-3 (as a quaternary carbon) is part of an azepane as well as a a cyclohexene moiety. It was shown to act as a potent inhibitor of cholinesterase and is applied in the therapy of Alzheimer disease [23a].

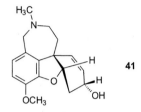

41

7.4   Diazepines

The 1,2-, 1,3- and 1,4-diazepines **1-3** are the parent compounds.

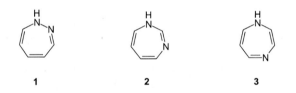

| 1 | 2 | 3 |

The 1,2- and 1,4-diazepines are of great importance, especially the 1,4-diazepine derivatives 2,3-dihydro-1,4-diazepine **4**, 1,5-benzodiazepine **5** and 1,4-benzodiazepine **6**:

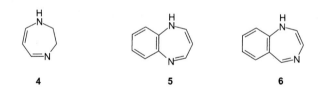

| 4 | 5 | 6 |

1,2-Diazepines

A 1,2-Diazepines occur in various tautomeric forms, e.g. 1*H*-, 3*H*- and 4*H*-1,2-diazepines, **1**, **7** and **8**. However, 5*H*-1,2-diazepines **9** exist only as the valence tautomeric 3,4-diazabicyclo-[4.1.0]hepta-2,4-diene **10**.

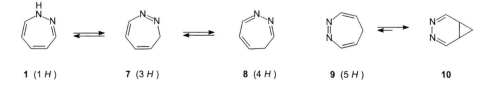

| **1** (1 *H*) | **7** (3 *H*) | **8** (4 *H*) | **9** (5 *H*) | **10** |

C The synthesis of 1,2-diazepines is carried out by various routes.

(1)   1,7-Electrocyclization of diazopenta-2,4-dienes followed by a 1,5 hydrogen shift leads to 3*H*-1,2-diazepines **11** [24]:

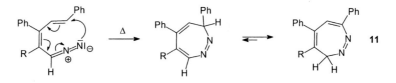

(2)   Photochemically induced isomerization of pyridine *N*-ylides **12** yields 1*H*-1,2-diazepines with electron-withdrawing N-substituents on N-1 **14** ([25], cf. p 290):

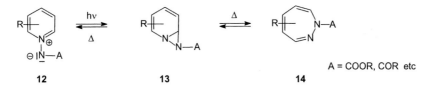

The initial products of the photoreaction are diaziridines **13** which yield the products **14** in a thermal reaction. The diaziridines can be thermally reconverted into the pyridine *N*-ylides **12**.

(3)   On reaction with hydrazine or methylhydrazine, pyrylium or thiinium salts produce 4*H*- or 1*H*-diazepines via a sequence of ring-opening-ring-closure ([26], cf. p 225), e.g.:

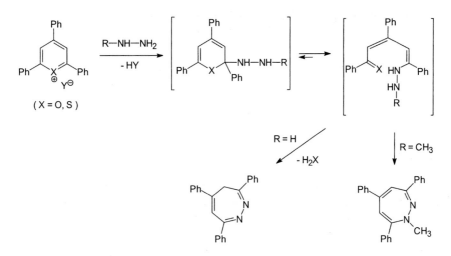

The tranquilizer 5*H*-2,3-benzodiazepine **16** (grandaxin) was commercially prepared from the benzopyrylium salt **15** by the above method:

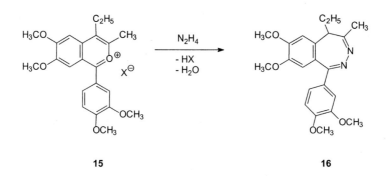

15   16

1,4-Diazepines

 2,3-Dihydro-1,4-diazepines 4 are formed from ethylenediamine and β-diketones by acid-catalysed cyclocondensation:

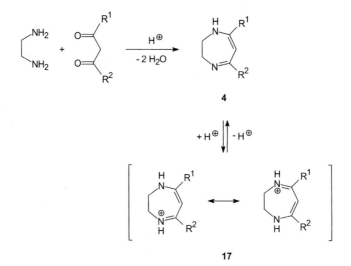

The dihydrodiazepines **4**, being vinylogous amidines, are strong bases ($pK_a \approx 13\text{-}14$). Their protonation leads to symmetrically delocalized cations **17** containing 6π-electrons corresponding to a trimethincyanine system. Its stabilization energy was estimated to be ca. 80 kJ mol^{-1}. The cations **17** possess a 'quasiaromatic' character because of their considerable resonance energy; they regenerate after electrophilic attack and thus undergo substitution reactions such as deuteriation, halogenation, nitration and azo-coupling at the 6-position:

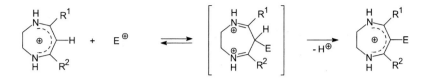

1,5-Benzodiazepines 5 are obtained by condensation of 1,2-diaminoarenes with 1,3-diketones [27]:

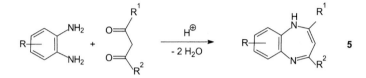

1,5-Benzodiazepines **5** are less basic ($pK_a \approx 4,5$) than their aliphatic analogues **4** because of the aromatic involvement of the amidine N-atoms. With acids, they form intensively coloured monocations **18**. When both N-atoms are protonated by a strong acid, colourless bisiminium ions **19** are formed.

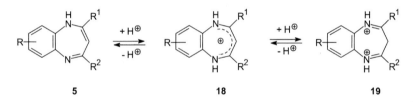

The end-groups of the cyanine moiety in the cations **18** are linked by a π-system. This system, in which the charge delocalization is somewhat higher than that of **17**, is referred to as a 'cyclocyanine'.

Dibenzo-1,5-diazepines were found to be useful building blocks for antipsychotic agents interacting with neuronal receptors [28], e.g. clozapine **20** which is effective for the treatment of schizophrenia:

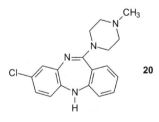

1,4-Benzodiazepines 6 and derived systems [29] are of considerable importance as pharmaceuticals, especially as tranquilizers and antidepressants. Several methods are available for their preparation.

(1)   2,3-Dihydro-1*H*-1,4-benzodiazepin-2-ones **21** are obtained from *o*-aminobenzophenones by a cyclocondensation with amino esters under base catalysis or by *N*-acylation with halo acid chlorides followed by cyclization with ammonia:

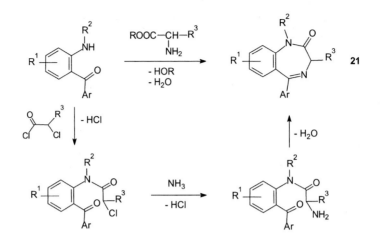

This reaction principle has considerable scope, as shown by the synthesis of various heterocyclic annulated 1,4-benzodiazepines. [1,2,3]Triazolo[1,5-*a*][1,4]benzodiazepines **23** are prepared from triazole derivatives **22** which are accessible by a 1,3-dipolar cycloaddition of *o*-azidobenzophenones with (aminomethyl)alkines; pyrrolo[2,1-*c*][1,4]benzodiazepines **25** are formed from 1,2-dihydro-3,1-benzoxazinediones **24** and proline:

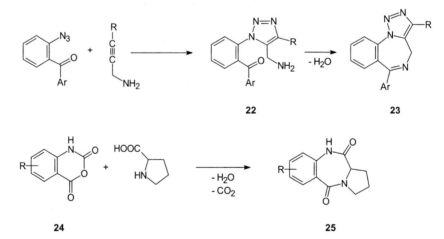

(2)   2-Amino-1,4-benzodiazepine 4-oxides **27** are prepared from 2-(chloromethyl)quinazoline 3-oxides **26** by interaction with NH_3 or primary amines (STERNBACH 1961). The ring-expansion is brought about by addition of the nucleophile to C-2 of **26**. It leads to elimination of the halide with a 1,2-shift in the nitrone **28** giving a mesomerically favoured amidinium system **29**, which on deprotonation yields **27**. The starting material **26** is obtained from *o*-aminobenzophenone oximes **30** by *N*-acylation with chloroacetyl chloride giving **31** followed by acid-catalysed cyclodehydration.

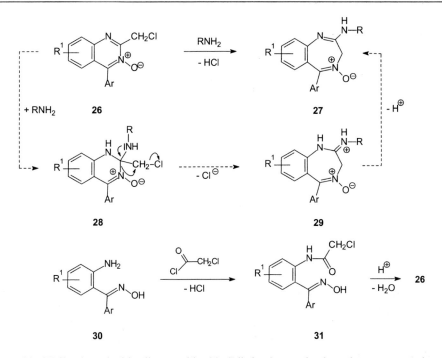

Diazepam **32** (Valium) and chlordiazepoxide **33** (Librium) are the best known examples of 1,4-benzodiazepines. They are prepared from 5-chloro-2-aminobenzophenone [30]:

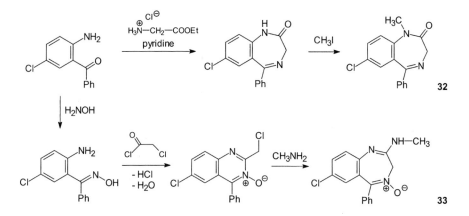

Chlordiazepoxide yields the nitrone **34** on mild alkaline hydrolysis. Treatment of the nitrone with acetic anhydride functionalizes the CH$_2$ group yielding the acetate **35** by a POLONOVSKI reaction. Reaction with p-toluenesulfonyl chloride causes ring contraction, affording tetrahydroquinoxaline **36** analogous to a BECKMANN rearrangement. However, the acid-catalysed rearrangement of 3-hydroxybenzazepin-2-one **37** leads to quinazoline-2-carbaldehyde **38** [31]:

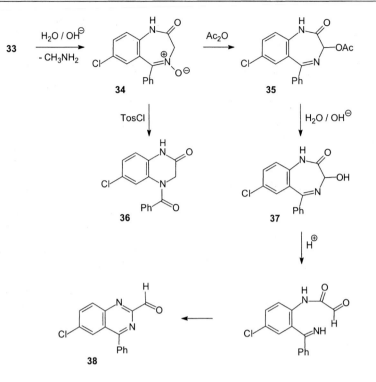

References

[1] R. Wehner, H. Günther, *Chem. Ber.* **1974**, *107*, 3179.

[2] D. M. Hayes, S. D. Nelson, W. A. Garland, P. A. Køllman, *J. Am. Chem. Soc.* **1980**, *102*, 1255.

[3] W. Tochtermann, C. Franke, *Angew. Chem. Int. Ed. Engl.* **1969**, *8*, 68.

[4] E. Vogel, H. Günther, *Angew. Chem., Int. Ed. Engl.* **1967**, *6*, 385.

[5] G. S. Shirwaiker, M. V. Bhatt, *Adv. Heterocycl. Chem.* **1984**, *37*, 68.

[6] M. N. Akhtar, D. R. Boyd, J. D. Neill, D. M. Jerina, *J. Chem. Soc., Perkin Trans. 1*, **1980**, 1693.

[7] E. Vogel, M. Biskup, W. Pretzer, W. A. Böll, *Angew. Chem., Int. Ed. Engl.* **1964**, *3*, 642; cf. also: B. Ganem, G. W. Holbert, L. B. Weiss, K. Ishizumi, *J. Am. Chem. Soc.* **1978**, *100*, 6483.

[8] A. Kleve, F. Bohlmann, *Tetrahedron Lett.* **1989**, *30,* 1241.

[9] K. Nishino, S. Yano, Y. Kokashi, K. Yamamoto, I. Murata, *J. Am. Chem. Soc.* **1979**, *101*, 5059.

[10] K. Satake, *J. Chem. Soc., Chem. Commun.,* **1991**, 1154.

[11] W. L. F. Armarego, *Stereochemistry of Heterocyclic Compounds*, Wiley, New York 1977.

[12] E. Vogel, U. Brocker, H. Junglas, *Angew. Chem., Int. Ed. Engl.* **1980**, *19*, 1015.

[13] H. Prinzbach, D. Stusche, J. Markert, H.-H. Limbach, *Chem. Ber.* **1976**, *109*, 3505; A. Hassner, D. J. Anderson, *J. Org. Chem.* **1974**, *39*, 3070.

[14] P. Casagrande, L. Pellacani, P. A. Tardella, *J. Org. Chem.* **1978**, *43*, 2725; N. R. Ayyangar, R. B. Bambal, A. G. Lugade, *J. Am. Chem. Soc., Chem. Commun.* **1981**, *790*.

[15] L. A. Paquette, D. E. Kuhla, J. H. Barrett, R. J. Haluska, *J. Org. Chem.* **1969**, *34*, 2866.

[16] C. Wentrup, *Adv. Heterocycl. Chem.* **1981**, *28*, 231.

[17] I. R. Dunkin, P. C. P. Thomson, *J. Chem. Soc., Chem. Commun.,* **1980**, 499.

[18] S. Batori, R. Gompper, J. Meier, H. U. Wagner, *Tetrahedron Lett.* **1988**, *29,* 3309.

[19] S. Klötgen, E.-U. Würthwein, *Tetrahedron Lett.* **1995**, *36*, 7065;
S. Klötgen, R. Fröhlich, E.-U. Würthwein, *Tetrahedron* **1996**, *52*, 14801.

[20] A. Fürstner, *Angew. Chem. Int. Ed. Engl.* **2000**, *39*, 3012.

[21] M. Arisawa, E. Takezawa, A. Nishida, M. Tori, M. Nakagawa, *Synlett* **1997**, 1179;
see also: C. A. Tarling, A. B. Holmes, R. E. Markwell, N. D. Pearson, *J. Chem. Soc., Perkin Trans. 1* **1999**, 1695;
J. Pernerstorfer, M. Schuster, S. Blechert, *Synthesis* **1999**, 1695.

[22] O. Sterner, B. Steffen, W. Steglich, *Tetrahedron,* **1987**, *43*, 1075.
The synthesis of optically active 2*H*-azepines by cyclization of aminoacid-derived building blocks is also described: D. Hamprecht, W. Josten, W. Steglich, *Tetrahedron* **1996**, *52*, 10883.

[23] P. J. Parsons, N. P. Camps, J. P. Camps, J. M. Underwood, D. M. Harvey, *Tetrahedron* **1996**, *52*, 11637.

[23a] B. Paruche, M. Schulz, *Pharm. Ztg.* **2002**, *147*, 490.

[24] G. Zecchi, Synthesis **1991**, 181.

[25] V. Snieckus, J. Streith, *Acc. Chem. Res.* **1981**, *14*, 348.

[26] D. J. Harris, G.Y.-P. Kan, T. Tschamber, V. Snieckus, *Can. J. Chem.* **1980**, *58*, 494;
J. Körösi, T. Láng, *Chem. Ber.* **1974**, *107*, 3883.

[27] L. F. Tietze, Th. Eicher, *Reactions and Synthesis in the Organic Chemistry Laboratory*, p 374, University Science Books, Mill Valley, CA **1989**.

[28] T. D. Penning et al., *J. Med. Chem.* **1997**, *40*, 1347.

[29] J. B. Bremner, *Prog. Heterocycl. Chem.* **2001**, *Vol. 13*, p 357;
L. H. Sternbach, *J. Med. Chem.* **1979**, *22*, 1;
M. Williams, *J. Med. Chem.* **1983**, *26*, 619.

[30] Th. Eicher, H. J. Roth, *Synthese, Gewinnung und Charakterisierung von Arzneistoffen*, p 204, Thieme, Stuttgart **1986**.

[31] J. T. Sharp in: A. R. Katritzky, C. W. Rees, W. Lwowski (Eds.), *Comprehensive Heterocyclic Chemistry,* Vol. 7, p 617, Pergamon Press, Oxford **1984**.

8   Larger Ring Heterocycles

In this chapter, eight-membered (e.g. azocines **1**), nine-membered (e.g. oxonin, thionin and azonine **2-4**) and larger ring heterocycles, as well as porphyrins and other tetrapyrrole systems, will be discussed.

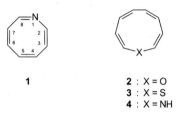

1

2 : X = O
3 : X = S
4 : X = NH

8.1   Azocine

A,C   Azocine **1** is the aza analogue of cyclooctatetraene. It was obtained by vacuum flash pyrolysis of diazabasketene and removal from the pyrolysate by freezing at -190°C [1]:

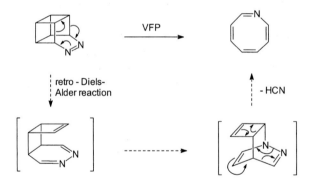

Azocine is thermally labile and decomposes at -50°C with polymerization. In its ^{1}H NMR spectrum, it exhibits protons in the olefinic region [δ = 7.65 (H-2), 7.0 (H-8), 5.0 (H-7); the remaining protons are at $\delta \approx 6$]. Therefore, azocine is likely to be an atropic 8π-system comparable to cyclooctatetraene.

D   **2-Methoxyazocine 9** and analogous systems are readily accessible (PAQUETTE 1971). 2-Methoxyazocine is synthesized by a [2+2] cycloaddition of chlorosulfonyl isocyanate to 1,4-cyclohexadiene. The chlorosulfonyl group in the adduct **5** is removed by treatment with benzenethiol, and the β-lactam **6** is converted into the imidic ester **7** by alkylation with trimethyloxonium tetrafluoroborate. Addition of bromine and two-fold dehydrohalogenation afford the azabicyclooctatriene **8**, which yields the yellow 2-methoxyazocine **9** by a thermal electrocyclic ring-opening reaction [2]:

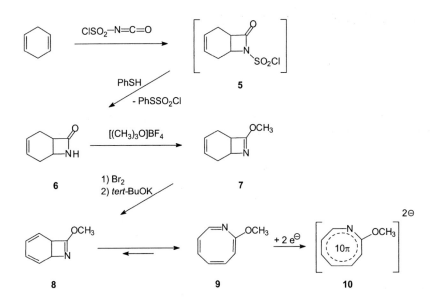

2-Methoxyazocine [^{1}H NMR: δ = 6.54 (H-8), 5.12 (H-7), 5.75-6.05 (H-3–H-6)] on reduction with potassium in liquid NH$_3$ or THF, takes up two electrons (as observed also for **1**) forming a red azocine dianion **10**. Its proton signals are significantly downfield shifted ($\approx$ 1 ppm) relative to **9**. This is ascribed to induction of a diamagnetic ring current and demonstrates the analogy of **10** with the aromatic 10π-dianion of cyclooctatetraene.

Reactions of 2-methoxyazocine proceed via its bicyclic valence isomer **8** [3]. Although it is the minor constituent of the equilibrium mixture, it is more reactive owing to its higher ring strain. Dienophiles (e.g. 4-phenyl-4,5-dihydro-3*H*-1,2,4-triazole-3,5-dione) undergo DIELS-ALDER reactions to give **11**. Demethylation with HBr yields the β-lactam **12**. Acid or base hydrolysis leads to ring-opening of the four-membered ring producing methyl benzoate and benzonitrile:

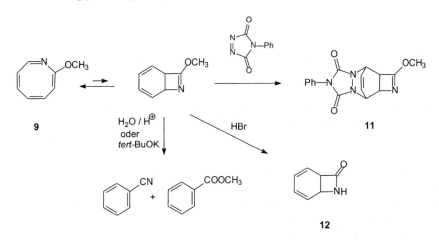

Eight-membered ring heterocycles with two heteroatoms (diheterocines) are derived from *cis*-benzene dioxides and *cis*-benzene diimines by a process of valence isomerization (PRINZBACH, VOGEL 1972) [4]:

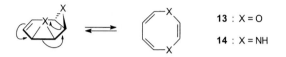

13 : X = O

14 : X = NH

1,4-Dioxocine 13, a colourless, rapidly polymerizing liquid, behaves like a polyalkene [^{1}H NMR: δ = 6.00 (H-2/H-3), 6.59 (H-5/H-8), 5.12 (H-6/H-7)] without any sign of aromatic stability [5].

1,4-Dihydro-1,4-diazocine 14, mp 135°C, forms colourless crystals on sublimation. It is almost planar and, according to X-ray analysis, its bond lengths are largely evened out. ^{1}H NMR data [δ = 6.25 (H-2/H-3), 6.80 (H-5/H-8), 5.72 (H-6/H-7)] indicate a diamagnetic ring current comparable to that of furan or thiophene and, therefore, characterize **14** as an aromatic 10π-system [6]. The diazocine **14** is prepared from the *cis*-benzene dioxide **15** according to the following sequence:

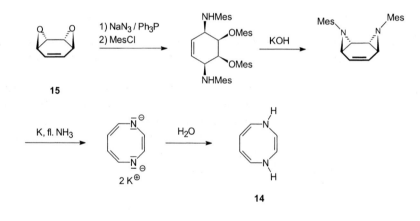

8.2 Heteronines and larger-membered heterocycles

A-D   Oxonin **2** and azonines **16** with electron-accepting N-substituents are thermally unstable. They are strongly folded π-localized polyene structures; their ring protons have chemical shifts of atropic 8π-systems without ring current effects in the ^{1}H NMR spectra. Azonine **4** and its anions **17** possess relatively high thermal stability and a largely planar delocalized structure. Thus the downfield shifts of their ring protons in the ^{1}H NMR spectrum meet the criteria for diatropic aromatic 10π-systems [7].

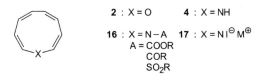

2 : X = O   4 : X = NH

16 : X = N−A   17 : X = N I$^{\ominus}$ M$^{\oplus}$
     A = COOR
     COR
     SO$_2$R

Therefore, 8π-heteronines **2** and **16** engage in pericyclic reactions observing the WOODWARD-HOFFMANN rules; photochemical $_\pi 8_\sigma$-isomerizations lead to bicyclo[6.1.0]nonatrienes **18**, thermal $_\pi 6_\sigma$ isomerizations yield *cis*-bicyclo[4.3.0]nonatrienes **19**:

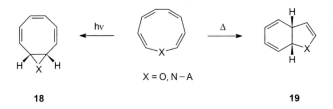

18   19

The principle developed by VOGEL of methano- or oxy-bridging of medium-sized annulenes [8] to strengthen the perimeter has proved advantageous in the synthesis of larger-membered heteroannulenes of various structures.

The methano-bridged aza[10]annulene **20** [9], a 10π-analogue of pyridine, is a stable, yellow compound with a quinoline-like odour and a flattened perimeter (UV: λ_{max} = 364 nm). The ^{1}H NMR data confirm the aromatic character of **20**. The perimeter protons combine the features of α-substituted 1,6-methano[10]annulenes and quinoline (see p 318); the upfield shift of the bridge H-atoms (δ CH$_2$ = -0.40 / +0.65) is characteristic. It is caused by the shielding effect of the diatropic heteroarene system. The heteroannulene **20** is less basis (pK_a = 3.20) than pyridine (pK_a = 5.23) or quinoline (pK_a = 4.94).

20   21 (X = O, S)

In contrast, the π-excessive hetero[11]annulenes **21** [10] and the pyridine-like hetero[12]annulene **22** [11] are atropic molecules. As shown by their ^{1}H NMR spectra, they avoid antiaromatic destabilization of a paratropic 12π-perimeter by assuming nonplanar structures.

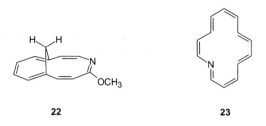

22   **23**

Unbridged analogues of pyridine containing 14π- and 18π-electrons are known (SCHRÖDER 1979). The aza[14]annulene **23** [12] is a thermally stable, violet compound (UV: λ_{max} = 620 nm) and shows signals in the ¹H NMR spectrum which are due to conformationally mobile intra- and extraannular protons with a shift difference of 9.5 ppm. This indicates the presence of a flat heteroannulene perimeter with a delocalized, strongly diatropic π-system.

   The blackish-green aza[18]annulene **24** [13] is stable (UV: λ_{max} = 682 nm) and shows a chemical shift difference of some 11 ppm between the intra- and extrannular protons in the ¹H NMR spectrum. The ¹H NMR spectrum of **24**, contrary to **23**, is temperature independent, i.e. it behaves like a 1,2-disubstituted [18]annulene [14]. Its violet hydrochloride **25** occurs in two rapidly equilibrating forms (**a** and **b**, in the proportion of 1:4). Their NH signals differ by ca. 20 ppm (**25a**: δ_{NH} = -0.81; **25b**: δ_{NH} = +19) because of their intra- and extrannular position. This supports the diatropic character of a HÜCKEL-aromatic 18π-system.

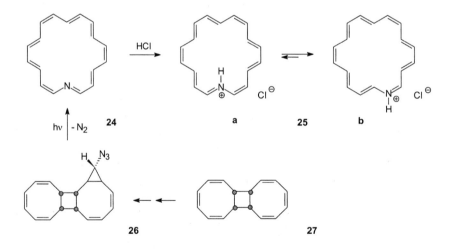

The synthesis of the aza[18]annulene **24** is carried out by photolysis of the tetracyclic azide **26**. This is obtained from the [2+2] dimer **27** of cyclooctatetraene by a conventional method (cyclopropanation with diazoacetic ester followed by conversion of the ester function into an azide via $CO_2Et \rightarrow CON_3$ $\rightarrow N=C=O \rightarrow NH–CO–NH_2 \rightarrow O=N–N–CO–NH_2 \rightarrow N_3$).

Crown ethers and cryptands [15] also belong to the group of larger ring heterocycles, but will not be discussed further in this book.

8.3   Tetrapyrroles

A The parent compound is the porphyrin ring system **1*** in which four pyrrole units are linked by four sp^2 methine bridges forming a cyclic conjugated C_{20} structure [16]. Partially hydrogenated porphyrins, such as chlorin **2** (17,18-dihydroporphyrin), phlorin **3** (15,24-dihydroporphyrin), bacterio-chlorin **4** (7,8,17,18-tetrahydroporphyrin) and porphyrinogen **5** (5,10,15,20,22,24-hexahydropor-phyrin), are also important:

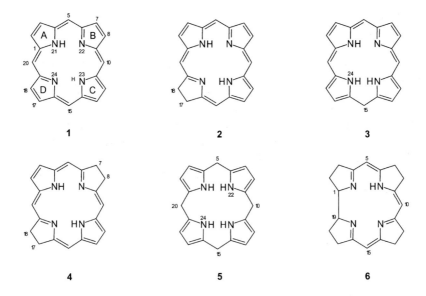

1   **2**   **3**

4   **5**   **6**

All pyrrole units of the porphyrin system are partially hydrogenated in the C_{19} structure of corrin **6**. Two pyrrole rings are directly linked at the α-positions and not via a methine group.

Porphyrins form a planar macrocycle that contains a conjugated, cyclic delocalized system of 22 π-electrons [17]. 18 Electrons can be assigned to the perimeter of a 1,16-diaza[18]annulene, an arrangement which is also found in the chlorin and bacteriochlorin systems. Porphyrins and chlorins are intensely coloured (**1**: red to violet, $\lambda_{max} \approx$ 500-700 nm; **2-4**: green, $\lambda_{max} \approx$ 600-700 nm).

B The reactions of porphyrins arise from the aromatic annulene character and the amphoteric behaviour of the pyrrole units which exist as 1*H*- or 2*H*-structures. Accordingly, porphyrins form tetradentate chelate complexes **7** with many cations, and as weak bases ($pK_{a1} \approx 7$, $pK_{a2} \approx 4$) they can be protonated to give dications **8**.

---

* The parent substance **1** is also known as porphin, see *Pure Appl.Chem.* **1987**, *59*, 779.

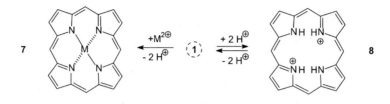

Porphyrins and their metal complexes can be substituted by electrophiles, e.g. by deuteriation, nitration and VILSMEIER formylation at the methine and pyrrole C-atoms.

Reduction of porphyrins by catalytic hydrogenation leads to phlorins and with diimides, chlorins are obtained; porphyrinogens are formed with sodium borohydride, Na/Hg or by catalytic hydrogenation under drastic conditions.

Oxidation of porphyrins was used to elucidate its structure. It leads (e.g. with $KMnO_4$) to the formation of pyrrole-2,5-dicarboxylic acids with retention of the methine C-atoms, or (e.g. with CrO_3) to maleimides with loss of the methine C-atoms:

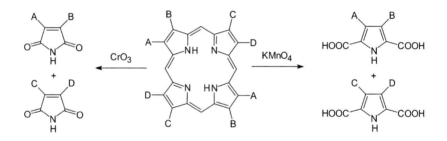

Single-electron oxidation of the metalloporphyrins **9** (e.g. by halogens or electrochemically) leads to the formation of radical cations. These undergo addition with nucleophiles at a methine bridge and, after acid demetalation, yield monosubstituted porphyrins **10** [18]:

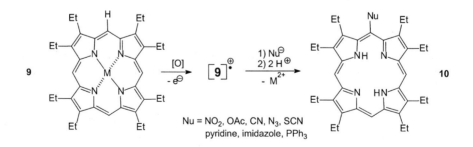

Nu = NO_2, OAc, CN, N_3, SCN
pyridine, imidazole, PPh_3

C Three main approaches are convenient for the synthesis of porphyrins [19], starting from appropriately substituted pyrrole derivatives.

(1)   Symmetrically substituted porphyrins are produced by joining four monopyrrole building blocks by cyclization (cyclotetramerization). For instance, the 5,10,15,20-tetraarylporphyrins **12** are obtained by cyclocondensation of pyrrole with arenecarbaldehydes catalysed by acid (BF_3–etherate, CF_3CO_2H etc.; see p 90) followed by dehydrogenation (chloranil, O_2 etc.) of the initially formed porphyrinogens **11** [20]. Octaethylporphyrin **14** ($R^1 = R^2 = Et$) is synthesized by cyclocondensation of 3,4-diethylpyrrole with formaldehyde followed by dehydrogenation with O_2 in 66% yield [21], alternatively from 2-(dimethylaminomethyl)pyrrole **13** ($R^1 = R^2 = Et$) by heating in acetic acid [22]:

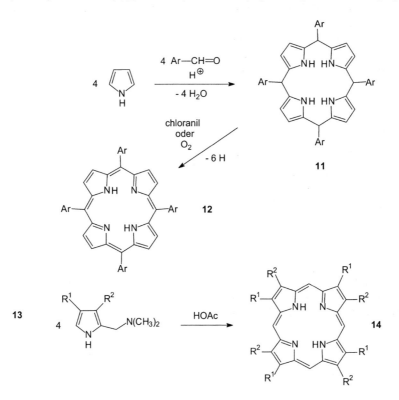

Porphyrin formation from **13** occurs as the result of a repeated cationic domino process [23]. This also applies to the biosynthesis of haemin, chlorophyll and vitamin B_{12} from 5-amino-4-oxobutanoic acid via porphobilinogen (**13**; $R^1 = CH_2CO_2H$, $R^2 = CH_2CH_2CO_2H$, NH_2 in place of NMe_2) [24].

(2)   Dipyrroles (dipyrrylmethenes, dipyrrylmethanes) are suitable building-blocks for convergent porphyrin syntheses. For instance, the acid-catalysed cyclocondensation of 5-bromo-5'-methyldipyrryl-methenes **15** yields the porphyrins **16** directly with double elimination of HBr:

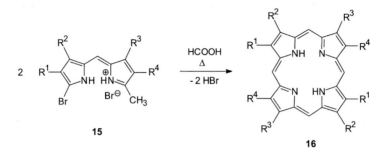

15   **16**

5,5'-Diformyl-substituted dipyrrylmethanes **17** cyclocondense with 5,5'-disubstituted dipyrrylmethanes **18** yielding the dications **19** which are dehydrated to the porphyrins **20**:

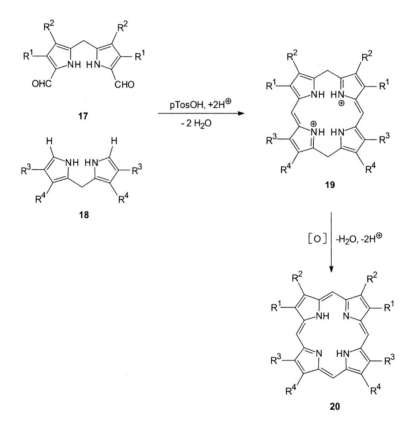

This variant has preparative advantages compared with the direct method because it proceeds under milder conditions and the starting materials are easily accessible.

(3)   Cyclization of open-chain, suitably substituted tetrapyrroles affords the possibility of a linear synthesis of unsymmetrically substituted porphyrins. 1-Bromo-19-methylbiladienes **23** are particularly suitable. They cyclize thermally via the terminal methyl group forming porphyrins **24** with elimination of HBr:

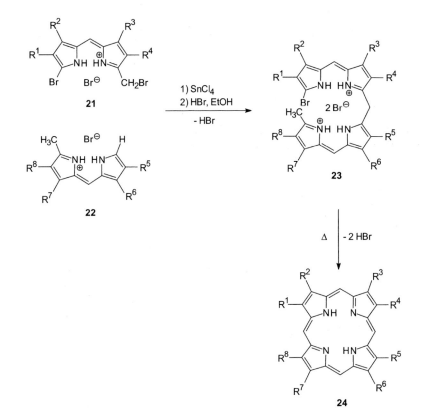

The biladienes **23** are prepared in high yield by a Friedel–Crafts alkylation of 5'-unsubstituted 5-methyldipyrrylmethenes **22** with 5-bromo-5'-(bromomethyl)dipyrrylmethenes **21** in the presence of SnCl₄ [25].

> **D** **Porphyrin 1** forms dark red crystals (dec temp. ca. 360°C) sparingly soluble in most solvents.
>
> **Haemin 25** ($C_{34}H_{32}ClFeN_4O_4$) is obtained from haemoglobin, the red pigment of vertebrate red blood cells, by separation of the protein component (globin) with acetic acid and sodium chloride (from which the Cl⁻ present in haemin is derived). Removal of the Fe(III) central ion from the chelate complex **25** (demetalation) by acid yields the 'protoporphyrin' **26**. Its hydrogenation and decarboxylation leads to 'aetioporphyrin' **28**; oxidation of haemin produces the haemin acid **27**. The first synthesis of haemin was achieved in 1929 by H. FISCHER [26].

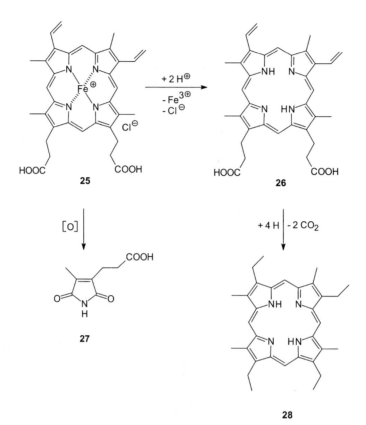

25   **26**   **27**   **28**

$+2 H^{\oplus}$
$-Fe^{3\oplus}$
$-Cl^{\ominus}$

[o]

$+4 H$ $-2 CO_2$

Haemoglobin transports oxygen from the lungs to the tissues. At normal pressure, 1 g of haemoglobin binds 1.35 cm³ of O_2, i.e. in the ratio $Fe:O_2$ of 1:1. In haemoglobin, iron has an oxidation level of +2. The globin-free iron(II) complex of protoporphyrin is known as haem. Oxidation of haemoglobin by O_2 leads to methaemoglobin containing iron at an oxidation level of +3. The globin-free iron(III) complex is known as haematin.

Biological degradation of haemoglobin to bilirubin (see p 98) starts (expressed in a simplified manner) with oxidation at a methine C-atom forming iron oxyphlorin which exists as enol **29** in an acidic and as ketone **30** in a basic medium. Further oxidation of **30** leads to biliverdin with extrusion of the methine bridge which, on reduction, yields bilirubin.

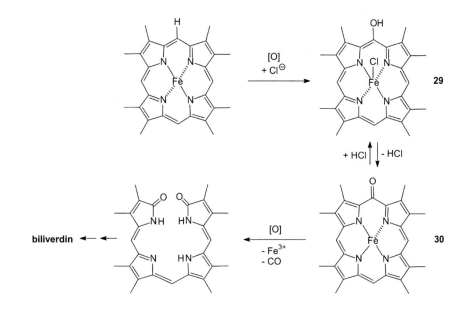

Catalase and peroxydase are enzymes which are structurally related to haemoglobin. Catalase catalyses the decomposition of hydrogen peroxide, which occurs in some metabolic processes, into water and oxygen. Peroxidase catalyses the oxidation of certain substrates by H_2O_2.

Cytochromes are porphyrin iron complexes bound to proteins which, as enzymes, play an important role in the oxidative phosphorylation in the respiratory cycle.

Chlorophyll is the green colouring matter in higher plants. It acts as catalyst in the energy uptake from sunlight for the photosynthesis of carbohydrates from H_2O and CO_2. It consists of chlorophyll a and chlorophyll b (**31** and **32**) in a proportion of 5:2. These are optically active and can be separated by chromatography (TSWETT 1906).

Both chlorophylls contain a central magnesium ion in a chlorin chelate complex. Acid treatment removes the magnesium to give phaeophytin a and b, which can be hydrolysed yielding phytol ($C_{20}H_{39}OH$) and phaeophorbide a and b (**33** and **34**), respectively:

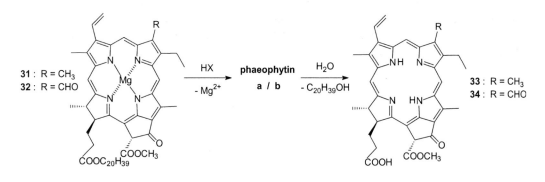

Phaeophorbide has been converted into aetioporphyrin **28**, which confirmed the structural relationship between blood and leaf pigments (H. FISCHER 1929-40).

Phaeophytins and phaeophorbides isomerize readily to porphyrins, the vinyl group accepting the H-atoms of the dihydropyrrole ring. The synthesis of chlorophyll a was carried out by WOODWARD in 1960 [27].

Cyanocobalamin 35 (vitamin B_{12}) was isolated from liver extract and from *Streptomyces griseus*. It is of great importance as the antipernicious anaemia factor. Its complex chemical structure was elucidated by X-ray investigation (CROWFOOT-HODGKIN 1957). Its central unit is a cobalt complex of a highly substituted corrin system (see p 177). The total synthesis of vitamin B_{12} was achieved by ESCHENMOOSER and WOODWARD in 1971 [28].

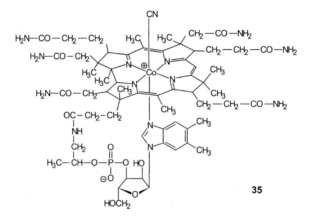

Tetraazaporphyrin contains in its porphyrin framework four N-atoms in place of the methine groups. Its purple magnesium complex **36** is formed by heating maleinonitrile with magnesium propanolate:

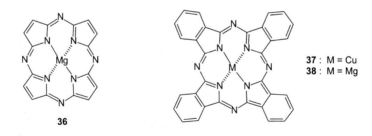

37 : M = Cu
38 : M = Mg

Phthalocyanines are tetrabenzotetraazaporphyrins. Copper(II) phthalocyanine **37** (LINSTEAD 1934) is prepared from phthalonitrile and copper powder/copper(I) chloride. Magnesium phthalocyanine **38** is obtained from Mg and 2-cyanobenzamide by thermal reaction in nitrobenzene [29], the free ligand system is produced by demetalation with acid of the magnesium complex.

Copper phthalocyanine is a dark blue, extremely stable compound which sublimes at 500°C without decomposition. It is not affected by heating in hydrochloric acid or potassium hydroxide. Copper phthalocyanine can be chlorinated or sulfonated. Its chlorination products (replacement of 14–16H by Cl) are green. Copper phthalocyanine and its substitution products are important as pigments in the dyestuff industry.

Verteporfin is a second-generation photosensitizer developed for photodynamic therapy of eye diseases, non-melanoma skin cancer and psoriasis [30]. It is constituted as a mixture of the two regio-isomeric dihydrobenzoporphyrine acids **39/40** obtained from non-selective hydrolysis of the corresponding (synthetic) tetraester.

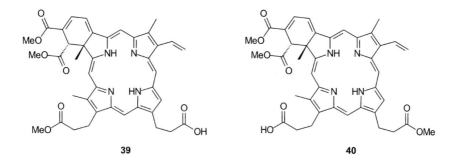

Porphyrinoids (porphyrin analogues, porphyrin homologues and porphyrin vinylogues [31]), like porphyrins, are of interest as sensitizers in photodynamic tumour therapy [32]. Some of these systems, e.g. **41 - 43** have been synthesized (FRANCK, VOGEL 1990) [33]. Their nomenclature takes into account the cyclic conjugation of the annulene perimeter and the number of methine groups between the four pyrrole units, e.g. porphyrin **1** is described as [18]porphyrin(1.1.1.1); porphycin **43**, which is isomeric with porphyrin, is known as [18]porphyrin(2.0.2.0).

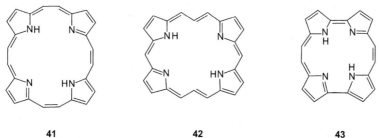

References

[1] D. W. McNeil, M. E. Kent, F. Hedaya, P. F. D'Angelo, P. O. Schissel, *J. Am. Chem. Soc.* **1971**, *93*, 3171; phenyl-substituted and benzannelated azocines are stable: B. M. Adgar, C. W. Rees, R. C. Storr, *J. Chem. Soc., Perkin Trans 1* **1975**, 45.

[2] L. A. Paquette, *Angew. Chem. Int. Ed. Engl.* **1971**, *10*, 11.

[3] L. A. Paquette, T. Kakihana, J. F. Kelly, *J. Org. Chem.* **1971**, *36*, 435; L. B. Anderson, J. F. Hansen, T. Kakihana, L. A. Paquette, *J. Am. Chem. Soc.* **1971**, *93*, 161.

[4] H. D. Perlmutter, *Adv. Heterocycl. Chem.* **1989**, *45*, 185.

[5] E. Vogel, H.-J. Altenbach, D. Kremer, *Angew. Chem. Int. Ed. Engl.* **1972**, *11*, 1013.

[6] H.-J. Altenbach, H. Stegelmeier, M. Wilhlem, B. Voss, J. Lex, E. Vogel, *Angew. Chem. Int. Ed. Engl.* **1979**, *18*, 962; M. Breuninger, B. Gallenkamp, K.-H. Müller, H. Fritz, H. Prinzbach, J. J. Daly, P. Schönholzer, *Angew. Chem. Int. Ed. Engl.* **1979**, *18*, 964.

[7] A. G. Anastassiou, *Acc. Chem. Res.* **1972**, *5*, 281; A. G. Anastassiou, *Acc. Chem. Res.* **1976**, *9*, 453; A. G. Anastassiou, H. S. Kasmai, *Adv. Heterocycl. Chem.* **1978**, *23*, 55.

[8] E. Vogel, *Isr. J. Chem.* **1980**, *20*, 215.

[9] M. Schäfer-Ridder, A. Wagner, M. Schwamborn, H. Schreiner, E. Devrout, E. Vogel, *Angew. Chem. Int. Ed. Engl.* **1978**, *17*, 853.

[10] E. Vogel, R. Feldmann, H. Düwel, H.-D. Kremer, H. Günther, *Angew. Chem. Int. Ed. Engl.* **1972**, *11*, 217.

[11] L. A. Paquette, H. C. Berk, S. V. Ley, *J. Org. Chem.* **1975**, *40*, 902.

[12] H. Röttele, G. Schröder, *Angew. Chem. Int. Ed. Engl.* **1980**, *19*, 207.

[13] W. Gilb, G. Schröder, *Angew. Chem. Int. Ed. Engl.* **1979**, *18*, 312.

[14] R. Neuberg, J. F. M. Oth, G. Schröder, *Liebigs Ann. Chem.* **1978**, 1368.

[15] A. D. Hamilton in: A. R. Katritzky, C. W. Rees, W. Lwowski (Eds.), *Comprehensive Heterocyclic Chemistry,* Vol. 7, p 731, Pergamon, Oxford **1984**.

[16] D. Dolphin (Ed.), *The Porphyrins* (several volumes), Academic, New York **1978**;

Porphyrine, Phlorine und Corrine, in: G. Habermehl, P. E. Hammann, *Naturstoffchemie*, p 487ff, Springer, Berlin **1992**.

[17] Aromaticity of porphyrins: P. v. Rague Schleyer et al., *Angew. Chem. Int. Ed. Engl.* **1998**, *37*, 177.

[18] K. M. Smith, G. H. Barnett, B. Evans, Z. Martynenko, *J. Am. Chem. Soc.* **1979**, *101*, 5953.

[19] For a review on the synthesis of chlorines and related systems see: F.-P. Montforts, M. Glasenapp-Breiling, *Progr. Heterocycl. Chem. 10*, **1998**, 1

[20] J. S. Lindsey, I. C. Schreiman, H. C. Hsu, P. C. Kearney, A. M. Marguerettaz, *J. Org. Chem.* **1987**, *52*, 827.

[21] J. L. Sessler, A. Mozaffari, M. R. Johnson, *Org. Synth.* **1992**, *70*, 68.

[22] H. W. Whitlock, R. Hanauer, *J. Org. Chem.* **1968**, *33*, 2169.

[23] L. F. Tietze, U. Beifuss, *Angew. Chem. Int. Ed. Engl.* **1993**, *32*, 131.

[24] B. Franck, *Angew. Chem. Int. Ed. Engl.* **1982**, *21*, 343; A. R. Battersby, *Pure Appl. Chem.* **1989**, *61*, 337.

[25] K. N. Smith in *Compr. Heterocycl. Chem.* (Eds. A. R. Katritzky, C. W. Rees) Vol. 4, p 416, Pergamon Press, Oxford 1984; for a modification of this principle of porphyrine synthesis see: C.-B. Wang, C. K. Chang, *Synthesis* **1979**, 548.

[26] H. Fischer, K. Zeile, *Liebigs Ann. Chem.* **1929**, *468*, 98.

[27] R. B. Woodward, *Angew. Chem.* **1960**, *72*, 651; R. B. Woodward, *Tetrahedron* **1990**, *46*, 22.

[28] R. B. Woodward, *Pure Appl. Chem.* **1973**, *33*, 145; *Pure Appl. Chem.* **1971**, *25*, 283; A. Eschenmoser in: 23rd *IUPAC Congress Boston 1971*, Vol. 2, p 69, Butterworth, London **1971**.

[29] L. F. Tietze, Th. Eicher, *Reactions and Synthesis in the Organic Chemistry Laboratory*, p 405, University Science Books, Mill Valley, CA **1989**.

[30] L. J. Scott, K. L. Goa, *Drugs and Aging* **2000**, *16*, 139; F. Sell, *Arch. Ophthalmol.* **1999**, *117*, 1400.

[31] E. Vogel, *J. Heterocycl. Chem.* **1996**, *33*, 1461; J. L. Sessler, S. J. Weghorn, *Expanded, contracted and isomeric porphyrins*, Pergamon Press, Oxford **1997**.

[32] *Photosensitizing Compounds: Their Chemistry, Biology and Clinical Use* (Ciba Foundation Symposium 146), Wiley, New York **1989**.

[33] B. Franck, H. König, C. Eickmeier, *Angew. Chem. Int. Ed. Engl.* **1990**, *29*, 1393;
E. Vogel, M. Jux, E. Rodriguez-Val, J. Lex, H. Schmickler, *Angew.Chem. Int. Ed. Engl.* **1990**, *29*, 1387;
J. L. Sessler, A. K. Burrell, *Top. Curr. Chem.* **1992**, *161*, 177;
E. Vogel, *Pure Appl. Chem.* **1993**, *65*, 143;
Th. Wessel, B. Franck, M. Möller, U. Rodewald, M. Läge, *Angew. Chem. Int. Ed. Engl.* **1993**, *32*, 1148;
porphycin [1.1.0.2]: E. Vogel et al., *Angew. Chem. Int. Ed. Engl.* **1997**, *36*, 1651.

9   Problems and Their Solutions

The following problems are organized according to the contents of the foregoing chapters. Thus, 1-11 refer to chapters 3/4, 12-62 to chapter 5, 63-109 to chapter 6 and 110-122 to chapters 7/8; 123-143 cover miscellaneous topics.

The purpose of these problems is twofold. First, they may serve as an exercise to deepen and to extend the facts and materials of heterocyclic chemistry presented in this book. Second, the reader should be enabled to apply his knowledge to the understanding of results and developments in the modern literature on reactivity, synthesis and application of heterocycles. Therefore, the explicit problem solutions throughout are n o t given in detail: It is up to the reader to elaborate the solutions for himself by studying the references and literature results from which the problems have been selected.

The problems are arranged in concise and standardized form as follows:

| P | Formulation of the **problem**, reactant(s), experimental conditions and other data relevant for the transformations described and for the formation of products;

| Q | **questions** to be answered and considered such as structure and/or stereochemistry of products, reaction types, principles and names, reaction intermediates and mechanisms (in most cases summarized by the term "mode of formation"), systematic nomenclature of products;

| S | literature references, notes and additional information relevant for the **solution** of the problem.

Problem 1

| P | 1,2-Diaminocyclohexane [(*R*,*R*)- and (*S*,*S*)-enantiomer] forms an imine (SCHIFF base) with 2,5-di-*tert*-butylsalicylaldehyde, which gives a chiral Mn(III) (salen) complex with Mn(II)acetate and oxygen. In contrast to the SHARPLESS-KATSUKI protocol (p 20), this complex effects the stereoselective oxygen transfer (from oxidants, e.g. monopersulfate or NMO) to **un**functionalized alkenes (JACOBSEN epoxidation [1], extended by KATSUKI [2]) giving rise to enantiomeric oxiranes with 90-98% *ee*.

| Q | Suggest a reasonable mechanism for this asymmetric epoxidation reaction.

| S | The mechanism of the JACOBSEN epoxidation is discussed on the basis of MO calculations in ref. [3].

[1] E. N. Jacobsen et al., *J. Org. Chem.* **1994**, *59*, 1939; *J. Am. Chem. Soc.* **1991**, *113*, 7063.
[2] K. Miura, T. Katsuki, *Synlett* **1999**, 783; P. Pietikäinen, *Tetrahedron Lett.* **2000**, *40*, 1001.
[3] L. Cavallo, E. N. Jacobsen, *Angew. Chem. Int. Ed. Engl.* **2000**, *39*, 589.

Problem 2

P Starting with cycloheptene (**X**) the following sequence of reactions is carried out:

(a) **X**, *m*-chloroperbenzoic acid → **A** (70%) ;

(b) **A**, C_6H_5SH, hexafluoroisopropanol (solvent) → **B** (stereochemistry ?) ;

(c) **B**, H_2O_2 → **C** (80%, two steps) ;

(d) **C**, K_2CO_3, 170 °C (vacuum) → **D** (90%).

Q What are the structures of the products **A** – **D** and their modes of formation ?

S Consult [4] and compare other ring-opening reactions of oxiranes, e.g. the β-cyclodextrin-catalyzed aminolysis [5] and problem 3. For an enantioselective ring-opening of oxiranes utilizing the JACOBSEN catalyst (problem 1) see [6].

[4] V. Kesavan, D. Bonnet-Delpon, J. P. Bégué, *Tetrahedron Lett.* **2000**, *41*, 2895.
[5] L. R. Reddy, M. A. Reddy, N. Bhanumathi, K. R. Rao, *Synlett* **2000**, 339.
[6] J. L. Leighton, E. N. Jacobsen, *J. Org. Chem.* **1996**, *61*, 389.

Problem 3

P Perform the following sequence of reactions:

(a) 2-(*O*-Benzyl)glycerol, allyl bromide, base → **A** ;

(b) **A** + *m*-chloroperbenzoic acid (excess) → **B** ;

(c) **B** + H_2N-CH_2-CH_2-CH_2-NH_2 + H_2O → **C** (90%).

Q What are the structures of the products **A** – **C** and their modes of formation ?

S Consult ref. [7] and compare with problem 2.

[7] M. F. Sebban, P. Vottero, A. Alagui, C. Dupuy, *Tetrahedron Lett.* **2000**, *41*, 1019.

Problem 4

P 2-Homopropargyl cyclohexanone reacts with the S-ylide from trimethyl sulfoxonium iodide in a COREY synthesis (cf. p 21). The resulting spiroepoxide **A** can be subjected to a 5-exo-cyclization catalyzed by titanocene (Cp_2TiCl_2) in the presence of a stoichiometric amount of Mn yielding a product **B** in 90% yield.

Q Suggest the structures of **A** and **B**, the stereochemistry of **B** and a suitable mechanism for its formation.

S According to ref. [8], a radical cyclization process has to be envisaged, in which redoxactive Cp_2TiCl participates; Mn plays the role of a reductant regenerating the Ti(III) species in the catalytic cycle.

By this cyclization tetrahydrofuran derivatives have also been prepared. Compare the LEWIS-acid induced tandem oxacyclization with oxepane formation in ref. [9].

[8] A. Gansäuer, M. Picrobon, *Synlett* **2000**, 1357.
[9] F. E. McDonald, X. Wang, B. Do, K. I. Hardcastle, *Org. Lett.* **2000**, *2*, 2917.

Problem 5

 P The methyl ester of 2-methyloxirane-2-carboxylic acid reacts with acetonitrile in the presence of BF$_3$-etherate and gives in 90% yield a product **A**, which on hydrolysis in aqueous acid leads to CH$_3$COOH and a tertiary α-aminoalcohol **B**.

Q Suggest structures for **A/B** and a reasonable mechanism for the formation of **A**.

S Consult ref. [10] and p 134.

[10]   J. L. García Ruano, C. García Paredes, *Tetrahedron Lett.* **2000**, *41*, 5357.

Problem 6

P When isatoic anhydride (**A**) is subjected to flash vacuum pyrolysis at 550°C, a product C$_7$H$_5$NO (**B**) is obtained in 80% yield which is stable only at temperatures below -20°C.

Q (a) Which is the structure of **B**; which second product is observed ?
(b) Give a short synthesis of **A** starting from phthalimide.

S Consult for (a) ref. [11] and for (b) p 377.

[11]   S.-J. Chiu, C.-H. Chou, *Tetrahedron Lett.* **1999**, *40*, 9271.

Problem 7

 P In the so-called aza-PAYNE rearrangement [12], (aminomethyl)oxiranes **A** are transformed to (hydroxymethyl)aziridines **B**. However, if compounds **A** are treated with magnesium bromide in dioxane, isomers of **B** (of different ring size) are obtained (**C**).

Q Discuss the structure of **C** and explain its mode of formation from **A**.

 S MgBr$_2$ effects a regioselective oxirane ring-opening by bromide and recyclization by S$_{Ni}$-displacement of Br via amino nitrogen according to ref. [13]. The reaction proceeds with a net retention of configuration.

[12]   P. C. Bulman Page, C. M. Rayner, I. O. Sutherland, *J. Chem. Soc. Chem. Commun.* **1988**, 356.
[13]   M. Karikomi, K. Arai, T. Toda, *Tetrahedron Lett.* **1997**, *38*, 6059.

Problem 8

 P (*N*-Benzyl)benzaldimine reacts with ethyl diazoacetate in the presence of lanthane(III)-trifluoroacetate in EtOH to give a product **A** in good yield.

Q Discuss the structure of **A**, its stereochemistry and its mode of formation.

S Consult ref. [14]. For this synthetic protocol, an enantioselective version has been elaborated [15] using chiral boron LEWIS acids.

[14]   W. Xie, J. Fang, J. Li, P. G. Wang, *Tetrahedron* **1999**, *55*, 12929.
[15]   J. C. Antilla, W. D. Wulff, *J. Am. Chem. Soc.* **1999**, *121*, 5099.

Problem 9

P 2-Phenyl-1-(phenylsulfonyl)aziridine is treated (a) with NaI, (b) with phenylisothiocyanate to give a product **A** in good yield.

Q Discuss the structure of **A** and its mode of formation.

S According to ref. [16] the reaction is believed to proceed via initial attack of iodide. Instead of iodide, the reaction can be initiated by (PhCN)$_2$PdCl$_2$ [17].

[16] U. K. Nadir, N. Basu, *J. Org. Chem.* **1995**, *60*, 1458.
[17] J.-O. Baeg, C. Bensimon, H. Alper, *J. Am. Chem. Soc.* **1995**, *117*, 4700.

Problem 10

P Benzaldehyde and styrene undergo cyclization on irradiation.

Q Which reaction occurs ? Discuss the stereochemical outcome and the possibilities for stereo-control in this photocyclization.

S Consult ref. [18] and p 39. For stereoselectivity by use of trimethylsilylether substitutents at the styrene moiety see ref. [19].

[18] T. Bach, *Liebigs Ann. Chem.* **1997**, 1627.
[19] T. Bach, F. Eilers, K. Kather, *Liebigs Ann. Chem.* **1997**, 1529.

Problem 11

P Starting with the monomethylester of methyl(*tert*-butoxymethyl)malonic acid **X** the following sequence of reactions is carried out :

(a) **X** + SOCl$_2$ → **A** ;

(b) **A**, CH$_2$Cl$_2$, rfl. (- HCl) → **B** (54%, two steps).

Q What are the structures of the products **A/B** and their modes of formation ?

S Consult for (a) ref. [20], for (b) p 40.

[20] F. R. Alexandre, S. Legoupy, F. Huet, *Tetrahedron* **2000**, *56*, 3921.

Problem 12

P Starting with sodium *o*-aminobenzenesulfinate (**X**) the following sequence of reactions is performed :

(a) **X**, NaNO$_2$, H$_2$SO$_4$/H$_2$O, -6°C; then Et$_2$O, -20°C → **A** (65%, thermally unstable) ;

(b) **A**, furan (excess), rfl. (under N$_2$) → **B** (54%) ;

(c) **B**, HCl/CH$_3$OH, Δ → **C** (91%, isomeric to **B**).

Q What are the structures of products **A** – **C** and their modes of formation ? Formulate inter-mediates and the systematic name of **A**.

S Consult ref. [21] and p 56. For further information see ref. [22].

[21] G. Wittig, R. W. Hoffmann, *Chem. Ber.* **1962**, *95*, 2718.
[22] R. W. Hoffmann, *Dehydrobenzene and Cycloalkynes,* p 78, Verlag Chemie, Weinheim **1967**.

Problem 13

| P | Starting from 1,1,3,3-tetrabromoacetone the following reactions are performed :
(a) X + $Fe_2(CO)_9$ → A ;
(b) A + furan; then Zn/HCl → B ;
(c) H_2, Pd-C; then FSO_3H → C ;
(d) C, DDQ → D.

| Q | What are the structures of the products A – D and their modes of formation ? Which intermediates are formed?

| S | Consult refs. [23]/[24]. A is a reactive intermediate, which is formed from tetrabromoacetone and Fe_2CO_9 and trapped by furan to give dibromo-B, Zn/HCl effects reductive debromination.

[23] R. Noyori, T. Sato, Y. Hayakawa, *J. Am. Chem. Soc.* **1978**, *100*, 2561.
[24] L. S. Hegedus, *Organische Synthese mit Übergangsmetallen*, p 121, VCH, Weinheim **1995**.

Problem 14

| P | With furan-2-carboxylic acid (X) as reactant, the following reactions are carried out :
(a) X, 2-amino-2-methylpropanol, TosOH (cat.) → A ;
(b) A, *sec*-BuLi, THF, -78°C → B ;
(c) B, CH_3I, then H_2O/HX → C.

| Q | What are the structures of the products A – C and their modes of formation ?

| S | Consult ref. [25] and p 136. Product C is soluble in aqueous Na_2CO_3 solution.

[25] D. J. Chadwick, M. V. McKnight, R. I. Ngochindo, *J. Chem. Soc. Perkin Trans. 1* **1982**, 1343.

Problem 15

| P | The furanosesquiterpene tsitsikammafuran $(X, C_{15}H_{18}O)$ was recently isolated from the South African sponge *Dysidea* and its structure was proven by the following synthetic sequence :
(a) 2-Isopropyl-5-methylphenol (thymol), PBr_5, Δ → A (53%) ;
(b) A, *n*-BuLi, TMEDA, THF, -78°C; then 3-furaldehyde, then H_2O → B (60%) ;
(c) B, Me_3SiCl, NaI → X (100%).

| Q | Which structure possesses X on the basis of its independent synthesis ?

| S | Consult ref. [26] for structure and synthesis of X. Reaction (c) is used as a dehydroxylation procedure [27].

[26] K. L. McPhail, D. E. A. Rivett, D. E. Lack, M. T. Davies-Coleman, *Tetrahedron* **2000**, *56*, 9391.
[27] P. J. Perry, V. H. Pavlidis, I. C. G. Coutts, *Synth. Commun.* **1996**, *26*, 101.

Problem 16

P Furfuryl alcohols are converted to pyranone derivatives by the ACHMATOWICZ oxidative ring expansion methodology, which is frequently applied in carbohydrate chemistry [28]. Starting from furfuryl alcohol (**X**), the following reactions are performed:

(a) **X**, Br$_2$, CH$_3$OH, -40°C ; then NH$_3$ (gas), -40°C to r.t. → **A** (90%) ;

(b) **A**, trifluoromethanesulfonic acid, H$_2$O → **B** ;

(c) **B**, Ac$_2$O, NaOAc → **C** (69%, 2 steps).

Q What are the structures of the products **A – C** and their modes of formation ?

S Consult ref. [29], for (a) compare p 55.

[28] J. M. Harris, M. D. Keranen, G. A. O'Doherty, *J. Org. Chem.* **1999**, *64*, 2982;
     D. Balachari, G. A. O'Doherty, *Org. Lett.* **2000**, *2*, 863.
[29] D. Caddick et al., *Tetrahedron* **2000**, *56*, 8953.

Problem 17

P The oxidative cycloaddition of 1,3-dicarbonyl compounds to cyclic enol ethers [30] can be conveniently mediated by tetra-*n*-butylammonium peroxidisulfate as exemplified by the following reaction :

Cyclohexane-1,3-dione, 2,3-dihydrofuran, [*n*-Bu$_4$N]$_2$S$_2$O$_8$, KOAc, CH$_3$CN, r.t. → **A** (90%).

Q What is the structure of product **A** and its mode of formation ?

S Consult ref. [31]. For an alternative approach to fused cyclic acetals see ref. [32].

[30] S. C. Roy, P. K. Mandal, *Tetrahedron* **1996**, *52*, 12495.
[31] F. E. Chen, H. Fu, G. Meng, Y. Cheng, Y.-L. Hu, *Synthesis* **2000**, 1091.
[32] M. C. Pirrung, J. Zhang, A. T. McPhail, *J. Org. Chem.* **1991**, *56*, 6269.

Problem 18

P Perform the following sequence of reactions :

(a) (CH$_3$)$_3$C-CO-CH$_2$-CN + HO-CH$_2$-COOEt, Ph$_3$P, EtOOC-N=N-COOEt, THF
     → **A** (80%, conditions of a MITSUNOBU reaction) ;

(b) **A**, NaH, THF → **B** (88%) .

Q What are the structures of the products **A/B** and their modes of formation ?

S Consult ref. [33].

[33] A. M. Redman, J. Dumas, W. J. Scott, *Org. Lett.* **2000**, *2*, 2061.

Problem 19

| P | Perform the following reactions:

(a) *tert*-Butyl acetoacetate, phenacyl bromide, NaH, THF → **A** (95%) ;

(b) **A**, CF_3COOH, CH_2Cl_2/THF → **B** (67%) ;

(c) **B**, THF, NaH; then Br-CH_2-$COOCH_3$ → **C** (89%) ;

(d) **C**, CF_3COOH, CH_2Cl_2/THF → **D** (67%) .

| Q | What are the structures of the products **A** – **D** and their modes of formation ?

| S | Consult ref. [34], compare with p 59 (FEIST-BENARY synthesis).

[34] F. Stauffer, R. Neier, *Org. Lett.* **2000**, *2*, 3535.

Problem 20

| P | Starting with 3-bromothiophene-2-carbaldehyde (**X**) the following sequence of reactions is carried out :

(a) **X**, HS-CH_2-$COOCH_3$, K_2CO_3, DMF, r.t. → **A** (80%) ;

(b) **A**, LiOH, H_2O → **B** (90%) ;

(c) **B**, Cu, quinoline (solvent), +260°C → **C** (88%).

| Q | What are the structures of the products **A** – **C** and their modes of formation ? Give the systematic name of **C**.

| S | Consult ref. [35] and compare p 76. Outline a 2-step synthesis of **X** starting from 3-bromothiophene.

[35] L. S. Fuller, B. Iddon, K. A. Smith, *J. Chem. Soc. Perkin Trans. 1* **1997**, 3465. .

Problem 21

| P | Starting with cyclohexane-1,3-dione (**X**) the following reactions are performed :

(a) **X**, K_2CO_3, (1) CS_2, (2) Br-CH_2-COOEt, then CH_3I → **A** ;

(b) **A**, $POCl_3$, DMF → **B** (61%; VILSMEYER-HAACK reaction, cf. p 76) ;

(c) **B**, Na_2S, Cl-CH_2-COOEt, then CH_3ONa/CH_3OH → **C** (63%) .

| Q | What are the structures of the products **A** – **C** and their modes of formation ? Which intermediates are formed ?

| S | Consult ref. [36].

[36] D. Prim, D. Joseph, G. Kirsch, *Liebigs Ann. Chem.* **1996**, 239; D. Prim, G. Kirsch, *Synth. Commun.* **1995**, *25*, 2449.

Problem 22

P *N,N*-Diethylacetoacetamide (**X**) is subjected to the following reactions :

    (a) **X**, 2.3 equiv. LDA, THF (solvent), -78°C to +20°C ;

        then Cl-CH$_2$-CH$_2$-Br → **A** (intermediate) ;

    (b) **A**, reflux → **B** (80%, > 98:2-mixture of stereoisomers).

Q What are the structures of the products **A/B** and their modes of formation ? Which configuration possesses **B** ?

S Consult ref. [37]; compare ref. [38], where a similar reaction principle is reported.

    [37] P. Langer, I. Karimé, *Synlett* **2000**, 743.
    [38] M. Makosza, J. Przyborowski, R. Klayn, A. Kwast, *Synlett* **2000**, 1773.

Problem 23

P Starting from diethylketone (**X**) the following sequence of reactions is performed :

    (a) **X**, HCOOEt, NaH, EtOH/benzene → **A** (55%) ;

    (b) **A**, ethyl glycinate hydrochloride, Et$_3$N, EtOH → **B** (85%) ;

    (c) **B**, EtONa, EtOH, refl. → **C** (45%).

Q What are the structures of the products **A** – **C** and their modes of formation ?

S Consult ref. [39] and compare p 98 (Kenner synthesis).

    [39] H. K. Hombrecher, G. Horter, *Synthesis* **1990**, 389 ;
         G. H. Walizei, E. Breitmaier, *Synthesis* **1989**, 337.

Problem 24

P In a one-pot three-component cyclocondensation, but-2-en-1-yl ethyl ketone, *n*-butylamine and nitromethane form a product **A** in 47% yield in THF at 60°C in the presence of SmI$_2$ as catalyst.

Q Suggest the structure for **A** and discuss its mode of formation.

S Consult ref. [40].

    [40] H. Shiraishi, T. Nashitani, S. Sakaguchi, Y. Ishii, *J. Org. Chem.* **1998**, *63*, 6234.

Problem 25

P Starting from methyl *N*-tosyl(phenylpropargyl)glycinate (**X**) the following reactions are carried out :

    (a) **X**, I$_2$, K$_2$CO$_3$, 0°C to r.t., acetonitrile (solvent) → **A** (74%) ;

    (b) **A**, DBU, DMF (solvent), r.t. → **B** (90%).

Q What are the structures of the products **A/B** and their modes of formation ?

S Consult ref. [41].

    [41] D. W. Knight, A. L. Redfern, J. Gilmore, *J. Chem. Soc. Chem. Commun.* **1998**, 2207.

Problem 26

P│ Perform the following reactions :

(a) 1-Allyl-1,2,3-benzotriazole, n-BuLi; then $(4\text{-ClC}_6\text{H}_4)\text{-CH=N-Ph}$ → **A** (80%) ;

(b) **A**, Pd(OAc)$_2$, Ph$_3$P, CuCl$_2$, K$_2$CO$_3$ → **B** (76%) .

Q│ What are the structures of the products **A/B** and their modes of formation ?

S│ Consult ref. [42], for comparison see p 208 and p 225 (KATRITZKY's benzotriazole-mediated methodology). For a further instructive example see ref. [43].

[42] A. R. Katritzky, L. Zhang, J. Yao, O. V. Denisko, *J. Org. Chem.* **2000**, *65*, 8074.
[43] A. R. Katritzky, T.-B. Huang, M. V. Voronkov, M. Wang, H. Kolb, *J. Org. Chem.* **2000**, *65*, 8819.

Problem 27

P│ Starting with methyl cinnamate (**X**), the following reactions are performed :

(a) **X**, TosMIC, NaH, DMSO/Et$_2$O (solvents) → **A** (70%) ;

(b) **A**, KOH, MeOH/H$_2$O; then HCl/H$_2$O → **B** ;

(c) **B**, HO-CH$_2$-CH$_2$-NH$_2$ (solvent), Δ → **C** (82%, 2 steps) .

Q│ What are the structures of the products **A** – **C** and their modes of formation ?

S│ Consult ref. [44], for comparison see p 128/129, for an application see problem 28.

[44] N. P. Parvi, M. L. Trudell, *J. Org. Chem.* **1997**, *62*, 2649.

Problem 28

P│ Starting with the methylester of pyrrole-2-carboxylic acid (**X**), the following sequence of reactions is carried out :

(a) **X**, CH$_3$O-CHCl$_2$, AlCl$_3$, CH$_2$Cl$_2$ (solvent) → **A** (80%) ;

(b) **A**, TosCl, NaH → **B** ;

(c) **B**, (EtO)$_2$PO-CH$_2$-COOBn, NaH, THF (solvent) → **C** (80%, 2 steps) ;

(d) **C**, TosMIC, LiN(SiMe$_3$)$_2$ → **D** (65%) ;

(e) **D**, (COCl)$_2$, CH$_2$Cl$_2$ (solvent), 0°C → **E** ;

(f) **E**, SnCl$_4$, 0°C → **F** (72%, 2 steps).

Q│ What are the structures of the products **A** – **F** and their modes of formation ?

S│ Consult ref. [45] for this sequence. For a review on the subject of the sequence given see [46].

[45] P. Carter, S. Fitzjohn, S. Halazy, P. Magnus, *J. Am. Chem. Soc.* **1987**, *109*, 2711.
[46] D. C. Boger, C. W. Boyce, R. M. Garbacchio, J. A. Goldberg, *Chem. Rev.* **1997**, *97*, 787.

Problem 29

P Perform the following reactions :

(a) Br-CN + HN(CH$_3$)$_2$ → **A** (94%) ;

(b) **A** + Cl$_2$C=N$^+$(CH$_3$)$_2$Cl$^-$ → **B** (96%, salt !) ;

(c) **B** + [C$_5$H$_5$]$^-$Na$^+$ → **C** (82%, red crystals).

Q What are the structures of the products **A** – **C** and their modes of formation ?

S Consult refs. [47] and [48] .

[47] H.-J. Gais, K. Hafner, *Tetrahedron Lett.* **1974**, 771. .
[48] L. F. Tietze, Th. Eicher, *Reactions and Synthesis in the Organic Chemistry Laboratory*, University Science Books, Mill Valley, CA **1989**.

Problem 30

P When D,L-alanine and α-chloroacrylonitrile are heated at +100°C in an excess of acetic anhydride, CO$_2$ is evolved and 3-cyano-2,5-dimethylpyrrol (**A**) is isolated as sole reaction product in 70% yield.

Q Explain the formation of **A** by a reasonable mechanism and formulate the intermediates of this reaction.

S Consult ref. [49] and compare p 129 and 192.

[49] I. A. Benages, S. M. Albonico, *J. Org. Chem.* **1978**, *43*, 4273.

Problem 31

P Starting from *o*-nitrotoluene (**X**) the following reactions are carried out :

(a) **X**, (H$_3$C)$_2$N-CH(OCH$_3$)$_2$ → **A** (not isolated) ;

(b) **A**, (H$_3$C)$_2$NH, H$_2$C=O, AcOH/CH$_3$OH → **B** (86%, 2 steps) ;

(c) **B**, O=CH-NH-CH(COOEt)$_2$ (**Y**), NaOEt → **C** (78%) ;

(d) **C**, H$_2$/Raney-Ni; then Δ (- H$_2$O) → **D** (96%) ;

(e) **D**, NaOH, H$_2$O; then AcOH, Δ → **E** (80%) .

Q What are the structures of the products **A** – **E** and their modes of formation ?

S Consult refs. [48] and [50], compare p 104. For the synthesis of **Y** see ref. [48].

[50] U. Hengartner, A. D. Batcho, J. F. Blount, W. Leimgruber, M. E. Larscheid, J. W. Scott, *J. Org. Chem.* **1979**, *44*, 3748.

Problem 32

P Perform the following reactions :

(a) H₂C=CH-CH=O, CH₃CO-NH-CH(COOEt)₂, NaOEt → **A** (not isolated) ;

(b) **A**, phenylhydrazine, AcOH → **B** (67%; 2 steps) ;

(c) **B**, H₂O/H₂SO₄ → **C** (72%) ;

(d) **C**, NaOH/H₂O; then AcOH → **D** (80%).

Q What are the structures of the products **A** – **D** and their modes of formation ?

S Consult ref. [51] and compare problem 31.

> [51] Th. Eicher, H. J. Roth, *Synthese, Gewinnung und Charakterisierung von Arzneistoffen*, Thieme, Stuttgart **1986**.

Problem 33

P When *o*-iodoaniline, pyruvic acid and DABCO are heated to 105°C in DMF in the presence of a catalytic amount of Pd(OAc)₂, a product **A** is obtained in 82% yield after the usual workup.

Q What is the structure of **A** and its mode of formation ? Formulate the intermediates of this reaction.

S Consult ref. [52] and compare p 105.

> [52] C. Chen, D. R. Lieberman, R. D. Larsen, T. R. Verhoeven, P. R. Reider, *J. Org. Chem.* **1997**, *62*, 2676.

Problem 34

P Starting with *o*-iodoaniline (**X**) the following reactions are performed :

(a) **X**, H₂C=CH-CH₂-N(Boc)₂, catalytic amounts of Pd(OAc)₂ and P(*o*-tolyl)₃, Et₃N, acetonitrile (solvent), rfl. → **A** (73%) ;

(b) **A**, H₂/CO (1 : 1), catalytic amount of HRh(CO)(PPh₃)₃, PPh₃, 70°C → **B** (69%), (the formation of **B** proceeds via an intermediate **C**) ;

(c) **B**, CF₃COOH → **D** (80%) .

Q What are the structures of the products **A** – **D** and their modes of formation ? Formulate the intermediates of the reactions (a) – (c).

S Consult ref. [53] and compare problem 33.

> [53] Y. Dong, C. A. Busacca, *J. Org. Chem.* **1997**, *62*, 6464.

Problem 35

P Starting from indole (**X**) the following sequence of reactions is carried out:

(a) **X**, Na[H₃BCN], AcOH → **A** (85%) ;

(a) **X**, Na[H$_3$BCN], AcOH → **A** (85%) ;

(b) **A**, HCOOH, toluene (solvent), rfl. → **B** (94%) ;

(c) **B**, (1) (COCl)$_2$, (2) HÜNIG's base, (3) Br$_2$, CHCl$_3$, (4) H$_2$O, HCl → **C** (79%) .

Q What are the structures of the products **A – C** and their modes of formation ? Formulate the intermediates in reaction (c).

S Consult ref. [54] and compare the SANDMEYER isatine synthesis in ref. [48].

[54] O. Meth-Cohn, S. Goon, *Tetrahedron Lett.* **1996**, *37*, 9381;
      Y. Kumar, L. Florvall, *Synth. Commun.* **1983**, *13*, 489.

Problem 36

P Indigo is subjected to the following transformations :

(a) oxidation with HNO$_3$;

(b) dehydrogenation with PbO$_2$;

(c) reaction with Na$_2$S$_2$O$_4$ in aqueous NaOH ;

(d) thermolysis at 460°C.

Q What are the products of the reactions (a) – (d) and their modes of formation ?

S Consult refs. [55]/[48] and p 110.

[55] G. Haucke, G. Graness, *Angew. Chem. Int. Ed. Engl.* **1995**, *34*, 67.

Problem 37

P When phenylhydrazine and 4-chlorobutanal are reacted in refluxing ethanol, a 70% yield of tryptamine is obtained after the usual workup (GRANDBERG synthesis).

Q Explain the mechanism of this indole synthesis and formulate its intermediates.

S Consult ref. [56] and compare p 106.

[56] I. I. Grandberg, *Chem. Heterocycl. Compounds* (Engl. Translation) **1974**, *10*, 501.

Problem 38

P Starting from *o*-aminobenzophenone (**X**) the following reactions are carried out :

(a) **X**, benzoyl chloride, pyridine → **A** (90%) ;

(b) **A**, titanium on graphite (TiCl$_3$/2 C$_8$K), THF, rfl. → **B** (90%).

Q What are the structures of the products **A/B** and their modes of formation ?

S Consult ref. [57]. Outline an alternative synthesis for **B** based on phenylhydrazine.

[57] A. Fürstner, B. Bogdanovic, *Angew. Chem. Int. Ed. Engl.* **1996**, *35*, 2442; A. Fürstner, A. Hupperts, A.
      Ptock, E. Janssen, *J. Org. Chem.* **1994**, *59*, 5215.

Problem 39

P Starting from aniline (**X**) the following sequence of reactions is performed :

(a) **X**, acrylonitrile, Cu(OAc)$_2$ → **A** (85%) ;

(b) **A**, hydroxylamine hydrochloride, NaHCO$_3$, H$_2$O/EtOH → **B** (70%) ;

(c) **B**, ethyl benzoate, NaOEt → **C** (85%) ;

(d) **C**, *n*-butanol (solvent), rfl. → **D** (95%, isomer of **C**) .

Q What are the structures of the products **A** – **D** and their modes of formation ?

S Consult refs. [58] and [48].

[58] D. Carbonits, E. M. Bako, K. Horvath, *J. Chem. Res. (S)* **1979**, 64.

Problem 40

P Starting from ethyl 3-phenyl-3-oxopropanoate (**X**) the following reactions are performed :

(a) **X**, hydrazine hydrate, EtOH/H$_2$O → **A** (72%) ;

(b) **A**, Cl$_2$, nitromethane (solvent) → **B** (95%) ;

(c) **B**, NaOH, H$_2$O → **C** (89%) + N$_2$.

Q What are the structures of the products **A** – **C** and their modes of formation ? Formulate the intermediates for the reactions (a) – (c).

S Consult refs. [59] and [48].

[59] L. A. Carpino, *J. Am. Chem. Soc.* **1958**, *80*, 599.

Problem 41

P Starting from *N*-benzyl-*N*-formylglycine ethylester (**X**) the following sequence of reactions is performed :

(a) **X**, HCOOEt, EtONa → **A** (40%) ;

(b) **A**, KSCN, HCl, H$_2$O → **B** (97%) ;

(c) **B**, HNO$_3$ → **C** (83%) ;

(d) **C**, LiAlH$_4$, diethylether → **D** (82%) ;

(e) **D**, SOCl$_2$ → **E** (99%) ;

(f) **E**, CH$_3$CO-NH-CH(COOEt)$_2$ → **F** (83%) ;

(g) **F**, HCl, H$_2$O → **G** (45%) .

Q What are the structures of the products **A** – **G** and their modes of formation ? Give the systematic name of **G**.

S Consult ref. [60] and compare ref. [51], p 143.

[60] R. G. Jones, *J. Am. Chem. Soc.* **1949**, *71*, 644; R. G. Jones, K. McLaughlin, *J. Am. Chem. Soc.* **1949**, *71*, 2444; T. B. Stensboel et al., *J. Med. Chem.* **2002**, *45*, 19.

Problem 42

P   Starting with phenyloxirane (**X**) the following sequence of reactions is performed :

(a) **X**, $H_2N-CH_2-CH_2-OH$, Δ → **A** (50%) ;

(b) **A**, HCl (gas), toluene (solvent); then $SOCl_2$, then H_2O → **B** ;

(c) **B**, H_2O, thiourea; then TosOH → **C** (66%, 2 steps, TosOH salt) ;

(d) **C**, $SOCl_2$; then $NaHCO_3/H_2O$ → **D** (54%) .

Q   What are the structures of the products **A – D** and their modes of formation ?

S   Consult ref. [51], p 222. rac-**D** (tetramisol) is used as anthelminticum in veterinary medicine, the (*S*)-enantiomer (levamisol) has found use as an immunostimulant.

Problem 43

P   Perform the following reactions :

(a) Ethyl (ethoxymethylene)cyanoacetate, hydrazine hydrate, EtOH → **A** (97%) ;

(b) **A**, formamide, Δ → **B** (99%) .

Q   What are the structures of the products **A/B** and their modes of formation ? Give the systematic name of **B**.

S   Consult ref. [51], p 219. **B** (allopurinol, cf. p 416) is medicinally applied for treatment of gout.

Problem 44

P   Starting from chloroacetonitrile (**X**) the following sequence of reactions is performed :

(a) **X**, S_2Cl_2, CH_2Cl_2 (solvent) [via Cl_2CH-CN] → **A** ($C_2Cl_3NS_2$, salt, 85%) ;

(b) **A**, 4-methoxyaniline, pyridine (solvent) → **B** (85%) ;

(c) **B**, 200°C (for 1 min, - HCl, - S_2) → **C** (60%) ;

(d) **C**, pyridinium hydrochloride → **D** (40%) ;

(e) **D**, (*S*)-cysteine, MeOH/H_2O → **E** (94%) .

Q   What are the structures of the products **A – E** and their modes of formation ?

S   Consult ref. [61] and p 157. Compare the approach to the analogous benzimidazole derivative in ref. [62].

[61] R. Appel, H. Janssen, M. Siray, F. Knock, *Chem. Ber.* **1985**, *118*, 1632;
V. Beneteau, T. Besson, C. W. Rees, *Synth. Commun.* **1997**, *27*, 2275;
E. H. White, F. McCapra, G. F. Field, *J. Am. Chem. Soc.* **1963**, *85*, 337.
[62] O. A. Rakitim, C. W. Rees, O. G. Vlasova, *Tetrahedron Lett.* **1996**, *37*, 4589.

Problem 45

 2,2-Dimethyl-3-dimethylamino-2*H*-azirine (**X**) was reacted   (a) with BF$_3$-etherate at –78°C in CH$_2$Cl$_2$,   (b) with benzamide (**Y**) in the presence of   sodium hexamethyldisilazane in THF at –78°C. After workup, a new heterocylic product (**A**) was isolated in 49% yield (together with 49% of unreacted **Y**).

Q   Suggest a structure for **A** and its mechanism of formation.

S   Consult ref. [63] and p 26. **A** contains all structural elements of **X**.

[63] F. Arnhold, S. Chaloupka, A. Linden, H. Heimgartner, *Helv. Chim. Acta* **1995**, *78*, 899.

Problem 46

 In a one-pot synthesis, 2-fluoro-4-methoxybenzonitrile is subjected to reaction with acetohydroxamic acid (CH$_3$-CO-NH-OH) in the presence of *tert*-BuOK in DMF. After treatment with NaCl/H$_2$O/ethyl acetate a product **A** can be isolated in 56% yield.

Q   What is the structure of **A** and its mode of formation ? Formulate intermediates and the systematic name of **A**.

S   Consult ref. [64] and compare ref. [65].

[64] M. G. Palermo, *Tetrahedron Lett.* **1996**, 37, 2885.
[65] D. M. Fink, B. E. Kurys, *Tetrahedron Lett.* **1996**, *37*, 995.

Problem 47

 N-Tosyl-*p*-chlorbenzaldimine (**X**) and CN-CH$_2$-CO-N(C$_5$H$_{10}$) (**Y**, isocyanoacetopiperidide) react   (a) in the absence of any catalyst/acid/base to give a product **A** (81%) containing **X** and **Y** in the ratio 2 : 1,   (b) in the presence of NaH to give a product **B** (90%) containing **X** and **Y** in the ratio 1 : 1.

Q   What are the structures of the products **A/B** and their modes of formation ?

S   Consult ref. [66] and compare p 128/129.

[66] X.-T. Zhou, Y.-R. Lin, L.-X. Dai, J. Sun, *Tetrahedron* **1998**, *54*, 12445.

Problem 48

P   5-Methoxy-3-phenylisoxazole (**X**) was allowed to react with FeCl$_2$ · 4 H$_2$O in acetonitrile to give a product **A** in 95% yield, which is isomeric to **X**. When **A** was treated with anhydrous FeCl$_2$ in acetonitrile, 80% of a product **B** was obtained, which formally results from reductive ring opening of **X**.

Q   What are the structures of the products **A/B** and their modes of formation ?

S   Consult ref. [67] and compare p 140 and 144. Suggest a simple synthesis for **X**.

[67] S. Auricchio, A. Bini, E. Pastormerlo, A. M. Truscello, *Tetrahedron* **1995**, *53*, 10911.

Problem 49

P | Perform the following reactions :
(a) 2-Butanone, isopropylamine, concd HCl (catalyst) → **A** (52%) ;
(b) **A**, (1) LDA, THF (solvent), - 10°C, (2) methyl benzoate, then H_2O/HCl
    → **B** (not isolated) ;
(c) **B**, hydroxylamine hydrochloride, H_2O → **C** (not isolated) ;
(d) **C**, H_2SO_4/H_2O → **D** (80%, 3 steps).

Q | What are the structures of the products **A** – **D** and their modes of formation ?

S | Consult ref. [68] and compare p 141 and 143.

[68] W. H. Bunnelle, P. R. Singam, B. A. Narayanan, C. W. Bradshaw, J. S. Liou, *Synthesis* **1997**, 439.

Problem 50

P | Starting from *o*-aminobenzonitrile (**X**) the following sequence of reactions is carried out:
(a) **X**, 2,5-dimethoxytetrahydrofuran, P_4O_{10}, toluene → **A** (85%) ;
(b) **A**, $H_2C=O$, dimethylamine hydrochloride, EtOH (solvent), refl. → **B** (90%) ;
(c) **B**, methyl iodide, diethylether (solvent), r.t. → **C** (92%) ;
(d) **C**, NaN_3, [18]crown-6, dioxane (solvent), rfl. → **D** (72%) ;
(e) **D**, xylene (solvent), rfl. → **E** (88%) .

Q | What are the structures of the products **A** – **E** and their modes of formation ? Give the systematic name of **E**.

S | Consult ref. [69] and compare p 58 and 215. Compounds **D** and **E** are isomers.

[69] D. Korakas, A. Kimbaris, G. Varvounis, *Tetrahedron* **1996**, *52*, 10751;
       D. Korakas, G. Varvounis, *Synthesis* **1994**, 164.

Problem 51

P | Starting with acetone (**X**) the following sequence of reactions is performed :
(a) **X**, semicarbazide (H_2N-CO-NH-NH_2), NaOAc, H_2O (solvent) → **A** (90%) ;
(b) **A**, $POCl_3$, DMF → **B** (93%) ;
(c) **B**, diethyl succinate, *tert*-BuOK, *tert*-BuOH (solvent) → **C** (85%) ;
(d) **C**, acetic anhydride, NaOAc → **D** (85%) .

Q | What are the structures of the products **A** – **D** and their modes of formation ? Give the systematic name of **D**.

S | Consult ref. [70] and compare p 186.

[70] P. G. Baraldi et al., *Synthesis* **1997**, 1140.

Problem 52

P   Perform the following reactions :

(a) 2,2-Dimethylpropanoyl chloride (pivaloyl chloride), ethyl isocyanoacetate, CH$_2$Cl$_2$ (solvent), rfl. → **A** (not isolated) ;

(b) **A**, triethylamine, - 10°C to 0°C → **B** (75%, 2 steps) .

Q   What are the structures of the products **A/B** and their modes of formation ?

S   Consult ref. [71] and compare p 129.

[71] W.-S. Huang, Y.-X. Zhang, C.-Y. Yuan, *Synth. Commun.* **1996**, *26*, 1149.

Problem 53

P   *N*-Benzoyl-L-alanine was reacted with oxalyl chloride in CH$_2$Cl$_2$/benzene at room temperature. The solvents were replaced by toluene and triethylamine was added, followed by methanol. Workup by chromatography gave 51% of a product **A**.

Q   What is the structure of **A** and its mechanism of formation ?

S   Consult ref. [72] and compare p 136 and 137.

[72] T. Cynkowski, G. Cynkowska, P. Ashton, P. A. Crooks, *J. Chem. Soc. Chem. Commun.* **1995**, 2335.

Problem 54

P   Starting with 2-bromopyridine (**X**) the following reactions are carried out :

(a) **X**, NaN$_3$, HCl/H$_2$O, EtOH, rfl. → **A** (69%, shows in the IR spectrum no absorption at ca. 2000 cm^{-1}) ;

(b) **A**, dimethyl acetylenedicarboxylate, +80°C → **B** (83%) .

Q   What are the structures of the products **A/B** and their modes of formation ? Give the systematic name of **A**.

S   Consult refs. [73a] and [73b], compare p 213.

[73a] J. H. Boyer, D. I. McCane, W. J. McCarville, A. T. Tweedle, *J. Am. Chem. Soc.* **1953**, *75*, 5298; R. Huisgen, K. von Fraunberg, H. J. Sturm, *Tetrahedron Lett.* **1969**, *10*, 2589.
[73b] R. Huisgen, *Angew. Chem. Int. Ed. Engl.* **1980**, *19*, 947.

Problem 55

P   Reaction of phenyl azide and benzyl cyanide (in EtOH in the presence of EtONa at r.t.) surprisingly does not lead to the expected product of a 1,3-dipolar cycloaddition (**A**), but to 5-amino-1,4-diphenyl-1,2,3-triazole (**B**, 80%). Triazole **B** isomerizes readily on heating in pyridine solution to give 5-anilino-4-phenyl-1,2,3-triazole (**C**).

Q   Discuss the formation of **B** instead of **A** and the isomerization of **B** to **C** in terms of reasonable reaction mechanisms. Which general reaction principles account for both transformations ?

S | Consult refs. [74a]/[74b] and compare with p 202. For additional information see ref. [73b].

[74a] O. Dimroth, *Chem. Ber.* **1902**, *35*, 4041;
         E. Lieber, T. S. Rao, C. N. R. Rao, *Org. Synth.* **1957**, *37*, 26 (87% yield).
[74b] E. Lieber, T. S. Rao, C. N. R. Rao, *J. Org. Chem.* **1957**, *22*, 654;
         D. R. Sutherland, G. Tennant, *J. Chem. Soc. (C)* **1971**, 706.

Problem 56

P | 1,1-Diacetylcyclopropane (**X**) reacts with hydrazine hydrochloride in H_2O solution to give a product **A** in 80% yield. **A** possesses the molecular formula $C_7H_{11}N_2Cl$ and does no more contain a cyclopropane structural unit according to its spectroscopic data.

Q | Assign a structure to **A** and explain its formation from **X** by a reasonable reaction mechanism.

S | Consult ref. [75]. Develop a short synthesis of **X** from readily available reactants.

[75] N. S. Zefirov et al., *Tetrahedron* **1982**, *38*, 1693 ; *Tetrahedron* **1986**, *42*, 709.

Problem 57

P | (*O*-Mesitylenesulfonyl)hydroxylamine (**X**) is used as a reagent for electrophilic NH_2 transfer to the "pyridine-like" N of azoles and azines (cf. p 273). The resulting N-aminated systems can be subjected to acylation, deprotonation (to give dipolar N-imides) and other transformations [76]. The following sequence of reactions may serve as an example:

(a) Thiourea, chloroacetaldehyde dimethylacetal, H_2O, H_3PO_4 → **A** (60%) ;

(b) **A**, formic acid → **B** (80%) ;

(c) **B**, **X**, CH_2Cl_2 (solvent), r.t. → **C** (70%) ;

(d) **C**, polyphosphoric acid, 100 – 110°C, then H_2O → **D** (90%) .

Q | What are the structures of the products **A** – **D** and their modes of formation ? Give the systematic name of **D**.

S | Consult refs. [76]/[77] and ref. [51], p 153.

[76] Y. Tamura, M. Ikeda, *Adv. Heterocycl. Chem.* **1981**, *29*, 71.
[77] Y. Tamura et al., *J. Heterocycl. Chem.* **1974**, *11*, 459; *J. Heterocycl. Chem.* **1973**, *10*, 947.

Problem 58

P | Starting with 2-nitro-*N*-methylformanilide (**X**) the following sequence of reactions is performed :

(a) **X**, $NaBH_4$, Pd (catalyst) → **A** (not isolated) ;

(b) **A**, - H_2O → **B** ;

(c) **B**, methyl propiolate, r.t. → **C** (not isolated) ;

(d) **C**, thermal isomerization → **D** .

Q | What are the structures of the products **A** – **D** and their modes of formation ?

S   Consult ref. [78].

[78] T. L. Gilchrist, *Heterocyclic Chemistry*, 2nd edition, p 96 and 328, Longman, Burnt Mill/England
     1993; P. N. Preston, *Chem. Rev.* **1974**, *74*, 279.

Problem 59

P   Piracetam (**B**) is therapeutically used as a cerebrotonicum (cf. p 115). For its synthesis the
following 2 step-sequence has to be carried out :

(a) 2-Pyrrolidone, NaH, benzene (solvent); then ethyl chloroacetate, rfl. → **A** (70%) ;

(b) **A**, NH_3 (gas), MeOH, +40°C → **B** (89%) .

Q   What are the structures of the products **A/B** and their modes of formation ?

S   Consult ref. [51], p 134. Outline an alternative approach to **B** using glycine as
starting material.

Problem 60

P   Starting with cyclopentene oxide (**X**) the following transformations are performed :

(a) **X**, $H_2C=CH-CH_2-MgBr$, diethylether; then H_2O → **A** (95%) ;

(b) **A**, (1) phthalimide, Ph_3P, EtOOC-N=N-COOEt, THF, (2) hydrazine, MeOH, rfl.,
    (3) HCl/H_2O, rfl., (4) conc. HCl/H_2O, then $NaOH/H_2O$ to pH 13 → **B** (76%) ;

(c) **B**, TosCl, pyridine → **C** (69%) ;

(d) **C**, $(CH_3CN)_2PdCl_2$ (10 mol%), LiCl, Na_2CO_3, 1,4-benzoquinone, THF, rfl.
    → **D** (86%) .

Q   What are the structures of the products **A** − **D** ? What is the stereochemistry of **A/B** and the
mechanism of reaction (d) ?

S   Consult ref. [79a] and ref. [24], p 187. For additional information see ref. [79b].

[79a] L. S. Hegedus, J. M. McKearin, *J. Am. Chem. Soc.* **1982**, *104*, 2444.
[79b] L. S. Hegedus, *Angew. Chem. Int. Ed. Engl.* **1988**, *27*, 1113.

Problem 61

P   *o*-Iodo-(*N*-allyl)aniline (**X**) is subjected to reaction with $Pd(OAc)_2$ (2 %), Na_2CO_3,
[*n*-Bu₄N]Cl, DMF (solvent) at +25°C. After the usual workup a product **A** (C_9H_9N) is isolated in 97% yield.

Q   Suggest a structure for **A** and explain its formation by a reasonable mechanism. The same
product (**A**) is obtained from **X** on reaction with $(CH_3CN)_2PdCl_2$ (2 %), 2 equiv. of 1,4-
benzoquinone, LiCl, in refluxing THF.

S   Consult ref. [80a], compare p 105 and ref. [80b].

[80a] L. S. Hegedus, G. F. Allen, J. J. Bozell, E. L. Waterman, *J. Am. Chem. Soc.* **1978**, *100*, 5800;
      cf. ref. [79b] and ref. [24], p 188.
[80b] Reductive N-heteroannulation of 2-nitrostyrenes: B. C. Söderberg, J. A. Shriver, *J. Org. Chem.*
      1997, *62*, 5838;
      Pd(II)-catalyzed cyclization of olefinic tosylamides: R. C. Larock, T. R. Heitauer, L. A. Hasvold,
      K. P. Peterson, *J. Org. Chem.* **1996**, *61*, 3584.

Problem 62

P   The reductive cyclization of imides is a well-established method for the construction of five- and six-membered heterocycles and works often in a stereoselective fashion [81a]. An early example utilizes the following reactions :

(a) N-[2-(3,4-dimethoxyphenyl)ethyl]succinimide (**X**), NaBH$_4$, EtOH/H$_2$O (9 : 1), 0 to +5°C, pH 8-10; then HCl/H$_2$O → **A** (not isolated) ;

(b) **A**, TosOH, benzene → **B** (73%, two steps) .

Q   What are the structures of the products **A/B** and their modes of formation ?

S   Consult ref. [81b].

[81a] W. N. Speckamp, H. Hiemstra, *Tetrahedron* **1985**, *41*, 4367;
       T. A. Blumenkopf, L. E. Overman, *Chem. Rev.* **1986**, *86*, 857.
[81b] J. C. Hubert, W. Steege, W. N. Speckamp, H. O. Huisman, *Synth. Commun.* **1971**, *1*, 103.

Problem 63

P   2,4,6-Trimethylpyrylium tetrafluoroborate (**X**) serves as reactant in the following transformations:

(a) **X**, NaBH$_4$, H$_2$O/diethylether → **A** (91%) ;

(b) **X**, NaCN, H$_2$O → **B** (95%) ;

(c) **X**, CH$_3$NO$_2$, *tert*-BuOK, *tert*-BuOH → **C** (77%) ;

(d) **X**, NH$_3$, H$_2$O → **D** (100%) .

Q   What are the structures of the products **A** – **D** and their modes of formation ?

S   Consult refs. [82] – [84] and ref. [48], compare p 229 and 230.

[82] E. N. Marvel, T. Gosink, *J. Org. Chem.* **1972**, *37*, 3036.
[83] A. T. Balaban, C. D. Nenitzescu, *J. Chem. Soc.* **1961**, 3566.
[84] K. Dimroth, K. H. Wolf, *Newer Methods of Preparative Organic Chemistry*, Vol. 3, Academic Press, New York **1964**.

Problem 64

P   Starting with pyridine (**X**) the following reactions are carried out :

(a) **X**, 2,4-dinitrochlorobenzene → **A** ;

(b) **A**, NaOH, H$_2$O → **B** (orange) ;

(c) **B**, HClO$_4$, H$_2$O → **C** (red) → **D** (salt, colourless) .

Q   What are the structures of the products **A** – **D** and their modes of formation ?

S   Consult ref. [48] and p 228.

Problem 65

P 2-Methoxy-5,6-dihydro-2*H*-pyran-5-one (**X**) is subjected to the following transformations :

(a) **X**, 1-acetoxybutadiene, 14 d, r.t. → **A** (80%) ;

(b) **A**, triethylamine → **B** (99%) ;

(c) **B**, Pd-C, toluene (solvent), 2.5 h, rfl. → **C** (50%) + **D** (34%, separable) ;

(d) **C**, Ac$_2$O, AcOH, 1 h, rfl. → **E** (63%) ;

(e) **E**, bicyclo[2.2.1]hepta-2,5-diene (norbornadiene), triethylamine → **F** (75%) .

Q What are the structures of the products **A** – **F** and their modes of formation ?

S Consult ref. [85a], compare p 241 and p 313. For additional information see ref. [85b].

[85a] P. G. Sammes, R. B. Whitby, *J. Chem. Soc. Perkin Trans 1* **1987**, 195.
[85b] A. R. Katritzky, N. Dennis, *Chem. Rev.* **1989**, *89*, 827; application in natural product synthesis:
    P. A. Wender, J. L. Mascarenas, *J. Org. Chem.* **1991**, *56*, 6267.

Problem 66

P Starting from benzaldehyde (**X**) the following reactions are carried out :

(a) **X**, acetophenone (2 equiv.), BF$_3$-etherate → **A** (80%) ;

(b) **A**, Na$_2$CO$_3$, I$_2$ → **B** (70%) ;

(c) **B**, *tert*-BuOK, dioxane (solvent) → **C** (94%, HALLER-BAUER-type cleavage) .

Q What are the structures of the products **A** – **C** and their modes of formation ?

S Consult ref. [86] and compare p 224 and p 228.

[86] I. Francesconi, A. Patel, D. W. Boykin, *Synthesis* **1999**, 61.

Problem 67

P Starting with 3-chloro-1,3-diphenyl-1-oxopropane (**X**) the following sequence of reactions is carried out:

(a) **X**, benzonitrile, SnCl$_4$, CHCl$_3$ (solvent) → **A** (64%) ;

(b) **A**, NH$_3$, H$_2$O → **B** (82%) ;

(c) **B**, trityl perchlorate, acetonitrile (solvent) → **C** (86%) ;

(d) **C**, malononitrile, Et$_3$N (10% excess), CH$_3$CN (solvent) → **D** (60%) + **E** (33%) .

Note: **E**, Δ → **D** (100%).

Q What are the structures of the products **A** – **E** and their modes of formation ?

S Consult ref. [87] and compare p 377 and 378.

[87] R. R. Schmidt, *Synthesis* **1972**, 333.

Problem 68

P   Starting from cyclopentene (**X**) the following reactions are carried out :

(a) **X**, phenylmalonodialdehyde, CH_2Cl_2 (solvent), hv   →   **A** (51%) ;

(b) **A**, acetyl chloride, pyridine, toluene (solvent)   →   **B** (73%) .

Q   What are the structures of the products **A/B** and their modes of formation ? Give the name of reaction (a).

S   Consult ref. [48].

Problem 69

P   1-Phenyl-1,3-dioxobutane (**X**) is the starting material for the following transformations :

(a) **X**, KNH_2 (2 equiv.), liquid NH_3 →   **A** (intermediate);

(b) **A**, CO_2 (solid), diethylether   →   **B** (58%, 2 steps) ;

(c) **B**, polyphosphoric acid, 100°C, then H_2O →   **C** (98%) .

Q   What are the structures of the products **A – C** and their modes of formation ?

S   Consult ref. [88]. Outline an alternative formation of **C** starting from ethyl acetoacetate.

[88] C. R. Hauser, T. M. Harris, *J. Am. Chem. Soc.* **1958**, *80*, 6360.

Problem 70

P   5-Bromo-2*H*-pyran-2-one (**X**) is the starting material for the following reactions :

(a) **X**, $H_2C=CH-SnMe_3$, $(Ph_3P)_4Pd$, toluene (solvent)   →   **A** (45%) ;

(b) **A**, maleic anhydride, benzene (solvent), rfl.   →   **B** (48%) .

Q   What are the structures of the products **A/B** and their modes of formation ?

S   Consult ref. [89a] and compare p 235. Reactant **A** was prepared from 5,6-dihydro-2*H*-pyran-2-one by bromination with NBS and dehydrobromination with triethylamine [89b].

[89a] Z. Liu, J. Meinwald, *J. Org. Chem.* **1996**, *61*, 6693.
[89b] K. Afarinkia, G. H. Posner, *Tetrahedron Lett.* **1992**, *33*, 7839.

Problem 71

P   (*o*-Hydroxy)acetophenone can be reacted with acetic anhydride in the presence of sodium acetate to give (in competition) two different oxygen-containing benzoheterocycles **A** and **B**.

Q   Discuss the structure of these products (**A/B**) and their mechanism of formation together with the possible intermediates.

S   This reaction has been widely used as KOSTANECKY-ROBINSON synthesis and should be discussed with regard to p 263/264.

Problem 72

P   (3,5-Dimethoxy)benzyl methyl ketone reacts with acetic anhydride in the presence of per-chloric acid at 0°C to give a product **A** (as perchlorate salt) in 74% yield.

Q   Discuss the structure of **A** and its mode of formation.

S   Consult ref. [90] and compare ref. [91].

[90] G. Bringmann, J. R. Jansen, *Liebigs Ann. Chem.* **1985**, 2116.
[91] B. K. Blount, R. Robinson, *J. Chem. Soc.* **1933**, 555.

Problem 73

P   The same product **A** is obtained in high yield by the following reactions :
    (a) (*o*-Allyl)benzoic acid, $(CH_3CN)_2PdCl_2$, Na_2CO_3, THF;
    (b) (*o*-iodobenzyl) methyl ketone, $(Ph_3P)_2PdCl_2$, Et_3N, THF, 600 psi CO .

Q   What is the structure of **A** and its mechanism of formation ?

S   Consult ref. [92] and compare ref. [93].

[92] For reaction (a): D. E. Korte, L. S. Hegedus, R. K. Wirth, *J. Org. Chem.* **1977**, *42*, 1329;
      ref. [24], p 96 and 187.
      For reaction (b): I. Shimoyama, J. Zang, G. Wu, E.-I. Negishi, *Tetrahedron Lett.* **1990**, *31*, 2841.
[93] The same type of heterocyclization takes place with *o*-iodobenzoates and alkynes :
      H.-Y. Liao, C.-H. Cheng, *J. Org. Chem.* **1995**, *60*, 3711.

Problem 74

P   Perform the following reactions :
    (a) *o*-Hydroxybenzaldehyde, acetophenone, KOH, EtOH (solvent) → **A** (56%) ;
    (b) **A**, HCl, CH_3COOH, rfl. → **B** (90%) .

Q   What are the structures of the products **A/B** and their modes of formation ?

S   Consult ref. [94] and compare p 252 and 253.

[94] R. J. W. Le Fevre, *J. Chem. Soc.* **1929**, 2771 ; D. D. Pratt, R. Robinson, *J. Chem. Soc.* **1923**, 745.

Problem 75

P   Starting from *o*-hydroxy-*ω*-formylacetophenone (**X**) the following reactions are carried out:
    (a) **X**, ethyl orthoformate, acetic acid → **A** (70%) ;
    (b) **A**, ethyl vinyl ether, Δ → **B** (76%) ;
    (c) **B**, H_2SO_4/H_2O/ethanol → **C** (49%) .

Q   What are the structures of the products **A** – **C** and their modes of formation ? What is the stereochemistry of **B** ?

S   Consult ref. [95] and compare p 263.

[95] C. K. Ghosh, M. Tewari, A. Bhattacharyya, *Synthesis* **1983**, 614;
      C. K. Ghosh, *J. Heterocycl. Chem.* **1983**, *20*, 1437.

Problem 76

P Perform the following reactions :

(a) *o*-Hydroxyacetophenone, diethyl carbonate, NaH, toluene (solvent) → **A** (90%) ;

(b) **A**, benzalacetone, H$_2$O → **B** (60%).

Q What are the structures of the products **A/B** and their modes of formation ?

S Consult ref. [96] and compare p 251.

[96] R. C. Hayward, *J. Chem. Educ.* **1984**, *61*, 87; cf. ref. [51], p 160.

Problem 77

P 2-Pyridone (**X**) is the starting material for the following sequence of reactions :

(a) **X**, oleum, H$_2$SO$_4$/HNO$_3$ → **A** (72%) ;

(b) **A**, PCl$_5$ → **B** (80%) ;

(c) **B**, MeONa, MeOH, rfl. → **C** (90%) ;

(d) **C**, H$_2$N-O-CH$_3$, ZnCl$_2$, *tert*-BuOK, DMSO, r.t. → **D** (87%) + CH$_3$OH .

Q What are the structures of the products **A** – **D** and their modes of formation ?

S Consult ref. [97] and compare p 274 and 312.

[97] S. Seko, K. Miyake, *Chem. Commun.* **1998**, 1519;
A. G. Burton, P. J. Halls, A. R. Katritzky, *Tetrahedron Lett.* **1971**, *24*, 2211;
H. L. Friedman et al., *J. Am. Chem. Soc.* **1947**, *69*, 1204.

Problem 78

P When 3-cyano-1-methylpyridinium iodide (**X**) is heated in 2 N aqueous NaOH together with methylamine hydrochloride, 2-(methylamino)pyridine-3-carbaldehyde (**A**) is formed in 60% yield.

Q Explain the transformation **X** → **A** by a reasonable mechanism and formulation of intermediates.

S Consult ref. [98] and compare p 280.

[98] J. H. Blanch, K. Fretheim, *J. Chem. Soc. (C)* **1971**, 1892.

Problem 79

P Starting with 4-aminopyridine (**X**) the following sequence of reaction is carried out :

(a) **X**, di-*tert*-butyl carbonate → **A** (80%);

(b) **A**, *n*-BuLi (2 equiv.), THF, - 30°C → **B** (in situ) ;

(c) **B**, CH$_3$I, - 70°C → **C** (54%, 2 steps) ;

(d) **C**, *tert*-BuLi (2 equiv.), THF, - 30°C; then DMF → **D** ;

(e) **D**, 6 N HCl/H$_2$O, + 50°C → **E** (79%, 2 steps) .

Q What are the structures of the products **A – E** and their modes of formation ? Give the systematic name of **E**.

S Consult ref. [99] and compare p 279 and p 282.

[99] D. Hands, B. Bishop, M. Cameron, J. S. Edwards, I. F. Cottrell, S. H. B. Wright, *Synthesis* **1996**, 877.

Problem 80

P Pyridine (**X**) is the starting material for the following reactions:
(a) **X**, benzaldehyde, benzoyl chloride → **A** (78%) ;
(b) **A**, KOH, H$_2$O, 0°C → **B** ;
(c) **B**, isopropylamine, benzene (solvent), 0°C → **C** (62%, 2 steps) + **D** .

Q What are the structures of the products **A – D** and their modes of formation ?

S Consult ref. [100] and compare p 280/281.

[100] D. Reinehr, T. Winkler, *Angew. Chem. Int. Ed. Engl.* **1981**, *20*, 881.

Problem 81

P When pyridine-1-oxide is reacted with *N,N*-dimethylcarbamoyl chloride [(CH$_3$)$_2$N-COCl] and trimethylsilylcyanide in CH$_2$Cl$_2$ at r.t., a product **A** is obtained in 94% yield, which does not contain oxygen.

Q Suggest the structure of **A** and a mechanism accounting for its formation.

S Consult ref. [101] and compare p 286/287.

[101] W. K. Five, *J. Org. Chem.* **1983**, *48*, 1375; H. Vorbrüggen, K. Krolikiewicz, *Synthesis* **1983**, 316.

Problem 82

P Starting from phenylacetaldehyde (**X**) the following reactions are carried out :
(a) **X**, ethyl acetoacetate, concd NH$_3$, EtOH (solvent) → **A** (71%) ;
(b) **A**, sulphur (elemental) → **B** (70%) ;
(c) **A**, NaNO$_2$, HOAc → **C** (95%) .

Q What are the structures of the products **A – C**, their modes of formation and the by-products of reaction (c) ?

S Consult ref. [102a] and compare the instructive example in [102b].

[102a] B. Loev, K. M. Snader, *J. Org. Chem.* **1965**, *30*, 1914; cf. ref. [48].
[102b] C. Seoane, M. Suárez et al., *J. Heterocycl. Chem.* **1995**, *32*, 235.

Problem 83

P   Perform the following reactions :

(a) (±)-Alanine ethylester hydrochloride, diethyl oxalate, Et₃N, EtOH → **A** (90%) ;

(b) **A**, COCl₂/Et₃N 1 : 2, chloroform → **B** (70%) ;

(c) **B**, 2-isopropyl-4,7-dihydro-1,3-dioxepine (**X**), Δ → **C** (66%, heterobicyclic) .

Q   What are the structures of the products **A** – **C** and their modes of formation ? What is the mechanism of reaction (c) and the systematic name of **C** ?

S   Consult ref. [103] and compare p 129 und p 131.

[103] I. Maeda et al., *Bull. Soc. Chem. Jap.* **1969**, *42*, 1435;
R. Lakhan, B. Ternai, *Adv. Heterocycl. Chem.* **1974**, *17*, 99;
D. L. Boger, *Tetrahedron* **1983**, *39*, 2869.

Problem 84

P   Ethionamide (**X**, cf. p 306) is used as a tuberculostatic. For its synthesis the precursor **C** is prepared by the following steps :

(a) Diethyl oxalate, ethylmethylketone, *tert*-BuOK, diethylether → **A** (66%) ;

(b) **A**, cyanoacetamide, piperidine, EtOH (solvent) → **B** (60%) ;

(c) **B**, concd HCl → **C** (81%) .

Q   What are the structures of the products **A** – **C**, their modes of formation and the names of the reactions used ? Which further transformations are required to prepare **X** from the precursor **C** ?

S   Consult ref. [51], p 163, and compare p 314.

Problem 85

P   Starting from *m*-chloroaniline (**X**) the following sequence of reactions is performed:

(a) **X**, diethyl ethoxymethylenemalonate, 100°C → **A** ;

(b) **A**, 250°C → **B** (85 %, 2 steps) ;

(c) **B**, NaOH, H₂O; then HCl → **C** (90%) ;

(d) **C**, 250-270°C → **D** (70%) ;

(e) **D**, POCl₃, Δ → **E** (90%) ;

(f) **E**, H₂N-CH(CH₃)-(CH₂)₃-N(C₂H₅)₂, Δ → **F** (90%) .

Q   What are the structures of the products **A** – **F** and their modes of formation ?

S   Consult ref. [104] and compare p 330 and 335. Compound **F** is used as an antimalarial.

[104] *Org. Synth.* Coll. Vol. III, 272; for comparison see :
A. R. Surrey, H. F. Hammer, *J. Am. Chem. Soc.* **1946**, *68*, 113.

Problem 86

P   In a recent modification of the FRIEDLÄNDER synthesis the following reactions are performed in a one-pot procedure starting from m-methoxy-N-(2,2-dimethylpropanoyl)aniline (**X**) :

(a) **X**, sec-BuLi (2 equiv.), THF, 0°C → **A** (intermediate) ;

(b) **A**, DMF (1 equiv.), 0°C → **B** (intermediate) ;

(c) **B**, 3-pentanone, potassium hexamethyldisilazane, 0°C; then NH_4Cl/H_2O → **C** (82%) .

Q   What are the structures of the products **A** – **C** and their modes of formation ?

S   Consult ref. [105] and compare p 328.

[105] J. I. Ubeda, M. Villacampa, C. Avendano, *Synlett* **1997**, 285.

Problem 87

P   Starting from o-iodoacetanilide (**X**) the following reactions are carried out :

(a) **X**, 3,3-diethoxy-1-propyne, Et_2NH, CuI (catalyst), $(Ph_3P)_2Pd(OAc)_2$ (catalyst), DMF (solvent) → **A** (85%);

(b) **A**, iodobenzene, $Pd(OAc)_2$, HCOOK, DMF (solvent) → **B** (75%) + **C** (4%) ;

(c) **B**, TosOH, EtOH (solvent) → **D** (89%) .

Q   What are the structures of the products **A** – **D** and their modes of formation ? Formulate detailed mechanisms for reactions (a) and (b).

S   Consult ref. [106] and compare p 327.

[106] S. Cacchi, G. Fabrizi, F. Marinelli, L. Moro, P. Pace, *Tetrahedron* **1996**, *52*, 10225; for comparison see : S. Cacchi, G. Fabrizi, F. Marinelli, *Synlett* **1999**, *4*, 401.

Problem 88

P   Isoquinoline reacts with CH_3I to give a product **A**. The following transformations are carried out with **A** :

(a) **A**, $LiAlH_4$, diethylether (solvent) → **B** (70%) ;

(b) **A**, CH_3MgI, diethylether (solvent) → **C** (86%) ;

(c) **A**, NaOH, H_2O; then $K_3[Fe(CN)_6]$ → **D** (70%) .

Q   What are the structures of the products **A** – **D** and their modes of formation ? Give the systematic name of **A**.

S   Consult ref. [107] and compare p 280, 337 and 340.

[107] W. J. Gensler, K. T. Shamasundar, *J. Org. Chem.* **1975**, *40*, 123.

Problem 89

P   In the course of an investigation on the synthesis of azapolycycles containing an octahydro-isoquinoline moiety the following reactions were carried out :

(a) Isoquinoline, H_2, Pd-C, CF_3COOH → **A** (70%) ;
(b) **A**, KNH_2, liquid NH_3; then $Ph-CH_2-CH_2-I$ → **B** (56%) ;
(c) **B**, CH_3I, diethylether (solvent) → **C** (70%) ;
(d) **C**, $NaBH_4$, EtOH → **D** (80%) .

Q What are the structures of the products **A** – **D** and their modes of formation ?

S Consult ref. [108] and compare p 282, 293 and 340.

[108] G. L. Patrick, *J. Chem. Soc. Perkin Trans 1* **1995**, 1273.

Problem 90

P Starting from β-(3,4-dimethoxyphenyl)ethylamine (**X**) the following reactions were carried out to synthesize the natural product (±)-carnegine (**E**) :
(a) **X**, HCOOH, Δ → **A** (80%) ;
(b) **A**, $POCl_3$ → **B** (80%) ;
(c) **B**, ClCOOEt → **C** ;
(d) **C**, CH_3MgI, diethylether (solvent) → **D** (58%, 2 steps) ;
(e) **D**, $LiAlH_4$, diethylether (solvent) → **E** (93%) .

Q What are the structures of the products **A** – **E** and their modes of formation ? Give the systematic name of **E**.

S Consult ref. [109] and compare p 338 and 343.

[109] A. P. Venkov, S. M. Statkova-Abeghe, *Synth. Commun.* **1995**, *25*, 1824.

Problem 91

P *N*-(2-Iodobenzylidene)-*tert*-butylamine is reacted with phenylacetylene
(a) in the presence of $(Ph_3P)_2PdCl_2$, CuI and Et_3N at 55°C, then (b) with CuI in DMF at 100°C. After the usual workup a product **A** is isolated in 91% yield.

Q What is the structure of **A** ? Formulate detailed mechanisms for the reactions (a) and (b).

S Consult ref. [110] and compare other Pd-mediated heterocyclizations reported there.

[110] K. R. Roesch, R. C. Larock, *Org. Lett.* **1999**, *1*, 553.

Problem 92

P Perform the following reactions :
(a) 1,3-Diphenylisobenzofuran, 3-bromopyridine, $[(CH_3)_3Si]_2NLi$, THF (solvent), - 15°C to rfl. → **A** (60%) ;
(b) **A**, $LiAlH_4$, diethylether (solvent) → **B** (60%).

Q What are the structures of the products **A/B** and their modes of formation ?

S Consult ref. [111] and compare p 66 and problem 12.

[111] C. Bozzo, M. D. Pujol, Synlett **2000**, *4*, 550; *Heterocycl. Commun.* **1996**, *2*, 163.

Problem 93

P Starting from 3-bromopyridine (**X**) the following sequence of reactions is performed :

(a) **X**, LDA, THF, - 78°C; then DMF, - 78°C to r.t.; then H_2O → **A** (92%) ;

(b) **A**, phenylacetylene, $(Ph_3P)_2PdCl_2$, CuI, Et_3N, DMF, r.t. → **B** (67%) ;

(c) **B**, NH_3, EtOH (solvent), 80°C → **C** (83%) ;

(d) **B**, hydroxylamine hydrochloride, AcOH, H_2O, r.t. → **D** (97%) ;

(e) **D**, K_2CO_3, EtOH (solvent), 80°C → **E** (80%) .

Q What are the structures of the products **A** – **E** and their modes of formation ? Give the systematic name of **C**.

S Consult ref. [112] and compare p 279.

[112] A. Numata, Y. Kondo, T. Sakamoto, *Synthesis* **1999**, 306.

Problem 94

P Methyl 7-oxoocten-2-oate (**X**, CH_3-CO-CH_2-CH_2-CH_2-CH=CH-$COOCH_3$) can be transformed to the corresponding 2,6-disubstituted piperidine derivative (**A**) by reductive amination with ammonium acetate/Na[H_3BCN] and subsequent intramolecular MICHAEL addition. Interestingly, (*E*)-**X** gives *cis*- and (*Z*)-**X** gives *trans*-2,6-disubstituted **A** as exclusive reaction product.

Q Give an explanation for this stereochemical outcome of piperidine formation.

S Consult ref. [113].

[113] M. G. Banwell, C. T. Bui, H. T. T. Pham, G. W. Simpson,
J. Chem. Soc. Perkin Trans 1 **1996**, 967.

Problem 95

P Perform the following reactions :

(a) (*R*)-*N*-Boc-(*N*-allyl)-(*C*-allyl)glycine methyl ester, $(Boc)_2O$ → **A** (85%) ;

(b) **A**, ruthenium alkylidene catalyst [114], CH_2Cl_2 (solvent), rfl. → **B** (93%) ;

(c) **B**, CF_3COOH → **C** (90) .

Q What are the structures of the products **A** – **C** and their modes of formation ? What is the name of reaction (b) ?

S Consult ref. [115] and compare p 470.

[114] R. H. Grubbs et al., *J. Am. Chem. Soc.* **1993**, *115*, 9858;
Angew. Chem. Int. Ed. Engl. **1995**, *34*, 2039.
for comparison see the PAUSON-KHAND cyclization of *N*-allyl-propargylglycine ester:
G. L. Bolton, J. C. Hodges, J. R. Rubin, *Tetrahedron* **1997**, *53*, 6611.

[115] F. P. J. T. Rutjes, H. E. Schoemaker, *Tetrahedron Lett.* **1997**, *38*, 677.

Problem 96

P   In a synthesis of the alkaloid (-)-pseudoconhydrine [(-)-**G**] the following reactions were carried out starting from (*S*)-proline (**X**):

(a) **X**, benzoylchloride, NaOH, H$_2$O → **A** ;

(b) **A**, K$_2$CO$_3$, CH$_3$I → **B** (88%, 2 steps) ;

(c) **B**, anodic oxidation in CH$_3$OH in the presence of [Et$_4$N]OTos
    → **C** (95%, 2 diastereomers 1 : 1) ;

(d) **C**, *n*-PrMgBr, (Me$_2$S)CuBr, BF$_3$-etherate, diethylether (solvent)
    → **D** (86%, 2 diastereomers 9 : 1, not separable) ;

(e) **D**, LiAlH$_4$, THF (solvent) → **E** (96%, 2 diastereomers 9 : 1, separable) ;

(f) (-)-**E**, CF$_3$COOH; then Et$_3$N followed by NaOH/H$_2$O → (-)-**F** (57%) ;

(g) (-)-**F**, H$_2$, Pd(OH)$_2$ → (-)-**G** (60%) .

Q   What are the structures of the products **A** – **G** and their modes of formation ?

S   Consult ref. [116].

[116] J. Cossy, C. Dumas, D. G. Pardo, *Synlett* **1997**, 905.

Problem 97

P   In a recent approach to 1,2,3,4-tetrahydroisoquinoline derivatives the following reactions were performed starting from Ph$_2$C=N-CH$_2$-COOEt (**X**) :

(a) **X**, 3-bromopropyne, K$_2$CO$_3$, acetonitrile (solvent) → **A** (86%) ;

(b) **A**, 1 N HCl, diethylether → **B** (88%) ;

(c) **B**, TosCl, Et$_3$N, CH$_2$Cl$_2$ (solvent) → **C** (78%) ;

(d) **C**, allylbromide, K$_2$CO$_3$, CH$_3$CN (solvent) → **D** (98%) ;

(e) **D**, GRUBBS catalyst [117], toluene, rfl. → **E** (70%) ;

(f) **E**, dimethyl acetylenedicarboxylate; then DDQ → **F** (85%) .

Q   What are the structures of the products **A** – **F** and their modes of formation ?

S   Consult ref. [118] and compare problem 95. For another modern approach to the 1,2,3,4-tetrahydroisoquinoline system see ref. [119].

[117] R. H. Grubbs, S. Chang, *Tetrahedron* **1998**, *54*, 4413.
[118] S. Kotha, N. Sreenivasachary, *J. Chem. Soc. Chem. Commun.* **2000**, 503.
[119] S. Kotha, N. Sreenivasachary, *Bioorg. Med. Chem. Lett.* **2000**, *10*, 1413.

Problem 98

P   In a synthesis of the local anesthetic cinchocaine (**E**) the following steps were used starting from isatine (see p 108) :

(a) Isatine, Ac$_2$O → **A** (80%) ;

(b) **A**, NaOH, H$_2$O → **B** (80%) ;

(c) **B**, PCl$_5$ → **C** (100%) ;

(d) **C**, Et$_2$N-CH$_2$-CH$_2$-NH$_2$, H$_2$O/benzene → **D** (99%) ;

(e) **D**, *n*-BuONa, *n*-BuOH → **E** (94%) .

Q   What are the structures of the products **A** – **E** and their modes of formation ?

S   Consult ref. [51], p 179.

Problem 99

P   In a synthesis of the chemotherapeutic trimethoprim (**D**) the following steps were used starting from methyl 3,4,5-trimethoxybenzoate (**X**) :

(a) **X**, H$_3$C-SO-CH$_3$, NaNH$_2$ → **A** (79) ;

(b) **A**, NaBH$_4$, H$_2$O/EtOH → **B** (96%) ;

(c) **B**, β-anilinopropionitrile, *tert*-BuOK, DMSO → **C** (63%) ;

(d) **C**, guanidinium nitrate, MeONa, MeOH → **D** (93%).

Q   What are the structures of the products **A** – **D** and their modes of formation ? What is the mechanism of reaction (c) ?

S   Consult ref. [51], p 184.

Problem 100

P   *N,N'*-Dimethylurea (**X**) is the starting material for the following sequence of reactions :

(a) **X**, cyanoacetic acid, Ac$_2$O → **A** (83%) ;

(b) **A**, KOAc, H$_2$O → **B** (84%) ;

(c) **B**, NaNO$_2$, HCOOH; then Na$_2$S$_2$O$_4$, then formamide, Δ → **C** (55%) ;

(d) **C**, EtONa, EtOH; then CH$_3$I → **D** (80%) .

Q   What are the structures of the products **A** – **D** and their modes of formation ? What are the (systematic) names of (c) and (d)

S   Consult ref. [51], p 213. Concerning the principle of heterocycle formation in (b) compare the example in ref. [120].

[120] M. Butter, *J. Heterocycl. Chem.* **1992**, *29*, 1369.

Problem 101

P   Methyl (*N*-cyano)acetimidate [**X**, (CH$_3$)(OCH$_3$)C=N-CN] is the starting material for the following sequence of reactions :

(a) **X**, CH$_3$-NH-CH$_2$-CN–hydrochloride, Et$_3$N, EtOH → **A** (68%) ;

(b) **A**, NaOEt, EtOH → **B** (79%) ;

(c) **B**, H$_2$, Pd-C, H$_2$O/H$_2$SO$_4$ → **C** (80%) ;

(d) **C**, CH$_3$CH$_2$CN, HCl (gas), dioxane; then H$_2$O/NH$_3$ → **D** (64%) ;

(e) **C**, malononitrile, MeONa, MeOH; then H$_2$O → **E** (80%) .

Q   What are the structures of the products **A** – **E** and their modes of formation ?

S   Consult refs. [121]/[122] and compare p 411.

[121] F. Perandones, J. L. Soto, *J. Heterocycl. Chem.* **1997**, *34*, 107; **1997**, *34*, 1459.
[122] A. Edenhofer, *Helv. Chim. Acta* **1975**, *58*, 2192.

Problem 102

P   Phenylglyoxalmonooxime (**X**, Ph-CO-CH=N-OH) is subjected to the following transformations :

(a) **X**, semicarbazide hydrochloride, NaOAc, EtOH/H$_2$O, 50°C → **A** (70%) ;

(b) **A**, 5% HCl/H$_2$O, rfl. → **B** (56%) ;

(c) **B**, DMF, POCl$_3$, CHCl$_3$, rfl. → **C** (47%) ;

(d) **C**, phenylacetonitrile, *tert*-BuOK, *N,N*-dimethylacetamide; then H$_2$O → **D** (86%) ;

(e) **D**, NH$_3$, acetone/H$_2$O → **E** (95%) :

Q   What are the structures of the products **A** – **E**, their modes of formation and the mechanisms of reactions (d) and (e) ?

S   Consult ref. [123]. The mechanistic considerations should refer to p 418.

[123] A. Rykowski, E. Wolinska, *Tetrahedron Lett.* **1996**, *37*, 5795;
        R. I. Trust, J. D. Albright, F. M. Lovall, N. A. Perkinson, *J. Heterocycl. Chem.* **1979**, *16*, 1393.

Problem 103

P   In a recent investigation on azaindolizines the following sequence of reactions was carried out starting from pyrrole-2-carbaldehyde (**X**) :

(a) **X**, TosMIC, DBU, THF → **A** (82%) ;

(b) **A**, Na-Hg, Na$_2$HPO$_4$, THF/MeOH → **B** (51%);

(c) **B**, phenacyl bromide, acetone (solvent) → **C** (78%) ;

(d) **C**, K$_2$CO$_3$, acetonitrile (solvent); then acrylonitrile → **D** (62%) ;

(e) **D**, DDQ (2 equiv.), CH$_2$Cl$_2$ → **E** (95%) .

Q   What are the structures of the products **A** – **D** and their modes of formation ? Give the systematic name of **E**.

S   Consult ref. [124] and compare p 129 and 284.

[124] J. M. Minguez, J. J. Vaquero, J. Alvarez-Builla, O. Castano, J. L. Andres,
        J. Org. Chem. **1999**, *64*, 7788. Compare the application of reaction (d) in :
        J. Zhou, Y. Hu, H. Hu, *Synth. Commun.* **1998**, *28*, 3397.

Problem 104

$\boxed{P}$ Barbituric acid (**X**) is the starting material for the following sequence of reactions :

(a) **X**, POCl$_3$ (excess), DMF; then H$_2$O → **A** (55%) ;

(b) **A**, N$_2$H$_4$ (anhydrous), 2-methoxyethanol → **B** (79%) ;

(c) **B**, benzaldehyde, DMF → **C** (85%) ;

(d) **C**, 70% HNO$_3$, DMF → **D** (85%) ;

(e) **D**, POCl$_3$, rfl. → **E** (80%) ;

(f) **E**, thiourea, *sec*-BuOH → **F** (81%).

$\boxed{Q}$ What are the structures of the products **A – F** and their modes of formation ?

$\boxed{S}$ Consult ref. [125].

[125] T. Nagamatsu, T. Fujita, *J. Chem. Soc. Chem. Commun.* **1999**, 1461;
         F. Yoneda, Y. Sakuma, S. Mizumoto, R. Ito, *J. Chem. Soc. Perkin 1* **1976**, 1805.

Problem 105

$\boxed{P}$ Starting with acetonitrile (**X**) the following sequence of reactions is performed :

(a) **X**, ethyl benzoate, EtONa, EtOH → **A** (65%) ;

(b) **A**, CH$_3$-NH-NH$_2$, EtOH, rfl. → **B** (60%) ;

(c) **B**, 2,4,6-tris(ethoxycarbonyl)-1,3,5-triazine (**Y**), DMF, Δ

   → **C** (70%) + NC-COOEt + NH$_3$.

$\boxed{Q}$ What are the structures of the products **A – C** and their modes of formation ? Formulate mechanism and intermediates in reaction (c).

$\boxed{S}$ Consult ref. [126] and compare the unusual reactivity of **Y** with p 452.

[126] Q. Dang, B. S. Brown, M. D. Erion, *J. Org. Chem.* **1996**, *61*, 5204.

Problem 106

$\boxed{P}$ In a novel 3-component one-pot diazine synthesis 4-iodonitrobenzene (**A**) was reacted in the presence of (Ph$_3$P)$_2$PdCl$_2$, CuI and Et$_3$N in THF with 1-phenylpropyn-1-ol (**B**) followed by (4-bromophenyl)amidinium chloride (**C**). Workup by flash chromatography gave a product **F** in 56% yield, the formation of which is likely to proceed via the following steps :

(a) **A** + **B** → **D** ;

(b) **D** → **E** (isomerization) ;

(c) **E** + **C** → **F** .

$\boxed{Q}$ What are the structures of the products **D/E/F** and their modes of formation ?

$\boxed{S}$ Consult ref. [127]. For reaction (c) an additional dehydrogenation has to take place. Compare ref. [128].

[127] T. J. J. Müller, R. Braun, M. Ansorge, *Org. Lett.* **2000**, *2*, 1967.
[128] T. J. J. Müller, M. Ansorge, D. Aktah, *Angew. Chem. Int. Ed. Engl.* **2000**, *39*, 1253.

Problem 107

P   Starting with 1,2-diaminobenzene (**X**) the following sequence of reactions is carried out:
(a) **X**, anthranilic acid, polyphosphoric acid → **A** (72%) ;
(b) **A**, CS$_2$, KOH, MeOH, rfl., microwave irradiation → **B** (95%) ;
(c) **B**, CH$_3$I, NaH, DMF (solvent), r.t. → **C** (95%);
(d) **C**, anthranilic acid, graphite, microwave irradiation → **D** (95%) .

Q   What are the structures of the products **A** – **D** and their modes of formation ?

S   Consult ref. [129] and compare p 176.

[129] T. Besson et al., *Tetrahedron Lett.* **2000**, *41*, 5857.

Problem 108

P   Isatoic anhydride (**X**, see p 377) is used as starting material for the following sequence of reactions :
(a) **X**, NH$_3$, H$_2$O → **A** (80%) ;
(b) **A**, 3-chloropropionyl chloride, dioxane (solvent) → **B** (77%) ;
(c) **B**, Na$_2$CO$_3$, H$_2$O/MeOH → **C** (100%) ;
(d) **C**, piperidine, AcOH/MeOH → **D** (100%) .

Q   What are the structures of the products **A** – **D** and their modes of formation ?

S   Consult ref. [130] and compare p 433.

[130] A. Witt, I. Bergman, *Tetrahedron* **2000**, *56*, 7245.

Problem 109

P   In a study on diazines with potential applications in nonlinear optics the following sequence of reactions was performed for the synthesis of compound **H** :
(a) *o*-Iodoaniline, trimethylsilylacetylene, (Ph$_3$P)$_2$PdCl$_2$, CuI, Et$_3$N → **A** (95%) ;
(b) **A**, NaNO$_2$, HCl, H$_2$O; then Δ → **B** (73%) ;
(c) **B**, POCl$_3$ → **C** (77%) ;
(d) **C**, (4-methoxyphenyl)boronic acid, (Ph$_3$P)$_4$Pd, K$_2$CO$_3$, DME/H$_2$O → **D** (86 %) ;
(e) **D**, LDA; then (CH$_3$)$_3$SiCl → **E** (67%) ;
(f) **E**, LiTMP; then I$_2$ → **F** (72%) ;
(g) **F**, (3-nitrophenyl)boronic acid, (Ph$_3$P)$_4$Pd, K$_2$CO$_3$, toluene/H$_2$O → **G** (72%) ;
(h) **G**, TBAF, H$_2$O/THF → **H** (76%) .

Q   What are the structures of the products **A** – **H** and their modes of formation ?

S   Consult ref. [131] and compare p 105, 279, 432 and ref. [180] of chapter 6.

[131] V. Gautheron Chapoulaud, N. Plè, A. Turck, G. Queguiner, *Tetrahedron* **2000**, *56*, 5499.

Problem 110

P Clonazepam (**E**) is medicinally used as an anticonvulsant. For its synthesis the following sequence of reactions is described :

(a) 2-Chloro-2'-nitrobenzophenone, H_2, RANEY-Ni → **A** ;

(b) **A**, Br-CH_2-CO-Br → **B** ;

(c) **B**, liquid NH_3 → **C** ;

(d) **C**, pyridine → **D** ;

(d) **D**, KNO_3, H_2SO_4 → **E** .

Q What are the structures of the products **A** – **E** and their modes of formation ? Give the systematic name of **E**.

S Consult ref. [132] and compare p 476.

[132] L. H. Sternbach et al., *J. Med. Chem.* **1963**, *6*, 261.

Problem 111

P Cloxazolam (**B**), a tranquilizer of the benzodiazepine type, is prepared by the following two-step synthesis :

(a) 2-Amino-2',5-dichlorobenzophenone, Br-CH_2-COCl → **A** ;

(b) **A**, ethanolamine, CH_3COONa → **B** (possesses a tricyclic structure).

Q What are the structures of the products **A**/**B** and their modes of formation ?

S Consult ref. [133] and compare problem 110.

[133] E. Schulte, *Dtsch. Apoth. Ztg.* **1975**, *115*, 1253 and 1828.

Problem 112

P Clozapine (**E**, cf. p 475) is medicinally used as a neuroleptic. For its synthesis the following sequence of reactions starting from 4-chloro-2-nitroaniline (**X**) is performed :

(a) **X**, methyl 2-chlorobenzoate, Cu → **A** ;

(b) **A**, 1-methylpiperazine, Δ → **B** ;

(c) **B**, H_2, RANEY-Ni → **C** ;

(d) **C**, $POCl_3$ → **D** .

Q What are the structures of the products **A** – **D** and their modes of formation ?

S Consult ref. [134] and compare problem 110.

[134] F. Hunzicker et al., *Helv. Chim. Acta* **1967**, *50*, 1588.

Problem 113

P For the synthesis of the heterobicyclic compound **D** the following sequence of reactions was carried out :

(a) Methyl (*N*-cyano)acetimidate, (*ω*-methylamino)acetophenone hydrochloride,
     triethylamine → **A** (51%) ;

(b) **A**, EtONa, EtOH → **B** (85) ;

(c) **B**, Br-CH₂-CO-Br, benzene (solvent) → **C** (77%) ;

(d) **C**, NH₃ → **D** (57%) .

Q What are the structures of the products **A – D** and their modes of formation ?

S Consult ref. [122] and compare problems 101 and 110.

Problem 114

P Perform the following sequence of reactions :

(a) Diaminomaleonitrile, triethyl orthoformiate → **A** (80%) ;

(b) **A**, hydrazine hydrate, dioxane (solvent), r.t. → **B** (100%) ;

(c) **B**, acetic anhydride, acetonitrile (solvent), r.t. → **C** (83%) ;

(d) **C**, KOH, H₂O → **D** (79%) ;

(e) **D**, NaNO₂, AcOH/H₂O → **E** (94%, contains an azide group) ;

(f) **E**, CH₃-CO-CH₂-COOCH₃, K₂CO₃, DMSO/EtOH (solvents)
     → **F** (68%, salt, heterobicyclic) ;

(g) **F**, HCl, THF → **G** (58%, contains two directly joined heterocyles) .

Q What are the structures of the products **A – G** and their modes of formation ?

S Consult refs. [135] and [136].

[135] M. J. Alves, B. L. Booth, A. P. Freitas, M. F. J. R. P. Proenca,
      J. Chem. Soc. Perkin Trans 1 **1992**, 913.
[136] A. P. Freitas, M. F. J. R. P. Proenca, B. L. Booth, *J. Heterocycl. Chem.* **1995**, *32*, 457.

Problem 115

P 1,2-Diamino-4-methylbenzene reacts with two equivalents of 4-nitroacetophenone in the presence of H₂SO₄ in CH₃OH to give a 77 : 23 mixture of two isomers **A/B** in 89% yield.

Q Suggest structures for **A** and **B** and discuss their mode of formation.

S Consult ref. [137] and compare ref. [138].

[137] B. Insuasty, R. Ahonia, J. Quiroga, H. Meier, *J. Heterocycl. Chem.* **1993**, *30*, 229.
[138] B. Insuasty, M. Ramos, J. Quiroga, A. Sanchez, M. Nogueras, N. Hanold, H. Meier,
      J. Heterocycl. Chem. **1994**, *31*, 61.

Problem 116

  $\boxed{\text{P}}$ Starting with 4-methylpyridine (**X**) the following reactions are performed :

       (a) **X**, 5-bromo-3-methylpenta-1,3-diene → **A** (100%) ;

       (b) **A**, *tert*-BuOK, THF/CH$_3$CN, Δ → **B** (63%) .

  $\boxed{\text{Q}}$ What are the structures of the products **A/B** and their modes of formation ?

  $\boxed{\text{S}}$ Consult ref. [139].

       [139] K. Marx, W. Eberbach, *Tetrahedron* **1997**, *53*, 14687.

Problem 117

  $\boxed{\text{P}}$ Perform the following reactions :

       (a) 4-Methoxybenzaldehyde, ethyl chloroacetate, *tert*-BuOK,
          tert-BuOH (solvent) → **A** (60%) ;

       (b) **A**, 2-aminothiophenol, toluene (solvent), microwave irradiation (390 watts),
          20 min → **B** (75%, *cis/trans* 9 : 1) .

       Note: When reaction (b) is performed in acetic acid with microwave irradiation at 490 watts for 10 min, diastereoselectivity of product formation (84%) was reversed to a *cis/trans* ratio of 1 : 9.

  $\boxed{\text{Q}}$ What are the structures of the products **A/B** and their modes of formation ?

  $\boxed{\text{S}}$ Consult ref. [140]. Compound **B** is a key intermediate in the preparation of the calcium channel blocker diltiazem.

       [140] J. Alvarez-Builla et al., *Tetrahedron Lett.* **1996**, *37*, 6413.

Problem 118

  $\boxed{\text{P}}$ Furfural reacted with diethyl azodicarboxylate (EtOOC-N=N-COOEt) on heating to 100°C to give a product **A** in 40% yield, which was shown to possess the structure of a *N*-substituted 6,7-diaza-8-oxabicyclo[3.2.1]oct-3-en-2-one by means of its spectroscopic data.

  $\boxed{\text{Q}}$ Discuss the structure of product **A** and its mechanism of formation.

  $\boxed{\text{S}}$ Consult ref. [141].

       [141] J. Sepulveda-Arquez et al., *Tetrahedron* **1997**, *53*, 9313.

Problem 119

  $\boxed{\text{P}}$ Starting with 1,2-dibromocyclohex-1-ene (**X**) the following sequence of reactions was performed :

       (a) **X**, *tert*-butyl acrylate (2 equiv.), Pd(OAc)$_2$, Ph$_3$P, Et$_3$N, DMF (solvent), 90-100°C
          → **A** (57%) ;

(b) **A**, CF$_3$CO$_3$H, Na$_2$HPO$_4$, CH$_2$Cl$_2$ (solvent) → **B** (74%) ;

(c) **B**, dioxane (solvent), 80°C → **C** (83%) .

Q What are the structures of the products **A** – **C** and their modes of formation ? What is the stereochemistry of **A** – **C** ?

S Consult refs. [142] and [143].

[142] A. de Meijere et al., *Eur. J. Org. Chem.* **1998**, 1521.
[143] A. de Meijere et al., *J. Org. Chem.* **1999**, *64*, 3806.

Problem 120

P 2-Chloropyridine (**X**) is the starting material for the following sequence of reactions :

(a) **X**, LiTMP, THF (solvent), -60 to -70°C; then *N*-formylpiperidine, -60°C to r.t.; then NH$_4$Cl, H$_2$O → **A** (65%) ;

(b) **A**, benzylamine, H$^+$ (cat.), Δ → **B** (100%) ;

(c) **B**, LDA, THF (solvent), -78°C; then phenyl-*O*-ethylthiocarboxylate (Ph-CS-OEt), -78°C to r.t.; then rfl. 2 h → **C** (65%) .

Q What are the structures of the products **A** – **C** and their modes of formation ? Give the systematic name of **C**.

S Consult ref. [144] and p 279.

[144] A. Couture, E. Deniau, P. Grandclaudon, C. Simion, *Synthesis* **1996**, 986.

Problem 121

P Starting from succinic acid (**X**) the following sequence of reactions was performed to synthesize compounds **D** and **E** :

(a) **X**, propionyl chloride, AlCl$_3$, CH$_3$NO$_2$ (solvent) → **A** (75%) ;

(b) **A**, CH$_3$I, MeONa, MeOH → **B** (85%) ;

(c) **B**, hν/300 nm, benzene (solvent), benzophenone (sensibilisator) → **C** (92%) ;

(d) **C**, hydrazine hydrate, EtOH, rfl. → **D** (93%) ;

(e) **C**, H$_2$N-CH$_2$-CH$_2$-NH$_2$, EtOH, rfl. → **E** (79%) .

Q What are the structures of the products **A** – **E** and their modes of formation ? Give the mechanism of reaction (c) and the systematic name of **D/E**.

S Consult refs. [145] and [146].

[145] H. Schick, G. Lehmann, G. Hilgetag, *Chem. Ber.* **1969**, *102*, 3238.
[146] R. S. Reddy, K. Saravanan, P. Kumar, *Tetrahedron* **1998**, *54*, 6553.

9   Problems   533

Problem 122

P Perform the following sequence of reactions :

   (a) *N*-Ethyl-*p*-toluenesulfonamide, *n*-BuLi (2 equiv.), THF (solvent), 0°C to r.t.; then B(OMe)₃, -78°C to r.t.; then H₂O₂, 0°C to r.t. → **A** (90%) ;

   (b) **A**, NaH (1 equiv.), DMF (solvent), 0°C; then allyl bromide (1 equiv.); then subsequently 1 equiv. NaH and allyl bromide → **B** (84%);

   (c) **B**, 10 mol% GRUBBS catalyst [114], CH₂Cl₂ (solvent) → **C** (82%) .

Q What are the structures of the products **A – C** and their modes of formation ?

S Consult ref. [147] and compare problem 95.

[147] C. Lane, V. Snieckus, *Synlett* **2000**, 1294.

Problem 123

P Pyrvinium iodide (**1**) is structurally related to the cyanine dyes and was used (in the form of the embonate) as an anthelmintic. It is obtained from the building blocks **2** and **3**, the synthesis of which is performed by the following sequences **I** (starting from aniline **A**) and **II** (starting from 4-amino-*N,N*-dimethylaniline **B**).

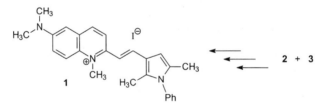

  I :   (a) **A**, hexane-2,5-dione → **C** (62%) ;

        (b) **C**, DMF, POCl₃ → **D** (83%),   = building block **2** ;

  II :   (c) **B**, CH₃-CH=CH-CH=O, HCl/H₂O → **E** (50%) ;

        (d) **E**, CH₃I, *i*-PrOH (solvent) → **F** (63%),   = building block **3** ;

    (e) convergence: **D** + **F**, MeOH (solvent), piperidine → **1** (92%) .

Q What are the structures of the products **C – F** and the mechanisms for reactions (a) – (e) ?

S Consult ref. [51], p 174.

Problem 124

P Clavicipitic acid (**1**) is an unusual ergot alkaloid biosynthesis derailment product and occurs in nature as a mixture of *cis*- and *trans*-diasteromers (**1a/1b**). The synthesis of the (*N*-acetyl)methylesters of **1a** and **1b** has been achieved by the following route starting with 2-bromo-6-nitrotoluene (**X**) :

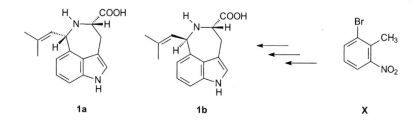

1a   1b   X

(a) **X**, Br$_2$, CCl$_4$ (solvent), $h\nu$ → **A** ;

(b) **A**, Ph$_3$P → **B** ;

(c) **B**, formaldehyde (gas), Et$_3$N → **C** (80%, 3 steps) ;

(d) **C**, Fe, AcOH; then TosCl, pyridine → **D** ;

(e) **D**, (CH$_3$CN)$_2$PdCl$_2$ (5%), 1,4-benzoquinone → **E** (80%, 2 steps) ;

(f) **E**, Hg(OAc)$_2$, HClO$_4$ (catalyst); then I$_2$ → **F** (97%) ;

(g) **F**, H$_2$C=C(NHAc)(COOCH$_3$), Pd(OAc)$_2$ (5%), Et$_3$N, CH$_3$CN
      → **G** (60%, Z-isomer) ;

(h) **G**, H$_2$C=CH-C(CH$_3$)$_2$OH, Pd(OAc)$_2$ (8%), Et$_3$N, P(o-tolyl)$_3$, CH$_3$CN → **H** (83%) ;

(i) **H**, (CH$_3$CN)$_2$PdCl$_2$ (15%), CH$_3$CN → **I** (95%) ;

(j) **I**, NaBH$_4$, Na$_2$CO$_3$, $h\nu$, MeOH/H$_2$O/DME
      → (N-acetyl)methylesters of **1a/1b** 1 : 1 (74%) .

Q What are the structures of the products **A** – **I** and their modes of formation ?

S Consult ref. [148] and ref. [24], p 96.

[148] P. J. Harrington, L. S. Hegedus, K. F. McDaniel, *J. Am. Chem. Soc.* **1987**, *109*, 4335.

Problem 125

P Starting from 1,2-diaminobenzene (**X**) the following sequence of reactions was performed to synthesize the polyheterocyclic compound **D**:

(a) **X**, cyanoacetic acid, Δ → **A** (70%) ;

(b) **A**, ethyl 2-oxocyclopentane-1-carboxylate, ammonium acetate (2 equiv.), Δ
      → **B** (92%, tetracyclic) ;

(c) **B**, POCl$_3$, Δ → **C** (99%) ;

(d) **C**, n-octylamine (excess), Δ → **D** (58%, main product) + **E** (minor product) .

Q What are the structures of the products **A** – **E** and their modes of formation ? Give the systematic name of **C**.

S Consult ref. [149] and compare p 176, 277 and 312.

[149] R. K. Russel, C. E. van Nievelt, *J. Heterocycl. Chem.* **1995**, *32*, 299.

Problem 126

P   Ningalin B (**1**) belongs to a class of recently identified marine natural products. Its synthesis was achieved by the following sequence of reactions starting from 2-bromo-4,5-dimethoxybenzaldehyde (**X**) :

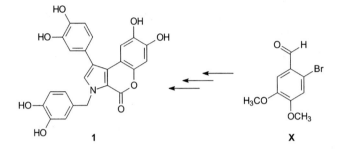

(a) **X**, (3,4-dimethoxyphenyl)acetylene, (Ph₃P)₂PdCl₂, CuI, Et₃N, DMF (solvent), Δ
  → **A** (87%) ;

(b) **A**, *m*-chloroperbenzoic acid; then KOH/MeOH, then MOM-Cl/DIPEA
  → **B** (67%) ;

(c) **B**, dimethyl 1,2,4,5-tetrazine-3,6-dicarboxylate, mesitylene (solvent), Δ
  → **C** (92%) ;

(d) **C**, Zn, AcOH → **D** (62%) ;

(e) **D**, (3,4-dimethylphenyl)ethyl bromide, K₂CO₃, DMF (solvent), Δ → **E** (94%) ;

(f) **E**, HCl, ethyl acetate (solvent) → **F** (95%) ;

(g) **F**, LiI, DMF (solvent), rfl. → **G** (80%) ;

(h) **G**, Cu₂O, quinoline, Δ → **H** (70%) ;

(i) **H**, BBr₃, hexane (solvent) → **1** (98%) .

Q   What are the structures of the products **A – H** and their modes of formation ?

S   Consult ref. [150]. For reaction (d) 1,4-reduction and *N,N*-fission should be considered. Compare also p 452.

[150] D. L. Boger, D. R. Soenen, C. W. Boyce, M. P. Hedrick, Q. Jin, *J. Org. Chem.* **2000**, *65*, 2479.

Problem 127

P   Methoxatin (**1**, cf. p 336) is a dehydrogenase cofactor in methylotrophic bacteria (also called PQQ = pyrroloquinoline quinone). The synthesis of its triester **J** is achieved by the following sequence based on 2-methoxy-5-nitroaniline (**X**) :

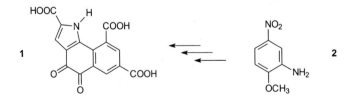

(a) **X**, formic acid, acetic anhydride → **A** (92%) ;

(b) **A**, H₂, Pd-C → **B** (80%) ;

(c) **B**, NaNO₂, HCl; then HBF₄ → **C** ;

(d) **C**, ethyl 2-methylacetoacetate, NaOAc, EtOH/H₂O (solvents) → **D** ;

(e) **D**, pH 6.5 to 5.2 → **E** (70%, 3 steps) ;

(f) **E**, formic acid, EtOH (solvent) → **F** (62%) + isomeric indole (separable) ;

(g) **F**, HCl/H₂O/acetone → **G** (85%) ;

(h) **G**, dimethyl 2-oxoglutaconate (CH₃OOC-CO-CH=CH-COOCH₃), CH₂Cl₂ (solvent)
→ **H** (73%, *cis*-position of COOCH₃) ;

(i) **H**, Cu(OAc)₂, HCl (gas), O₂, CH₂Cl₂ (solvent) → **I** (95%) ;

(j) **I**, Ce(NH₄)₂(NO₃)₆, H₂O → **J** (70%) = triester of **1** .

Q What are the structures of the products **A** – **J** and their modes of formation ?

S Consult refs. [151]/[152] and compare p 106 and 331.

[151] E. J. Corey, A. Tramontano, *J. Am. Chem. Soc.* **1981**, *103*, 5599.
[152] P. Martin, E. Steiner, K. Auer, T. Winkler, *Helv. Chim. Acta* **1993**, *76*, 1667.

Problem 128

P A new access to the hydrogenated indolo[2.3-a]quinolizine skeleton **1** was achieved by utilizing a sequence of transformations, in which the following reactions were used :

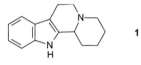

(a) **X**, H₂C=CH-CH=O, vinyl ethyl ether, Δ → **A** (cf. p 241) ;

(b) **A**, HO-N=C(Cl)COOEt, Et₃N, THF (solvent), r.t.
→ **B** (61%, mixture of diastereomers) ;

(c) **B**, HCl/H₂O → **C** ;

(d) **C**, tryptamine, CF₃COOH, CH₂Cl₂ (solvent) → **D** (65%, 2 steps) ;

(e) **D**, KOH, EtOH/H₂O; then HCl → **E** ;

(f) **E**, DMF (solvent), Δ → **F** (not isolated) ;

(g) **F**, Δ → **G** (61%, 3 steps) .

Q What are the structures of the products **A** – **G** and their modes of formation ?

S Consult ref. [153] and compare p 141, 241 and 346.

[153] C. Perez, Y. L. Janin, D. S. Grierson, *Tetrahedron* **1996**, *52*, 987.

Problem 129

P Grossularine-1 (**1**) is a natural product isolated from the tunicate *Dendrodoa grossularia* (Stylidae) and possesses interesting antitumor properties. In a recent approach the tetracyclic ester **2** as key intermediate for the synthesis of **1** was constructed from the heterocycles **X** and **Y** performing the following transformations :

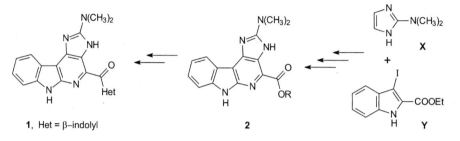

1, Het = β–indolyl   **2**   **Y**

(a) **X** [154], NaH, DMF; then SEM-Cl, 0°C → **A** (83%) ;
(b) **A**, *tert*-BuLi, diethylether, -78°C; then *n*-Bu₃SnCl → **B** ;
(c) **B**, **Y**, (Ph₃P)₄Pd, DMF (solvent) → **C** (82%, 2 steps) ;
(d) **C**, Na₂CO₃, H₂O; then HCl → **D** ;
(e) **D**, diphenyl phosphorazidate, Et₃N, benzene (solvent), 60°C → **E** (86%, 2 steps) ;
(f) **E**, *o*-dichlorobenzene (solvent), 170°C → **F** (88%) ;
(g) **F**, trifluoromethanesulfonic anhydride, pyridine → **G** (94%) ;
(h) **G**, carbon monoxide (1 atm), MeOH, Et₃N, Pd(OAc)₂, dppf → **H** (77%) ;
(i) **H**, 1-TIPS-3-lithioindole, diethylether; then 6 N HCl/H₂O → **I** (63%) = **1** .
Notes: SEM   = [(trimethylsilyl)ethoxy]methyl
          dppf   = 1,1-bis(diphenylphosphino)ferrocene
          TIPS   = triisopropylsilyl

Q What are the structures of the products **A** – **H** and their modes of formation ? Give the systematic name of **H**.

S Consult ref. [155].

[154] A. Dalkafouki et al., *Tetrahedron Lett.* **1991**, *32*, 5325.
[155] S. Hibino et al., *J. Org. Chem.* **1995**, *60*, 5899.

Problem 130

P The alkaloid physostigmin (**1**) and its analogues bearing pyrrolo[2.3-b]pyrrolidine structural units have attracted attention because of their activity as cholinesterase inhibitors. A short synthesis of compound **2** possessing a pyrrolopyrrolidine skeleton includes the following reactions, in which 4-hydroxycoumarine (**X**, cf. problem 76) is used as starting material :

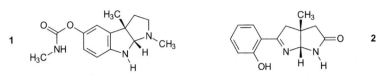

(a) **X**, CH₃-CO-CH(OCH₃)₂, CH₂(CN)₂, piperidine, benzene (solvent), rfl.
  → **A** (92%) ;

(b) **A**, 2N HCl/H₂O, rfl. → **B** (77%) ;

(c) **B**, NH₃/H₂O, 80°C (sealed tube) → **2** (63%) .

Q What are the structures of the products **A/B** and their modes of formation ?

S Consult ref. [156]. Each of the reactions (a) – (c) has to be considered as a cascade of several transformations (domino reaction, cf. p 487).

[156] T. Okawara et al., *J. Chem. Soc. Perkin Trans. 1* **1997**, 1323.

Problem 131

P In a study on biomimetic alkaloid synthesis the pentacyclic indole alkaloid vincadifformin (**1**) was synthesized in two steps starting from tryptamine hydrochloride (**X**):

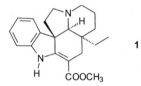

(a) **X**, methyl pyruvate, MeOH (solvent), rfl. → **A** (72%) ;

(b) **A**, 5-chloro-2-ethylpentanal (**Y**), TosOH (catalyst), toluene, rfl. → **1** (84%) .

Q What is the structure of **A** ? What are the modes of formation for **A** and **1** ? Formulate the intermediates in reaction (b).

S Consult ref. [157]. Transformation (b) is an example of a domino reaction (cf. problem 130). Aldehyde **Y** is obtained in 2 steps from (*N*-cyclohexyl)butane-1-imine and 1-bromo-3-chloropropane. Formulate this synthesis.

[157] M. E. Kuehne, J. A. Huebner, T. M. Matsko, *J. Org. Chem.* **1979**, *44*, 2477.

Problem 132

P Epirizole (**F**) possesses analgesic and anti-inflammatory properties. Its synthesis is possible according to the following sequence of reactions :

(a) Ethyl acetoacetate, urea, EtONa/EtOH → **A** ;

(b) **A**, POCl₃ → **B** ;

(c) **B**, MeONa, MeOH → **C** ;

(d) **C**, hydrazine hydrate → **D** ;

(e) **D**, 1,3-but-3-enolide (diketene) → **E** ;

(f) **E**, dimethyl sulfate → **F** .

| Q | What are the structures of the products **A** – **F** and their modes of formation ? Give the systematic name of **F** ?

| S | Consult ref. [158]/[159] and compare p 188, 400 and 402.

[158] H. Vanderhaeghe, M. Claesen, *Bull. Soc. Chim. Belg.* **1959**, *68*, 30.
[159] A. Kleemann, J. Engel, *Pharmaceutical Substances*, p 709, Thieme Verlag, Stuttgart-New York **1999**.

Problem 133

| P | Etizolam (**G**), a benzodiazepine tranquilizer with anxiolytic and sedative effects, is synthesized utilizing the following sequence with 2-chloro(cyanoacetyl)benzene (**X**) as starting material :

(a) **X**, butanal, sulfur (elemental) → **A** ;

(b) **A**, C_6H_5-CH_2-O-CO-NH-CH_2-CO-Cl → **B** ;

(c) **B**, HBr, AcOH → **C** ;

(d) **C**, pyridine, AcOH → **D** ;

(e) **D**, P_4S_{10} → **E** ;

(f) **E**, hydrazine hydrate; then acetic anhydride → **F** ;

(g) **F**, AcOH → **G** .

| Q | What are the structures of the products **A** – **G** and their modes of formation ? Give the systematic name of **G**.

| S | Consult ref. [160] and compare p 77/475 and problems 110/111.

[160] T. Tahara et al., *Arzneim.-Forsch.* **1978**, *28*, 1153; ref. [159], p 761.

Problem 134

| P | Irbesartan (**F**), an antihypertensive acting as angiotensin II antagonist, is prepared by the following route starting with cyclopentanone (**X**):

(a) **X**, NaCN, NH_4Cl, NH_3, MeOH/H_2O → **A** ;

(b) **A**, H_2SO_4, H_2O → **B** ;

(c) **B**, pentanoyl chloride, Et_3N, THF → **C** ;

(d) **C**, KOH, NH_4Cl, MeO/H_2S → **D** ;

(e) **D**, 4'-(bromomethyl)-2-cyanobiphenyl [161], NaH, DMF (solvent) → **E** ;

(f) **E**, NaN_3; then HCl/H_2O → **F** .

| Q | What are the structures of the products **A** – **F** and their modes of formation ? Give the systematic name of **F**.

| S | Consult ref. [162] and compare p 178 and 215.

[161] For the synthesis of this building block see ref. [159], p 1116.
[162] C. A. Bernhart et al., *J. Med. Chem.* **1993**, *36*, 3371.

Problem 135

P The analgesic (-)-levorphanol is obtained by resolution of racemic (±)-3-hydroxy-*N*-methyl-morphinane (**G**) with (+)-tartaric acid. The synthesis of **G** is accomplished by the following sequence of reactions :

(a) Cyclohexanone, cyanoacetic acid, [NH₄]OAc, Δ → **A** (double bond endocyclic) ;

(b) **A**, H₂, RANEY-Ni → **B** ;

(c) **B**, (4-methoxyphenyl)acetyl chloride → **C** ;

(d) **C**, POCl₃ → **D** ;

(e) **D**, H₂, RANEY-Ni → **E** (one mole of H₂ is consumed) ;

(f) **E**, formaldehyde, H₂, RANEY-Ni → **F** ;

g) **F**, H₃PO₄ → **G** .

Q What are the structures of the products **A** – **G** and their modes of formation ? Give the systematic name of **F**.

S Consult ref. [163] and compare p 343.

[163] O. Schnider, J. Hellerbach, *Helv. Chim. Acta* **1950**, *33*, 1437;
     O. Schnider, A. Grüssner, *Helv. Chim. Acta.* **1951**, *34*, 2211; ref. [159], p 1096.

Problem 136

P The anti-inflammatory meloxicam (**D**), a cyclooxygenase-Z inhibitor, is synthesized starting with saccharin-Na (**X**) :

(a) **X**, methyl chloroacetate → **A** (60%) ;

(b) **A**, MeONa, DMSO → **B** (77%) ;

(c) **B**, CH₃I, NaOH, H₂O/EtOH → **C** (88%) ;

(d) **C**, 2-amino-5-methylthiazole (**Y**), xylene (solvent), rfl. → **D** (70%) .

Q What are the structures of the products **A** – **D** and their modes of formation ? Outline syntheses of **X** and **Y**.

S Consult ref. [51], p 198, and ref. [159], p 1156/1699. Compare p 152.

Problem 137

P The gyrase inhibitor ofloxacin (**F**, cf. p 335) is medicinally used as an antibiotic. Its synthesis is accomplished from 1,2,3-trifluoro-4-nitrobenzene (**X**) utilizing the following sequence of transformations :

(a) **X**, KOH, MeOH → **A** ;

(b) **A**, chloroacetone, K₂CO₃, KI → **B** ;

(c) **B**, H₂, RANEY-Ni → **C** (bicyclic !) ;

(d) **C**, diethyl ethoxymethylenemalonate → **D** ;

(e) **D**, ethyl polyphosphate; then HCl/H₂O → **E** ;

(f) **E**, *N*-methylpiperazine → **F** .

Q What are the structures of the products **A** – **F** and their modes of formation ? Formulate detailed mechanisms for reactions (a), (c) and (f).

S Consult ref. [159], p 1380/1381.

Problem 138

P The radioprotector methoxsalen (**F**) is synthesized by performing the following sequence of reactions :

(a) Pyrogallol, chloroacetic acid, POCl$_3$ → **A** ;

(b) **A**, NaOAc → **B** ;

(c) **B**, H$_2$, Pd-C → **C** ;

(d) **D**, hydroxybutanedioic acid (malic acid), H$_2$SO$_4$ → **D** ;

(e) **D**, dimethyl sulfate, K$_2$CO$_3$ → **E** ;

(f) **E**, DDQ → **F** .

Q What are the structures of the products **A** – **F** and their modes of formation ?

S Consult ref. [159], p 1210, and compare p 251.

Problem 139

P Nedocromil (**G**), an antiallergic and antiasthmatic, is synthesized according to the following route starting with 4-(acetylethylamino)-2-hydroxyacetophenone (**X**) :

(a) **X**, allyl bromide, K$_2$CO$_3$ → **A** ;

(b) **A**, *N*,*N*-diethylaniline (solvent), Δ → **B** (isomer of **A**) ;

(c) **B**, H$_2$, Pt; then HBr/H$_2$O → **C** ;

(d) **C**, dimethyl acetylenedicarboxylate → **D** ;

(e) **D**, polyphosphoric acid → **E** ;

(f) **E**, diethyl oxalate, NaH, DMF (solvent) → **F** ;

(g) **F**, NaHCO$_3$, EtOH → **G** .

Q What are the structures of the products **A** – **G** and their modes of formation ?

S Consult ref. [164] and compare p 263 and 330.

[164] H. Cairns et al., *J. Med. Chem.* **1985**, *28*, 1832; ref. [159], p 1313.

Problem 140

P The cardiotonic pimobendan (**D**) is synthesized according to the following sequence starting with methyl-3-(4'-amino-3'-nitrobenzoyl)butyrate (**X**) :

(a) **X**, 4-methoxybenzoyl chloride, chlorobenzene (solvent), Δ → **A** ;

(b) **A**, hydrazine hydrate, AcOH, 0°C → **B** ;

(c) **B**, H_2, Pd-C, EtOH (solvent) → **C** ;

(d) **C**, AcOH → **D** .

Q   What are the structures of the products **A** – **D** and their modes of formation ? Outline a synthesis for **X**.

S   Consult ref. [159], p 1521; compare p 176 and 395.

Problem 141

P   Olprinone hydrochloride (**E**), a cardiotonic and vasodilator acting as PDE III-inhibitor, is synthesized by the following route starting with 2-amino-5-bromopyridine (**X**) :

(a) Bromoacetaldehyde diethylacetal, HCl/H_2O; then $NaHCO_3$, **X** → **A** (71%) ;

(b) **A**, EtMgBr, THF, 0° to 10°C; then 1-chloro-2-methylpropene; then NH_4Cl, H_2O
   → **B** (71%) ;

(c) **B**, O_3, $HCl/H_2O/EtOH$ → **C** (71%) ;

(d) **C**, dimethylformamide dimethylacetal, DMF, 80°C → **D** (75%) ;

(e) **D**, cyanoacetamide, DMF, 80° to 90°C; then HCl/H_2O → **E** (51%) .

Q   What are the structures of the products **A** – **E** and their modes of formation ?

S   Consult ref. [165], compare p 284 and 314.

[165] M. Yamanaka et al., *Chem. Pharm. Bull.* **1991**, *39*, 1556; **1992**, *40*, 1486.

Problem 142

P   In a study of endothelin antagonists a potent compound (EMD 122801, **H**) was synthesized by the following route starting from (4-aminophenyl)acetic acid (**X**) :

(a) **X**, acetic anhydride → **A** ;

(b) **A**, HNO_3, AcOH → **B** ;

(c) **B**, EtOH, HCl, rfl. → **C** (68%, 3 steps) ;

(d) **C**, H_2, Pd-C, EtOH (solvent) → **D** (90%) ;

(e) **D**, $SOCl_2$, DMF (catalyst) → **E** (75%) ;

(f) **E**, 4-methoxyphenacyl bromide, *tert*-BuOK, *N*-methylpyrrolidone (solvent)
   → **F** (75%) ;

(g) **F**, 3,4,5-trimethoxybenzaldehyde, MeONa, MeOH, rfl. → **G** (86%) ;

(h) **G**, NaOH, MeOH → **H** (100%) .

Q   What are the structures of the products **A** – **H** and their modes of formation ?

S   Consult ref. [166].

[166] W. W. K. R. Mederski et al., *Bioorg. Med. Chem. Lett.* **1998**, *8*, 17.

Problem 143

P Porantherin (**1**) is an alkaloid from the Australian plant *Poranthera corymbosa* and possesses an interesting tetracyclic skeleton of the 9b-azaphenalene type [167]. In an instructive biomimetic synthesis of **1** developed by COREY [168] the key intermediate (**X**) was subjected to the following reactions (a) and (b) leading via **A** to **B**, which was further transformed to the target molecule **1** :

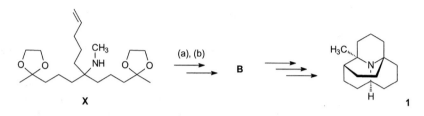

(a) **X**, 10% HCl/H$_2$O → **A** (90%) ;

(b) isopropenyl acetate, TosOH, benzene (solvent), Δ → **B** (45%) .

Q Suggest structures for **A/B** and their mechanisms of formation. Develop a short synthesis of **X** starting from 5-chloropentan-2-one.

S Consult ref. [168].

[167] M. Hesse, *Alkaloide*, p 210, Wiley-VCH, Weinheim **2000**.
[168] E. J. Corey, R. D. Balanson, *J. Am. Chem. Soc.* **1974**, *96*, 6516.

10 Indices

The contents of the foregoing chapters have been distributed among t w o indices.

The "**General Subject Index**" (Chapter 10.1) mainly deals with
- the heterocyclic compounds presented in this book including natural products and systems of pharmaceutical, medicinal and technical relevance;
- characteristic unnamed (!) reactions, reaction types and applications of heterocycles;
- important general principles of heterocyclic chemistry and nomenclature of heterocycles.

This index does not list basic transformations such as cycloaddition, cyclocondensation, thermolysis, photolysis, ring-opening hydrolysis, oxidation, reduction, decarboxylation, valence tautomerism, metalation etc.. They are listed under the appropriate parent heterocyclic system and the subsections on reactions and synthesis.

The "**Index of Named Reactions**" (Chapter 10.2) covers all the named reactions, reaction types and principles for synthesis of the aforementioned heterocyclic systems. The heterocycle resulting from a named synthesis is shown in brackets, e.g. Fischer synthesis (indole).

10.1 General Subject Index

10.2 Index of Named Reactions